Comparative Cognition

Integrating developments from [illegible] neuroscience, this is an undergraduate introduction to cognitive pr[illegible] across species. The authors [illegible] classic studies and contemporary research to give students the full picture of the evolving field of comparative cognition.

Engaging students in the discipline from its roots in animal learning and evolutionary biology through to current research, the chapters cover both controlled laboratory and comparative cross-species studies in the natural environment. This approach provides students with complementary ethological and neurobiological perspectives on cognition. Feature boxes encourage engaged learning, giving a deeper understanding of topics discussed in the main text. These are supported by end-of-chapter questions to help check understanding of key theories and concepts.

Online resources include links to sites of interest, further reading, PowerPoint lecture slides and additional questions, all available at www.cambridge.org/cognition.

Mary C. Olmstead is Professor of Psychology and Neuroscience at Queen's University, Ontario. Her research is directed towards understanding the neural and psychological interface between motivation and cognition, or how rewarding stimuli influence learning.

Valerie A. Kuhlmeier is Associate Professor of Psychology and Canada Research Chair in Cognitive Development at Queen's University, Ontario. Her research program explores cognition from a developmental and evolutionary perspective.

Both the authors are research psychologists who have been teaching at Canadian, US, and European institutions for the last 15–20 years.

Comparative Cognition

MARY C. OLMSTEAD

AND

VALERIE A. KUHLMEIER

Queen's University, Ontario, Canada

CAMBRIDGE
UNIVERSITY PRESS

University Printing House, Cambridge CB2 8BS, United Kingdom

Cambridge University Press is part of the University of Cambridge.

It furthers the University's mission by disseminating knowledge in the pursuit of education, learning and research at the highest international levels of excellence.

www.cambridge.org
Information on this title: www.cambridge.org/9781107011168

First published 2015

Printing in the United Kingdom by TJ International Ltd. Padstow Cornwall

A catalogue record for this publication is available from the British Library

Library of Congress Cataloguing in Publication data
Olmstead, Mary C.
Comparative cognition / Mary C. Olmstead, Valerie A. Kuhlmeier.
pages cm
ISBN 978-1-107-01116-8 (Hardback) – ISBN 978-1-107-64831-9 (Paperback) 1. Psychology, Comparative. 2. Cognition. I. Kuhlmeier, Valerie A. II. Title.
BF671.O46 2014
156′.3–dc23 2014021238

ISBN 978-1-107-01116-8 Hardback
ISBN 978-1-107-64831-9 Paperback

Additional resources for this publication at www.cambridge.org/cognition

TABLE OF CONTENTS

PREFACE

Comparative cognition is a highly interdisciplinary field that arose from a synthesis of evolutionary biology and experimental psychology. In its modern form, researchers from a variety of scientific backgrounds (e.g. neuroscience, behavioral ecology, cognitive and developmental psychology) come together with the common goal of understanding the mechanisms and functions of cognition. Over the past 10 years, we have taught both undergraduate and graduate courses that covered the subject matter of comparative cognition, although frequently in a course of another name. Like many instructors, we put together course material that included scientific articles, chapters in other textbooks, and our own writings as an attempt to represent the evolving field of comparative cognition. This was not ideal as the presentation of material from different sources is uneven, and undergraduate students often have difficulty conceptualizing the fundamentals of a discipline without the framework provided by a solid textbook. We realized that our experience was not unique when we spoke to colleagues teaching similar courses at other universities.

Our textbook provides an introduction to the field of comparative cognition. It begins with an historical view of the field, emphasizing the synergy of different disciplines, both in terms of theoretical foundations and methodological tools. This lays the groundwork for later chapters in which controlled laboratory studies are presented alongside comparative studies in the natural environment. The first half of the text covers topics that reflect the influence of behavioral psychology on comparative cognition. This material, therefore, overlaps with traditional animal learning texts. The distinguishing feature of our text is an increased emphasis on the evolutionary function and underlying neural mechanisms of cognition. In addition, issues that are central to cognitive psychology (e.g. attention, episodic memory, and cognitive maps) are interwoven throughout these chapters. The second half of the book focuses on what are often described as 'higher cognitive processes,' describing recent research on topics such as tool use and causal reasoning, imitation, and functionally referential communication.

Although different cognitive processes are discussed in separate chapters, the point is continually made that none of these functions in isolation. Even seemingly simple behaviors, like foraging for food, depend on an interaction between many cognitive processes. For example, spatial navigation is required to locate the food and then return to home base, perceptual discrimination is necessary to distinguish what is a good food source from what is not, emotional memory is used to recall dangers that were encountered in previous foraging expeditions, and decision making allows animals to make cost/benefit calculations of the energy output in searching for food versus the energy gained in food consumption. Foraging may also involve more complex processes such as communication with conspecifics about where the food is located, cooperation in hunting prey or retrieving food, as well as knowledge of social hierarchies (e.g. whether a dominant animal is likely to steal the food).

The text also reflects the fundamental importance of cross-species analyses in comparative cognition research. Each chapter includes descriptions and analyses of scientific studies conducted with humans and a variety of non-human animals. (In the interest of succinctness, from hereon we use the term 'animal' to refer to all non-human animals and 'human' to refer to *Homo sapiens*, and scientific nomenclature for species' names will be presented only the first time a species is mentioned.) The relative weighting of material drawn from research using humans versus animals (or indeed one animal species versus another) depends on the topic under discussion. For example, scientific studies reviewed in the early chapters often include multi-species comparisons because research that informs these topics (e.g. perception, long-term memory) spans the entire animal kingdom; in contrast, topics covered in the later chapters are frequently defined by a comparison to human abilities (e.g. theory of mind). In some cases, we know more about how a specific cognitive process operates in certain animals simply because it has been studied more frequently in that species (e.g. operant conditioning in rats). As research in comparative cognition progresses, both the function and mechanisms of cognition are being studied in an increasing number of species. Future editions of this text, therefore, will undoubtedly include an expanded list of cross-species comparisons.

Comparative cognition research is characterized by an ongoing dialogue between scientists with different areas of specialization. This component of the field is represented in the text by an integration of material across disciplines. Importantly, each topic includes up-to-date research on neural mechanisms of cognition and on cognitive development, two of the most fruitful areas of research that inform and complement comparative cognition research. Because each topic is covered from different perspectives, the structure of the text emphasizes how complementary empirical studies lay the groundwork for theoretical positions. Many of the chapters finish with a description of ongoing controversies in the field, with no apparent 'conclusion' to a specific topic. This is a reflection of the dynamic and productive field of comparative cognition: like many scientific endeavors, there is always more work to be done. The book is timely in that the field is advancing rapidly; a textbook that presents fundamentals of this discipline will be a foundation for further scientific and scholarly investigations.

Using the book

This book is organized in such a way that it may be flexibly adapted to courses with different program restrictions and pedagogical requirements. One instructor may opt to cover all of the chapters in a single term, another to divide these by presenting basic (Chapters 2 through 7) and higher order (Chapters 8 through 13) cognitive processes in separate terms. We adopt the latter approach in our own teaching and assign Chapter 1 (Introduction to Comparative Cognition) to both courses. Those interested in using the text (or single text chapters) for upper level undergraduate or graduate courses can take advantage of information provided in the instructor's manual. Among other tools, this on-line resource includes a list of Supplementary Material for each chapter that presents specific topics at a more advanced level of pedagogy. For most chapters, this includes at least one 'origin of the field' paper, such as an original excerpt from one of Darwin's or Tinbergen's publications. Depending on the chapter topic, the supplementary material could also include a recent high-impact paper in the field. Each entry is accompanied by a short commentary specifying its importance and pointing out controversies that were stimulated or resolved by the publication. The section concludes with suggestions for short assignments based on this material.

Finally, the textbook includes a number of features designed to enhance student engagement. Text boxes within each chapter complement material presented in the main text by outlining extensions of comparative cognition research. Some of these discuss scientific discoveries in related disciplines such as anthropology, computer science, artificial intelligence, linguistics, or philosophy. Others present real-world applications of knowledge gained from comparative cognition studies to fields such as clinical psychology or environmental studies. Each chapter profiles one contemporary researcher in the field of comparative cognition. As interested students (not so long ago!), we often wondered how scientists began their careers, especially in a field like comparative cognition that has so many diverse influences. An overview of these biographies reveals that career and education paths vary widely, that research interests evolve, and that many different scientific perspectives converge in the field of comparative cognition. Each chapter also includes a list of 6–10 questions that cover both factual and theoretical information. Students can use these to gauge their own understanding of basic and conceptual issues related to the chapter topic. Each chapter concludes with a Further Reading section, an annotated list of 8–10 selections of supplementary readings. These include scientific articles (primarily reviews or 'classic' papers in the field), book chapters, or entire books. A description of each entry along with its relevance to the chapter material is provided such that students and instructors can select the readings based on their own interest and/or the course curriculum.

Despite teaching comparative cognition courses for close to 15 years, we never cease to be surprised (and delighted) by the insightful questions and thought-provoking comments of our students. These were the original source of inspiration for this text, for which we are eternally grateful. We were fortunate to have supportive Department Heads, Merlin Donald, Vernon Quinsey, and Richard Beninger, who encouraged and facilitated the writing of this text. In addition, many of our colleagues generously offered suggestions on individual chapters, including Hans Dringenberg, Kristen Dunfield, Barrie Frost, Megan Mahoney, Janet Menard, Niko Troje, Kristy vanMarle, and Ron Weisman, as well as the members of the Infant Cognition Group and Moticog Lab. Ruxandra Filip and the staff at CUP were invaluable during the final stages of text editing.

We are particularly indebted to the collegiality displayed by Francesco Bonadonna and his research team at the Centre d'Ecologie, Functionelle et Evolutive in Montpellier, France. Along with Anne Gorgeon, they provided a stimulating environment and ensured that nothing was lost in translation. We are grateful to a set of international reviewers who provided feedback on earlier versions of this text, and to the Natural Sciences and Engineering Research Council (NSERC) of Canada for ongoing research funding.

This text would not have materialized without the unfailing enthusiasm and support of Rodney Birch and Leslie Farquharson. We also thank Bradley, Darrell, Kermit, Digger, Bobby, Sarah, Abby, and Sheba who taught us invaluable lessons about animal communication and interspecific mutualism. Finally, Catia and Lucas opened our eyes to the wonders of the natural world and graciously put up with an absentee mother on far too many nights and weekends.

This book is dedicated to Patricia Jean Olmstead and to James and Alba Kuhlmeier.

1 History of comparative cognition

Background

Observing animals in their natural habitat often calls attention to interesting and unusual behaviors. As an example, early in the twentieth century, inhabitants of the British Isles noticed that some birds were using their beaks to poke through the lids of milk bottles that were delivered to houses. By puncturing the lid, the birds were able to get the cream that floated to the top of the bottle, no doubt annoying the humans who had purchased the milk. The bottles were often attacked by swarms of birds as soon as they were left at the door, and there were even reports of flocks following milk carts on their delivery route (Fisher & Hinde, 1949). Because this habit started in a small village and spread across a large geographical region, many people believed that the birds were learning the behavior by watching other birds open bottles. In other words, social learning *could* explain how the behavior was transmitted so quickly in a population.

The only way to verify that social learning explains the transmission of bottle opening by birds is to examine the behavior under controlled experimental conditions. This may occur either in the lab or in the natural environment, although many researchers favor the former because it is easier to control factors that may influence or bias the results in a lab setting (although that in itself is debatable). Either way, researchers examining the role of social learning in bottle-opening behavior of birds would need to consider alternative explanations. For example, did one bird have to see another bird open a bottle to be able to do this? Did the birds have to interact in order for the behavior to be transmitted? Is it possible that individual birds were following other birds to a milk bottle, tasting the cream, and then approaching new milk bottles on their own which they learned to open through trial-and-error? If the birds were learning from other birds, what conditions made this learning possible?

This list is only a subset of the possible questions that scientists could ask regarding social learning in birds and there is no guarantee that any one researcher, or even any group of researchers, will be able to answer all of them. Often new questions arise as research progresses and, in some cases, scientists return to the natural environment for more detailed observations of the behavior under study. This general approach of combining field observations and controlled laboratory experiments exemplifies the subject matter of this text: comparative cognition.

Chapter plan

This chapter provides a brief overview of the historical and intellectual influences that led to the contemporary field of comparative cognition. An interest in animal cognition has been documented for much longer than is reported here: theories of animal cognition were presented by early philosophers such as Aristotle (384–322 BC) and later by Rene Descartes (1596–1650), among others. The approach in this chapter is to first define the hallmarks of research in comparative cognition and then to focus on three modern, scientific perspectives that, together, influenced the emerging field of comparative cognition. In doing so, this chapter outlines the intellectual traditions that laid the groundwork for the discipline.

Students and researchers in any field must be aware of the intellectual and social influences that led to the development of their field. This allows them to place current scientific advances within historical context and to interpret the significance of research findings that are presented through published journals, research conferences, academic courses, or the popular media. An inherent component of any scientific investigation, therefore, is the understanding that research advances are built on prior work; understanding these influences will facilitate further progress in that knowledge of the past can direct future endeavors.

For the field of comparative cognition, this tradition emerged from the work of early experimental psychologists who devised carefully controlled experiments to examine behavioral responses to environmental events, the work of early biologists who were interested in the evolution of animal behavior and conducted studies in naturalistic settings, and the work of early cognitive psychologists who considered the underlying mental representations that might guide behavior. After detailing these historical influences, the chapter concludes by characterizing the modern interdisciplinary trends in comparative cognition and suggests some likely future directions for the field.

1.1 Hallmarks of comparative cognition

Researchers in the field of comparative cognition study a wide diversity of species, employ a variety of methodologies, and conduct their work within many different academic and research departments. Nonetheless, the field is characterized by three unifying hallmarks:

1. *Examination of cognitive processes.* **Cognition**, translated from the Latin cognosco, means *knowledge* or *thinking*. It is frequently equated with information processing in that the study of cognition examines how humans and animals acquire, store, and process information[1]. A more specific definition of cognition is the "mental processes and activities used in

[1] Life on earth is classified according to a biological taxonomy, originally developed in the eighteenth century by Carl Linnaeus (see Chapter 9 for details). The system divides living organisms into progressively smaller groups, based on shared physical traits, according to the following scheme: Domain; Kingdom; Phylum; Class; Order; Family; Genus; and Species. Humans would be classified as follows: Eukarya (Domain); Animalia (Kingdom); Chordata (Phylum); Mammalia (Class); Primates (Order); Hominidae (Family); Homo (Genus); and sapiens (Species). An organism's scientific name, also called the binomial name, consists of the genus and species which are usually derived from a Latin or Greek root. These are written in italics with the genus name capitalized. Thus, humans are *Homo sapiens*. According to this classification, humans are part of the Kingdom Animalia but, for succinctness, this book will use the term 'animal' to refer to all non-human animals and 'human' to refer to *Homo sapiens*. In addition, the scientific name will be provided in parentheses the first time that a species is mentioned in each chapter.

perceiving, remembering, thinking, and understanding and the act of using these processes" (Ashcraft & Klein, 2010, p. 9). Any cognitive process, therefore, is internal: it happens inside the organism's brain. Although scientists have developed some very clever tools and tests to examine these internal processes, cognition is typically inferred from some behavioral measure. For example, in studying memory, a researcher may test whether humans can repeat a list of words or whether rats will return to a location where they found food after different delay intervals. The researcher then uses the behavioral data (i.e. verbal recitation or number of visits to a food location) to make inferences about time-dependent effects on memory retention.

2. *Experimental procedures*. Research in comparative cognition typically involves some experimental manipulation. Behavior that was initially observed in the wild is often 'moved' into the laboratory for controlled empirical study. In other instances, researchers conduct experiments in an animal's natural habitat; one of the best-known examples is the study of European honeybee (*Apis mellifera*) communication by von Frisch. Von Frisch determined how honeybees indicate to other bees where to find food by moving food sources to different locations and then observing how the forager bees 'dance' when they return to the hive. There are many other examples of outstanding research using this naturalistic approach, despite the fact that it is more difficult to control extraneous variables outside of the lab.
3. *Evolutionary framework*. A guiding principle in comparative cognition is that cognitive abilities emerge through the same evolutionary process that shapes physiological traits (i.e. natural selection). Some researchers examine whether a given cognitive ability in humans is present in animals (e.g. counting, planning for the future, or understanding social relationships among group members), with the goal of understanding more about human cognition (e.g. if humans have unique cognitive abilities, what does this tell us about how this species processes information?). In other cases, researchers compare cognitive processes across species with the goal of understanding how and why these processes evolved. It is important to note that not all researchers in comparative cognition study a variety of species; in reality most focus on one or two. Yet, their findings are interpreted from an evolutionary perspective in that they reveal how a particular cognitive process may function in relation to certain environments.

These hallmarks of comparative cognition are exemplified in research by Galef and his colleagues that aimed to explain why some male grouse end up with a disproportionately large number of matings. In the natural environment, grouse mate on 'leks,' locations where males gather and display by fluttering their wings, rattling their tails, and vocalizing. Females observe the males during these displays and then select the male grouse with whom they will mate. It *appeared* that females were making their mate decisions based on the males' behavior but, despite extensive examination, observers could find no clear difference between males that would explain why some of these animals attracted more females. This led to the idea that females simply observe the mate choice of other females and then copy them (e.g. Bradbury & Gibson, 1983).

Galef tested this social learning hypothesis by studying mate-choice copying of another ground-dwelling bird, the Japanese quail (*Corturnix japonica*). In a series of tightly controlled lab experiments, he and his collaborators recorded the reactions of female quail to male quail that they had observed interacting with other females. For example, in one experiment they determined which of two males a female preferred. Then, the female observed her 'nonpreferred' male interacting with another female quail. This observation session was enough to

Researcher profile: **Dr. Bennett Galef**

Figure 1.1
Dr. Bennett Galef.

As an undergraduate at Princeton, Bennett Galef had planned to major in chemistry with a minor in art history and then begin a career as a forensic art historian. After attending courses in chemistry, physics, and calculus, he ended up majoring in psychology – not because he particularly liked it, but because he held the all-too-common misconception that it would be an easy major. His senior thesis examined concept formation in humans, and he graduated with honors standing and the requisite coursework under his belt, but only two courses in animal learning (Galef, 2009). It was only in preparation for his comprehensive exams and PhD dissertation that Galef began reading ethology texts, including Tinbergen's classic studies of animal behavior in the natural environment (Tinbergen, 1951, 1958). Through these texts, he found his passion for the burgeoning field of comparative cognition.

Upon graduating, Galef was hired as an assistant professor at McMaster University in Ontario, Canada, where he stayed until his retirement in 2008. At McMaster, Galef began the research that would make him a prominent member of his field. In 1971, he reported the finding that rat pups learned to choose particular foods based on the food choice patterns of adults. The adults had been trained previously to avoid a tasty, yet toxic food and instead to eat a far less palatable food. The pups, even without personal exposure to the toxin, also favored the less palatable, safe food, but only when they observed adults making this food choice (Galef & Clark, 1971). This was the first systematic demonstration of social learning across generations in a controlled laboratory setting. (Social learning will be detailed in Chapter 13.)

Since then, social transmission of behavior, particularly food choice and mating, has been the focus of Galef's research. From his perspective, social learning lies at the intersection of biology and psychology: Galef has approached the topic as such, spending time conducting both field and laboratory research. He has served in important roles that span the fields of psychology and biology, including editor of the journal *Animal Behaviour* and president of the Animal Behavior Society. His interdisciplinary framework has enabled Galef to develop new hypotheses and theories regarding the 'hows' and 'whys' of social transmission, evidenced by his numerous empirical papers, book chapters, and edited volumes.

According to Galef (2009), "As I have told my students, probably far too often, I see a life in science as a marathon, not a sprint. Ask simple questions arising from clearly stated hypotheses. Use simple experimental designs and transparent statistical analyses. One step at a time, experiment after experiment, frequently replicating your main effect, until you understand what you set out to understand and can be quite sure that, when others attempt to repeat your procedures, they will get the same results that you did. And if not, you will know why not." (p. 304).

increase her preference for this male bird (White & Galef, 1999). In other words, watching a male bird interacting with another female makes him more attractive. Moreover, when females mate with these males, they lay more fertilized eggs than when they mate with males that they have not observed interacting with other females (Galef, 2008; Galef & White, 1998; White & Galef, 1999). This work helped to determine that social learning has a profound impact on reproductive behavior (at least in this avian species).

1.2 Influence: theory of evolution by natural selection

Prior to the end of the nineteenth century, most people believed that all animals (including humans) were distinct and that each species had arrived on earth in its current form. This 'arrival' was explained, most commonly, by divine creation. These ideas were challenged by Darwin's theory of evolution, detailed in *On the Origin of Species* (Darwin, 1859) and *The Descent of Man and Selection in Relation to Sex* (Darwin, 1871). Other scientists and philosophers had discussed evolution, the idea that one type of animal could descend from another type of animal, but Darwin's hypothesis of how this occurred was novel. Alfred Russel Wallace independently came to similar conclusions (Wallace, 1870) although it was Darwin's published works that engendered excitement, inquiry, and controversy.

1.2.1 Tenets of the theory of evolution by natural selection

The basic tenets of Darwin's theory can be simplified to the following three points:

1. *Variation*: Individuals within a species display variability in both physiological and behavioral traits. Many of these variations reflect random mutations of genetic material, although Darwin elaborated his theories without benefit of this knowledge.
2. *Heritability:* Offspring inherit traits from their parents. Again, Darwin had no understanding of the mechanisms explaining trait heritability, which occurs through transmission of genetic material.
3. *Survival and reproduction:* If a certain trait promotes survival or reproduction, individuals possessing this trait will have a greater chance of transmitting traits to their offspring.

According to Darwin, this cycle continues across generations such that individuals with the trait promoting survival and reproduction will begin to outnumber those without the trait. This principle is often paraphrased as 'survival of the fittest' in which **fitness** refers to the ability to survive and reproduce. The process by which inherited traits become more (or less) prominent in a population, due to differences in fitness, is called **natural selection**.

Darwin's description of evolution by natural selection provided a theoretical framework for many fields of science, including modern-day comparative cognition. Put another way, if physiological traits are shaped by evolutionary pressures, so are cognitive traits. Moreover, if there are physical similarities between species due to common ancestry, there are likely to be similarities in behavior, emotion, and thought processes (i.e. cognition) as well. Finally, the idea that humans evolved, through variation and selection, from other animals, does not mean that humans are 'descended from chimpanzees' or any other species of animal that is alive today. Rather, humans and chimpanzees both evolved from a common ancestor that lived approximately 4–6 million years ago. This common ancestor was neither a chimpanzee nor a human, and there is no reason to believe that it was more chimp-like than human-like.

1.2.2 Adaptations

To examine cognitive processes within an evolutionary framework is to consider how a cognitive trait might improve fitness in a particular environment. Evolution produces **adaptations**, byproducts, and random effects (Buss, 2004), but only adaptations are the result of natural selection.

That is, only adaptations provide some evolutionary advantage to the individual. Adaptations, therefore, can be defined as traits that improve fitness; these have been selected for because they increase survival and reproduction. Some of an organism's adaptive traits can already be seen at birth (e.g. reflexes such as sucking); in other cases, input from the environment may be necessary for the adaptive trait to develop (e.g. language). Finally, some adaptive behavioral traits may not appear until later in life because they depend on developmental processes (e.g. human bipedal walking).

It is sometimes difficult to ascertain whether a particular trait is an adaptation because the fitness advantage it provided in the past may not be the same across evolution. In other words, a particular characteristic may be providing a valuable service now, without having been selected for that function in the past (i.e. without being an adaptation). These adaptations to one environmental problem that can be co-opted to solve another are called **exaptations**. Lewens (2007) provides an illustrative example: many people find that a screwdriver is very good at lifting lids from paint cans, but that is not what a screwdriver was originally designed to do. Some physiological traits fit the same pattern. Bird feathers probably evolved for thermoregulation but then served the important function of early flight. As another example, primate hands probably have evolved for manual dexterity, but humans also use them to hold a wedding ring which symbolizes a monogamous relationship. This symbol arguably helps with social bonding and co-parenting, leading to increased fitness of offspring. Yet, one would not propose that the structure of fingers is an evolved trait for pair-bonding and offspring survival.

The difficulty of defining and identifying adaptive traits led Williams (1966) to conclude that "evolutionary adaptation is a special and onerous concept" (1966, p. vii). The 'onerous' problem is compounded with cognitive traits because, even with contemporary species, cognition must be inferred through other measures. Fossil records and artifacts provide minimal information on the behavior of extinct animals, although they can be used to infer what animals were capable of doing (e.g. pterodactyls could fly and the early human ancestor *Homo habilis* created basic tools). In some cases, scientists compare cognitive abilities of contemporary species that share a common ancestor as a means to understand whether a particular cognitive trait is an adaptation. It is important to remember, however, that different animals will often employ different solutions to the same problem. Mate and offspring recognition is a perfect example: this is a survival problem that must be solved by birds and mammals, particularly for species in which both parents contribute to rearing the young. Most animals rely on sensory abilities, such as vision or olfaction, to recognize their mate and offspring, but the particular mechanisms that accomplish this task vary across species. The female emperor penguin (*Aptenodytes forsteri*), for instance, has the amazing ability to use vocal calls to identify her mate, who has been incubating their egg for over 2 months, within a colony containing a cacophony of hundreds of vocalizing males (Figure 1.2).

Finally, to fully understand adaptations, it is helpful to also consider the other products of evolution that are not the result of natural selection. **Byproducts** are side effects of adaptations (e.g. a belly button is the byproduct of the adaptive umbilical cord; Buss, 2004) and **random effects** (sometimes called noise) are chance mutations (e.g. the particular curvature of one's belly button) that do not provide any survival or reproductive advantage. Both byproducts and random effects may disappear across evolution, particularly if there is an evolutionary cost to maintaining these traits (e.g. a chance mutation that produces brightly colored feathers on a bird may make them more conspicuous to predators). On the other hand, environmental conditions may change such that a trait which provided no fitness advantage when it emerged may be adaptive in the future, in which case it would be more likely to be passed on to the next generation.

Figure 1.2 Emperor penguin colony with cubs in Antarctica.

1.2.3 Speciation

The concept of adaptation helps to explain the emergence of different species across evolution. To Darwin and his contemporaries, a species was a group of animals that resembled each other; a primary task of biologists during this period was to classify different animals into groups, based on shared physical characteristics. By the mid-twentieth century, with advanced knowledge of molecular biology, species were defined based on genetic similarities which map very closely to physiological attributes. On a general level, a species is a group of animals capable of interbreeding and producing fertile offspring. That is, different species cannot breed together, although there are a few notable exceptions (e.g. dogs and wolves). It is unlikely that humans and chimpanzees could produce new baby organisms (even with current reproductive technologies) because the eggs and sperm would not join properly. But the common ancestor of humans and chimpanzees was a species that bred together, very effectively, until approximately four to five million years ago.

Wallace (1870) explained how more than one species can evolve from a common ancestor based on the idea that subpopulations of a single species may display a trait that becomes adaptive when environments change. An animal's environment may change due to extraneous factors, such as the meteorite impact that led to the extinction of dinosaurs. A more common scenario is that small groups of animals move to a new habitat that provides better foraging, shelter, or reproductive opportunities. Some of these animals may possess a physiological or behavioral trait that had no fitness benefit in the previous habitat, but now provides a reproductive or survival advantage in the new conditions. These traits are more likely to be passed on to the next generation, but *only* in the group of animals that moved to the new environment. The separated groups, originally members of the same species, may diverge to the point that they can no longer breed together, a process called **speciation**. This tight link between physiological traits and local environmental conditions is illustrated by one of Darwin's best-known examples, shown in Figure 1.3.

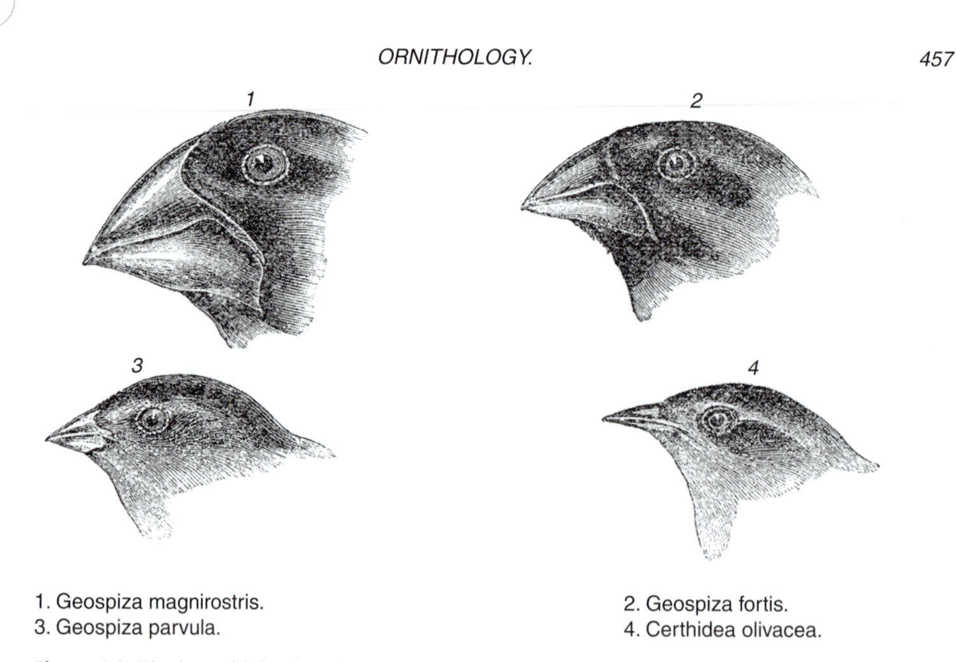

Figure 1.3 Finches with beaks adapted to different diets observed by the Charles Darwin in the Galapagos Islands during his voyage on the Beagle.

The same principle applies to cognitive processes, even if these are more difficult to observe and track across evolution. In other words, certain environmental conditions in the past must have favored specific types of information processing. Language is a prime example: many people believe that this form of communication evolved in humans when the complexity of our social environment increased. The question of whether other animals have language and whether social complexity explains the evolution of human language is not resolved (see Chapter 12 for details). In contrast, there is clear evidence that heritable differences in another cognitive function (visual processing) have resulted in species divergence (Boughman, 2002). A subpopulation of cichlid fish living in Lake Victoria in East Africa carry genes that make them more sensitive to red light. (Cichlid fish belong to the family *Cichlidae*, which contains a number of species.) Because light colors are absorbed differently as they pass through the water, blue light is present near the surface whereas just a few meters below the surface, red light dominates. Fish sensitive to red light will be better able to find food in deeper water and red-colored mates. Fish lacking these genes are better adapted to the blue light conditions of shallow water. Although there is no physical barrier between the shallow and deep areas of the lake, fish with different light sensitivity slowly separated into two groups. With continued environmental pressures favoring one type of light sensitivity over the other, these neighboring populations became two separate species of fish within the last few hundred thousand years (a relatively short time in evolutionary history). This example makes the point that differences in cognitive processing, like physiological or behavioral traits, may contribute to evolutionary change and speciation.

1.2.4 Continuity hypothesis

The basic tenets of the theory of evolution by natural selection led to Darwin's **continuity hypothesis**, which is the idea that trait differences between animals and humans will be quantitative, not qualitative. In *The Descent of Man and Selection in Relation to Sex* (Darwin, 1871), he wrote the following:

There can be no doubt that the difference between the mind of the lowest man and that of the highest animal is immense. An anthropomorphous ape, if he could take a dispassionate view of his own case, would admit that though he could form an artful plan to plunder a garden – though he could use stones for fighting or for breaking open nuts, yet the thought of fashioning a stone into a tool was quite beyond his scope. Still less, as he would admit, could he follow out a train of metaphysical reasoning, or solve a mathematical problem, or reflect on God, or admire a grand natural scene... Nevertheless, differences in mind between man and the higher animals, great as it is, is certainly one of degree and not of kind. (pp. 104–105)

In this passage, Darwin acknowledges that humans and animals will differ – sometimes greatly – on many traits and abilities, but he believed that the difference is not in the trait, only in how it is expressed. According to Darwin, therefore, animals possess to some degree many, if not all, of the cognitive and emotional traits that humans display, even if only at an incipient level. Modern researchers, however, do not simply automatically attribute human characteristics and traits to animals (referred to as **anthropomorphism**); rather, they test these ideas experimentally. Yet, this experimental framework was not available to Darwin, who instead relied on anecdotes from people who had frequent interactions with animals, including pet owners, zookeepers, and hunters. Each provided compelling accounts of what appeared to be highly intelligent behavior in animals. Darwin was not alone in using this technique; in the early 1880s his friend and colleague, George Romanes, compiled a book of similar anecdotes, entitled *Animal Intelligence*. Here is an example of a typical story:

As a beautiful instance of the display of sympathy, I may narrate an occurrence which was witnessed by my friend Sir James Malcolm – a gentleman on the accuracy of whose observation I can rely. He was on board a steamer where there were two common East India monkeys, one of which was older and larger than the other, though they were not mother and child. The smaller monkey one day fell overboard amidships. The larger one became frantically excited, and running over the bulwarks down to a part of the ship which is called 'the bend', it held on to the side of the vessel with one hand, while with the other it extended to her drowning companion a cord with which she had been tied up, and one end of which was fashioned around her waist. The incident astonished everyone on board, but unfortunately for the romance of the story the little monkey was not near enough to grasp the floating end of the cord. The animal, however, was eventually saved by a sailor throwing out a longer rope to the little swimmer, who had sense enough to grasp it, and so to be hauled on board. (Romanes, 1883; pp. 474–475)

1.2.5 Anecdotal method

As many critics noted at the time, the fact that Darwin and Romanes relied so heavily on anecdotes to develop their theories of continuity was problematic. In the case of Romanes' reports, the observations were usually made by one person, and in many cases relayed to him second, or even third, hand. In all likelihood, the stories were embellished with each retelling and few of these anecdotes were later re-examined more carefully. Even if it were presumed that the animals *did* engage in the behaviors depicted in the stories, it is difficult to judge whether the actions reflected the complex cognitive processes the writer often assumed. In Romanes' story, it is not clear that the larger monkey threw out the rope because she understood the smaller monkey's desire to be pulled out of the water. Nor is it obvious, as Romanes assumed, that the monkey's frantic excitement reflects a 'display of sympathy.' The problem with attributing cognitive abilities to animals, based on a single subject, is often illustrated using the Clever Hans story (see Box 1.1).

Box 1.1 Clever Hans

Figure 1.4 Clever Hans, circa 1904.

Clever Hans was a horse owned by a German schoolteacher, Wilhelm von Osten, who noticed that Hans seemed to have quite an intellect. When given math problems, for example, Hans would answer by tapping on the ground with his hoof until reaching the right answer. He could even reply to questions with German words by tapping out the numbers of the appropriate rows and columns on an alphabet table. It is easy to see why he was the toast of Berlin in the early 1900s (Figure 1.4).

In 1904, a group of academics and animal trainers known as the September Commission was created to examine Hans' abilities. There was suspicion that von Osten was giving Hans signals for the correct answer. Much to their surprise, even when other individuals posed questions to Hans in the absence of von Osten, the horse answered correctly. In the end, the commission found no indication of trickery underlying Hans' performance (Candland, 1993).

Despite this, one member of the committee, psychology professor Carl Stumpf, felt that surely something must be going on. He asked his student, Oskar Pfungst, to do an investigation, focusing on the possibility that somehow Hans was being cued. After several experiments, Pfungst found that when the tester was aware of the question or answer himself, Hans did well. But when the tester did not know, Hans got the answer wrong, for example by overestimating the correct number in an arithmetic problem. Pfungst concluded that a knowledgeable tester would enact subtle body movements just as Hans reached the correct answer, effectively cuing Hans to stop tapping.

Pfungst never claimed that von Osten was a charlatan; the owner and the experimenters were unaware of their cuing and von Osten was completely convinced that his horse could solve arithmetic problems. In this way, Hans offers an important lesson on the potential biasing of results for those studying animals and humans. Indeed, most experimental protocols now ensure that 'Clever Hans Effects' are eliminated through a 'double blind' procedure in which neither the subject nor the tester knows the expected outcome.

The Clever Hans affair occurred during a period in which many such stories of great cognitive feats by animals were spreading. Many of these cases were likely the result of intentional trickery on the part of the owners, although others might have been as innocent as the actions of von Osten. Pfungst's findings served to appropriately temper the general excitement and speculation about the animal mind (Roberts, 1998), and in turn became a factor in the development of a controlled, experimental field of study.

Even without Clever Hans, many scientists were skeptical of evidence derived from anecdotes. In addition to the problem of reliability, they noted that personal recollections of a specific incident may not reflect typical or general trends in a population. In psychology, this is sometimes referred to as the 'person who' fallacy in reference to the fact that individuals often reach conclusions based on their own personal experience (e.g. "I know a person who..."). This is not to say that anecdotes have no place in scientific research; indeed, they are often the impetus for new scientific discoveries. For example, an astute observer may note that a particular animal displays a behavior that is unique or different from the rest of their group, such as avoiding a particular food that other animals are eating. This could lead them to investigate whether this avoidance response is due to the animal's previous experience with the food, observing another animal not eating the food, or a food aversion that was inherited from a parent. In this way, anecdotes can inform scientific research but they cannot, on their own, be used as scientific evidence.

1.3 Influence: experimental psychology and behaviorism

One of the most vocal critics of the anecdotal method (and of Darwin himself) was the British psychologist, C. Lloyd Morgan (1842–1936). Morgan's disbelief of Darwin's claim for animal intelligence stemmed from the fact that all of the evidence was anecdotal and, therefore, difficult to replicate. His book *An Introduction to Comparative Psychology* (Morgan, 1894) described a number of simple experiments in which he presented animals (often his own dog) with a problem to solve. The animal's ability to find the solution was an indication of its intelligence. For example, anecdotal evidence suggested that dogs spontaneously figure out how to open gates but when Morgan tested this idea systematically, he concluded that the animals learn this through trial-and-error. Morgan's interpretation of his data followed a straightforward rule: keep it simple. In other words, there is no need to attribute complex thought processes to animals (or indeed to humans) if their behavior can be explained by simple or basic mechanisms. This directive, now referred to as **Morgan's canon**, had a significant impact on the development of experimental psychology. Even today, many researchers attempt to rule out, or control for, simple explanations of behavior prior to attributing higher order cognitive functions to their subjects (either human or animal).

The growth of experimental psychology was advanced further by Edward L. Thorndike (1874–1949), who advocated a systematic analysis of animal behavior using experimental manipulations. His 1911 publication, *Animal Intelligence*, was a direct rebuttal to Romanes' earlier work of the same name (Thorndike, 1911). Thorndike conducted hundreds of experiments with chicks, rats, cats, dogs, fish, monkeys, and humans. Much of his work centered on trial-and-error learning in which cats learned a particular response, most commonly escape from a box, over a series of test trials. Thorndike argued that escape times decrease because animals form associations between environmental stimuli and their own responses (e.g. pulling a rope that opened a hatched door in the box). Another way to state this is to say that behavior is shaped by an animal's reinforcement history: specifically how frequently it is rewarded or punished for its actions. An important premise of Thorndike's work was that behavioral tests must provide objective, quantifiable, and replicable measures of learning. Although Thorndike's theoretical positions on learning were later challenged, this tightly controlled methodology helped to establish comparative psychology as an experimental science.

1.3.1 The rise of behaviorism

John B. Watson (1878–1958) championed Thorndike's approach in the early twentieth century, arguing that psychologists must focus on observable events (both stimuli and responses) without speculating on the inner processes that may control behavior. Many of Watson's ideas were a reaction to nineteenth-century mentalistic psychology which proposed theories of thought processes that could not be tested scientifically. Watson's influential paper, *Psychology as a Behaviorist Views It* (Watson, 1913), expounded the following principles, which guided the field for almost 50 years:

1. Behavior, and only behavior, should be the subject matter of psychology.
2. Mentalistic concepts such as consciousness and imagery should not be the topics of an objective science of psychology.
3. Both human and animal behavior is modified by experience, specifically the learning of stimulus–response associations.

Watson pointed out that the contents of the mind could not be observed directly, so if psychology was to gain respect as a science, it must focus on objective and quantifiable measures of behavior. The idea that behavior (not thoughts or ideas or any other cognitive process) is the only justifiable object of study in psychology is associated with a subdiscipline of the field known as **behaviorism**. The extreme position, that mentalistic states have no role in behavioral change, is now referred to as **radical behaviorism**. In truth, many psychologists at the time, even those who aligned themselves with behaviorism, recognized that mental processes exist and no doubt contribute to behavioral change. The problem is that these internal processes are difficult, if not impossible, to measure; thus, many researchers practiced **methodological behaviorism** in that their research involved quantifiable measures of behavioral output and tight control of extraneous variables. Many described their findings in terms of reinforcement history (i.e. stimulus–response associations) even if they accepted that internal, unobservable states could influence behavior. Today, researchers who are guided by the premise that behavior should be described in terms of observable stimuli and responses (i.e. methodological behaviorism) are simply called behaviorists.

1.3.2 Animal thought

Although Watson's ideas had a huge impact on the field, many researchers continued to study thought processes in animals and to interpret the animals' behavior as the output of complex, cognitive functions. The most prominent of these, Edward Tolman (1886–1959), agreed that stimuli should be controlled in each experiment and that the unit of analysis should be behavior. Nonetheless, he hypothesized that animals formed mental representations of their environment that were more complex than simple stimulus–response associations. Although Tolman did not use the term 'cognition,' the ideas put forward in his 1932 publication, *Purposive Behavior in Animals and Men*, fit clearly with the idea that behavior is controlled by information processing that occurred inside the animal's head. Tolman is just one example of many researchers who were studying 'higher processes' during a period in which behaviorism dominated psychology (Dewsbury, 2000). (See Box 1.2 for another prominent example).

Today, it is common for researchers to describe animal behavior in terms of thought (or cognitive) processes. Nonetheless, the legacy of behaviorism is still apparent in that this work relies on observable, quantifiable, and replicable measures of behavior. Thus, the growing interest in the experimental study of animal behavior during the early part of the twentieth century laid the foundation for work that continues to the present day.

Box 1.2 Wolfgang Köhler: insight learning

Figure 1.5 Kohler's chimps.

In 1913, German psychologist Wolfgang Köhler was invited to direct a primate research station on Tenerife in the Canary Islands. The facility held a colony of captive chimpanzees (*Pan troglodytes*) that had access to an outdoor pen, and Köhler set out to "test the intelligence of higher apes" (*The Mentality of Apes*, 2nd Edition, 1927, p. 1). The research aims were largely based on the theory of continuity of psychological processes, with Köhler noting that the "chemistry of their bodies... and the structure of their most highly developed organ, the brain" are similar to our own. Thus, he wondered "whether they do not behave with intelligence and insight under conditions which require such behavior" (1927, p. 1).

The chimpanzees were provided with a variety of objects such as sticks, poles, boxes, and other objects. Köhler developed 'problems' for the chimps to solve, typically involving obtaining food that was out of reach (Figure 1.5). In what is likely the most popular situation, a chimpanzee jumps and reaches unsuccessfully at bananas that have been hung out of reach. Usually, after a time, the chimpanzee attends to the objects in the pen, then back at the food, and then at the objects again. Ultimately, the chimpanzee utilizes an object (e.g. a long pole or stacked boxes) to enable the retrieval of the food.

Across many of these situations, the behaviors Köhler observed appeared to involve insight and planning. These seemingly sudden solutions were evidence to Köhler that the associative processes detailed by the behaviorists did not describe all learning events. Here, learning did not seem to be the result of the development of stimulus–response connections by repetitive reinforcement (Köhler, 1927, also see Spence, 1938); chimpanzees were not coming upon the solution *by chance* and then repeating the behavior because it was reinforced. That said, Köhler did suggest that experience has an important role in insight. In *The Mentality of Apes*, he noted that the chimpanzees played with the sticks and other objects, often using them as substrates for climbing prior to applying them for problem-solving. He proposed that this experience set the groundwork for success in the insight tasks, perhaps by creating the opportunity to learn about relationships between placement of the objects and certain motoric actions.

However, not everyone agreed with Köhler's interpretation that the solutions reflected insight. For example, Schiller (1952) suggested that problem-solving by the chimpanzees was based on innate patterns of motion that were used during the tasks and thus reinforced. That is, the fact that chimpanzees often played with the objects in manners that would also allow success in the tasks (e.g. stacking boxes or climbing poles) suggests that perhaps 'solutions' were the result of engaging in an innate motor pattern that, in a task, happened to lead to food reward. The criticisms of Köhler's work are well taken; if insightful problem-solving requires completely novel actions, his work does not provide clear evidence for this capacity in chimpanzees. Of course, given this criterion, it might also be difficult to demonstrate the phenomenon in humans (Gould & Gould, 1994).

1.4 Influence: ethology and behavioral ecology

During the same period that psychologists were examining learned responses in the lab, many European scientists were studying behavior by observing animals in their natural environment. This approach can be traced back to the naturalists who catalogued and described species-typical behaviors in great detail. These include observations of honeybee hive construction by Jean-Henri Fabre (1823–1915) and beaver dam building by Thomas Morton (1579–1647). John James Audubon (1785–1851) documented a variety of natural behaviors in animals, primarily birds, although he is better known for his engravings depicting his subjects in naturalistic settings. The early naturalists were often amateurs, but their work provided the foundation for scientific studies in **zoology**, the branch of biology that relates to the animal kingdom.

1.4.1 Ethology

Naturalistic observations of animal behavior became more systematic and directive during the eighteenth and nineteenth centuries. A turning point in the field was the publication of the first volumes of *Histoire Naturelle* (1749) by Georges-Louis Leclerc, Comte de Buffon (1707–1788). In this work, Buffon recommended that the classification of species should include behavioral (not just morphological) traits such as social organization, habitat selection, and foraging patterns. Fabre is also considered to be a pioneer in this emerging discipline (Danchin *et al.*, 2008), particularly for his focus on naturalistic behaviors that were *not* the product of learning. Both Buffon and Fabre emphasized that understanding the behavior of one species would be enhanced by comparing it to the behavior of similar species. This comparative approach foreshadowed the emergence of **ethology**, the scientific study of animal behavior, as a branch of zoology distinct from comparative anatomy.

Ethology, which flourished in Europe in the early to mid-twentieth century, is based on the premise that the study of behavior should be naturalistic. Lab experiments may be justified due to practical constraints (including better control of observation) but, even in these studies, researchers should attempt to recreate natural conditions and to measure responses that are part of the animal's behavioral repertoire. In describing a particular behavior, ethologists focus on evolutionary explanations, suggesting that behavioral traits must be understood in terms of natural selection. Put another way, ethology promotes the idea that behavior is an inherited trait; just like physiological traits, such as body size or coloration, behavior can be described in terms of random variation, adaptations, and fitness.

Ethologists working at the beginning of the twentieth century, mostly notably Konrad Lorenz, Nikolaas Tinbergen, and Karl von Frisch, focused on **instincts** or behavioral patterns that appear in

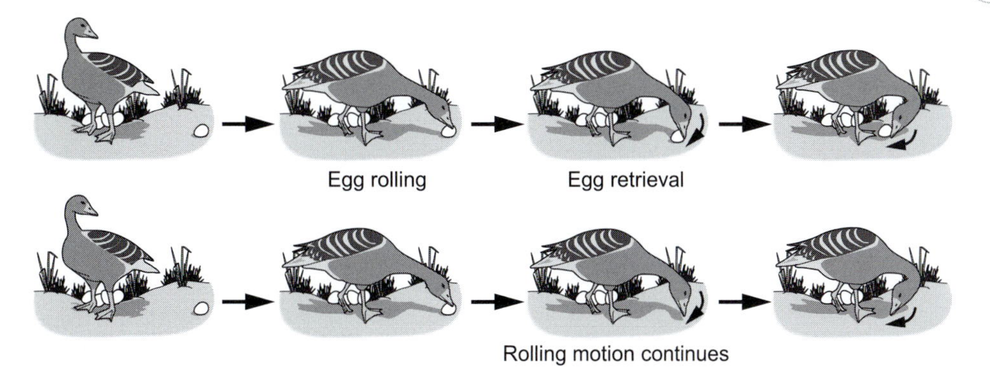

Figure 1.6 Fixed action pattern in graylag geese. Top panel: When a goose sees an egg outside of its nest, it retrieves the egg by rolling it back into the nest with its beak. Bottom panel: If the egg is removed part way through the retrieval, the goose continues to engage in the rolling behavior as it retrieves a phantom egg.

full form the first time they are displayed, even if the organism has no previous experience with the stimuli that elicit this behavior. Instincts are often described as **innate** responses, those that are inherited or 'hard-wired' (i.e. not learned). In studying instincts, Lorenz (1950) and Tinbergen (1951) developed the concept of **fixed action patterns (FAPs)**: stereotyped, species-typical behaviors that occur in a rigid order and are triggered by specific stimuli in the environment. Once initiated, these behavioral sequences continue to completion, sometimes to comical effect. For example, when a graylag goose (*Anser anser*) observes an egg-sized object near the nest, it triggers a FAP that retrieves the object by rolling it back into the nest in a specific, predictable series of actions. If an experimenter removes the object in the middle of the goose's rolling, the bird will continue to engage in rolling motions *as if* it were bringing a phantom egg back into the nest (see Figure 1.6) (Tinbergen, 1951).

Ethologists characterized many naturalistic behaviors, such as sexual responses, aggression, and hunting or foraging as FAPs, but it quickly became apparent that these behaviors were not as 'fixed' as they originally appeared. Tinbergen (1951) was particularly vocal in pointing out that many apparently innate responses, including FAPs, could be modified by experience. He emphasized that behavior must be understood as an interaction between learning and inherited mechanisms, demonstrated most clearly in Lorenz's study of **imprinting** (Lorenz, 1952). Imprinting is a particular type of learning in which exposure to specific stimuli or events, usually at a young age, alters behavioral traits of the animal. In a familiar case of filial imprinting, young birds learn the characteristics of the first moving object they encounter, and then follow it around. The behavior develops rapidly very early in life and remains intact, more or less, throughout the organism's life. Lorenz noted a critical period of 13 to 16 hours in which newly hatched geese would imprint on the first suitably sized, moving object they encountered, most famously Lorenz himself (Figure 1.7). The innate aspect of imprinting is the tendency to attend to a moving visual stimulus soon after birth. But the specific environmental stimulus may vary so that individual birds learn to follow different 'mothers,' depending on what they observed during this critical period.

Similar criticisms arose regarding other FAPs, specifically that these were not entirely inflexible and mechanistic. Even the goose egg rolling response can be modified by experience (within limits). Moreover, some FAPs require environmental feedback to be expressed in their entirety and the behavioral sequence of many FAPs varies, even if the same stimulus is presented repeatedly. Because of these problems, some authors suggest that 'modal action pattern' is a more appropriate

Figure 1.7 Konrad Lorenz with some of his geese (1903).

description of the phenomenon (Domjan, 2010), although it is still used less commonly than FAP. The point to remember is that few (if any) behaviors are absolutely fixed, even those that appear to show little variability between individuals and across situations. Thus, rather than separating innate from learned behaviors, ethologists now view these as traits that vary along a continuum. Inherited mechanisms set up a behavioral response that is then modified by experience; this modification may be minimal or extreme, making it difficult or easy, respectively, to detect any effect of learning.

1.4.2 Four questions of ethology

As the influence of ethology broadened and more researchers turned their attention to studying animals in their natural environment, Tinbergen recognized the need to develop an organized approach to the study of behavior. The presentation of his 'four questions' (Tinbergen, 1963) provided a framework for the growing field of ethology and a means by which researchers could focus their research. More specifically, Tinbergen proposed that behavior should be analyzed in terms of four separate causes, each of which can be framed as a scientific question. These are:

1. Adaptive value. *What is the function of the behavior?*
2. Evolution. *How did the behavior develop across evolution and how does it compare to the behavior of closely related species?*
3. Ontogeny. *How does the behavior change across the lifespan of the organism?*
4. Immediate causation. *What are the internal mechanisms that produce the behavior?*

The first two questions, explaining behavior in terms of adaptive value and evolution, are referred to as **ultimate causes** of behavior. The second two, explaining behavior in terms of development and physiology, are referred to as **proximate causes** of behavior. Ultimate explanations focus on the animals' evolutionary lineage as well as the ecological pressures, both past and current, that affect reproductive success and survival.

Proximate explanations deal with the building and operation of an animal that enables it to behave in a particular way. It is important to remember that these separate levels of analysis are not mutually exclusive: one explanation does not preclude another. Rather, the four causes (or questions) are complementary in that they attempt to explain the same behavior from different perspectives.

The application of Tinbergen's four questions to a specific behavior is often illustrated by the simple observation that a male American robin (*Turdus migratorius*) will start to sing in the spring. The function of the behavior is mate attraction or territorial defense (Question 1). Singing evolved in some, but not all, birds because it provided a fitness advantage in certain environments. One possibility is that bigger and stronger birds were capable of emitting more complex vocalization so male birds displaying this trait were more attractive to females (Question 2). The characteristic of a male song depends on exposure to other male songs during development, particularly its own father (Question 3). Finally, male birds begin to sing in the spring when longer periods of daylight stimulate hormone production that acts on specific brain areas responsible for vocalization (Question 4).

1.4.3 Behavioral ecology

Lorenz, Tingergen, and von Frisch shared the Nobel Prize in Physiology or Medicine in 1973, attesting to the impact and importance of the discipline which they founded. Even so, both the methodologies and theories of this research area were criticized in the mid to late twentieth century. Many scientists were uncomfortable with the ethologists' focus on FAPs and the fact that behavior of individual organisms was frequently examined in isolation. Many animals live in social groups so any theoretical account of behavior should include social interactions. These ideas were promoted by Wilson (1975), the founder of **sociobiology** (a field of study in which the principles of population biology and evolutionary theory are applied to social organizations), who explained many human activities, such as cooperation, altruism, and sibling rivalries, as adaptive traits. Sociobiologists made the point that humans, like animals, are always in competition for limited resources (e.g. food, territories, or mates); behaviors that provide a competitive advantage in these situations will increase an individual's chance of surviving and reproducing. Put another way, behavior facilitates the preservation of an individual's genes within a population, an idea that was put forward in the highly successful book, *The Selfish Gene* (Dawkins, 1976).

The rise in sociobiology coincided with scientific analyses of animal behavior within an economic framework. Several researchers in **ecology** (the scientific study of interactions among organisms and their environment) proposed that activities, such as foraging or hunting, reflected an internal process that weighed the costs and benefits of a particular behavior (Emlen, 1966). The costs associated with foraging include energy output, potential dangers that animals may encounter during their search, as well as time away from other fitness-related activities, such as reproduction. Behaviors that maximized gains and minimized costs were adaptive; as with physiological traits, these were more likely to be passed on to the next generation.

The increasing focus on the evolution of social behavior combined with an economic analysis of behavior led to the development of a new discipline. **Behavioral ecology** is defined as the scientific study of the evolutionary basis for animal behavior due to ecological pressures (Danchin *et al*., 2008). Behavioral ecologists study the interaction between organisms and their environment, and how these interactions result in differential survival and reproduction. The field solidified as a separate discipline with the 1978 publication of *Behavioural Ecology: An Evolutionary Approach*. Three more editions of this multi-author work appeared over the next 20 years. Behavioral ecology is also distinguished by the use of genetic markers to trace heritability of behavioral traits, a technique that was not available to early ethologists. Importantly, these tools have provided new insight into social interactions. For example, co-parenting was often taken as evidence for paternity, but genetic analyses revealed that socially monogamous species often engage in extra-pair copulations. That is, males of many different species help to raise non-biological offspring, a fact that has implications for understanding the evolution of social behaviors.

Today, behavioral ecology is viewed as an offshoot of ethology with an increased emphasis on ultimate explanations of behavior (Bolhuis & Giraldeau, 2005). The field arose, partially, as a reaction to the focus on proximate causes of behavior that characterized many early studies in ethology (e.g. imprinting by chicks). Although Tingergen's four questions of ethology gave equal weight to both ultimate and proximate explanations of behavior, behavioral ecology assigns a higher priority to the former. According to a contemporary text (Danchin *et al.*, 2008), physiological and developmental mechanisms underlying a behavior can be understood *only* within the context of evolution and function. Despite a different theoretical focus, ethologists and behavioral ecologists investigate the same phenomenon (natural behaviors) and frequently work within the same research unit, so the distinction between the two disciplines is often difficult to establish.

1.5 Emergence of comparative cognition

Comparative cognition grew out of two traditional sciences: psychology and zoology or, more specifically, behaviorism and behavioral ecology. The two disciplines developed and worked (more or less) independently for over 50 years. In many cases, scientists in each field would be interested in exactly the same question (e.g. how do animals transmit information to each other?), but there was little cross-talk between them in the initial part of the twentieth century. Communication increased in the 1960s and 1970s when it became apparent that researchers in the two areas were frequently studying the same phenomenon. The book *Animal Behavior: A Synthesis of Ethology and Comparative Psychology* (Hinde, 1970) provided clear indication that the goals and interests of the two disciplines were complementary. In some cases, experimental psychology and behavioral ecology were merging into a single field of behavioral research that was no longer distinguished by methodological or theoretical differences. Many psychologists began to conduct fieldwork and behavioral ecologists frequently brought their studies into controlled laboratory settings. More importantly, there was a growing recognition that a complete understanding of any behavior or cognitive process requires explanations in terms of both proximate and ultimate causes.

This resulting synergy of theoretic perspectives is reflected in the contemporary field of comparative cognition. A guiding premise in this discipline is that understanding why a behavior occurs is informed by understanding how it occurs and vice versa. Put another way, cognitive processes are controlled by an interaction between factors in an animal's environment and biological predispositions to behave in a particular way. For example, the way in which experience may alter innate spatial navigation abilities will be discussed in Chapter 7, and the role of biological predispositions in learned associations will be discussed in Chapter 4.

1.5.1 A focus on cognitive processes

Comparative cognition, as a distinct discipline, arose in the mid-twentieth century with Darwin's theory of evolution by natural selection providing a theoretical framework, and both behaviorism and behavioral ecology providing methodological tools. The emphasis on cognition arose within the field of psychology, which underwent an important shift in emphasis during the mid-1900s. The impetus for these changes came from two other disciplines: **linguistics** (the study of the nature and structure of human language) and **computer science** (the study of computer design and computational theories). Linguists had obvious links to psychologists, particularly those who were studying language acquisition. According to Skinner, a renowned behavioral psychologist, learning to speak was no different than learning any other behavior: it could be explained by the same principles that governed rat lever pressing in the lab (Skinner, 1957). Not everyone agreed. Noam Chomsky, a prominent linguist at the time, wrote a scathing attack on Skinner's piece (Chomsky, 1959) in which he criticized Skinner's inconsistent and vague use of the term reinforcement and argued, further, that language acquisition could not be explained by stimulus–response learning. Chomsky pointed out that children all over the world master language and the rules of grammar at roughly the same age, with the exception of very rare and tragic cases involving social isolation (see Curtiss, 1977, for review). This amazing, universal ability could only be explained by innate, mental processes that structured language learning. Chomsky's criticism of Skinner's paper (and the field of behaviorism in general) was timely. Many people, including scientists working within the field, were disillusioned with behaviorism and its apparent inability to explain complex behaviors.

During the same period, computer science was developing as a discipline led by the pioneering work of Alan Turing. Turing designed and built some of the first computers, but his interest in these systems extended far beyond the mechanics of how they work. He wrote numerous papers on how information processing occurs in both humans and machines, pointing out the similarities between the two. One of his most influential articles (Turing, 1950) equated information processing with thinking and raised the question: "Can machines think?" Turing's influence was widespread. Over the next two decades, computer science and the study of artificial intelligence expanded rapidly at many academic institutions. Scientific advances in this area helped to establish the study of information processing as a credible endeavor. Psychologists who were frustrated by trying to explain all behavior without reference to hypothetical constructs quickly embraced computational models of thinking, reasoning, and decision making (topics that were seldom discussed by behaviorists).

Thus, the growing interest in cognition within psychology was both a reaction to the limitations of behaviorism and an acceptance of ideas put forth by linguists and computer scientists. Some texts describe this shift in the late 1950s and early 1960s as a cognitive revolution in that many tenets of behaviorism were rejected and replaced with discussions and theorizing of mental processes and representations. Others argue that the shift in research focus was an evolution, not a revolution (Leahey, 1992), pointing out that cognitive research continued throughout the early and mid-1900s, even if it took a backseat to the more dominant behaviorist work. For example, the first half of the twentieth century witnessed reports of tool use (e.g. Yerkes & Spragg, 1937), imitation (Warden & Jackson, 1935), and spatial maps (e.g. Tolman & Honzik, 1930) in animals, most of which were discussed as higher mental processes. Whether it is called a revolution or an evolution, psychology did undergo a change of emphasis, and the modern-day field of comparative cognition is steeped within this academic environment.

The legacy of behaviorism is still apparent in comparative cognition with an emphasis on quantifiable measures of behavior and an attempt to control extraneous variables (i.e. methodological behaviorism).

1.5.2 Cognition in an evolutionary framework

It should now be clear that the emergence of comparative cognition can be traced back to the synthesis of behaviorism and behavioral ecology, along with the growth of cognitive science. This convergence of ideas coincided with scientific discoveries in other fields, specifically molecular biology, that led to a widespread acceptance of evolutionary theory (see Box 1.3). This provided a theoretical framework for the burgeoning field of comparative cognition in that research in this area is guided by the premise that cognitive processes, like physical traits, are shaped by natural selection. Darwin, himself, promoted the idea; although he did not use the term cognition, he believed in the continuity of complex thought processes across species.

Box 1.3 A primer on genetics

When Darwin proposed his theory of evolution by natural selection, he had no means to explain how heritable traits are passed from parent to offspring. Nor did he know the work of his contemporary, Gregor Mendel (1822–1884), who had identified the laws of heredity by studying the transmission of observable characteristics, such as shape or color, in pea plants. Mendel suggested that these independent traits are passed from one generation to another in independent packages, which are now referred to as genes. Mendel's work was rediscovered decades after his death and his theories of inheritance were confirmed by advances in molecular biology, particularly the description of deoxyribonucleic acid (DNA) structure by the Nobel laureates, James Watson and Francis Crick (1953).

DNA forms long strands, called chromosomes, that are present in each cell in the body. Chromosomes contain four different nucleotides (adenine, thymine, guanine, and cytosine) that recur in various sequences across the entire strand. A gene is a specific section of the chromosome, made up of some combination of the four nucleotides. Genes have many forms (also called alleles), with a specific gene form being defined by the nucleotide sequence of the area it occupies on the chromosome. So, the gene is a particular location on the chromosome, and variations in the nucleotide sequence within that location are the gene alleles (Figure 1.8). Genes produce proteins, which build the physical body, and different alleles (i.e. nucleotide sequence) of the same gene account for variations in physical traits, such as the presence of dimples or freckles. Importantly, there is not a one-to-one mapping of gene to trait; even simple features like eye color are produced by a combination of three or more genes. Behavioral and cognitive differences are even more complex, even if these can sometimes be linked to a cluster of genes within a certain region of the chromosome.

All cells in the body with the exception of the sex cells (sperm and ova) have two sets of chromosomes. In the case of humans, each set has 23 but this number varies across other species. Offspring inherit one chromosome set from each parent so every individual possesses half of their mother's genetic code and half of their father's. If the same allele is inherited from both parents, the individual is homozygous for that particular gene; if not, they are heterozygous for that gene. Sometimes the nucleotide sequence on the chromosomes is altered during transmission, producing genetic variation

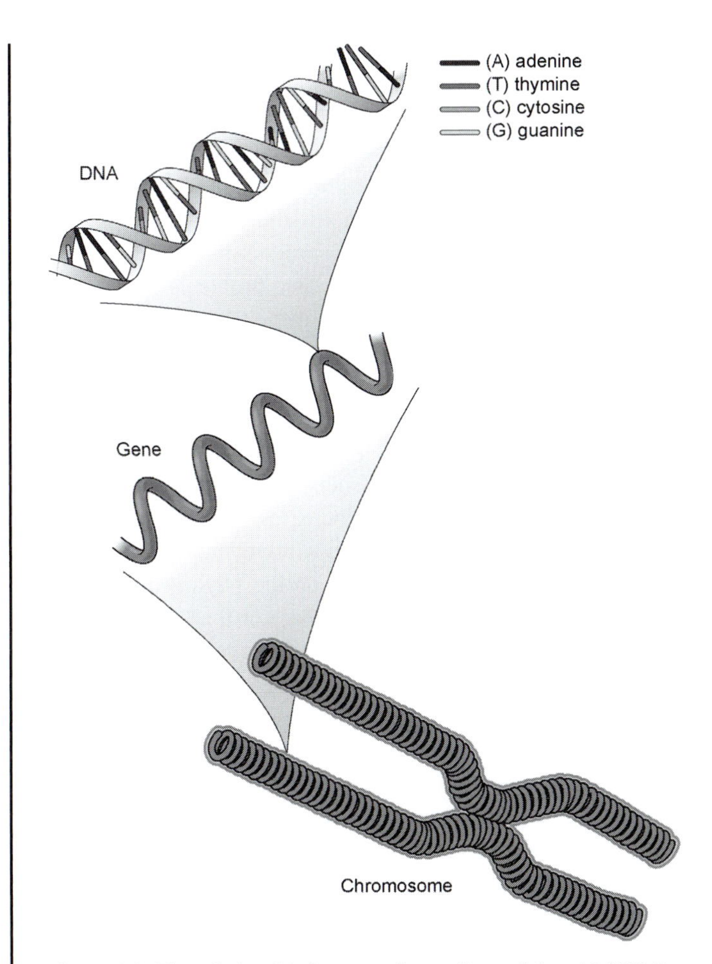

Figure 1.8 The relationship between deoxyribonucleic acid (DNA), genes, and chromosome. Genes are located on specific regions of the chromosome and are made up of a combination of DNA nucleotides (adenine, thymine, guanine, and cytosine).

or mutation. Most of these mutations are undetectable; in rare cases, the mutation is so severe that it causes death at an early stage of development. In other instances, some random genetic variation confers a fitness advantage to the individual (e.g. longer necks in giraffes that allow them to reach the best food), making it more likely that this individual will survive, reproduce, and pass on the gene to their own offspring. Thus, all three tenets of Darwin's theory of evolution by natural selection (Section 1.2.1) can be explained by genetic mutations and inheritance.

Examining cognition within an evolutionary framework presents a number of challenges. The first is determining how to classify cognition. This textbook provides one perspective, with each chapter presenting and then discussing a different cognitive process. Not everyone would agree with these divisions (or how the material is presented within each chapter), and future scientific advances will likely lead to a restructuring of these categories. Nonetheless, even if they disagree with the specific divisions, most researchers acknowledge that cognition should be viewed as a collection of processes that are characterized by distinct rules of operation, rather than a general purpose operating system. These individual processes interact to control both human and animal behavior, and researchers face the challenge of designing experiments that tap one process or the other. The task becomes even more complicated if the same cognitive process is to be examined in different species. These issues will arise repeatedly throughout the text, frequently with little or no resolution.

Once researchers identify a cognitive process that they wish to study, they face the added challenge of determining the adaptive value and evolutionary progression of this process. A premise of the work is that cognition evolved to solve fitness-related problems, but compared to many physical traits (e.g. large wings for migrating birds), it is difficult to determine how a particular cognitive process increased survival and reproduction in ancestral (or even contemporary) environments. One approach to this problem is to examine whether cognitive abilities of different animals match the ecological constraints of that species. For example, animals that live in social groups would be more likely to survive and reproduce if they cooperated with conspecifics in foraging, hunting, and defending territory. A corollary is that prosocial behaviors will be higher, or at least more complex, as the complexity of a social environment increases. Another approach is to assess whether closely related species exhibit the same (or similar) cognitive traits (MacLean *et al.*, 2012). Closely related species are represented as nearby branches on the phylogenetic tree (see Figure 1.9), indicating that they share a more recent evolutionary ancestor than distantly related species. If cognitive traits are shaped by natural selection, one would expect these to be comparable in closely related species that would have experienced similar evolutionary pressures.

Although this approach may be productive, inferring cognitive evolution through phylogenetic relatedness can be misleading. Even if different animals appear to possess similar cognitive abilities, these may be mediated by different mechanisms. Wasps, rodents, and birds all display spatial navigational skills when they return to a location where they found food in the past, but they may accomplish this feat in a variety of ways. Some animals may navigate by recreating a mental image of their environment whereas others may simply approach a cue that, previously, was in the vicinity of the food. In order to navigate effectively, some animals rely on visual cues, some on olfactory cues, and others on magnetic cues. Finding food is critical for survival so all animals must be able to do this, regardless of the physical traits that they possess (e.g. good or bad eyes, small or large brain, etc.). This process whereby individual species develop different strategies for dealing with the same fitness-related problem is called **common adaptation**. Common adaptation is also described as convergence on the same solution through independent evolutionary change.

The problem of common adaptation can be addressed, at least partially, by identifying biological mechanisms that mediate a particular cognitive ability. If the same brain processes control the same cognitive traits in different animals, it is more likely that these evolved from a common (perhaps recent) evolutionary ancestor. The rationale for this statement is that cognition is controlled by biological processes that are subject to the same evolutionary pressures as any other biological trait. Natural selection alters wing span in birds, making flight more or less efficient; the same process alters brain function, making cognitive processing more or less efficient. A less efficient cognitive process is one that uses more resources (primarily brain energy) and is not as accurate in solving an ecological problem (e.g. finding food). A number of other factors could decrease the efficiency of cognitive processing, many of which will be discussed throughout this text.

The idea that cross-species comparisons can inform studies of cognition raises the obvious question of how different species are ranked in terms of cognitive abilities. In the past, this question was often posed as "which animal is the most intelligent?" Darwin, himself, was interested in this question and comparative studies in the early to mid-twentieth century attempted to answer it by

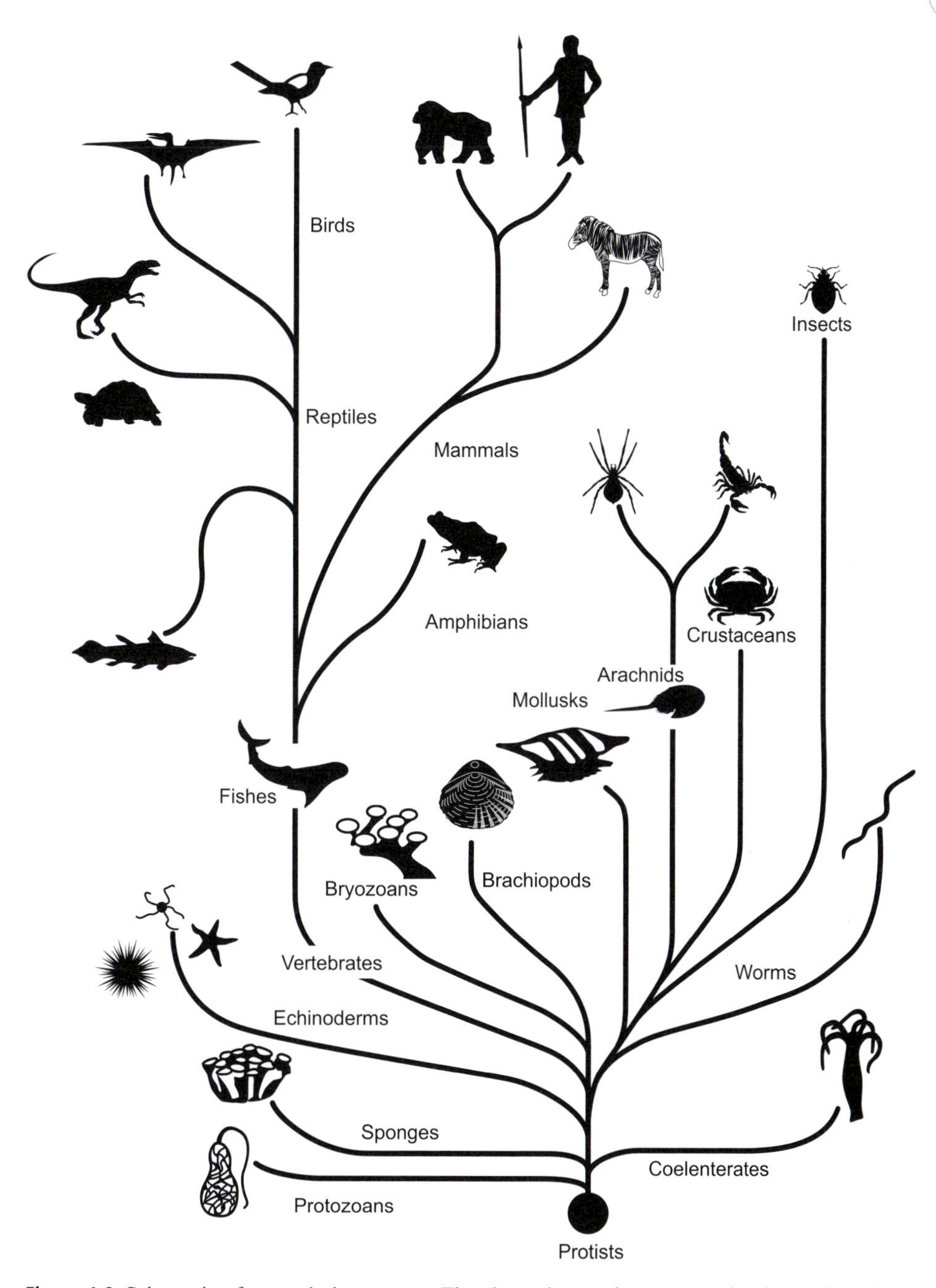

Figure 1.9 Schematic of an evolutionary tree. The closer the species are to each other on branches of the tree, the more traits they share.

assessing performance of different animals in the same behavioral paradigm (Bitterman, 1965; Harlow, 1953). The findings were generally inconclusive, partly because animals rely on different strategies to solve the same task (e.g. some animals have better visual abilities than others so learning to go through a maze with visual cues would be easier for this species). Also, it was never clear which dependent measure should be the most important (e.g. fewer errors, quicker task completion, or more rapid acquisition) in these studies. Other general problems of attributing intelligence to animals, based on what they do in an experiment, are outlined in the discussion of Köhler's work (see Box 1.2). Despite these challenges, certain researchers continued to study animal intelligence, promoting the idea that humans are not the only species to possess complex cognitive traits, such as higher order consciousness (see Box 1.4).

By the second half of the twentieth century, researchers began to recognize that intelligence, like cognition, is not a one-dimensional mechanism that varies quantitatively. Rather, it includes a number of attributes that must be assessed in different ways. For example, human intelligence is often assessed in terms of logic, abstract thought, communication, self-awareness, learning, and problem-solving, among others. A more productive approach to cross-species comparisons of intelligence, therefore, is to examine how a particular cognitive process varies across species and how this ability has allowed organisms to adapt to their environment. This, of course, is the approach of this textbook.

Box 1.4 Animal consciousness

Donald Griffin (1915–2003) conducted comparative cognition research, long before the term was coined, by integrating theories and techniques from biology, psychology, and cognitive science. His original research examined how bats navigate in the dark (Griffin, 1944). Griffin speculated that bats 'see' by emitting ultrasonic sounds and then listening to the echoes to reveal information about the objects around them. To test this idea, Griffin attached special microphones to the bats and set up vertical wires for them to fly around. By recording the ultrasonics, Griffin confirmed that bats were emitting bursts of sound as they flew (Figure 1.10). When he placed earplugs in the bats' ears, the bats were unable to avoid the objects, confirming that they were using echoes to locate and identify objects in their environment. Such information is so precise that bats can use it to find and eat tiny insects. Griffin coined the word 'echolocation' to describe this navigational skill in bats. The discovery was considered a major breakthrough, not only in biology, but also in the development of radar and sonar systems. Griffin even determined that blind people can use a similar strategy to navigate in familiar environments.

Despite the advances Griffin made in understanding auditory signaling, the question that intrigued him the most was that of animal consciousness. Griffin echoed Darwin's ideas when he noted that animals, like humans, engage in many complex and goal-directed behaviors. Given this similarity, it is entirely feasible that animals are also aware of their actions (assuming that one accepts humans have this awareness). In his numerous books and articles, Griffin provided examples of foraging behavior, predatory tactics, artifact construction, tool use, and many other abilities, concluding that it is simply impossible to explain what animals do without assuming that they are conscious of their own thoughts and behavior (Griffin, 1992).

Many researchers, even those examining cognitive processes in animals, regard Griffin's positions as extreme. Although many may believe that animals possess complex thought processes, they recognize the difficulty of finding an objective measure of consciousness in non-verbal subjects (i.e. animals). Despite the skeptics, Griffin continued to explore the topic, publishing *Animal Minds: Beyond Cognition to Consciousness* in 2001, just two years before his death (Griffin, 2001).

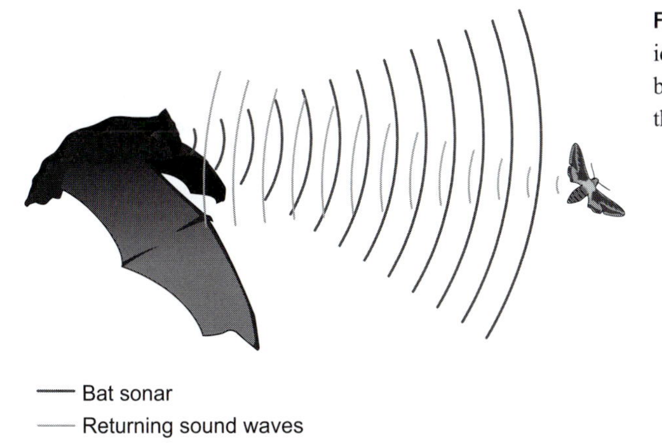

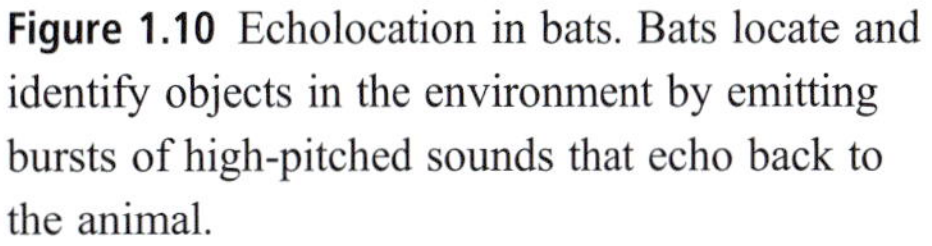

Figure 1.10 Echolocation in bats. Bats locate and identify objects in the environment by emitting bursts of high-pitched sounds that echo back to the animal.

1.6 An interdisciplinary approach

Comparative cognition is an interdisciplinary endeavor with scientists from different fields unified by the common goal of understanding how and why humans and animals process information as they do. This research objective can be approached by rephrasing Tinbergen's four questions in terms of cognition. That is, for any cognitive process (e.g. memory or communication), the following questions can be posed: What is its function? How did it develop across evolution? How does it change across the lifespan? And what are its underlying mechanisms? These last two questions are the purview of scientists in two disciplines that are intimately connected to comparative cognition: **developmental psychology** and **behavioral neuroscience**. Developmental psychologists study systematic changes that occur from birth to death in humans. Behavioral neuroscientists examine the relationship between brain function and behavior in all animals. The following two sections outline some of the ways that developmental psychology and behavioral neuroscience can inform research in comparative cognition (and vice versa).

1.6.1 Developmental psychology

Many people think of developmental psychology in terms of the changes that occur during infancy and childhood. Certainly much of the field is focused on these periods because the changes that humans undergo before puberty are truly remarkable. Anyone who has watched young children learn to walk, run, talk, and think for themselves will appreciate this statement. Developmental psychology, however, is not limited to these early ages; it extends across the entire lifespan to include adolescence, adulthood, and aging. Research in this field is often focused on understanding the interactions between nature (including possible innate predispositions) and nurture (experiences) that cause developmental changes.

Comparative cognition capitalizes on the fact that development psychology already has a long and established tradition of examining changes in cognition. Research in comparative cognition, in turn, builds on knowledge in developmental psychology by comparing the cognitive abilities of infants and children to those of other animals. These studies can be used to address a variety of questions that are the cornerstone of comparative cognition. For example, what cognitive capacities can exist in the absence of language? How do we define the biological predispositions and environmental conditions that influence what learning occurs? What is the evolutionary lineage of different cognitive processes?

Although research in animal cognition and human cognitive development has often delved into the same questions, only recently has there been an increase in collaborative work between the two disciplines. One problem was that developmental psychology and comparative cognition frequently employed different experimental procedures, making it difficult to compare results from the two sets of studies. This changed over the last decade with an increasing number of experimental studies that directly compare cognitive abilities of children and animals using very similar methodologies. For example, in the study of numerical cognition, a topic that will be discussed in Chapter 6, research focusing on counting processes has benefitted from dialogue and collaboration between those who traditionally study the cognitive development of human infants and those who examine animal cognition. In some cases, methodologies have been shared; infants have been tested using measures originally designed for rhesus monkeys (*Macaca mulatta*; e.g. Feigenson *et al.*, 2002) and, in other cases, monkeys have been examined with procedures that were previously developed for use with infants (e.g. Hauser & Carey, 2003). Through these studies, and many others like them, new theories on the cognitive mechanisms allowing for number representation have been developed. Importantly, the foundations of these abilities appear to be common across many different animal species.

1.6.2 Behavioral neuroscience

There are two factors that make comparative cognition and behavioral neuroscience complementary sciences. First, cognitive processes are mediated by biological systems, most notably the brain, and second, cognition must be measured, indirectly, through behavioral responses. Thus, the study of brain–behavior relationships (i.e. behavioral neuroscience) can inform comparative cognition research on a number of levels. As an example, neuroscience techniques are commonly employed to examine the mechanisms underlying specific cognitive abilities. These include manipulations of brain systems via lesions, electrical stimulation, and pharmacological activation. As well, brain activity may be monitored through brain imaging, electrophysiology, and measures of neurotransmitter release. Research using these techniques has provided detailed information on the proximate causes of many cognitive functions, such as perception and attention (see Chapter 2 for details). That is, behavioral neuroscientists have traced how sensory information is relayed from the environment to the nervous system, how it is processed in the brain, and how this information controls responding to incoming sensory stimuli.

Behavioral neuroscience can also address ultimate causes of cognition by comparing central nervous system function across species. This often occurs within the subfield of **neuroethology**, the study of how the nervous system controls natural behavior. A premise of this field is that the nervous system evolved to solve fitness-related problems so cross-species differences in brain morphology and function can be best understood in the context of adaptive traits. These comparative studies are informative because, despite differences in gross neuroanatomy (i.e. overall brain structure), there are widespread similarities in the nervous system of all animals. These include the structure and function of **neurons** (nerve cells), how neurons communicate with each other, and how proteins are formed within neurons. Moreover, developmental processes such as neuronal growth are preserved across species, at least in animals with an organized central nervous system.

If similar brain systems underlie the same cognitive process in different animals, it would suggest that the mechanisms that mediate this process have been preserved across evolution. Fundamental processes, such as the ability to associate environmental stimuli with motivationally significant events, may fall into this category. Other cognitive processes appear to be mediated by distinct physiological mechanisms in different animals. For example, most animals transmit information to conspecifics (discussed in Chapter 12), but the nature of the transmission varies dramatically. It can range from the production of auditory signals by the vocal cords in humans to the emission of **pheromones** by insects. Pheromones are chemicals released by one animal that are used as a scent signal by another animal, usually one of the same species. The ability to convey information, either about environmental dangers or resources, would enhance survival of a species, so communication evolved in many animals despite dramatic differences in body structure and neuronal architecture.

Not surprisingly, our knowledge of the neural basis of cognition varies widely, depending on the cognitive process under investigation. As outlined in Chapter 4, scientists have detailed the biological mechanisms involved in associative learning from brain systems to cellular changes to molecular signaling to genetic alterations that may mediate this process. In contrast, most have little understanding of how the brain controls social cognition, although research in this field (social cognitive neuroscience) is progressing rapidly.

In addition to uncovering the biological basis of cognitive processes, understanding how the brain operates can help scientists to refine cognitive theories of behavior. One may speculate on how cognition works by developing elaborate models of information processing, but these theories would be implausible if they did not account for the physical constraints of brain function. It is known, for example, that neurons transmit information in one direction only; a model of memory that proposes 'back propagation' to form associations would be untenable. Other brain properties such as neurotransmitter communication, anatomical connections, or neuronal firing patterns must be taken into account in any theory that provides either proximate or ultimate explanations of cognition.

1.6.3 Future directions

Comparative cognition is now firmly established as a discipline devoted to understanding cognition from a variety of perspectives. Since its inception in the late twentieth century, the number of species studied in comparative cognition research has expanded, a trend that is likely to continue in the future (Shettleworth, 2010). This represents a clear advance in the field, leading to an increased understanding of how specific cognitive traits vary across species. This information can then be used to develop theories of cognitive evolution, particularly as scientists continue to develop better tools for comparing phylogenetic relationships across species.

The influence of both developmental psychology and behavioral neuroscience on comparative cognition research will likely expand as communication between disciplines continues to grow. Originally, comparative cognition researchers adopted test paradigms from developmental psychology that allowed them to examine cognitive processes in nonverbal subjects. This cross-talk is now moving from methodological to theoretical questions as researchers begin to appreciate how developmental processes may be applied to phylogenetic questions. Behavioral neuroscience is growing as psychologists in related areas (e.g. social, cognitive, and clinical) incorporate more and more biological measures and theories into their work. Paralleling this shift, behavioral ecology is increasingly characterized by the study of brain–behavior relationships. Given these trends, comparative neurophysiology may become an important endeavor in the future (Danchin *et al.*, 2008), complementing work in comparative cognition. In all likelihood, the emerging discipline of neuroeconomics, studying neural systems that control mental computations, will have an important role to play in future advances in this field.

Finally, advances in comparative cognition are somewhat paradoxical in that they highlight cross-species commonalities in thought, emotion, and behavior. If other species are more similar to humans than originally thought, it raises the issue of how and when animals should be used in scientific research, even observational studies. Although the debate is ongoing, many people agree that the benefits accrued from this work, both for humans and animals, far outweigh the potential costs. Currently, research institutes must adhere to strict regulations on the ethical treatment of non-human subjects, both those that are tested in the lab and in the natural environment. In addition to medical advances based on animal models of disease states, research using animal subjects has improved the welfare and conservation of animals in both the natural environment and in captivity (Dawkins, 2006). Finally, like many fields, basic science research in comparative cognition often leads to practical applications that were not immediately apparent when the research was being conducted.

Chapter Summary

- Researchers in the field of comparative cognition study a wide diversity of species, employ a variety of methodologies, and conduct their work within many different academic and research departments. Nonetheless, the field is characterized by three unifying hallmarks: a focus on cognition, defined as the acquisition, storage, and processing of mental information; an experimental methodology, either in the lab or the natural environment; and an explanation of research findings within the framework of evolutionary theory.
- The basic tenets of Darwin's theory of evolution by natural selection can be simplified to the following three points: variation (i.e. individuals within a species display variability in both physiological and behavioral traits), heritability (i.e. offspring inherit traits from their parents), and survival and reproduction (i.e. individuals possessing a trait that promotes survival or reproduction will have a greater chance of transmitting traits to their offspring). A premise of comparative cognition research is that cognitive traits are shaped by the same evolutionary pressures (i.e. natural selection) that shape physiological traits.
- Despite the widespread influence of Darwin's theory of evolution by natural selection, many scientists criticized Darwin's consideration of animal thinking, noting that he relied heavily on anecdotal evidence to formulate these ideas, rather than tightly controlled experiments. Under this banner, a subdiscipline of psychology, behaviorism, flourished in the first half of the twentieth century and focused on behavior (e.g. learned responses) instead of mentalistic states.
- During the same period that psychologists were examining learned responses in the lab, many ethologists were studying behavior, primarily by observing animals in their natural environment. Tinbergen, who was one of the pioneers in the field, formulated the four questions of ethology as a means to understand a particular behavior. These can be divided into ultimate (functional, adaptive) and proximate (developmental, mechanistic) explanations. As a response to what was seen as a focus on proximate causes by ethologists, another discipline developed, behavioral ecology, which focused on ultimate explanations of behavior.
- Comparative cognition emerged out of two traditional sciences: psychology and zoology or, more specifically, behaviorism and behavioral ecology. Additionally, an emphasis on cognition arose within the field of psychology, as researchers reacted to the limitations of behaviorism and began to accept ideas put forth by linguists and computer scientists regarding mental processes and representations.
- The contemporary field of comparative cognition is an interdisciplinary endeavor with scientists from different fields unified by the common goal of understanding how and why humans and animals process information. The field has advanced through interactions with other researchers such as development psychologists and behavioral neuroscientists.

Questions

1. Imagine that you have birdfeeders in both your front and back garden. In the morning birds only visit one of these to feed, whereas in the evening they visit both. What factor(s) could explain this difference and how would you set up an experiment to test your hypothesis?

2. Many parents relate amazing cognitive abilities of their own children such as recognizing voices just after birth, using words at a very young age, or imitating behaviors that they observe in older children. Can these anecdotal reports ever be used to develop theories of cognitive development?
3. If scientists could determine that only humans have the ability to mentally represent numbers beyond 10, what would this tell us about our own evolutionary history? Look through the titles for the next 12 chapters in this book. Select one of these topics and try to explain why it would be so difficult to determine if the evolution of this process is an adaptation.
4. In the past, many psychologists who studied behavior in the lab measured responses, such as lever pressing in rats, that were not part of a species' behavioral repertoire in the natural environment. What are the advantages and disadvantages of this approach?
5. Given the definition of natural selection, what is artificial selection? When does it occur and what are the advantages and disadvantages of this process?
6. If damage to one small part of the human brain produced a deficit in the ability to find the way around, could one infer that this is the neural site of spatial navigation? Why or why not?
7. Although humans and chimpanzees are considered 'close' evolutionary relatives, a common ancestor linking the two goes back at least four to six million years. The spread is far greater for humans and rodents yet rats and mice are the most common animal model of human conditions, such as Alzheimer's disease. Does this cause problems for interpreting data from these studies? Why or why not?

FURTHER READING

Some of the early texts and historical reviews of comparative cognition provide the most informative and entertaining background reading. These include:

Darwin, C. (1909). *On the Origin of Species by Means of Natural Selection*. This edition has been reprinted many times, most recently by Arcturus Publishing, London, UK (2008), in celebration of the centenary of Darwin's original publication.

Dewsbury, D. D. (2000). Comparative cognition in the 1930s. *Psychonomic Bull Rev*, **7**, 267–283.

Houck, L. D., & Drickamer, L. C. (1996). *Foundations of Animal Behavior: Classic Papers with Commentaries*. Chicago, IL: University of Chicago Press.

Lorenz, K. Z. (1952). *King Solomon's Ring* (Wilson, M. K., trans). New York, NY: Thomas Y. Cromwell.

Tinbergen, N. (1958). *Curious Naturalist*. Garden City, New York, NY: Doubleday.

Tinbergen's four questions are presented in his 1963 paper, *On Aims and Methods of Ethology*. A reprint can be found in: Houck, L. D., & Drichamer, L. D. (eds.). (1996). *Foundations of Animal Behavior: Classic Papers with Commentaries*. Chicago, IL: University of Chicago Press.

Some of Darwin's most important ideas were based on his observation of finch beaks and how these varied in size and shape across the Galapagos Islands. A modern-day description of scientists documenting evolutionary changes in the same species in this region can be found in:

Weiner, J. (1994). *The Beak of the Finch: A Story of Evolution in Our Time*. New York, NY: Alfred A. Knopf Inc.

The evolutionary biologist and paleontologist, Stephen Jay Gould, has written many up-to-date accounts of evolutionary theory for the lay public. His writing often appeared as popular essays in *Natural History* magazine, some of which are reprinted in collected volumes such as:

Gould, S. J. (1977). *Ever Since Darwin*. New York, NY: W. W. Norton & Company.

Gould, S. J. (1980). *The Panda's Thumb*. New York, NY: W. W. Norton & Company.

The award-winning author and evolutionary biologist, Richard Dawkins, also presents compelling arguments for evolutionary theory in dozens of essays and books that are written, specifically, for non-specialists. These include:

Dawkins, R. (2010). *The Greatest Show on Earth: The Evidence for Evolution*. New York, NY: Free Press.
Dawkins, R. (2004). *The Ancestor's Tale: A Pilgrimage to the Dawn of Life*. Boston, MA: Houghton Mifflin.

Scientists and philosophers have been interested in animal minds since the beginning of antiquity. These ideas are presented in a comprehensive volume of books that discusses human relationships with animals across the last four centuries.

Kalof, L., & Resl, B. (eds.) (2007). *A Cultural History of Animals: Volumes 1–6*. Oxford, UK: Berg Publishers.

There are many excellent biology texts on animal behavior, such as:

Alcock, J. (2005). *Animal Behavior: An Evolutionary Approach*. Sunderland, NJ: Sinauer Associates Inc.
Bolhuis, J. J., & Giraldeau, L.-A. (eds.). (2004). *The Behavior of Animals*. Oxford, UK: Blackwell Publishing.
Danchin, E., Giraldeau, L.-A., & Cézilly, F. (eds.) (2008). *Behavioural Ecology*. Oxford, UK: Oxford University Press.
Dugatkin, L. A. (2004). *Principles of Animal Behavior*. New York, NY: W. W. Norton & Company.

Earlier textbooks and edited volumes on comparative cognition that inspired this current text include:

Dukas, R. (ed.) (1998). *Cognitive Ecology*. Chicago, IL: University of Chicago Press.
Roberts, W. A. (1998). *Principles of Animal Cognition*. Boston, MA: McGraw-Hill.
Shettleworth, S. J. (2010). *Cognition, Evolution, and Behavior*. New York, NY: Oxford University Press.
Wasserman, E. A., & Zentall, T. R. (2006). *Comparative Cognition: Experimental Explorations of Animal Intelligence*. Oxford, UK: Oxford University Press.

Many scientists agree that cognitive processes, like physiological traits, evolved through natural selection, but there is considerable debate on how this occurred. One intriguing hypothesis, presented in the text listed below, is that a symbiosis between biology and culture shaped human evolution, resulting in a combination of highly sophisticated cognitive abilities that are unique to this species.

Donald, M. W. (1993). *Origins of the Modern Mind: Three Stages in the Evolution of Culture and Cognition*. Boston, MA: Harvard University Press.

2 Sensory systems

Background

"What is it Like to be a Bat?" The title of Nagel's now famous philosophical paper is meant to conjure up images that are foreign to most humans, such as hanging upside down in a dark cave or swooping through the night to catch insects (Nagel, 1974). Although originally posed as a challenge to reductionist theories of consciousness, this hypothetical thought experiment raised an important question in comparative cognition research: is it possible for one individual to understand another individual's (or more accurately another species') experience? Nagel carefully selected bats for the title of his article because their primary interaction with the sensory environment is echolocation. Humans may understand how echolocation works but, with rare exceptions (see Box 1.4), they have never experienced this phenomenon. According to Nagel, this means that bats and humans have a different knowledge of the world. One is not necessarily better, more advanced, or even more intelligent than the other. They are simply different. The analogy extends beyond echolocation in that most animals possess some sensory ability that is absent in humans.

A fascinating example of these species-specific sensory abilities is provided by the mantis shrimp (*Odontodactylus scyllarus*). Mantis shrimp can detect a broader range of the visual spectrum than humans, whose vision is restricted to a narrow band of red, green, and blue wavelengths. Human vision is also limited to non-polarized light, sunlight that scatters randomly in all directions. Polarized light is created when sunlight is reflected off a surface, like water or snow, so that it is oriented in a linear direction (i.e. horizontally). Humans experience polarized light as a steady glare, which they minimize with polarized sunglasses. Other species, including many insects, can see polarized light but, to date, the mantis shrimp is the only animal known that can also detect circular polarized light (Chiou *et al.*, 2008). This visual experience is created underwater when light scatters up toward the surface and is reflected back down. Mantis shrimp detect circular polarized light via six rows of numerous small eyes which move independently and are arranged at exactly the right angle to convert circular light waves into signals that the other cells can detect (humans wearing special goggles that do this conversion see circular light as a bright steady glare). No one has determined the function of polarized light detection in the mantis shrimp although one possibility is that it allows secret communication within the species as reflections of polarized light from a mantis shrimp body would be invisible to other animals.

Chapter plan

According to Nagel, humans will never understand the perceptual experience of a bat or any other animal, but this has not inhibited scientific investigations in this area. This chapter presents an overview of that research, focusing on the basic properties of sensory system evolution, development, and function across species. An understanding of these topics is critical for the remainder of the text in that all cognitive processes are influenced, at least initially, by sensory input. In other words, sensory systems provide the first point of contact between an organism and its environment: information gathered about the sensory world is used when animals remember events, navigate through new or familiar territories, and communicate with conspecifics. These and other cognitive processes must function within the ecological constraints of individual species (e.g. communication will be different in aquatic and non-aquatic environments). Because of this, highly specialized sensory systems have evolved in many species. But this is only the first step: a fully functioning sensory system also depends on sensory experience that is acquired during an animal's lifetime. This chapter describes how these two processes – evolution and development – interact in a way that allows organisms to function effectively (i.e. to survive and reproduce) in their environment.

The chapter then moves on to a discussion of sensory system function, divided into detection and processing. Sensory detection begins when physical signals carrying information from the environment (e.g. light or sound waves) reach the animal. This information is transmitted to the central nervous system and on to different brain regions where it is processed. Despite differences in sensory abilities, a comparison of sensory detection and processing across species reveals a number of general principles of sensory system function. Sensory processing involves the segregation and organization of incoming information, which must then be recombined so that organisms can interpret the world around them. How this occurs is covered in the final sections of the chapter under the topics perception and attention.

2.1 Evolution of sensory systems

The bat and mantis shrimp examples in the chapter opening reveal that there are some types of energy (i.e. ultrasonic vibrations and circular polarized light) that humans cannot detect. The same is true of magnetic and electrical fields; humans must rely on technology to 'sense' these signals, whereas many birds and fish do so 'naturally' as they locate prey or return home from distant locations. Even for energy sources that humans can detect, they often do so within a much smaller range than other animals. For example, humans see vibrant colors in flowers, but not the ultraviolet reflections that guide many insects to their food. And, although they can detect odor cues from conspecifics, few humans could track another animal by following their scent on the ground (see Figure 2.1). These species-typical differences in sensory abilities allow animals to adapt to the particular ecological constraints of their environment. This may explain why the human auditory system is finely tuned to the range of sound frequencies in speech, an obvious advantage for a species that relies so heavily on verbal communication.

2.1.1 Visual adaptation

Sensory system adaptation has been studied extensively in terms of vision, an ability that varies widely across species. In all animals, the visual system works by absorbing light of particular wavelengths along the electromagnetic spectrum. Humans see within a 400–700 nm range (the

Figure 2.1 Coyote tracing prey in Death Valley National Park, California USA.

visible spectrum) because they have chemical pigments in the eye to absorb these wavelengths. Animals that inhabit environments with different patterns of light absorption, such as aquatic habitats, have evolved different visual sensitivities (Munz, 1958). For example, red, yellow, and orange wavelengths are absorbed at shallow water depths so many marine species are more sensitive to light within the blue spectrum. In addition, because little light reaches the bottom of the ocean, some deep-sea species are able to emit and detect chemical light, **bioluminescence**, which is created when organic compounds are mixed together. (Bioluminescence is a marine phenomenon although a similar process can be observed on land with fireflies.) In marine animals, bioluminescence occurs in the internal organs, which then glow in the dark. Fish in these environments have small eyes which are able to detect bioluminescence emissions, most of which appear as a blue hue to humans.

Evolution has also favored traits that help to enhance vision in land dwelling animals with different lifestyles. For example, many nocturnal animals have eyes with large pupils and lenses that allow them to pick up more light. Some of these differences are displayed in Figure 2.2. Compared to diurnal animals, nocturnal animals also have a much higher proportion of rods than cones. Rods and cones are the cells at the back of the eye that detect light and color: rods are 1000 times more sensitive to light than are cones so nocturnal animals can navigate very well in the dark but often have difficulty discriminating colors. The reverse is true for humans and other animals that function, primarily, during daylight hours.

Another factor that has a huge impact on what animals see in their environment is eye placement. Humans and other primates have frontally placed eyes, which means that they have very good

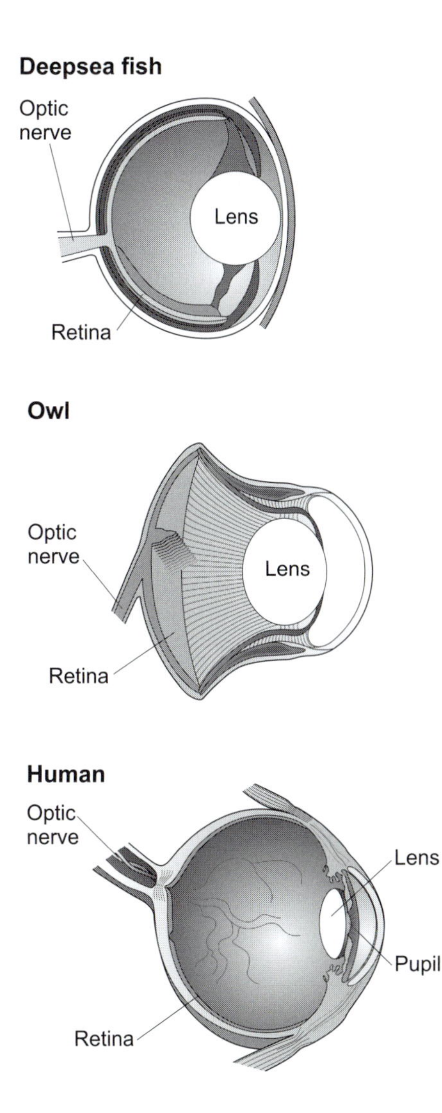

Figure 2.2 Eye structure of three different animals: deep-sea fish, owl, and human. Compared to diurnal humans, the nocturnal animals have larger lenses that can pick up more light.

binocular, but very poor lateral, vision. **Binocular vision** occurs when the two visual fields overlap, meaning that much of what is seen with the right eye is also seen with the left eye and vice versa. But the images are not exact: the small distance between them means that there is a slight disparity between the visual input that reaches the right and left eyes. The brain uses this information to produce depth perception, the ability to see the world in three dimensions including which objects are further away. Binocular vision is not the only cue for depth perception, but it is one of the most important. Animals that have eyes placed in different positions will have different levels of binocular vision, as represented in Figure 2.3.

Compared to binocular vision, laterally placed eyes permit a much wider field of view, even if depth perception is reduced. Animals that have eyes on the sides of their head, therefore, are able to see behind them, which has obvious advantages for avoiding predators. Indeed, a general rule of eye placement is that predators have frontally placed eyes, allowing them to track prey more effectively, whereas prey animals have laterally placed eyes, allowing them to detect these predators (see Figure 2.4). Given that vision is often directed toward other organisms, understanding visual

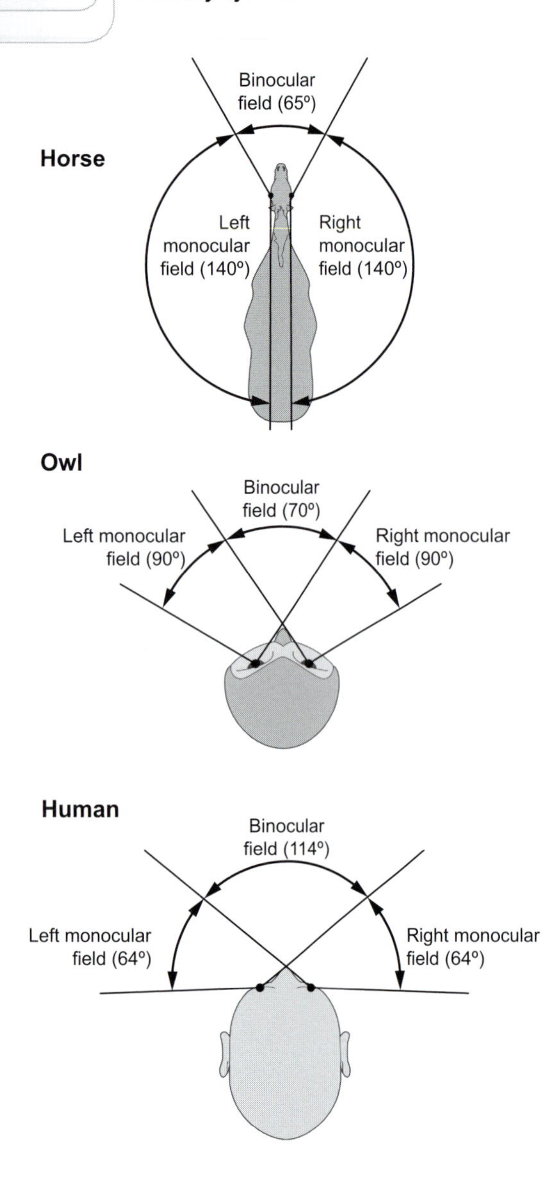

Figure 2.3 Visual fields of three different animals: horse, owl, and human. The laterally placed eyes in the horse provide a large monocular field on each side of the head and a small binocular field in which the two eyes can see the same image. In contrast, the frontally placed human eyes provide a large binocular field and smaller monocular fields of vision. The owl's eye placement and corresponding monocular and binocular fields are between the other two species.

system function in one species may provide functional explanations of physiological traits in another species (see Box 2.1).

2.1.2 Sensory adaptations in closely related species

The visual adaptations described above illustrate how dramatic differences in ecological circumstances alter sensory abilities. More subtle differences in sensory traits can be observed in closely related species that inhabit similar environments. For example, several snake species are able to detect chemical, visual, and thermal cues but they show greater sensitivity to one type of cue or the other, depending on their prey preference (Saviola *et al.*, 2012). Midland rat snakes (*Scotophis spiloides*), which feed primarily on rodents, respond to chemical cues allowing them to follow

(A) (B)

Figure 2.4 Eye placement of two different birds species. The predator, eagle owl (A), has frontally placed eyes whereas the prey, Victoria crowned pigeon (B), has laterally placed eyes.

Box 2.1 Why do zebras have stripes?

To most people, stripes are the defining characteristic of zebras, making them easy to distinguish from other animals of similar size and shape (see Figure 2.5). The stripes are not **sexually dimorphic**, meaning that the trait is similar in males and females, and there is no evidence that stripes make zebras more attractive to members of the opposite sex. With no apparent reproductive advantage, scientists struggled to formulate a functional explanation for this highly conspicuous trait.

One hypothesis is that body stripes provide protection against predators because the pattern helps zebras to blend in with the long grass around them. Lions (*Panthera leo*), the zebra's primary predator, are color blind so it doesn't matter that the color of the stripes does not match the grass. In addition, when zebras stick with their herd (which they usually do), lions may see a large mass moving in unpredictable ways, rather than individual animals. This makes it difficult for lions to single out and attack the weaker zebras in a herd. Stripes may also deter bloodsucking insects that deliver annoying bites and carry potentially dangerous germs. In an experimental test of this hypothesis, horseflies were less attracted to striped models of horses than to either solid black or solid white ones (Egri *et al.*, 2012). In fact, narrower stripes were the most effective insect repellant, which may explain why a zebra's thinnest stripes are on the face and legs, the body areas most vulnerable to insect bites. According to the authors, insects such as horseflies are attracted to polarized light because it is similar to light reflected off water surfaces. Black and white stripes disrupt the reflective signal of polarized light, making it less likely that the insects will land on these animals. Regardless of whether one or both hypotheses are correct, it seems likely that zebra stripes evolved as a mechanism of protection for these animals. But, it was only through studying the visual systems of other species (i.e. zebra predators and pests) that humans were able to reach this conclusion.

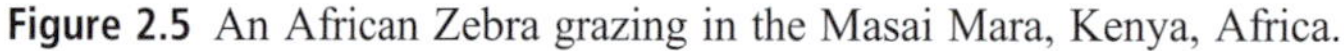

Figure 2.5 An African Zebra grazing in the Masai Mara, Kenya, Africa.

rodent odor trails. In contrast, western fox snakes (*Mintonius vulpina*) respond at higher rates to visual cues, matching their preference for ground-nesting birds. This suggests that snakes inherit an adaptive tendency to respond to sensory cues most likely to lead to prey capture. Moreover, as snakes age, their metabolism and ability to consume larger animals change, leading to altered responses to cues associated with particular prey (Saviola *et al.*, 2012).

Cross-species comparisons, such as Savliola's research with snakes, support (but do not *prove*) that natural selection plays a role in sensory system evolution. Evolution of any cognitive process is difficult to ascertain, partly because the evolutionary record of many species is incomplete. Even if fossil records are available, they provide little information beyond brain size, so sensory system adaptations must be inferred by comparing sensory traits in contemporary species that evolved from a common ancestor. One of the best examples of this approach is an analysis of auditory abilities in insects (Yack, 2004), specifically the detection of high pitched sounds by whistling moths (*Hecatesia exultans*) (Fullard & Yack, 1993). Unlike most moth species, whistling moths (and other moths in the noctuid family) have ears on either side of their thorax near the wings (Surlykke & Fullard, 1989). Moth species in the closely related saturniid family have a nearly identical anatomical structure in the same region that provides feedback on wing position. Noctuid and saturniid moths likely evolved from the same ancestor and the authors speculate that a small mutation in these wing position sensors allowed a predecessor of the whistling moth to detect ultrasound emissions from bat predators. The survival advantage conferred to these mutants led to a proliferation of this trait in a subpopulation of moths. With further adaptations, descendants of the early noctuid moths could then use their auditory sensors to detect the high pitched sounds of other moths. This evolution from wing sensor to sound detector likely occurred over many generations and may explain differences in intra-species communication that distinguishes noctuid from saturniid moths.

Other researchers have proposed similar models to explain intra-species differences in other sensory systems, with general agreement that sophisticated sensory abilities only emerge with high ecological demands (Niven & Laughlin, 2008). The processing of sensory information is costly in terms of metabolic energy, so sense capacities that are not 'needed' are often diminished or lost. In other words, selection pressures reduce sensory traits that are too costly to maintain. One example is a loss of visual function in species that inhabit environments with minimal light (Fong *et al.*, 1995). It is difficult to trace the evolution of this decline in the natural environment, but the phenomenon has been demonstrated in the lab (Tan *et al.*, 2005): lab-bred fruit flies (*Drosophila melanogaster*) show a reduction in eye size that is related to their time in captivity. Presumably this occurs over generations as the need for vision is reduced when animals are not required to locate food or mates.

2.1.3 Sensory drive hypothesis

The divergence of sensory abilities in different species fits well with the **sensory drive hypothesis**, an explanation for ecological divergence within species (Endler, 1992). Sensory drive describes how communication signals work effectively and how these coevolve with sensory systems. Basically, speciation by sensory drive is a special case of speciation by natural selection. According to this theory, when populations occupy new habitats with different sensory environments, natural selection favors adaptations that maximize the effectiveness of communication, both within and between species. In other words, the traits showing adaptations are those that improve the ability to send and receive signals in the new environment. The entire process could work in the following way: individuals of the same species become separated by a geographical barrier (e.g. a river or mountain range) partly because there is competition for resources, such as food or available mates. The new habitat has features that make the transmission of sensory signals more or less difficult than in the previously occupied territory. For example, if birds moved into a new area with denser trees, the light levels would be lower and it would be more difficult for females to detect the colored plumage on males. Females with better visual acuity (as well as males with brighter feathers) would have a reproductive advantage that was not apparent in the original habitat. This would promote the evolution of both traits in the isolated groups of animals. Other forms of communication, such as aggressive displays or prey–predator interactions, may also be subject to the forces of sensory drive. That is, sensory signals which are transmitted or detected more effectively in a new environment will be favored in terms of reproduction and survival. This topic is discussed in more depth in Chapter 12 (Communication).

If the sensory drive hypothesis is true, it should be possible to trace the heritability of sensory abilities across generations. Endler and colleagues conducted this experiment in a controlled laboratory setting using populations of guppies (*Poecilia reticulate*) that were selected for different visual sensitivity (Endler *et al.*, 2001). Guppies are able to see a range of colors across the visual spectrum although there is individual variability in how well they see each color. This sensitivity can be measured by placing fish in cylindrical tanks with black and white vertical stripes moving around the outside of the tank (see Figure 2.6). Guppies orient to the vertical stripes and maintain their position as they move around the tank. A colored filter is then placed in the light path to illuminate the stripes in red or blue. Fish continue to track the stripes as the light intensity is decreased. At a certain intensity, they stop tracking the stripes, indicating that they can no longer see them.

Using this approach, Endler and colleagues separated groups of guppies that were more sensitive to red light from those that were more sensitive to blue light and then interbred each color sensitive group. After the ninth generation, fish from each group (including a control

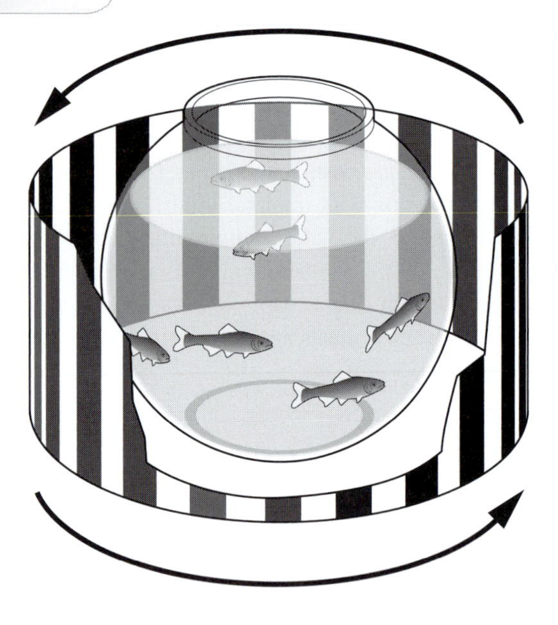

Figure 2.6 Experimental design to test visual sensitivity in guppies (Endler *et al.*, 2001). A cylindrical fish tank was surrounded by vertical black and white stripes that moved around the tank at a constant speed. Guppies swam around the perimeter of the tank in line with the movement of the stripes. The stripes were changed to red or blue by placing a colored filter in the light path and the intensity of the light was slowly decreased.

group that was cross-bred across the red and blue sensitive groups) were tested for color sensitivity. The two color sensitive groups showed an even greater difference in their visual sensitivity than the original parent groups, indicating that this sensory trait was passed from generation to generation. Endler's experiments were not the first to show heritability of sensory system function but they provide one of the best examples of how observational field studies can direct experimental research in the lab.

2.1.4 Sensory bias

One apparent paradox in the evolution of sensory systems is a phenomenon called **sensory bias**. Sensory bias refers to situations in which individuals of a species respond with increased vigor to stimuli that are exaggerated versions of naturally occurring stimuli. The phenomenon is often explained as a sensory adaptation that gets co-opted for sexual selection. Figure 2.7 shows a rather comical example of an experiment demonstrating sensory bias. Researchers attached abnormally large feathers of different colors to the heads of male birds, and then assessed female preference for these birds by measuring the time they spent with each decorated male (Burley & Symanski, 1998). White crested males were the most attractive to female birds, even though the females had never encountered such elaborate head ornaments on potential mates.

The paradox of sensory bias is that it cannot provide any evolutionary advantage because the exaggerated trait is not part of either animal's natural environment. The most likely explanation for sensory bias is that it is a byproduct of a sensory preference that had an evolutionary advantage in a different context. This idea, sometimes called **sensory exploitation**, implies that sensory signals which were important for one process have been co-opted by another. In the case of the female finches, their preference for large white headdresses on males may have been 'borrowed' from their nest building behavior, which uses white feathers. So the females may be hanging out with these males because they have an existing bias for white feathers, not because these feathers indicate anything about male fitness.

Figure 2.7 A male long-tailed finch (left) and a male zebra finch (right) decorated with artificial white crests. During mate choice trials, females preferred to spend time interacting with these birds than with undecorated males or males sporting red or green crests.

Sensory exploitation also describes situations in which a trait evolved to capitalize on an existing preference. This two stage process is revealed by tracing the phylogenetic development (i.e. evolutionary history) of a trait, and determining whether a preference for this trait predated its emergence. Basolo (1990) conducted this analysis by examining the evolution of the tail extension in swordtail fish (*Xiphophorus hellerii*). Swordtail fish are related to the swordless platyfish (*Xiphophorus maculatus*), and their common ancestor also lacked the sword-like extension of the tail. The characteristic fringe on the swordtail, therefore, evolved when the platyfish and swordtail fish diverged. Even though it never existed in their species, female platyfish prefer conspecific males with artificial extensions on their tails, suggesting that this sensory bias for long-tailed males was present in their evolutionary ancestors. When the two species diverged, male swordtail fish were at a reproductive advantage in that this female tendency created a feed forward loop in evolution: longer tailed males were more likely to be selected as mates, producing more males with long tails and more females who prefer them. The latent preference for long tails survived in female platyfish because it was not detrimental to survival or reproduction: it was only revealed when interfering scientists attached extensions to these unsuspecting males.

The principles of sensory exploitation may also explain heightened responses to exaggerated versions of natural stimuli (**supernormal stimuli**). Tinbergen documented this phenomenon in graylag geese (*Anser anser*) when they retrieved and attempted to incubate giant artificial eggs, even while neglecting their own normal sized eggs that sat nearby (Tinbergen, 1951). In other experiments, birds preferred to sit on dummy eggs with enhanced markings, such as more saturated versions of their own egg color. Similarly, stickleback fish (*Gasterosteus aculeatus*) ignore invading conspecifics to attack wooden models of fish with darker red undersides. Ethologists speculated that an animal's natural tendency to respond to the strongest stimulus in their environment leads to abnormal responding to stimuli that have no parallel in the natural world. For example, birds attend to the largest eggs in their brood as these are likely to produce the healthiest young: if big is better, biggest is best. Since oversized eggs only show up with roving ecologists, there is no evolutionary pressure to have an upper limit on this preference. If this is true, responding to supernormal stimuli can be explained as approximate rules of thumb in a world of normal variation.

In summary, a cross-species comparison of sensory systems reveals an enormous range of sensory abilities across the animal kingdom. All animals (and humans) experience only a limited portion of the sensory world, but these generally reflect the evolutionary history of that species. In other words, sensory system evolution, like other physiological traits, is an adaptation. Although most of this section focused on vision, the same analysis could be applied to any of the other senses in that the auditory, olfactory, and tactile systems of different animals have all evolved within the ecological constraints of particular environments.

2.2 Development of sensory systems

Evolution provides each organism with a genetic code or set of instructions for building proteins, but how this code is expressed in a fully developed individual varies widely. In some cases, the instructions provided by a genetic code are clear and exact: build two blue eyes. In others, there is a range of possible outcomes for each set of instructions: "build a device that can detect visual signals within a particular range of the light spectrum" could be accomplished in many ways. The difference in how these instructions are carried out is determined by both physiological and environmental factors. A person may be born with a genotype to produce a tall, lanky body but malnutrition or excessive consumption of fats early in life would alter the size and shape of their adult build.

2.2.1 Sensitive periods

Much of what is known regarding sensory system development comes from the pioneering work of Hubel and Wiesel. These scientists shared the Nobel Prize in Physiology or Medicine (1981) for their work describing how visual information is processed and organized in the brain. In the course of these investigations, Hubel and Wiesel uncovered a critical role for experience in visual development. It was already known that children who are born with cataracts never fully recover their sight if the cataracts are removed after the age of 3. In sharp contrast, cataracts that develop in adulthood and are removed have little or no impact on vision. Hubel and Wiesel created a parallel scenario in animals: if the eyes of a kitten or monkey were sewn shut during the first few weeks of life, normal vision never developed even if the eyes were later reopened. As little as 1 week of visual deprivation during the first 6 months of a monkey's life was enough to produce this deficit, whereas visual deprivation for weeks or even months in adulthood had no effect. Hubel and Wiesel also showed that cells in the eye which receive light signals respond normally after visual deprivation, whereas neurons in the cortex that receive signals from these cells in the eye do not. This suggested that the connection between the first and final stages in the visual pathway failed to develop under sensory deprivation conditions. The effect was so pronounced that even closure of one eye produced profound structural and functional changes in the cortical neurons that normally respond to visual input (Hubel & Wiesel, 1977). Hubel and Wiesel concluded that key components of the visual system develop after birth, and that the physiological changes underlying this development depend on visual experience. This period in which experience-dependent changes can have profound and enduring effects on development is called the **sensitive period**.

Knowledge of sensitive periods has improved the prognosis for a number of medical problems, including **strabismus**. Strabismus is a condition, often present at birth, in which the visual axes of the two eyes are misaligned. The misalignment makes it difficult to look at the same point in space with both eyes, thereby disrupting depth perception and binocular vision. Infants with strabismus have good vision but, as they grow, the images in the two eyes do not converge and the visual system fails to develop properly. Children tend to favor the stronger eye, often losing most of their vision in the neglected eye. Strabismus can be treated surgically and, in the past, surgery was often delayed until children were old enough to understand the medical procedures. Vision was improved in these cases, but binocular vision often remained impaired. Now that scientists have a greater understanding of sensitive periods and sensory system development, ophthalmologists treat strabismus at a much younger age. If the surgical intervention occurs as soon as the condition is diagnosed, children frequently develop normal binocular vision and depth perception.

Sensory deprivation is an extreme example that occurs, only rarely, in the natural world. But even subtle changes in sensory input during the sensitive period can alter development. For example, if cats are exposed to a particular visual stimulus during the first few weeks of life (e.g. horizontal versus vertical lines), their ability to detect the previously exposed pattern is enhanced in adulthood. This shift in sensitivity occurs because connections between neurons in the visual cortex are rearranged during development, and these cortical neurons become more responsive to the familiar pattern (Blakemore, 1976). Similar findings have been reported in both the auditory and tactile systems of other mammals in that alterations in early sensory experience modify neuronal connectivity and function in the adult brain. One of the most striking examples of this phenomenon is second language learning. Acquiring a foreign language is remarkably easy as a child and frustratingly difficult as an adult. One problem for late language learners is that they have difficulty distinguishing, and then reproducing, sounds that are not part of their native tongue. Exposure to these sounds during the first few years of life, like visual exposure to lines of different orientations, alters brain development and makes it easier to distinguish foreign sounds in adulthood (Kuhl *et al.*, 1992; Rivera-Gaxiola *et al.*, 2005). The concept of sensitive periods has been extended to other cognitive processes such as categorization (Chapter 9), social competence (Chapter 10), and communication (Chapter 12). Indeed the idea that early life experiences shape adulthood is the basis of the huge industry built around early childhood education and enrichment.

2.2.2 Compensatory plasticity hypothesis

Sensory deprivation reveals a critical role for experience in brain development. If vision or audition or any other sensory input is blocked during development, the functioning of that system may be disrupted in adulthood. Many people believe that these deficits in one sensory system lead to enhanced functioning in other sensory systems. Anecdotally, blind individuals can hear much better than people with normal vision, although this does not prove that the auditory system has compensated for visual deficits when people cannot see. It may simply be the case that those with normal vision have not learned to pay attention to auditory cues because they rely so heavily on vision. Careful laboratory experiments, however, confirmed that blind adults really do have superior auditory abilities, and the effect is confined to those with congenital blindness (Lessard *et al.*, 1998).

The question then arose whether people who are born blind have better hearing at birth, or whether this developed because visual input was restricted. Theoretically this issue could be resolved by testing audition in blind and sighted babies, but the experiment would be difficult to set up and may take years to complete. (Researchers would need to identify and test blind newborns with no other medical condition, and then match these to babies with normal vision). In contrast, an animal experiment to test this hypothesis can be designed with relative ease. Chapman *et al.* (2010) conducted this work by raising guppies under different light intensities, and then testing how they responded to visual or chemosensory (i.e. olfactory) cues. The prediction was that fish raised in environments with minimal visual input would develop improved olfaction. This is exactly what happened. When tested for their ability to find food, guppies reared at high light intensities responded better to visual cues, whereas those reared under low light intensity responded better to olfactory cues. Since the guppies were randomly assigned to each lighting condition at birth, the group differences must be due to their environmental experience early in life.

These results fit the **compensatory plasticity hypothesis** (Rauschecker & Kniepert, 1994) which states that a loss or deficit in one sense leads to a heightened capacity in another. Compensatory plasticity appears to be a developmental process, in that altered sensitivity to visual and olfactory cues only occurs in early life. Older guppies were unable to respond to similar changes in their sensory environment. That is, fish reared under low light conditions did not show improved visual responses when they were tested under high light as adults. It appeared that once the olfactory system became dominant in these animals, it could not be reversed.

The advantage of compensatory plasticity is that it allows animals to adapt, within their lifetime, to new sensory environments. Natural selection provides a mechanism for adaptation across generations, but this process is random and may be too slow to ensure survival. This is particularly true in aquatic systems that are easily altered by weather and human activity (dams, irrigation, pollution, etc.). Compensatory plasticity may be an adaptive mechanism to deal with these rapidly changing sensory conditions. How the process occurs is still not clear. The most likely explanation is that areas of the brain that were genetically programmed to respond to one sense (e.g. vision) are taken over by another sense (e.g. olfaction) when sensory input is reduced.

In sum, most animals do not have fully functional sensory systems at birth: both normal development and the fine tuning of sensory abilities depend on postnatal sensory experience. The remarkable changes in sensory function that occur with altered sensory input can provide clues to the evolution of sensory systems. Again, the visual system can be used to illustrate this point. Environments with poor visual conditions, such as caves, underground tunnels, or turbid waters, are often inhabited by animals with little or no visual acuity. Cave dwelling salamanders are an excellent example: they have eyes at the larval, but not the adult, stage and consequently no vision. Amazingly, if these animals are reared in light conditions, they develop normal eyes and low-level vision (Lythgoe, 1979). This suggests that the genetic code for building functional eyes is present in these animals, but is not expressed under their normal living conditions. It may be that vision provided some evolutionary advantage to the salamander's ancestors, but was no longer necessary when the animals' habitat changed. Perhaps vision remains viable in these animals because evolution has not selected against it. Like any other trait, if there is no disadvantage to vestigial vision (including that it does not require metabolic energy to be maintained), there may be no reason for it to disappear across evolution.

2.3 Sensory system function

Sensory system function can be divided into two stages: detection and processing. Sensory detection describes how animals acquire information about their sensory world; sensory processing describes how this information is organized and transmitted to other brain regions. Sensory detection and processing allow animals and humans to interpret the world around them and to respond appropriately to stimuli in their environment. For example, an animal that could not tell the difference between a mating call and an alarm call would be severely impaired in most social interactions.

2.3.1 Sensory detection

Sensory detection begins at the sense organs (eyes, ears, nose, skin, tongue), which each contain groups of specialized neurons called sensory receptors. Sensory receptors transmit information to the central nervous system through a process of neuronal communication. Neurons vary in size and shape, but all have four primary features that are necessary for interneuronal communication (see Figure 2.8). The **soma**, or cell body, contains the cell nucleus and other structures that aid cellular metabolism. Neuronal signals that are used for communication are generated in the cell body and then transmitted, by electrical impulses, along the **axon**. This long thin output fiber of the neuron has many branches, each of which swells at the tip to form a **presynaptic terminal**. When an electrical signal arrives at presynaptic terminals, it causes the neuron to release chemicals called neurotransmitters (dopamine, serotonin, and acetylcholine are all well known examples of neurotransmitters). This transmission of an electrical signal from the soma along the axon to the presynaptic terminal, resulting in the release of neurotransmitter, is called an **action potential**. In neuroscience terminology, neurons 'fire' an action potential and a neuron is 'firing' if action potentials are generated.

When a neurotransmitter is released from the presynaptic terminal, it crosses the **synapse** (a narrow gap between the axon tip of one neuron and the cell surface of another neuron) to act on postsynaptic receptors located on the soma and **dendrites** of other neurons. Dendrites are tiny fibers

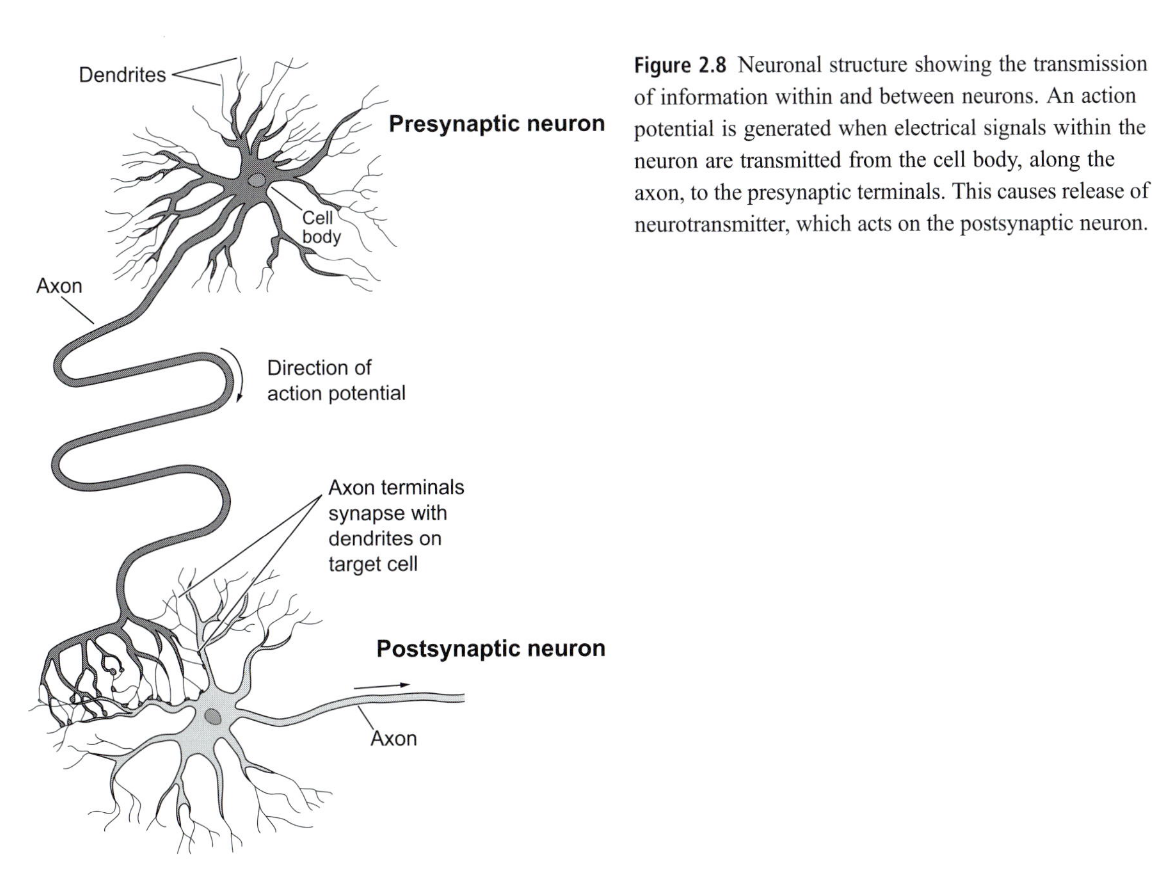

Figure 2.8 Neuronal structure showing the transmission of information within and between neurons. An action potential is generated when electrical signals within the neuron are transmitted from the cell body, along the axon, to the presynaptic terminals. This causes release of neurotransmitter, which acts on the postsynaptic neuron.

that branch out from the soma, increasing the surface area of a cell so that more information can be received from neighboring neurons. Activation of postsynaptic receptors causes biochemical changes in this neuron that increase or decrease the probability that it will fire an action potential. If the probability of neuronal firing is increased, the connection from the presynaptic to postsynaptic cell is excitatory; if the probability is decreased, the connection is inhibitory. A single neuron may receive input from dozens, hundreds, or even thousands of other neurons. The combined influence of all excitatory and inhibitory input determines whether an action potential will fire at any given moment.

Sensory receptors are unique in that they respond, not to neurotransmitter release, but to input from the environment. This input arrives as physical events such as sound waves (hearing), light waves (vision), or airborne chemicals (smell). When these events reach the sense organs, sensory receptors are activated (i.e. fire action potentials). These sensory receptors translate physical events into electrical signals, a process called **transduction**. In this way, sensory input is coded as neuronal messages that are transmitted to other parts of the central nervous system, including the brain.

Action potentials are all the same, so the brain must have some way to distinguish different types of sensory events. This is accomplished, first, at the level of the sensory receptors which respond to a specific type of physical energy. For example, visual receptors in the eye are activated by light waves, auditory receptors in the ear are activated by sound waves, olfactory receptors in the nose are activated by airborne chemicals, etc. This specialization ensures that action potentials in the visual system are interpreted as light whereas those in the auditory system are interpreted as sound. Gently pressing on closed eyelids produces an illusion of small dots of light because the brain interprets stimulation of sensory receptors in the eye as visual input. The effect is even more dramatic if stimulating electrodes are connected to nerve fibers that carry visual, auditory, or olfactory information. In these cases, individuals experience lights, sounds, and smells that are not present in the environment.

2.3.2 Sensory processing

Sensory information remains segregated by modality as it is transmitted from the sense organs to higher brain regions. The segregation occurs because there are separate neuronal pathways for vision, audition, olfaction, and other senses. Each pathway begins with a group of sensory receptors that sends axons to the spinal cord or brainstem where they connect to another distinct cluster of neurons. These neurons, in turn, connect to other groups of neurons with each pathway terminating in a specific region of the cerebral cortex. Thus, each sensory modality has a distinct pathway from the sensory receptors, where information is detected, to the cortex, where it is processed. This ensures that visual, auditory, and tactile messages do not get scrambled as they are being transmitted. This principle is illustrated in Figure 2.9, which shows the sensory pathways for vision in humans. Over human evolution, the number of 'stops' along this anatomical route from sensory detectors to higher brain regions has increased, allowing more complex processing of sensory information (Kass, 2008).

In the preceding figure showing sensory processing in the human brain, the visual cortex is much larger than the auditory cortex. This reflects a general principle of brain organization: the size of cortical area devoted to a function is a reflection of its importance. Humans rely much more heavily on vision than on audition so we need a larger area of cortex to process this information. In contrast,

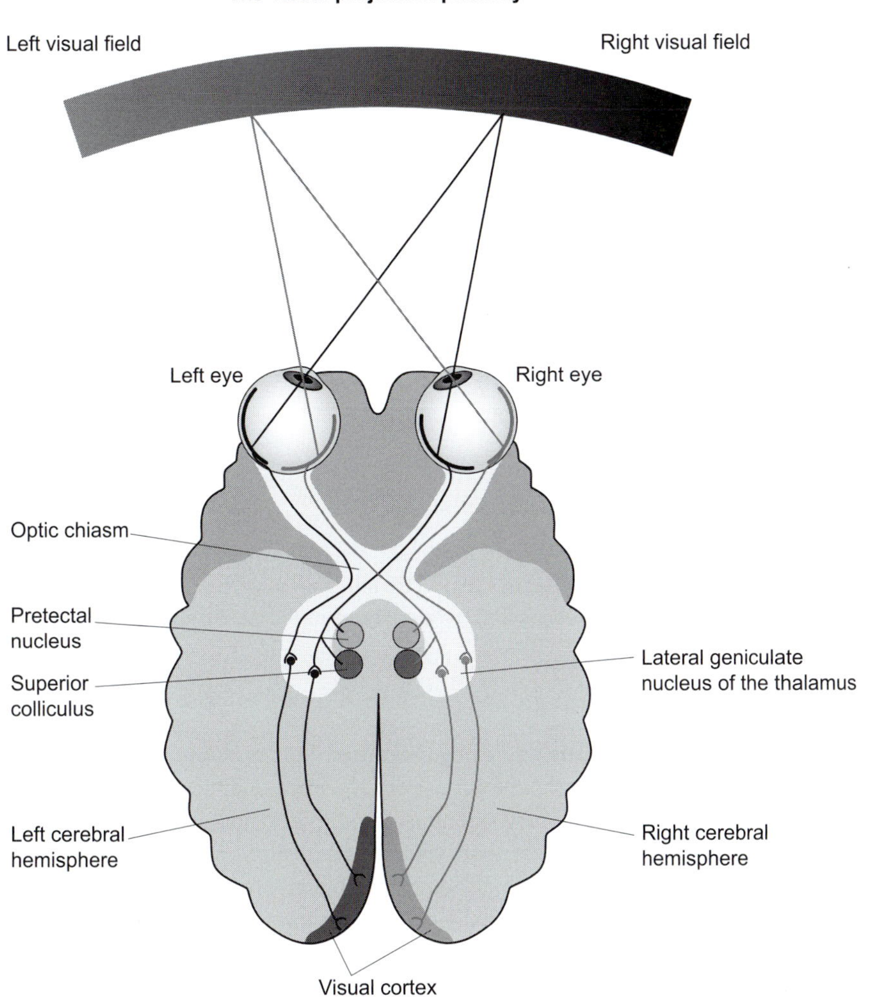

Figure 2.9 Neural pathway for transmission of visual information in the human brain.

animals that use other senses to gather important information about their environment should have larger cortical representations of these senses (see Figure 2.10).

Night hunting bats that use echolocation to identify prey have an exceptionally large auditory cortex (Suga, 1989). Other nocturnal animals, such as the platypus (*Ornithorhynchus anatinus*), spend much of their time searching through muddy waters for insects, shrimps, and other invertebrates. The platypus locates these prey by moving its large bill back and forth under the water. Sensory receptors on the bill detect mechanical ripples from other animals, indicating both their location and direction of movement. Accordingly, almost all of the sensory cortex of the platypus is devoted to the bill, with minimal representation of auditory and visual areas (Krubitzer *et al.*, 1995).

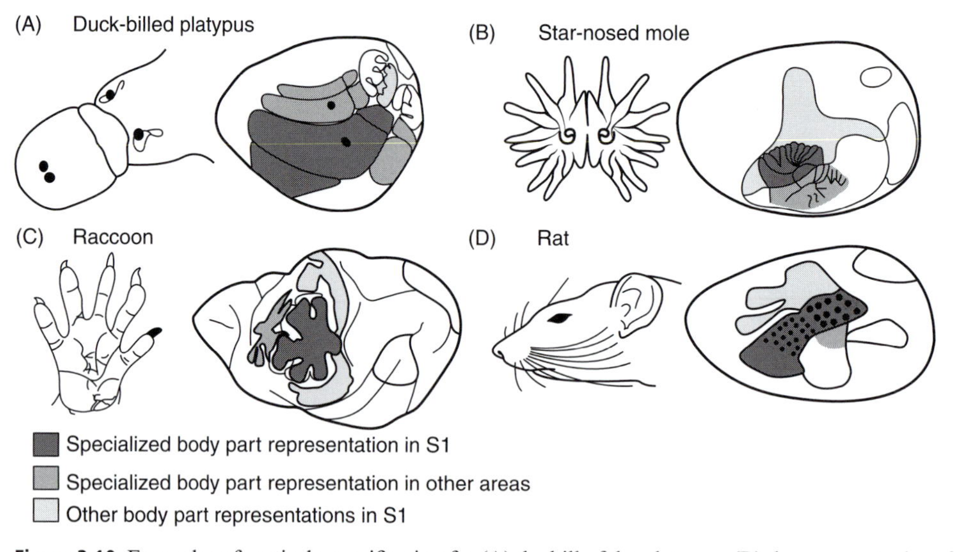

Figure 2.10 Examples of cortical magnification for (A) the bill of the platypus, (B) the nose tentacles of the star-nosed mole, (C) the hand of the raccoon, and (D) the whiskers of the rat. S1 = sensory cortex.

Coding stimulus dimensions

It is clearly important for organisms to distinguish visual from auditory from olfactory sensations, but this provides only a crude snapshot of the sensory world around them. Animals must extrapolate more detailed information about sensory stimuli in their environment if they are to use these cues to locate prey, attract mates, and care for their young. Information that is likely to be important to animals in these situations includes the attributes or quality of the stimulus, the intensity of the stimulus, and the length of time that it remains in the environment. This section outlines how the sensory system is set up to detect these differences.

Sensory quality refers to the type or category of stimulus within a modality. That is, different sounds, smells, or tactile sensations represent qualitative differences within a modality. These are coded in the central nervous system by the activation of different groups of receptors. For example, low and high pitch musical notes activate different combinations of auditory neurons. Similarly, sweet and sour tastes are distinguishable because the chemical qualities of these two tastes trigger different sets of neurons. The same is true of the somatosensory systems, with hot, cold, ticklish, and painful sensations producing action potentials in distinct sets of neurons. Sensory receptors often respond to more than one stimulus (i.e. taste receptors often respond to more than one chemical and auditory receptors often respond to more than one sound), although stimuli that activate a single sensory receptor typically share certain features. That is, one neuron in the visual system may respond to different lines of a similar orientation and a neuron in the olfactory system may respond to a number of chemicals of a similar structure.

The nervous system must also have a way to code the intensity of sensory stimuli so that organisms can detect differences between bright and dim lights, soft and loud sounds, and light and heavy touches to the skin. The first way that sensory intensity is coded is through the rate of cell firing: as the intensity of the physical stimulus increases, so does the frequency of action potentials. This principle is called **frequency coding** and it explains how the nervous system distinguishes weak from strong sensory input. But neurons can only fire so quickly because the cell membrane must recover after each action potential

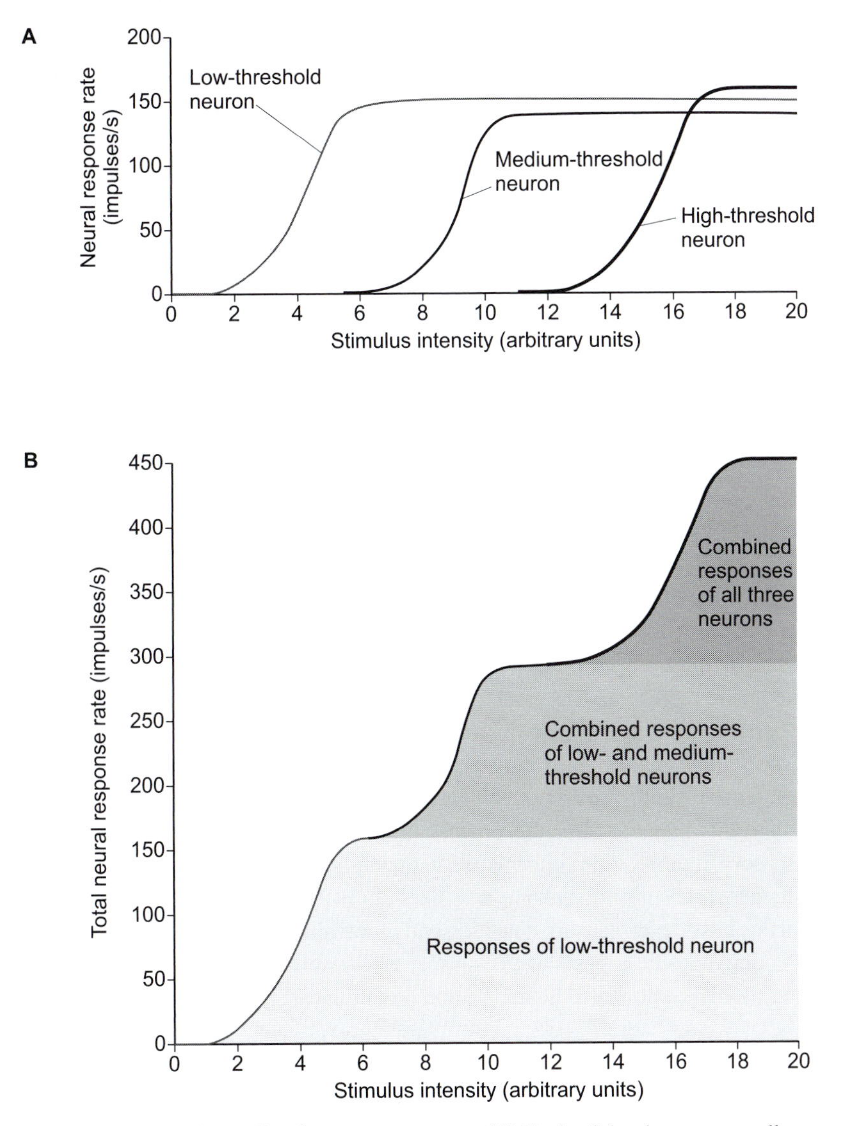

Figure 2.11 Intensity coding in sensory neurons. (A) Each of the three nerve cells represented has a different threshold – low, medium, or high – and thus a different rate of firing. Each cell varies its response over a fraction of the total range of stimulus intensities. (B) Although none of these nerve cells can respond faster than 150 times per second, the sum of all three can vary in response rate from 0 to 450 impulses per second, accurately indicating the intensity of the stimulus.

before another electrical impulse can be generated. Because of this, a change in stimulus intensity, beyond a certain threshold, will not be detectable by single neurons. This problem is overcome by having more than one neuron fire at a time. According to the principle of **population coding**, as the stimulus intensity increases, the number of sensory receptors firing action potentials also increases. Stimulus intensity is coded within each modality, therefore, as an increased rate of cell firing and an increased number of firing neurons. The relationship between frequency and population coding is illustrated in Figure 2.11.

Coding stimulus duration

The intensity of a stimulus indicates how much physical energy is reaching the sensory receptors. By coding stimulus intensity, animals can detect how bright lights are, how loud sounds are, and how pungent smells are. It seems almost trivial to say that animals also need to know how long these stimuli last. A sudden burst of noise and a persistent roaring probably mean something different, even if they are the same volume. This leads to the question "how does the central nervous system code stimulus duration?" The simplest solution may appear to be sustained firing: as long as a stimulus is present (and does not change), action potentials should continue at the same rate. The drawback to persistent cell firing is that it depletes energy and resources within the neuron. It may make more sense (at least metabolically) for the nervous system to respond to changes in sensory input. This also makes sense in terms of what animals need to know about their environment. Sights, sounds, and smells that remain constant probably have less significance than those that arrive or leave suddenly. The ability of animals and young children to keep track of temporal duration will be discussed in Chapter 6.

Sensory receptors code whether stimuli are constant or changing through altered patterns of cell firing. If sensory input is constant, the rate of cell firing declines. This decline may be rapid or slow; indeed different groups of sensory receptors can be categorized as slowly or rapidly adapting. Slowly adapting neurons fire a burst of action potentials when a stimulus arrives, then continue to fire at a slower rate while the stimulus is present. Rapidly adapting neurons fire a quick succession of action potentials, both at the onset and the offset of a stimulus. The duration of a sensory stimulus, therefore, is represented in clusters of sensory neurons that exhibit different patterns of adaptation to persistent stimuli. The combination of these two firing patterns allows animals to distinguish sensory input that remains constant from that which is changing.

For all organisms, including humans, sensory detection occurs automatically with little or no awareness of how this happens. But the process is far from simple: the segregation, coding, and organizing of sensory input provides information to the organism that is critical for survival. An animal that is unable to detect sensory information is unlikely to live, let alone reproduce and take care of its young. Beyond this basic level, sensory detection and processing are fundamental to all other cognitive processes. By coding specific aspects of the sensory environment, animals are able to find their way around (Chapter 5), make choices (Chapter 8), and communicate with other animals (Chapter 12).

2.4 Perception

Understanding how sensory information is detected, segregated, and transmitted is only the first step in understanding how animals are able to interact with their sensory environment. Sensory signals that travel through the central nervous system must be recombined in a way that is meaningful to the animal. This connection between sensory input and the analysis of this input describes the relationship between sensations and perception. **Sensations** are produced when physical stimuli (e.g. light or sound waves) activate sensory receptors, which then send neural signals through specialized circuits. **Perception** is the interpretation of these signals which occurs when sensory information is processed, organized, and filtered within the central nervous system. The distinction between sensation and perception is illustrated in Figure 2.12. Sensory receptors detect a series of black dots and random shapes on a white background, with little variation across the two-dimensional image. In contrast to these sensations, most people perceive a spotted Dalmatian with

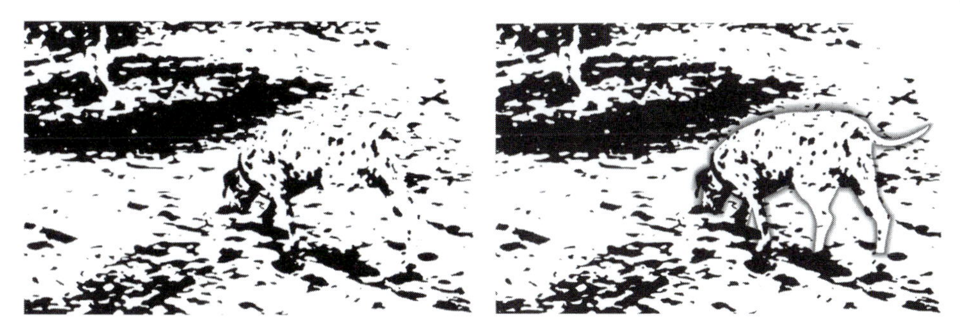

Figure 2.12 A series of black and white dots may be perceived as an image with identifiable shapes.

shadows in the background. This occurs because the visual system groups smaller elements into larger objects which are then recognizable to most humans.

2.4.1 Psychophysics

The study of the relationship between sensations and perception is called **psychophysics**. Psychophysics attempts to explain how the physical characteristics of a stimulus are interpreted by an organism. Psychophysical experiments in humans are conducted by asking a participant to indicate when they see a light, hear a sound, smell an odor, etc. The setup is familiar to anyone who has undergone a hearing test. In this clinical exam, tones of different volumes are played through earphones to each ear and the patient indicates when the sound is audible. The same basic setup is used in experimental labs that test responses to stimuli in other modalities. Naturally this is more difficult to study in animals which cannot make verbal responses. But experimenters have figured out clever ways to get yes/no answers from animals in the lab. The simplest way to do this is to train animals to make one response when a stimulus is present and another when it is absent. By varying the intensity of the stimulus, the researcher can determine the point at which an animal no longer perceives the stimulus (see Figure 2.13A). Blough (1956) set up one of the first psychophysical experiments using this procedure to test how well pigeons could detect light after spending different periods of time in the dark. As shown in Figure 2.13B, the threshold for detecting light dropped as the dark period increased. In other words, the birds were more sensitive to light the longer they stayed in the dark.

This lowered threshold for detecting light that occurs under reduced illumination is called **dark adaptation**. Dark adaptation takes a few minutes to develop because it involves a series of neurochemical reactions in the eye. The mechanism evolved in humans and some animals, providing these species with the ability to see during the day and at night. Dark adaptation provides clear evidence that perceptions are not a direct reflection of sensations: individuals see different aspects of their environment under altered lighting conditions. Indeed the same physical stimulus may be perceived in one context but not another. Like vision, perception in other sensory systems is also context dependent. For example, subtle flavors are easier to detect when an individual is hungry than when they are sated. Deciphering the soprano and bass lines may be simple in an unaccompanied choral piece, but near impossible in a large-scale opera with full orchestra. And a light brushing of the arm is likely to be interpreted (perceived) in very different ways if it comes from a close friend rather than a complete stranger.

The natural environment provides infinite examples that perception is altered under different stimulus conditions. Think about **camouflage**, in which structural adaptations allow a species to

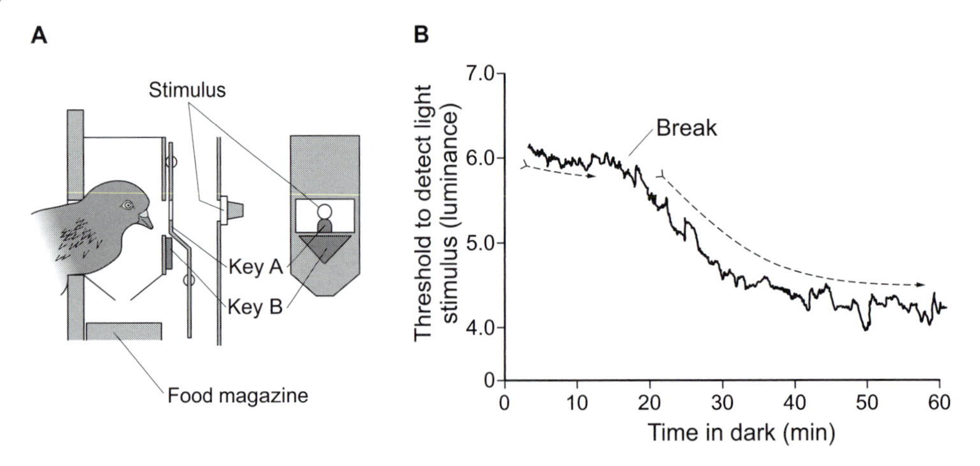

Figure 2.13 Experimental design (A) and data (B) from a study examining light perception in pigeons (Blough, 1956). Pigeons are placed in an operant chamber with a pecking key on either side of a food magazine. Over a series of trials, they learn to peck the A ('yes') key when a light is turned on above the food magazine and to peck the B ('no') key when no light is turned off. As the light intensity decreases, birds respond less on the 'yes' key, indicating that they can no longer detect the stimulus. This threshold to detect the light declines (i.e. birds are better able to see the light) when they are placed in a dark environment prior to testing.

blend in with its environment. The effectiveness of camouflage can be measured by testing whether the same animal is more difficult to detect in different environments. Kettlewell (1955) used this approach with the peppered moth (*Biston betularia*), which is either melanic (black) or lightly speckled (peppered). He placed each type of moth on dark or pale tree trunks and recorded how many became prey to local birds. As expected, melanic moths on light backgrounds and light moths on dark backgrounds were more conspicuous, being gobbled up quickly by the birds. Kettlewell's report helped to explain why the once abundant speckled moths were replaced, almost completely, by the rare melanic form in certain areas of Great Britain. From the mid-nineteenth to the mid-twentieth century, the treed areas in which the melanic form became dominant were darkened by industrial soot. Under these conditions, the light peppered moths became easy targets for predators, reducing the survival and reproductive success of this subgroup. Fortunately, Britain eventually reduced its use of coal and enforced stricter environmental laws so the speckled moth morphs increased once again (see Hooper (2002) for an alternative explanation).

Psychophysical experiments and the natural environment equivalent, camouflage, reveal that the ability to detect a sensory stimulus depends on the context. The same principle applies to detecting differences between stimuli. A **just noticeable difference (JND)**, the amount by which two stimuli must differ so that the difference can be detected, is not an absolute value, but a relative one. As the intensity of a stimulus increases, so does the intensity that produces a JND. Thus, if a dim light in a dark room is turned up a small amount, the change is immediately noticeable. In contrast, a very bright light needs a much larger illumination boost to produce a perceptual change. The ratio between current stimulus intensity and the intensity change required for a JND is called the Weber fraction in honor of Ernest Weber (1795–1878), the physiologist who characterized this relationship. JND makes sense if you consider that the function of perception is to interpret sensory input. What an organism needs to know is how particular stimuli stand out or differ from other stimuli. The best way to do this is to design a system that calculates the relationship between stimuli. In this way, the system can process sensory input that may vary over many orders of magnitude, without losing sensitivity in the lowest ranges. Most importantly, JND provides a striking example that the perceptual system does not reflect the absolutes of the sensory world.

2.4.2 Sensations transformed to perception

Given that sensations do not map directly onto perceptions, the central nervous system must transform sensory input into messages that are interpretable to the animal. In other words, sensory information that is segregated by modality (e.g. auditory or visual) and then coded for stimulus dimensions (e.g. intensity or duration) must be recombined into a perceptual whole. This translation of sensations into perceptions is accomplished in successive stages along the sensory pathways. Recall from Section 2.3.2 that sensory information is transmitted from receptors to cortical regions through specialized circuits. Each pathway has relay nuclei where signals are transmitted from one cluster of neurons to another. Sensory information is integrated at these relay points, and is also combined with signals arriving from other brain regions. This general organization of sensory processing is shown in Figure 2.14.

To understand how information is processed at each stage of the circuit, scientists implant electrodes in groups of neurons at each relay station and record how these neurons respond to different stimuli. The visual system again provides an excellent model for understanding this process because neuronal responses to visual stimuli have been characterized in a variety of animals. Most of the sensory receptors in the eye project to a region of the thalamus called the **lateral geniculate nucleus (LGN)**. Sensory receptors in the retina and LGN neurons both fire at rapid rates when small dots of light are presented to the eye. But this does not mean that they are performing the same function in terms of visual processing, or that visual signals are simply transferred from the eye to the cortex through the LGN. Rather, the LGN organizes incoming visual signals that arrive from the retina before transmitting this information to the cortex. First, input from the right and left eyes are separated into different layers of LGN neurons. More specifically, each LGN receives input from both eyes, but fibers from the ipsilateral (same side) and contralateral (opposite side) eyes project to defined regions of the LGN. This organization is even more detailed within layers of LGN

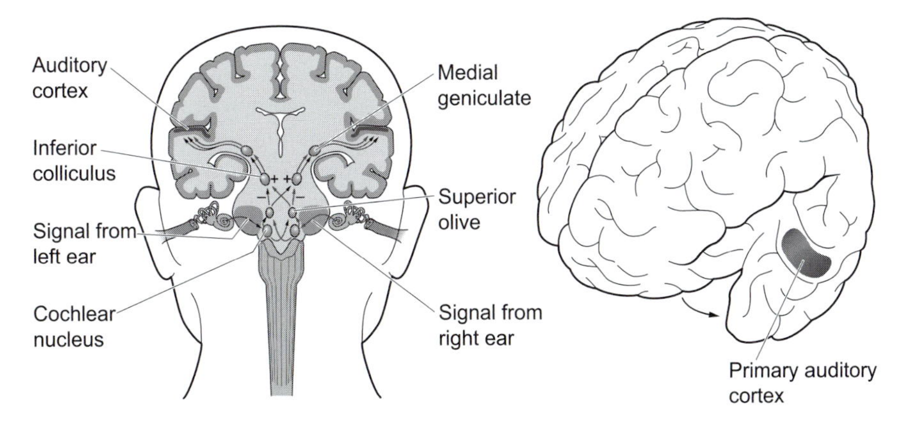

Figure 2.14 Route of auditory impulses from receptors in the ear to the auditory cortex of humans (left panel). The cochlear nucleus receives input from the ipsilateral ear (the one on the same side of the head). Information is then sent, bilaterally, to the superior olive, inferior olive, and medial geniculate nuclei. The medial geniculate nucleus (in the thalamus) sends auditory signals to the primary auditory cortex (right panel) located in the superior temporal cortex.

neurons, in that each point on the retina sends axons to a corresponding location on the LGN. Thus sensory neurons that are adjacent to each other in the retina are also adjacent to each other in the LGN. This produces a spatial mapping from the retina to the LGN, a truly amazing feat of organization given that each LGN receives input from approximately one million retinal fibers. Finally, the LGN regulates information flow from the retina by integrating these messages with signals from other brain regions, most notably the cortex. Indeed, there is a much larger flow of information from the cortex to the LGN than the other way around. This suggests that the LGN must be performing some kind of processing that modifies and refines sensory signals about the external world. A separate visual pathway, that bypasses the LGN, has a different role in visual perception (see Box 2.2).

Box 2.2 Blindsight

The primary visual cortex is the first cortical area to receive input from the retina (via the LGN), so it is not surprising that damage in this region produces blindness. The damage may be caused by stroke, brain injury, or surgery to remove a tumor. Patients with these conditions report no visual perception in the visual field that projects to the damaged area. Experimental lesions in animals appeared to confirm this phenomenon of cortical blindness, at least until the late 1960s. Around that time, researchers began to examine the function of visual pathways from the retina to the cortex that did not travel through the LGN. In a clever experiment with golden hamsters (*Mesocricetus auratus*), Schneider (1969) showed that lesions of the visual cortex rendered animals unable to discriminate between visual stimuli (lines of different orientation), but did not impair the ability to turn toward a food reward. Lesions of the retinal fibers that project to the superior colliculus produced the opposite pattern: these animals could not locate a food reward but were able to solve a visual discrimination task. The explanation for this dissociation of lesion effects was that hamsters (and probably many other animals) had two visual systems: one for identifying the 'where' and one for identifying the 'what' of visual stimuli.

Many people argued that the orienting response in humans had been lost or become dormant over evolution as cortical systems became dominant. A small portion (approximately 10%) of retinal fibers project to the superior colliculus in humans, but the function of this 'where' system appeared to be vestigial. The hypothesis was disproven in a dramatic demonstration of residual visual function in a patient with cortical blindness (Weiskrantz, 1976). D.B. had a large portion of his right visual cortex excised to remove a malformation. As expected, he reported no vision in his left visual field after the surgery. So it seemed outrageous to him that an experimenter would ask him to 'locate' an object in his left visual field by moving his eyes to the object's position. But trial after trial showed that D.B. performed significantly better than chance on this task.

Weiskrantz coined the term **blindsight** to describe this paradoxical ability of cortically blind individuals to locate a visual stimulus. Subsequent clinical reports confirmed this phenomenon in other patients, although the majority of individuals with cortical blindness do not demonstrate blindsight. Nonetheless, even isolated reports of blindsight helped to confirm a functional role for the superior colliculus in human vision. They also provided some of the first evidence for perception in the absence of knowledge of awareness. Blindsight patients repeatedly claim that they cannot see a stimulus; when pressed to make a choice, however, they can point to the stimulus location. These individuals are surprised, even incredulous, when their correct choices are revealed to them. Knowledge of perception of a stimulus, therefore, can be dissociated from the perception of that stimulus.

LGN signals that arrive in the primary visual cortex target one of three types of cells. Simple cells respond best to lines of a particular orientation, complex cells to a particular direction of movement, and end stopped cells to corners, angles, or bars of a particular length moving in a particular direction. In addition, the visual cortex organizes information into separate paths for color, shape, and motion. As sensory information is transmitted to higher brain regions, therefore, neurons at each stage respond to increasingly complex stimuli. The processing of sensory information does not stop in the primary sensory areas. Sensory signals continue to be organized, regulated, and integrated as information is transmitted to other parts of the cortex. In mammals, one of the most interesting, at least in terms of sensory perception, is the orbitofrontal cortex (OFC). The OFC receives direct and indirect connections from all primary sensory areas, suggesting that it has an important role in sensory integration. One proposal is that the OFC assigns motivational significance to sensory stimuli, allowing animals to make decisions about how to respond to these stimuli (Rolls & Grabenhorst, 2008). In sum, sensory processing occurs in stages with neurons responding to more and more complex aspects of the environment as information is transmitted to higher brain regions. This general plan of sensory processing is consistent across sensory modalities and across a variety of animals, providing strong evidence that the process is preserved over evolution.

2.4.3 Elemental versus ecological theories of perception

If neurons at each stage of the sensory system pathways respond to increasingly complex stimuli, it makes sense to think about sensory detection as taking information apart and perception as putting it back together. This idea was discussed by early psychologists, such as Wilhelm Wundt (1832–1920), but developed most thoroughly by his student Edward Titchener (1867–1927), who promoted a school of thought called **structuralism**. The basic idea behind structuralism is that perceptions are created by combining or adding up the elements of sensations. In terms of vision, the process occurs as follows: visual stimuli (e.g. shape, color, and size) are coded in different groups of neurons that convey this information, in parallel, to higher brain regions. When neural signals arrive at a new brain site (sometimes called a relay station), the information is recombined and transmitted to other parts of the brain. In this way, the elements of visual input are the building blocks for the recognition and interpretation of complex visual scenes.

The main evidence for this elemental theory of perception is that the time to process two different features of a stimulus (e.g. color and shape) is the sum of the time it takes to process each individually. Similarly, the probability of correctly identifying individual features when stimuli are flashed briefly on a screen is perfectly predicted by the combined probabilities of reporting each on their own. These effects, observed in both humans and pigeons (Cook, 1992), would not be expected if perception was synergistic. Theories that explain perception in terms of feature binding, or simply putting elements together (Katz *et al.*, 2010), are modern versions of structuralism. One of the most prominent is the **feature integration theory** (Treisman, 1988), which posits that elements of sensory input (e.g. shape, color, and motion for visual objects) are coded at the initial stages of processing, and then combined at higher levels to produce perceptual wholes. These and other elemental theories of perception are sometimes called bottom-up theories.

Elemental theories of perception appear to explain many experimental findings, but seem at odds with human experience in which the sensory world is perceived as a 'whole.' This criticism was formalized into alternative theories by the Gestalt psychologists, a group of European scientists (e.g. Werthemier) who proposed that individuals perceive sensory information in its entirety (e.g. a visual scene) and *then* divide it into elements only if further processing is required. In North America,

Figure 2.15 Video games that simulate rapid motion, such as fast driving, rely on optic flow to create the illusion of movement.

James J. Gibson (1904–1979) voiced a general disillusionment with elemental theories by pointing out that perception is not a passive documentation of features in the sensory world, but an active process of gathering ecologically relevant information. The information that an animal or human acquires, therefore, will vary depending on how it is moving through the environment. In studying this process, Gibson used the term **optic flow** to describe the movement of elements in a visual scene, relative to the observer (Gibson, 1979). Optic flow is one of the most important sources of information to an animal because it provides feedback on how its own behavior is altering sensory input. Gibson extended this idea further, noting that the primary function of perception is to allow animals to act upon and to interact with their surroundings. The principles of optic flow are an important component to many interactive computer games (Figure 2.15).

Gibson's ideas on optic flow and other ecologically relevant processes are closely tied to evolutionary theory. If optic flow is a critical component of visual perception, sensory and motor systems must have evolved together. Otherwise, the rate of information flow generated by movement through the environment may not match the rate at which sensory receptors can encode the moving information (Snyder *et al.*, 2007). This tight link between sensory and motor systems is illustrated in day- and night-active insects: the retinae of the night-active insects integrate visual information more slowly than their day-active relatives (Warrant *et al.*, 2004). Bipedal locomotion evolved in humans, so the rate at which you walk should be ideally suited for gathering visual information about your environment. Although it may seem that you miss many things, consider how much more you remember about a new city if you walk through it, compared to if you are driving or riding on a high-speed train.

Gibson's focus on perception as an interactive process helped to create a subdiscipline of psychology, called **ecological psychology**. According to this position, behavior (and cognition) can only be understood in the environmental context in which it takes place. Sensory perception, therefore, should be studied under real-world conditions or in laboratory experiments that simulate the natural environment. Despite the widespread influence of Gibson's theories, many researchers continued to study sensory systems in the lab and many contemporary theories explain sensory processing as the integration of elements or features. These two different approaches to studying perception (elemental versus ecological) should be considered as complementary rather than contradictory: each position explains the process from a different perspective. There is no doubt that sensory input is segregated into elements in the central nervous system, or that separate groups of neurons respond to different elements of the sensory environment. At the same time, at least some perception is accomplished via a computational process that depends on an animal's interaction with its natural environment. These two different types of perceptual processing (elemental and ecologically relevant) probably occur simultaneously in parallel pathways.

2.4.4 Stimulus filtering

Both elemental and ecological theories must account for **stimulus filtering**, the process of separating and extracting meaningful (i.e. biologically relevant) information from the abundance and diversity of sensory cues in the environment. Stimulus filtering occurs, first, at the level of the sensory receptors, which respond to a particular range of physical energy within a modality (see Section 2.3.1). For example, humans hear only a small spectrum of sound waves, but are acutely sensitive to those which transmit auditory signals from conspecifics (i.e. speech).

On a broader level, stimulus filtering is accomplished by **sign stimuli**, the essential features of a stimulus which are necessary to elicit a specific behavioral response. Responses to sign stimuli are typically species-typical behaviors that occur in a fixed order (although these may be modified with experience). Examples include courting, nest building, and defensive responses. Although these behaviors may be quite complex, they are often controlled by a small number of simple features (i.e. sign stimuli). Tinbergen (1951) documented one of the first descriptions of responses to sign stimuli in herring gull (*Larus smithsonianus*) chicks that peck a parent's bill. The pecking, which is directed to a red dot on the bill, causes the adult to regurgitate food for the youngster. By presenting artificial stimuli of various shapes and colors (e.g. cut-out cardboard, painted sticks, etc.), Tinbergen determined that the chicks were attending exclusively to the shape of the bill and the red dot. The pecking response, which is critical for the young bird's survival, therefore is elicited by a small but essential feature of the parent. Other well known examples of sign stimuli include the red belly of stickleback fish that elicits aggressive responses in other males, female pheromones that elicit ritual courting dances in male birds, and egg-shaped objects that elicit egg-rolling responses in nesting geese. Because they reliably produce species-specific behaviors that promote survival and reproduction, sign stimuli are sometimes called **releasers**.

Sign stimuli are an efficient means to interpret the sensory world because they provide meaningful information with minimal sensory input. An elegant demonstration of this principle comes from a classic study by Schiff *et al.* (1962). In an experimental setup like the one shown in Figure 2.16, researchers set subjects (rhesus monkeys; *Macaca mulatta*) on one side of an opaque screen and a light on the other. Placing a small circle in front of the light cast a shadow on the screen. Moving the circle away from the screen caused the shadow to expand – and the subjects to flee in terror. Moving the circle toward the screen, or simply changing the screen illumination, was ineffective. A range of animals from crabs to frogs to chicks to humans showed the same escape response, suggesting that these very different species were all interpreting a growing shadow in the same way (Schiff, 1965). According to Schiff and his colleagues, the rapid and symmetrical expansion of a visual stimulus is always perceived as an approaching object, a phenomenon they called

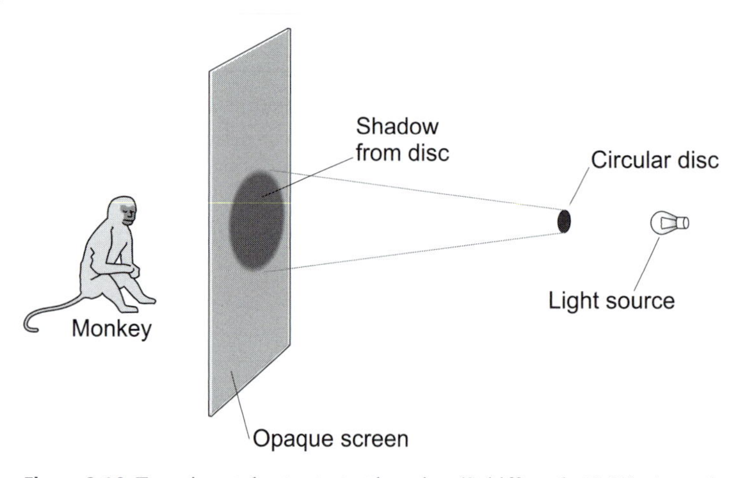

Figure 2.16 Experimental setup to test looming (Schiff *et al.*, 1962). A monkey sat facing one side of an opaque screen. A light behind a circular disc on the other side cast a shadow on the screen. Moving the disc caused the shadow on the screen to expand or contract.

'looming.' Even kittens and chicks that were reared in the dark (with no experience of approaching objects hitting them) were quick to avoid potential impact from the looming stimulus. This is a hard-wired response, no learning required. Indeed, the escape response is so strong that it is almost impossible to suppress, a fact that is exploited in video games, flight simulators, and IMAX theatres: most people cannot avoid ducking when an asteroid from outer space suddenly fills the screen.

An updated version of the looming experiments used computer simulated visual images to examine how the brain coordinates this rapid escape from approaching objects (Wang & Frost, 1992). It was already known that neurons in the visual area of a pigeon brain fire when a looming stimulus approaches. Wang and Frost varied the size and speed of these moving images on a computer screen to demonstrate that a small population of visual neurons compute, *exactly*, when an object will collide with the observer. The calculation in these 'time to collision' neurons is amazingly precise, taking into account how far away the object is, how big it is, and how fast it is moving. So two objects may be the same distance away, but if one is moving faster than the other, it will set the neurons firing at an earlier time point. Less than 100 milliseconds after the neurons begin to fire, heart rate increases and muscle activity peaks. This suggested to Wang and Frost that activity in the time to collision neurons is a signal for the animal to escape. In essence, these neurons are sending a very specific message to the rest of the body: 'get moving . . . NOW!'

The stimuli in both the looming and time to collision experiments were very simplistic: in the case of the former, just a growing shadow on a screen. But this information was enough to elicit consistent and reliable responses across animals. Organisms do not need to know much about a stimulus, what it is or where it comes from, to know that they should get out of the way if it is heading right toward them. Like other sign stimuli, the looming and time to collision stimuli allow animals to make cost-effective decisions with limited information and no time to waste. These are sometimes described as simple 'rules of thumb' for perception. It makes more sense to interpret a glimpse of moving fur in the bush as a predator, than to wait until the entire animal appears to judge whether or not it is dangerous. Errors that arise from these perceptual rules of thumb are usually rare and often inconsequential (i.e. erring on the

Researcher profile: **Dr. Barrie Frost**

Figure 2.17
Dr. Barrie Frost.

Barrie Frost's scientific investigations started very early in life. Growing up in New Zealand, he was an avid observer of the wildlife that surrounded him, often asking himself 'how' and 'why' animals behaved as they did. He took these questions to his formal education where he completed a BA and MA at the University of Canterbury (NZ), a PhD at Dalhousie University in Canada, and a postdoctoral fellowship at the University of California, Berkeley. He was appointed to the Department of Psychology at Queen's University (Kingston, Canada) in 1970 where he has continued to conduct world class research on a range of topics that cover neuroethology, animal navigation, auditory and visual neuroscience, as well as virtual reality in humans.

Frost's primary research interest is sensory processing. He has studied visual, auditory, somatosensory, and even olfactory function in a variety of animals ranging from butterflies to humans. One of his primary interests is how the brain distinguishes between movement caused by the head's own motion and the organisms' actual movement through the environment. Pigeons are an ideal model to study this phenomenon because they engage in constant head-bobbing as they walk. Frost was able to train pigeons to walk on a tiny treadmill and then varied their visual environment by moving films at different speeds around the birds. Based on this work, he and his team concluded that head-bobbing served an important function, allowing pigeons to take ongoing snapshots of the world around them. Frost has also looked at the role of this head-bobbing and other pigeon movements in reproduction by developing a virtual pigeon that elicits courting behavior in conspecifics. His work is cited in many introductory textbooks and his theories have led to practical applications such as an artificial ear for the profoundly deaf which allows users to 'feel' sound through their skin.

Frost's fascination with visual and auditory processing led him to investigate navigational skills of both birds and butterflies. How do these animals find their way part way around the world, sometimes travelling on their own for the first time? His work in this area has led him on many international excursions, clambering around on cliffs at night in the outer reaches of the South Pacific to track birds by equipping them with tiny radio transmitters that communicate their location to passing satellites. His younger colleagues call him the 'Indiana Jones of Neuroethology.'

Barrie Frost typifies all of the qualities that make a world class scientist. He is inquisitive, conscientious, and collaborative. He uses a range of techniques from behavioral observation to neurophysiological recordings, allowing him to address both proximate and ultimate explanations of behavior. In his words, "the technique is irrelevant, it is the research question that matters." Despite his retirement more than 10 years ago, Frost can still be seen racing madly through the halls of Queen's University, often with his butterfly net in hand!

side of caution when detecting a predator); thus, there is no evolutionary pressure for them to disappear.

2.5 Attention

Sensory detection and processing filters, organizes, and codes sensory signals in the environment, thereby reducing the amount of information that reaches higher brain regions. Even so, most organisms are exposed to a vast array of sensory input across modalities (e.g. visual, auditory, tactile, etc.). If there was no way to ignore irrelevant input, the sensory overload could create a constant state of scatterbrained confusion. **Attention** solves this problem: the mental process that selects which information will be processed further allows individuals to focus on particular stimuli or events.

Attention is difficult to quantify so it did not fit easily into models of stimulus–response learning promoted by behavioral psychologists. Nonetheless, many scientists in the early twentieth century discussed attentional mechanisms and how these may influence other cognitive processes such as memory. This increased with the rise of cognitive psychology and the realization that internal mental processes could be used to explain behavior. Even experimental psychologists studying animal behavior began to incorporate attentional mechanisms into S-R theories, noting that learning is affected by the amount of attention animals devote to a stimulus (Mackintosh, 1975; Pearce & Hall, 1980).

2.5.1 Selective attention

One of the first things researchers noted is that attention is not a unitary construct. This is apparent in the many descriptions of attention: 'pay attention,' 'short attention span,' and 'grabbing someone's attention' all refer to different processes. To deal with this complexity, researchers have categorized and defined different types of attention. One of the easiest to conceptualize is **selective attention**, the ability to attend to a limited range of sensory information while actively inhibiting competing input. Selective attention is sometimes called the cocktail party phenomenon in reference to the ability to converse with a small group of people while blocking out other conversations in the vicinity. In the visual system, selective attention is accomplished by moving the eyes to the point of interest. It can be measured in the lab using computerized eye tracking as shown in Figure 2.18.

Eye tracking experiments reveal an important aspect of selective attention in humans. When looking at another person's face, most people focus on the eyes, particularly when the facial expression is emotional. In general, females spend more time than males gazing at the eyes and are better at identifying emotions in other people (Hall *et al.*, 2010). In contrast, many autistic individuals have difficulty attributing emotional states to others, and those that do show reduced scanning of the eyes (Pelphrey *et al.*, 2002). Selective attention, allowing humans to focus on one feature (eyes) of a larger stimulus (face) is adaptive in that the ability to recognize an angry aggressor, a fearful opponent, or a friendly mate is critical for social species, including humans.

Even for species that do not spend time interpreting facial expressions in others, selective attention provides a number of advantages. Animals in their natural environment face the constant challenge of identifying food, potential mates, and threatening stimuli. The time and effort to search for these stimuli would be reduced if animals could focus on one item (e.g. a specific food) at a time. They may do this by forming a mental image of the target, and then scanning the environment for stimuli that match this image. Tinbergen (1960) was the first to put this hypothesis forward, suggesting that foraging animals form a **search image**, a mental representation of a target, which allows them to selectively attend to specific features in the environment. The idea came from his observation that birds did not eat certain types of insects when they first appeared in the spring, probably because they were difficult to detect on a cryptic background. As the birds had more experience with seeing these insects, however, they were able to locate (and quickly gobble up) the prey.

Pietrewicz and Kamil (1981) tested Tinbergen's idea that birds were forming search images by training blue jays (*Cyanocitta cristata*) to peck a screen when they saw a moth (see Figure 2.19).

Figure 2.18 Eye tracking software can provide a measure of selective attention by indicating which part of a visual image subjects spend the most time looking at.

The slide on the screen changed with each trial, and all slides contained one image of a moth on either a cryptic or a conspicuous background. The key point in the experiment is that there were two types of moths, distinguished in terms of the color and pattern on their wings. Birds were better at detecting one type of moth if it appeared in successive trials, suggesting that they were forming search images of the more plentiful prey. The results also fit with field research showing that common food items are foraged at higher rates in the natural environment. The explanation for both findings is that search images develop with experience, and selective attention allows animals to focus on stimuli that match these images. In that respect, both search images and selective attention are cognitive processes that provide an evolutionary advantage: they improve foraging and may aid searches for other biologically relevant stimuli.

Interestingly, a search image is maintained as a mental representation *only* when the target is difficult to find (i.e. it is cryptic) (Langley *et al.*, 1996). When food items are conspicuous and easy to locate, search images do not emerge or quickly disappear. This suggests that the cognitive resources required for developing and maintaining a search image are recruited only if they are necessary. This makes sense in that search images make scanning more efficient, but may cause organisms to 'miss' something unexpected in their environment if they are focusing only on stimuli that match their search image.

2.5.2 Sustained attention

Selective attention allows animals to focus on a small set of stimuli, but this is not enough to function effectively in the environment. Many situations require organisms to maintain a focus on one aspect of their surroundings for extended periods of time. This process is called **sustained**

Figure 2.19 A blue jay waits for an image of a moth to appear on one of the screens.

attention and is measured in both humans and animals using vigilance tasks. Rather than identifying a target (as in search tasks), subjects are required to monitor a particular location and to indicate when a target stimulus is presented. In humans this can be done on a computer screen with flashing lights: subjects are instructed to press a key as quickly as possible after a light appears somewhere on the screen. The light flashes are brief and the time between them is unpredictable so subjects must be vigilant in monitoring the screen. A similar setup is used for lab animals that peck a key (bird) or poke their nose in a hole (rodent) when they see a light flash. If humans or animals are distracted or their attention lapses, even briefly, they may miss the light and the opportunity to respond. Humans (and many animal species) exhibit wide variability in sustained attention (see Box 2.3).

The natural environment equivalent of sustained attention is continuously watching for predators or prey. Predators (as well as aggressive conspecifics) have a much higher chance of success if they can mount a surprise attack. So the ability to remain vigilant across time has a huge survival advantage. Of course, an individual animal that spends all of its time watching for attacks will have no opportunity to forage or hunt for itself, let alone reproduce or care for its young. Many species have solved this problem by living in social groups. Under these conditions, the task of keeping watch for approaching danger can be spread across the group. Think about our human ancestors taking turns on the 'night watch'; groups of animals employ the same tactic by having each member remain alert for short, alternating, periods. You can observe this in a flock of birds that each scan the environment between periods of searching for food. As soon as one member detects danger, they

Box 2.3 Attention deficit hyperactivity disorder

For most people selective and sustained attention are automatic processes that appear to require very little mental energy. For others, focusing on one thing at a time and sustaining this focus beyond a few minutes is almost impossible. These difficulties often lead to a diagnosis of attention deficit hyperactivity disorder (ADHD), which affects up to 12.3% of boys and 5% of girls in the USA (Pritchard *et al.*, 2012). Children with ADHD often perform poorly in school, even if they score at age level on standardized intelligence tests. The inability to focus on one source of information means that they often miss, or misinterpret, instructions and they have difficulty completing assigned tasks because they are easily distracted. The hyperactivity that accompanies these deficits makes classroom management of ADHD children particularly challenging.

The frequency of ADHD has increased dramatically in North America over the last 30 years, raising concerns that the disorder is simply overdiagnosed in a society that does not allow "boys to be boys." Other critics argue that young people today have difficulty focusing on one thing at a time because they are constantly surrounded by sensory stimulation from computer games, text messages, and electronic gadgets (Bailey *et al.*, 2010). One of the more intriguing hypotheses of ADHD etiology is that symptoms emerge when children have reduced opportunities to engage in rough-and-tumble play (Panksepp, 1998). In rat pups, play fighting is necessary for normal development of the frontal cortex; conversely, frontal lobe lesions in rats produce ADHD-like symptoms. Importantly, these deficits are alleviated if the brain-damaged pups are allowed to engage in vigorous play fighting with other young rats. Panksepp's ideas are not as far fetched as they sound: compared to 'neurotypical' controls of the same age, individuals with ADHD have smaller frontal lobes (Castellanos *et al.*, 1996) and reduced cortical thickness in the right frontal lobe that correlates with the severity of their illness (Almeida *et al.*, 2010).

The most common treatment for ADHD in North America is prescription medication. Methylphenidate (better known by its trade name, Ritalin) increases activity of the neurotransmitter dopamine, which then improves both selective and sustained attention (Swanson *et al.*, 2011). The downside to methylphenidate is that the drug is a stimulant, similar in structure and activity to amphetamine. Many people are concerned that long-term Ritalin treatment will produce additional problems, particularly in children. Clinicians and parents in favor of the drug point out that reduced misconduct and increased academic and social acceptance outweigh the potential risks of chronic drug use. Even without any treatment, symptoms often decline across childhood, although some people believe that affected adults who show minimal deficits have simply learned to cope with their disorder.

alert the other birds by emitting alarm calls and quickly escaping. Figure 2.20 shows the outcome of this process in a flock of coots all taking flight when a predator (in this case a bald eagle; *Haliaeetus leucocephalus*) is detected flying overhead. As with any cooperative endeavor, the workload for each individual goes down as the numbers in the group go up. Thus, the amount of time that birds spend scanning the environment decreases as the size of the flock increases (Bertram, 1980). This likely explains why guppies live in bigger schools as the risk of predation increases (Seghers, 1974), and why hawks are less successful in killing pigeons that live in large flocks (Kenward, 1978).

Figure 2.20 Flock of coots taking flight when a predator is detected.

2.5.3 Divided attention

Both selective and sustained attention allow organisms to focus on a limited part of their sensory world. **Divided attention** is the opposite: it describes the ability to process, simultaneously, sensory input from more than one source. Divided attention allows humans to do more than one thing at a time although, in most cases, one or both tasks suffer. Even relatively simple tasks, like reciting a nursery rhyme or calculating arithmetic sums, are disrupted when subjects must listen to a spoken message at the same time. Media multitasking is no different. In fact, individuals with more experience sharing their time between cell phones, texting, web surfing, and other activities perform *worse* on a series of cognitive tests (Ophir *et al.*, 2011). Contrary to popular belief, multi-tasking is NOT the most efficient way to get things done (see Figure 2.21).

Laboratory tests reveal that limits on divided attention are not restricted to humans. Using an experimental setup much like the one shown in Figure 2.19, Dukas and Kamil (2000) showed that blue jays are better at detecting prey when they can focus on one target as opposed to dividing their attention between two targets. Moreover, targets that are cryptic or difficult to detect require more attention, making it less likely that the birds will notice an approaching predator (Dukas & Kamil, 2000). This phenomenon can be observed in the natural environment: guppies that are feeding take longer to escape from an approaching cichlid than resting guppies, suggesting that the prey fish are less likely to notice a predator if they are paying attention to securing their own meal (Krause & Godin, 1996). Some animals (unlike many humans) appear to recognize their own limitations on attention because they modify their behavior when evaluating the tradeoff between feeding and predator detection. For example, stickleback fish prefer to feed on high density swarms of water fleas because these yield food at a high rate. But catching prey under these conditions demands more attention so the fish switch to low density swarms if a predator threatens (Milinski & Heller, 1978).

Figure 2.21 Comic depicting the negative outcomes of multitasking.

The fact that humans and animals have difficulty playing attention to more than one thing at a time illustrates an important principle of attention: it is limited. Limited attention means that the brain can only process a certain amount of information at any given time. Input that requires more attention uses more of this limited cognitive capacity, restricting other incoming information. One of the easiest ways to test this is to listen to a radio documentary while performing mental arithmetic. If the arithmetic is simple, information from the news report can be processed at the same time. As the difficulty increases, the ability to understand and remember the auditory information declines. This may simply reflect neural constraints on cognitive processing, which evolution has dealt with by allowing some processes to become automatic. For example, a sudden loud noise causes an orienting response in mammals that directs attention to a particular stimulus or location. This response, that occurs without awareness, focuses attention on potentially meaningful information (e.g. something that signals danger), but places minimal demands on cognition. The load on cognitive processes is reduced further if a behavior is repeated frequently enough to become automatic. Many highly accomplished musicians or athletes perform their skilled movements with little or no conscious awareness. This leaves enough attentional resources for other activities and helps to explain why some people really *do* seem capable of doing more than one thing at a time. In most cases, they are engaged in an over-rehearsed activity that requires minimal attention, leaving ample cognitive resources for other activities.

Chapter Summary

- There is an enormous range of sensory abilities across the animal kingdom, which often matches the ecological niche of a species. This suggests that evolutionary pressures have helped to shape sensory system function in different animals. The sensory drive hypothesis provides a specific explanation for the evolution of sensory systems by stating that natural selection favors adaptations that maximize the effectiveness of communication, both within and between species. Evolutionary adaptations can also explain the apparent paradoxes of sensory bias, sensory exploitation, and supernormal stimuli.
- Evolution provides the blueprint for sensory systems but the development and fine tuning of many sensory abilities depend on sensory experience during postnatal life. Hubel and Wiesel's work highlighted the importance of visual input in early life for normal visual development, with later work revealing the same phenomenon in other sensory systems. The specific timing of these sensitive periods varies across species and across sensory systems. Interestingly, a loss or deficit in one sense early in life may lead to a heightened capacity in another. This compensatory plasticity could allow animals to adapt, within their lifetime, to new sensory environments.
- Sensory system function can be divided into two stages: detection and processing. Sensory detection begins when physical signals (e.g. light or sound waves) activate the organs. This information is then segregated, organized, and transmitted to other brain areas for further processing. Despite the enormous variation in the type of sensory information that different organisms can detect, general principles of sensory system function are preserved across species. For example, stimulus properties such as intensity or duration are coded by the firing patterns of neurons in many different animals.
- Sensory information that is transmitted to higher brain regions is recombined into perceptions that allow organisms to make sense of the world around them. Psychophysics describes this relationship between the physical properties of stimuli (sensations) and how these are interpreted (perception). Elemental theories of perception focus on explaining how individual units of sensory input are combined, whereas ecological theories emphasize an organism's interaction with the environment as a determinant of perceptual experience. For both groups of theories, perception is an active process that involves filtering and extracting meaningful information from the environment.
- Perception is constrained, further, by attention, which allows an organism to focus on specific stimuli or events in the environment. Selective attention, the ability to attend to a limited range of sensory information, is aided by mental search images. Sustained attention, the ability to maintain focus on one aspect of the environment for an extended period, allows animals to remain vigilant in watching for prey or predators. Selective and sustained attention work in opposition to divided attention, the ability to process input from more than one source at the same time. Although many people do not believe it, almost all cognitive processes are compromised when attention is divided between too many sources.

Questions

1. The star-nosed mole (*Condylura cristata*) is a small, furry mammal (about the size of a hamster) that lives most of its life underground. It gets its name from the 22 wiggly tentacles that form a star shape on the front of its face. These tentacles are covered in extremely sensitive touch receptors that the animal uses to identify objects in the environment, including prey. The closest relatives of the

star-nosed mole are the shrew and hedgehog but neither of these animals have this unique anatomical feature. How and why do you think this adaptation evolved in the star-nosed mole?

2. Although they live in very different environments, both bats and dolphins use echolocation to locate food. Do you think this is an example of common adaptation? Why or why not? (See Chapter 1 for a definition of common adaptation.)
3. Describe the difference between frequency coding and population coding. How are these used to decipher stimulus intensity and/or duration?
4. The most common explanation for supernormal stimuli is that the sensory system evolved approximate rules of thumb in a world of normal variation. Can you think of examples of supernormal stimuli that would be disadvantageous? Do you think these are shaping evolutionary processes in the contemporary world?
5. Cochlear implants are a recent medical advance that allows deaf individuals to hear. The surgery cannot (yet) be performed on newborns so children with congenital deafness go through the first 1–2 years of life with no auditory input. There is an ongoing debate about whether these children should be taught sign language while they are waiting for the surgery. What are the advantages and disadvantages of each position?
6. What general principles of sensory system function are preserved across species?
7. According to the principles of ecological psychology, perception depends on the context in which it is studied. With this in mind, why would researchers continue to study and explain sensory system function in terms of elemental theories of perception?
8. Could you design an auditory equivalent of Gibson's visual looming experiment? If so, how would you do this?
9. Do you think it is possible for selective and divided attention to evolve simultaneously in the same species? What conditions would favor this?
10. How would you define acoustic ecology? What topics do you think would be discussed at a meeting of the Society for Acoustic Ecology?

FURTHER READING

One of the most compelling and readable descriptions of sensory system evolution is presented in:

Dawkins, R. (1996). *The Blind Watchmaker*. London, UK: Longman.

Many of the same arguments are presented in this edited work that focuses on aquatic vertebrates:

Thewissen, J. G. M., & Nummela, S. (eds.) (2008). *Sensory Evolution on the Threshold – Adaptations in Secondarily Aquatic Vertebrates*. Berkeley, CA: University of California Press.

Cross-species comparisons of visual function are combined with detailed and intricate drawings of eye structure in this updated version of Dawkin's original hypothesis:

Land, M. F., & Nilsson, D.-E. (2012). *Animal Eyes*. New York, NY: Oxford University Press.

This text emphasizes that comparative studies of 'simpler' organisms can be used to generate general principles of sensory system organization and processing:

Prete, F. R. (ed.) (2004). *Complex Worlds from Simple Nervous Systems*. Cambridge, MA: MIT Press.

Ultimate explanations for sensory system design are described with remarkable simplicity in the following articles:

Endler, J. A. (1992). Signals, signal conditions and the direction of evolution. *Am Nat*, **139**, 125–153.

Guildford, T., & Dawkins, M. S. (1991). Receiver psychology and the evolution of animal signals. *Anim Behav*, **42**, 1–14.

More than 30 years later, Gibson's influential theories on sensory system processing should not be overlooked:

Gibson, J. J. (1979). *The Ecological Approach to Visual Perception*. Boston, MA: Houghton Mifflin.

A number of popular science books examine human perception through a 'new lens,' focusing on cultural differences in how we see and interpret the world, explaining common visual illusions, or describing neurological disorders that alter perception. Many of these books are worth browsing through, if only for the beautiful (and sometimes comic) illustrations they contain.

Classen, C. (1993). *Worlds of Sense: Exploring the Senses in History and Across Culture*. London, UK: Routledge.

Gregory, R., Harris, J., Heard, P., & Rose, D. (eds.) (1995). *The Artful Eye*. Oxford, UK: Oxford University Press.

Sacks, O. (2010). *The Mind's Eye*. Toronto, Canada: Alfred A. Knopf.

Shepherd, G. (2011). *Neurogastronomy: How the Brain Creates Flavors and Why it Matters*. New York, NY: Columbia University Press.

Shepherd, R. (1990). *Mind Sights*. New York, NY: Freeman.

3 Memory

Background

I have a very vivid memory of the first time that I thought about memory, and wondered how it worked. As a teenager, I spent many nights and weekends babysitting two adorable little blond haired girls who lived around the corner. One winter evening, they decided to entertain me with Christmas carols they had learned at preschool. Before they made it through the first song, I was struck by an amazing difference: both had difficulty remembering the correct words but they made very different kinds of mistakes. Carolyn's rendition preserved the meaning of the text, but she had to adapt the words to fit the melody. Simone, who was two years older, remembered the phonetics of each line and adjusted her text accordingly. Although I don't recall a specific example (this really was a long time ago!), I can imagine "Oh Christmas tree, oh Christmas tree, how lovely are thy branches" being converted to "Oh Christmas tree, oh Christmas tree, you have beaut'ful boughs and leaves" (Carolyn) and "Oh Christmas tree, oh Christmas tree, who lively art by ranches" (Simone). Now that I have studied memory for many years, I have a better idea of why the two girls sang such different versions of the same songs. At the same time, I also know that I should question whether this event happened at all. (Cella Olmstead)

For many people, reading excerpts such as this one will remind them of their own past holidays. They are likely to remark that some festive events are easy to remember, whereas others are close to forgotten. In discussing these memories with old friends or relatives, it may become apparent that two individuals have dramatically different memories of the same event. In other situations, memories for these same occasions may be accessed unconsciously. Odors reminiscent of a freshly cooked holiday meal can induce a warm and pleasant feeling but it may take a few moments to realize that these emotions arose because the smells were previously associated with specific celebrations. At the same time that thoughts of past events are surfacing and receding, individuals may be using their memory in other ways as they strum a guitar, carry on a conversation, or study for an exam. This chapter helps to explain how and why memory operates in so many different ways.

Chapter plan

To say that this chapter provides a 'brief overview' of memory is a gross understatement. Entire textbooks as well as some full term college or university courses are devoted, exclusively, to this topic. A number of scientific journals only publish papers on memory research (e.g. *Learning and Memory*, *Neurobiology of Learning and Memory*) and even those that are not as limited in scope contain dozens of observational and experimental studies on memory. For obvious reasons, therefore, each section in this chapter provides only a synopsis of some aspect of memory research in humans or animals. The focus *across* the chapter is to understand these different components of memory within a comparative cognition framework.

The chapter begins with a discussion of how memory relates to other cognitive processes. The same issue came up in the previous chapter when it was noted that cognition must be understood in terms of sensory system function. Indeed, a recurring theme in this text is that cognitive processes constantly interact. In the case of memory, almost no other cognitive process functions in its absence. The fact that memory is critical for most cognitive functions is the basis for recent theories on the evolution of memory, which are described at the conclusion of the first section. This is followed by a description of how memory is processed over time through separate phases of encoding, consolidation, and storage. The next section describes the historical distinction between short- and long-term memory. Over the last two decades, these terms have gradually been replaced with working and reference memory, reflecting the fact that both involve active mental processing rather than passive transmission of information. Finally, an examination of the cellular and synaptic mechanisms of memory suggests that general biological principles of this cognitive process have been conserved across evolution.

3.1 Preliminary issues

Memory seems to be one of those terms that needs no definition. After all, most people would say that they know what memory is, even if they are not always sure how it works. But scientists must be very careful to define all of the terms that they use, even those (or particularly those!) that are common in everyday language. In the simplest sense, **memory** is the stored representation of past experiences. A more precise definition from cognitive psychology is "the mental processes of acquiring and retaining information for later retrieval" (Ashcraft & Klein, 2010, page 9).

3.1.1 Learning versus memory

Note that the cognitive definition of memory includes both the acquisition and retention of information. In the past, these two processes were separated into subcategories of **learning** (acquisition) and memory (retention) but the two are so tightly connected that they were often discussed as one topic. Although it may be possible to acquire information without retaining it (i.e. learning without memory), it is difficult to imagine how one could retain information that was not acquired (i.e. memory without learning). Given this tight association, it is surprising that, historically, learning and memory were studied in different ways. Scientific investigations of learning typically employed lab animals, such as rats or pigeons, to understand how stimulus–response associations are formed. Behavioral psychologists working with these models assumed that the principles of learning were constant across animals even when the type of information to be acquired varied dramatically (e.g. learning that a bell predicted food as opposed to learning how to find your way home). Insight from these studies, therefore, would help to build general theories of learning and, by extension, memory (see Chapter 1 for details, including the intellectual and scientific developments that led to the demise of this perspective). Few researchers today believe that all instances of learning can be explained by stimulus–response associations (Urcelay & Miller, 2010) although, as outlined in the next chapter, the study of associative processes continues to be a fruitful research endeavor.

In contrast to the study of learning, most memory research during the first half of the twentieth century tested the ability of human subjects to remember lists of words, often nonsense syllables. The technique was pioneered by the German psychologist Hermann Ebbinghaus (1850–1909) using himself as an experimental subject. Ebbinghaus purposefully employed typically meaningless sounds because he did not want his memory for these items to be influenced by prior experience. In other words, he wanted to study how information is acquired, not how it is modified by existing biases or tendencies. Ebbinghaus' results, as well as those of his many intellectual successors, were often interpreted within an associative framework. For example, two consecutive words in the list were more likely to be recalled together, presumably because subjects formed an association between them during list learning. With the rise of cognitive science in the 1960s, the theoretical framework for investigating memory shifted from an associative framework to an Information Processing Model.

Most contemporary researchers have adopted the cognitive definition of memory with the goal of understanding how information is represented and processed during acquisition, retention, and retrieval. Individual researchers may focus on one part of this equation (e.g. memory acquisition), but there is a growing recognition that learning and memory occur together and should be studied as one phenomenon. The current chapter adopts this approach, discussing all aspects of memory in a variety of subjects ranging from insects to birds to primates to fish. Indeed, because the subject is so vast and the interest in it so intense, memory is one of the few areas of comparative cognition research that spans the entire animal kingdom. The common thread in this work is an attempt to understand how organisms use new information to interact with their environment. Like other cognitive processes, memory is measured, most commonly, with some behavioral output. These can range from the modification of simple reflexes to complex verbal responses. A caveat to this definition is that not all behavioral change reflects memory. For example, muscle fatigue or loss of sensitivity to sensory input could alter behavior but neither would be described as memory. As always, researchers must be careful to eliminate these potential confounds when interpreting their data.

3.1.2 Relationship of memory to other cognitive processes

Given that memory is measured as behavioral change due to past experience, it is difficult to think of *any* cognitive process that is independent of memory. Read the remaining chapter titles in this book and try to imagine whether any of these could function in the absence of memory. As an example, identifying living things as plants versus animals may seem like an innate ability, but what about dogs versus cats, let alone Dalmations versus Irish Setters? This ability to categorize items correctly, covered in Chapter 9, obviously requires some element of memory. The same is true of all other cognitive processes discussed in this book. Of course, this makes it difficult to dissociate memory from functions such as decision making (Chapter 7), tool use (Chapter 8), or social learning (Chapter 12). The problem is not new: in the nineteenth century, language impairments that we now call aphasias were diagnosed as memory disorders or 'a failure to remember how to speak' (Squire & Wixted, 2011).

Despite these challenges, we and other authors believe that memory can (and should) be examined as a separate cognitive process. The rationale for this decision is that 'experience-dependent modification of behavior' does not provide a complete account of the cognitive processes discussed in later chapters. Memory may facilitate functions such as spatial navigation (Chapter 5), but it cannot explain the phenomenon entirely. For example, most students have no problem finding their way to a lecture hall on campus, but it is unlikely that they are doing so by memorizing the route from one building to the next (otherwise it would be impossible to get there from a different location). Like many animals, humans develop a spatial map of their surroundings that allows them to compute current position in relationship to global landmarks. Memory contributes to this process by using past experience to update spatial computations. In this way, it is parsimonious to consider memory as a necessary complement to other cognitive processes. According to contemporary authorities on the topic, one cannot understand cognition unless they understand memory (Ashcraft & Klein, 2010).

3.1.3 Evolution of memory

It seems almost trivial to say that memory improves an organism's fitness as the ability to retain information over long intervals has obvious advantages for survival and reproduction. In evolution, however, everything comes at a cost. Memory requires a lot of brain power, which uses resources that could be devoted to survival and reproduction. Moreover, the fact that behavior is modified by experience means that it starts out being less than perfect. All animals make mistakes, particularly in new situations, some of which are deadly (e.g. eating the wrong food or wandering into the wrong place). Why, then, would evolution favor a process of behavioral change that depends on experience? The answer lies in the ever-changing world around us. In a perfectly stable environment, memory would not be necessary: organisms could respond to the stimuli around them and those displaying the most adaptive responses would transmit these behaviors to their offspring. But no organism has ever lived under these conditions, so it is very likely that rudimentary memory emerged in the simplest forms of animal life millions of years ago. Indeed, this may be one thing that distinguishes plants from animals, although not everyone would agree (Trewavas, 1999).

If memory evolved through natural selection, memory abilities should vary across species that inhabit different environments. Although plausible, this hypothesis is difficult to test in cross-species comparisons of memory. For example, imagine the difficulty of designing an experiment to compare memory in worms and monkeys. Deciding what the animals should remember, let alone how to measure it, is complicated by dramatic differences in sensory and motor abilities, as well as more complex factors such as motivation. Even if it were possible to demonstrate that monkeys have longer lasting memories than worms, they also have a much longer lifespan: 1-year memory retention in monkeys may not be 'better' than a 4-hour memory in worms.

An alternative approach is to examine whether the *type* of information that animals remember matches the ecological constraints of their environment; or, more accurately, the environment of their evolutionary ancestors. The evidence is compelling in humans who show better memory for stimuli associated with snakes or spiders than with guns or electrical outlets (Őhman & Mineka, 2001), even though the latter pose a much greater threat in modern society.

Ecologically based memory is also evident in food caching animals, such as squirrels and some bird species, whose survival depends on remembering the location of hidden food. These animals display better and longer lasting memories for spatial information than animals who do not cache food in their natural environment. Some researchers have even speculated that the memory retention curve (i.e. how long a memory lasts) should mimic the time course of survival related events in the

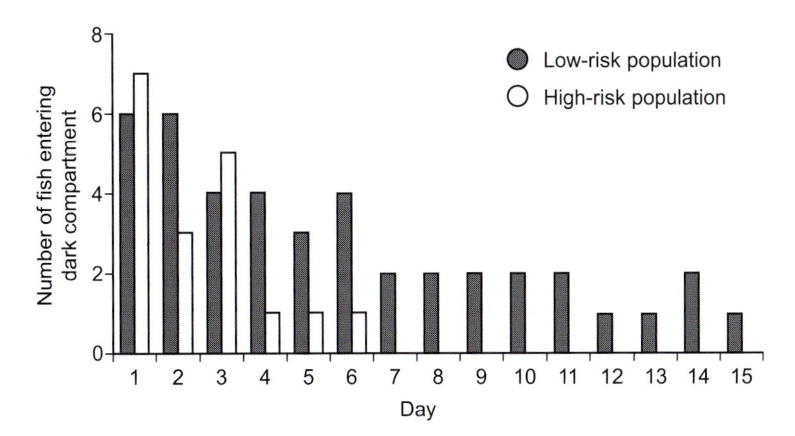

Figure 3.1 Inheritance of memory in stickleback fish. Offspring of fish from environments of high or low predation were tested in an avoidance learning paradigm (Huntingford & Wright, 1992). Each day, fish were given access to two compartments containing food. When they entered their preferred compartment (defined as the compartment they entered most frequently during training), they were presented with a simulated predatory attack. Offspring of fish from a high predation environment were quicker to learn to avoid this environment than were offspring of fish from a low predation environment.

environment (Anderson & Schooler, 1991). Indirect evidence for this hypothesis is provided by male stomatapod crustaceans (*Gonodactylus bredini*) which mate with females and then leave them to tend to the fertilized brood. If the pair interacts during the subsequent 4 weeks (approximately the time of brood development), the male is less aggressive with his former mate than he is with other females (Caldwell, 1992). Thus, the male remembers the female during the time that she is caring for his genetic offspring. Theoretically, his memory might decline after the period of brood care when it is no longer adaptive to keep a record of the information (unfortunately this was not tested).

Finally, the evolution of memory may also be examined by comparing memory abilities in members of the same species that inhabit different environments. Huntingford and Wright (1992) conducted this experiment by selectively breeding two populations of stickleback fish (*Gasterosteus aculeatus*): one from an environment with high predation risk and one from an environment with low predation risk. The first generation offspring of the two groups, reared in the lab, showed similar rates of foraging (finding food in one area of the fish tank). Differences emerged, however, when an aggressor was introduced to the environment: memory for events that predicted the aggressor was better in the group from the high predation area (see Figure 3.1). In the parent populations, fish in the high predation environment were more likely to survive and reproduce if they could modify their behavior (i.e. escape) based on experience, an ability that was passed to their offspring. In contrast, fish in the low predation environment were able to forage without interruption, so using valuable brain resources to remember when an aggressor may appear provided minimal evolutionary advantage. A similar phenomenon was demonstrated in fruit flies (*Drosophila melanogaster*) (Mery & Kawecki, 2002) in that successive generations of selective breeding produced a population of quick-learning flies with long-lasting memories (3 hours, which is a long time for a fly). As with the fish, it was the *ability* to remember, not a specific response to a stimulus, that was transmitted across generations because descendants of flies with the best memories were also better at learning (and remembering) completely novel associations.

These examples bring up an important point about memory: it is not a stored representation of past events but a cognitive *process* that allows organisms to adjust their behavior in an ever changing world (Klein *et al.*, 2002). Indeed, models of cognitive evolution argue that the transmission of fixed behavioral traits versus the ability to learn (and remember) is a function of the stability of an animal's environment (Stephens, 1991). As we noted previously, the ability to modify behavior based on experience would facilitate many cognitive functions, so it is very likely that memory evolved in parallel with the processes that will be discussed in subsequent chapters.

3.2 Stages of memory processing

The definition of memory includes a time-dependent component in that it involves mental processes of acquiring and retaining information for *later* retrieval. In truth, no cognitive process is a static event as each involves neural processing that occurs over time, even if this is only a few milliseconds. In the case of memory, the time lag between an experience and subsequent behavioral change (the operational definition of memory) can vary from a few seconds to several decades, depending on the species. For this reason, it is useful to think of memory as sequential stages of processing in which a memory is formed, stored, and then retrieved.

3.2.1 Encoding

This first stage of memory processing is **encoding**, defined as the conversion of incoming information into neural signals that will be used for later processing. As outlined in Chapter 2, incoming information arrives through the sensory system, but memory encoding is more than just sensory detection. In other words, not all sensations are encoded in memory. This fact should be obvious: few people recall every detail of a past experience even if their memory for the event is vivid. So, what determines which stimuli and events will be encoded? Attention is probably the biggest factor as this is the mental process that selects information for further processing. Organisms that are not attending to specific details may not remember them at all. Similarly, encoding is weaker if attention is divided due to distracting stimuli or events. Many people forget where they put their glasses, keys, or notebook because they set these down (absent mindedly) while they were doing something else. Conversely, salient or very surprising events that grab attention are more likely to be encoded and transferred to the next stage of memory processing.

Encoding is also enhanced by **elaboration**, the process of adding meaning, images, or other complex information to sensory input. A well known elaborative technique is to tag a visual image to a person's name (e.g. the name Mark may evoke an image of a large X on the individual's chest). In any elaboration technique, the greater the level of processing, the stronger and more durable the memory. This principle is apparent in human studies: when subjects are asked to identify whether letters were written in lower or upper case, they do not remember a word list as well as when they are asked to make a judgment about the meaning of the word (Craik & Lockhart, 1972). It is difficult to instruct animals to elaborate on information during memory encoding, so it is not clear whether this process is unique to humans. This seems unlikely, however, given that one of the most effective elaboration techniques is to process information in terms of its survival value (Nairne *et al.*, 2008). Indeed, humans have a better memory for words that they are asked to think of in terms of survival in an ancestral environment compared to modern-day scenarios (Nairne & Pandeirada, 2010) (see Figure 3.2). A critical feature of the study is that the words to be remembered were the same; the difference was the instructions of what to think about when they were being read. Because the

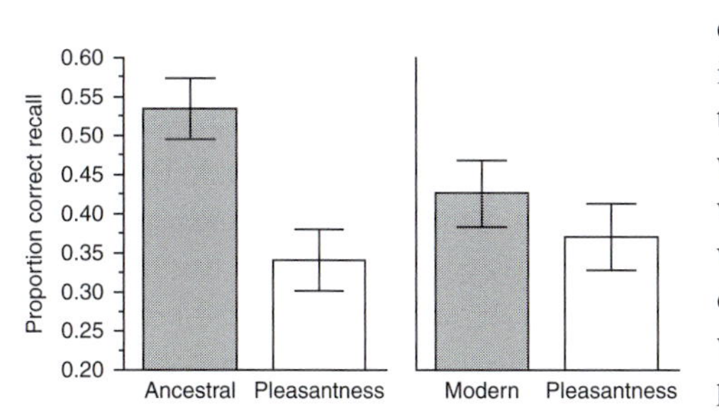

Figure 3.2 Memory for stimuli associated with survival in the natural environment. Participants were asked to read a list of words and imagine how relevant each of the items would be for surviving when they were stranded in the grasslands or a large city of a foreign land with no resources. They were also asked to rate other aspects of each word, including pleasantness. Importantly, the words to remember were the same in each condition, only the imagined scenario was different (ancestral or modern). In a later test, recall was greater for words in the ancestral condition but there was no difference in pleasantness between ancestral and modern conditions.

later memory test was a 'surprise,' it is unlikely that subjects were trying to memorize the words during reading.

One of the challenges to encoding (and to memory in general) is that there is a limit on the amount of information that the nervous system can process. A partial solution to this problem is to reorganize discrete elements of sensory input into larger units. Miller (1956) coined the term **chunking** to describe this process, suggesting that it increases memory capacity by reducing the amount of information that is encoded. For example, most people would have difficulty remembering this list of 39 letters:

LETTERSGROUPEDTOGETHERFORMCHUNKSOFWORDS

but could easily recall the string of words that they form. Although chunking reduces the number of items to be encoded, it increases the amount of information that is conveyed in each unit. This conversion from small to larger units requires cognitive processing; the associated mental effort decreases as experience with the chunking increases (e.g. reading becomes easier with practice). Beyond the obvious example of reading, combining information into meaningful units facilitates encoding in a variety of tasks: performance in memory tests is increased in pigeons (Terrace, 1991), rats (Dallal & Meck, 1990), and monkeys (D'Amato & Colombo, 1988) when the animals can chunk stimuli or responses into meaningful units. Indeed, chunking is used to explain a number of cognitive abilities in animals including serial list learning, covered in Chapter 6 (Timing and counting), and knowledge of kin relationships, covered in Chapter 10 (Social competence). The important point to remember is that the conversion of smaller elements into larger groups of elements provides evidence that encoding is not a passive transfer of information. Rather, it is an active mechanism that can be facilitated by cognitive processes such as attention, elaboration, and chunking. It should go without saying that better encoding leads to better memory.

3.2.2 Consolidation

The second stage of memory processing, **consolidation**, is defined as the process of modifying encoded representations so that they become more stable over time. The time-dependent property of memory consolidation was recognized by the early Greeks, and confirmed by clinical reports in the eighteenth century (Dudai, 2004). The latter described head trauma producing amnesia for events in the recent, but not the distant, past. This phenomenon has been reproduced repeatedly in the lab with electroconvulsive shock (Duncan, 1949), drugs (Flexner *et al.*, 1965), distracting stimuli (Gordon & Spear, 1973), and new learning (Boccia *et al.*, 2005) all of which disrupt the transfer of recently acquired information into stable memories. Conversely, drugs that increase neural activity

Box 3.1 Flashbulb memories

Figure 3.3 World Trade Center memorial, New York City.

"What were you doing when you learned that..." Depending on an individual's age and where they live, the completion to this sentence could read: John F. Kennedy had been shot; John Lennon was dead; Margaret Thatcher had resigned; or planes had crashed into the Twin Towers in New York City (see Figure 3.3). The commonality in all of these examples is that they elicited strong emotional reactions. Most people claim to have vivid memories for at least one of these events or, more precisely, for the circumstances surrounding the news about the event. Brown and Kulik (1977) coined the term 'flashbulb memories' to describe these very detailed and vivid 'snapshots' of the moment when an individual receives emotionally arousing news. They speculated that flashbulb memories differed from other, less reliable, memories in that they created a permanent record of the details and circumstances surrounding the event.

Animals cannot tell us 'what they were doing when...,' but hundreds of experiments using a variety of different species confirm that emotionally arousing events are more likely to be remembered in the future. The effect depends on the release of stress hormones which travel through the blood stream and into the brain, causing neurochemical changes in the amygdala, a small almond-shaped structure located in the temporal lobes (McGaugh & Roozendaal, 2009). Neuroimaging studies in humans confirmed that increased activity in the amygdala during consolidation improves accuracy in a later memory test (Cahill *et al.*, 2004). Importantly, the more shocking or emotionally arousing the event, the more likely it is to be remembered. Flashbulb memories are discussed, most frequently, in terms of unpleasant events, but highly emotional events that are pleasurable may also enhance memory (McGaugh, 2004). The evolutionary explanation for flashbulb memories is straightforward. Emotionally arousing stimuli are likely to signal important events so it is adaptive to remember the conditions under which these occurred.

Despite personal reports, not all research supports the phenomenon of flashbulb memories. The main point of contention is whether events surrounding highly emotional occurrences are remembered more accurately than other memories (Christianson, 1989). (In these studies, subjects are asked to remember what they were doing when they heard the news about an emotional event, not details of the event itself.) One thing is true across all of these studies: people are more *confident* that these memories are accurate, regardless of whether they actually remember more details (McCloskey *et al.*, 1988). This is perhaps what makes flashbulb memories unique: the certitude with which people recall these events. But why emotions would enhance confidence in memory without necessarily improving the memory itself is still a mystery.

associated with learning will improve memory if they are administered immediately following training (McGaugh & Krivanek, 1970). If the time between learning and drug administration is delayed (from a few minutes to a few hours), the facilitating effect is lost. The interpretation of these data is that consolidation takes time; memories are not stable until this process is complete.

Some researchers distinguish between consolidation and a later storage phase, listing four rather than three stages of memory processing (i.e. encoding, consolidation, storage, and retrieval). The problem with postulating a separate 'storage' phase is that it infers that representations of memory sit undisturbed in the brain for an indefinite period of time. This seems unlikely given the ongoing neural processing that characterizes cognition. It is more parsimonious, therefore, to think of consolidation as a process with two stages. The first occurs within minutes to hours and is dependent on protein synthesis. This early stage consolidation has been observed in a variety of species including worms, flies, and mice. In all of these studies, disruption of protein synthesis in the period immediately following learning disrupts memory formation in the same way that concussions produce amnesia for events preceding the trauma. The second stage of consolidation, which may take weeks to years, has only been characterized in mammals. This process involves structural changes in synaptic connections between neurons and a shift in the brain areas that store memory-related information (Milner *et al.*, 1998). It is very likely that this second consolidation phase is present in other animal classes (i.e. non-mammals) but the details have yet to be worked out.

Regardless of whether consolidation is a one- or two- stage process, it appears that all memories are labile before they become consolidated. If this is true, what is the evolutionary advantage of a memory process that takes time to stabilize? One answer is that the nervous system would quickly become overloaded with useless information and precious cognitive resources would be wasted if consolidation occurred instantaneously. A more direct explanation is that consolidation provides a timeframe in which the representation of a memory can be modified, taking into account both confirmatory and contradictory evidence. In this way, associations between events and generalizations across conditions can be incorporated into memory for a particular event (Dudai, 2004). Put simply, a snapshot of the environment provides less meaningful information than does a moving window of time. Conversely, it is possible that consolidation is *not* adaptive, but an artifact of brain biology. Both protein synthesis and the structural rewiring of neural circuits take time. Perhaps consolidation takes time simply because it relies on these pre-existing processes.

3.2.3 Retrieval

Information that is encoded and consolidated must be accessed if it is to guide behavior at a later time. **Retrieval**, the third phase of memory processing, is the mental activity of accessing this stored information. There are two mechanisms by which information is retrieved: recognition and recall. Recognition is the act of identifying information as familiar, even if the details of the memory that accompany it cannot be specified. Recall is a more cognitively demanding task of generating information from memory stores, in the absence of cues or 'reminders.' Recognition retrieval is much easier than recall retrieval: it may be difficult to generate the name of a previous head of state (prime minister or president), but most people can identify his or her name in a list.

Both recognition and recall are better if they occur in the same context in which the member was encoded. Context refers to the combination of cues, both internal and external, surrounding an event. These may include sensory signals such as specific sights or sounds as well as hormonal or emotional states (Bouton & Moody, 2004). One of the best examples, shown in Figure 3.4, is that memory is enhanced when encoding and retrieval occur during the same phase of the circadian cycle (Cain *et al.*, 2008). In this study, golden hamsters (*Mesocricetus auratus*) did not show any evidence that they remembered an environment where they had received a shock if they were trained and tested at different times of day. This suggests that time of day information may be coded as part of the memory formation such that circadian cues can facilitate memory retrieval. These cues may be particularly important in the natural environment because biologically relevant events often occur

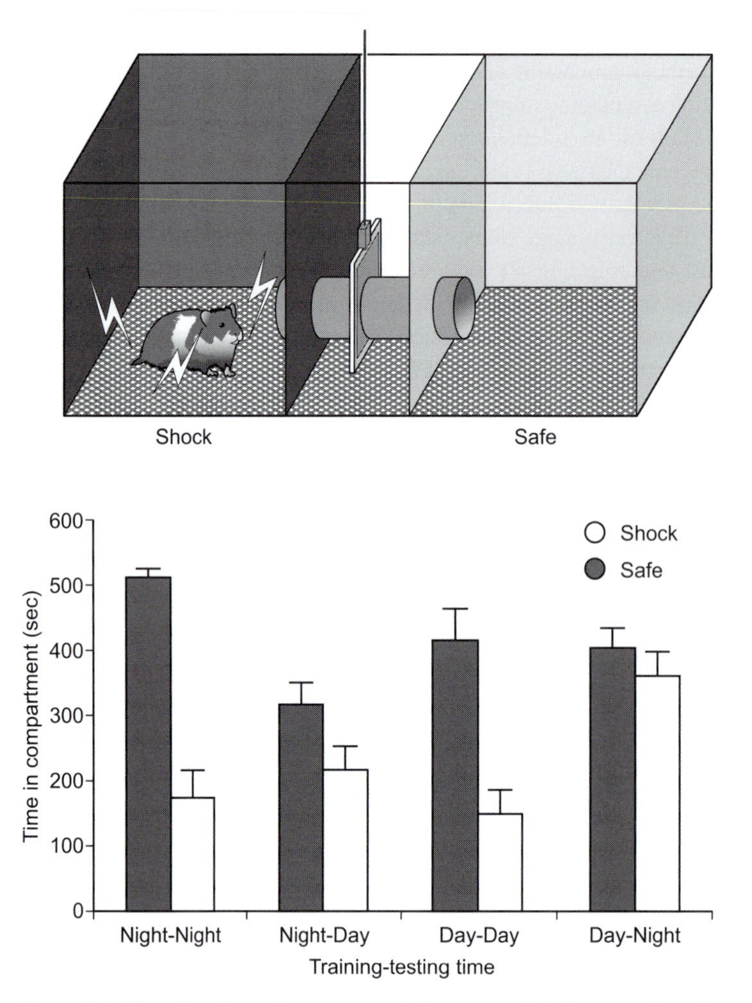

Figure 3.4 Circadian dependent memory in hamsters. (A) Over a series of conditioning trials, hamsters experienced shock (random presentation of 3–4 mild shocks over 5 min) in one side of a place preference apparatus. On alternate days, they were placed in the other side with no shock. Conditioning sessions always occurred at the same point in the circadian cycle. (B) No shocks were turned on during a subsequent test when animals had access to all compartments. Hamsters avoided the shock-paired compartment *only* if testing occurred during the same phase of the circadian cycle as training sessions.

at the same time of day. The fact that memories can sometimes be retrieved in one context but not another supports the idea that forgetting is often a failure of retrieval, not a decay of the memory trace.

The inability to retrieve a memory is not the only problem that arises during the third phase of memory processing. Just as consolidation does not produce a snapshot of past events, retrieval does not access a replica of the stored information. Bartlett (1932) was one of the first to propose that memory retrieval combines elements of the encoded information with existing knowledge. His subjects recounted stories that they had read previously, frequently altering the stories to match their perceptions or beliefs of how the stories should unfold. Subsequent work verified that retrieval is a reconstruction of past events, influenced by an individual's pre-existing knowledge as well as their current emotional state (Sulin & Dooling, 1974). Memory retrieval may also be distorted by suggestion (Schacter, 1999) when false or misleading information is introduced *after* a memory is encoded, but then influences how the original event is remembered (see Box 3.2).

If retrieval is an active process that modifies stored information, what happens to the memory trace after it is 'put back' in the brain? One suggestion is that retrieved memories, just like newly acquired information, are labile; in order to become stable, they must undergo a second consolidation phase, now called reconsolidation. The first systematic investigation of reconsolidation used a simple paradigm in which rats hear a tone and then experience a brief foot shock. The next day, the animals freeze when they hear the tone, indicating that they remember the sequence of events. If a

Box 3.2 False memories

Tales of dangerous or heroic deeds are often embellished in the retelling. Some of these distortions are just exaggerations or 'stretching the truth.' In other cases, people really believe their own (reconstructed) descriptions of the event. But, how far does this phenomenon extend? Is it possible to implant new information that is later retrieved as a 'memory'? For over 25 years, Elizabeth Loftus and her colleagues have addressed this question with an unequivocal response: YES! Loftus' original work was straightforward. She wanted to examine the reliability of eyewitness testimony so she showed film clips of traffic accidents to her students (Loftus & Palmer, 1974), and then asked them to respond to a series of questions. By altering the wording of questions (e.g. "how fast were the cars going when they hit each other?" versus "how fast were the cars going when they smashed into each other?"), researchers were able to modify subjects' reports of the incident. This suggestibility even extended to new events. That is, subjects 'remembered' details that did not occur when the questions provided misinformation about the previous event. For example, subjects who read that the cars smashed into each other were more likely to report seeing glass at the accident scene (there was none).

Loftus' most innovative work developed from her 'lost in the mall' study. In this experiment, students were asked to reflect on specific childhood events such as a wedding or a large celebration they had attended. One narrative, describing a day that they were lost in the mall, was fabricated (the students were told that the details of all incidents were provided by family members). One quarter of the students reported remembering the event, and many provided specific details of what happened that day. The finding has been replicated by other researchers using a variety of fabricated scenarios (e.g. personal injury from an accident or embarrassing events like spilling punch at a wedding reception) and participants ranging in age from young children to adults (Hyman & Pentland, 1996). Amazingly, the more subjects think about these 'memories,' the more detailed they become.

Loftus is considered one of the most influential psychologists of the twentieth century, but her work is highly controversial. She argues vehemently that eyewitness testimony is unreliable and has acted as an expert witness for a number of high profile legal cases. Her greatest challenge arose when she questioned the notion of recovered memories of sexual abuse, pointing out that suggestibility during therapeutic sessions can create false memories for events that did not occur. Not surprisingly, these statements inflamed sentiments across the globe. Loftus maintains that her research was not undertaken as a political or social statement, but a scientific investigation of factors that influence memory retrieval. Many scientists and clinicians are reluctant to take a stand on the sensitive issue of sexual abuse claims, but few can discount the overwhelming evidence that memories may be reconstructed through suggestibility.

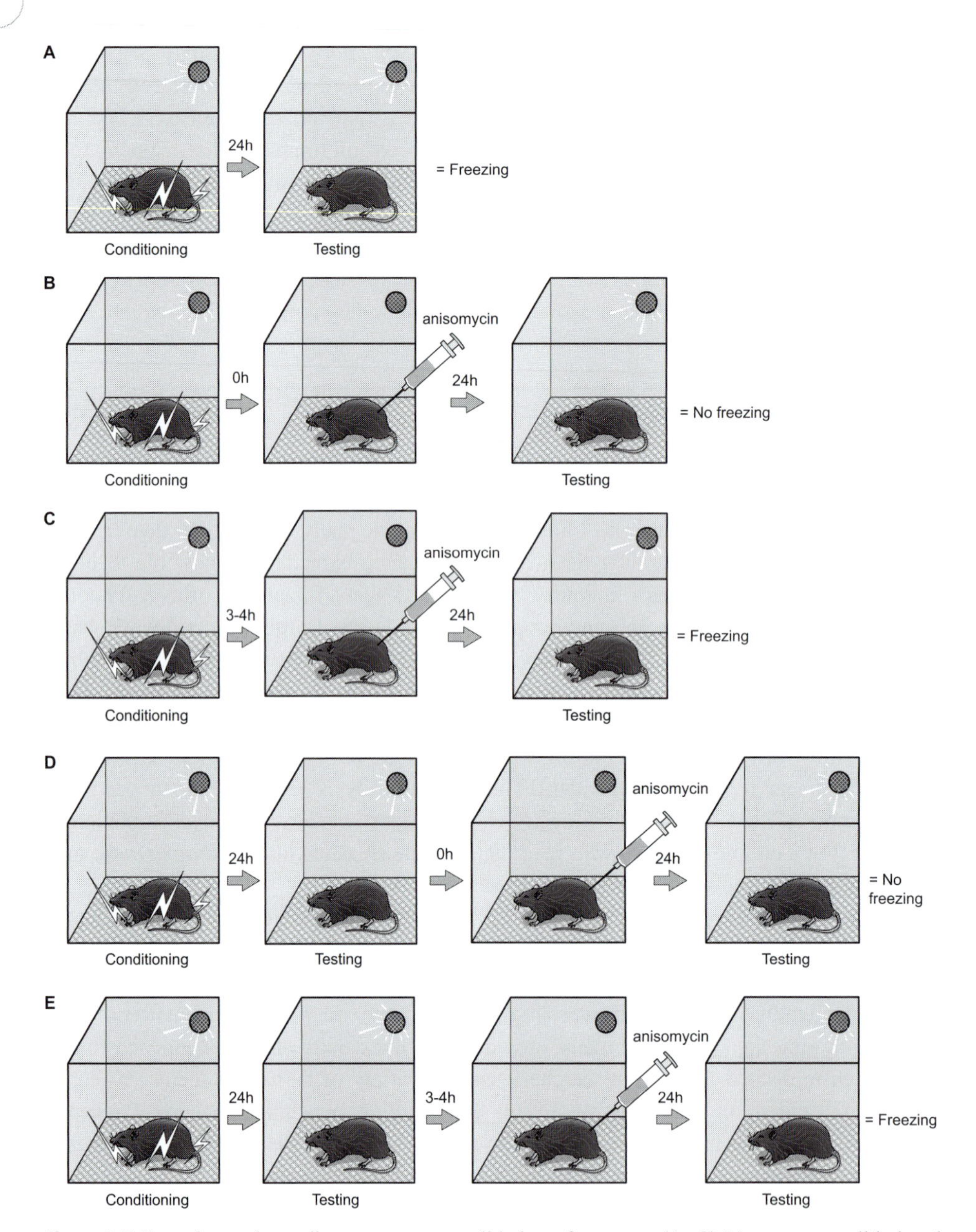

Figure 3.5 Experimental paradigm to test reconsolidation of memory. (A–C) Memory consolidation depends on protein synthesis. (A) In a single conditioning session, rats are presented with a light followed by a mild footshock. The next day, rats freeze when the light is turned on (no shock is presented). (B) Administration of the protein synthesis inhibitor, anisomycin, immediately following conditioning blocks the freezing response on day 2. (C) If anisomycin is administered 3–4 hours after conditioning, freezing remains intact. (D–E) Memory reconsolidation depends on protein synthesis. (D) Rats undergo conditioning and testing sessions as in A. Following testing (when animals freeze), they are injected with anisomysin. In a subsequent test, 24 hours later, rats do not freeze. (E) Rats freeze on the second test if anisomycin is injected 3–4 hours after the first test session.

protein synthesis inhibitor is administered immediately after the test tone, the freezing response disappears on subsequent tests (Nader *et al.*, 2000). Protein synthesis inhibitors disrupt the initial consolidation phase and these new data suggest that a similar process occurs during memory retrieval. Post retrieval modification of memory has been demonstrated in other species including worms, mice, and humans (Nader & Hardt, 2009) (see Figure 3.5). Previously consolidated memories, therefore, may undergo a second phase of instability when they are retrieved. Despite the apparent similarities, reconsolidation is not just a second consolidation as the two processes involve different biological mechanisms.

Not all evidence supports the concept of reconsolidation, although negative findings in these experiments can be difficult to interpret. At the very least, it may be too simplistic to assume that an established memory can be erased with protein synthesis inhibitors (Squire, 2009). No doubt there are restrictions on how and when memories are altered during retrieval, but this does not detract from evidence that information accessed from memory stores is subject to changes.

3.3 Working memory

Early descriptions of memory, some going back thousands of years, make a distinction between short- and long-term memory. Short-term memory was described as a time-dependent process with limited capacity. For humans, this is the information held in immediate consciousness, or the small number of items that can be remembered over a short period of time (i.e. seconds).

3.3.1 Definition and description

An interest in studying this process emerged in the 1950s with the rise of cognitive psychology and the recognition that keeping information in conscious awareness involves mental effort. Atkinson and Shiffrin (1968) formulated one of the first models explaining how incoming information is organized, in the short term, before being transferred to long-term memory. They used the term **working memory** to describe the temporary buffer that provides a gateway to permanent storage. The concept evolved over the next two decades with an increased emphasis on the active or dynamic aspects of working memory and how these contribute to cognition (Baddley & Hitch, 1974). Although there is still some disagreement on the specific components, a contemporary definition of working memory is the process of maintaining information in short-term store so that it can be used in other cognitive processes.

A primary distinction between the original and updated definitions of working memory is that the latter incorporates knowledge acquired during previous experiences. In other words, working memory is an ongoing mental computation that uses recently encoded information *and* information retrieved from long-term store to make decisions, navigate in space, or communicate with conspecifics. Because working memory holds information 'on-line' as source material for other operations, it is sometimes described as the brain version of random access memory (RAM) in a computer.

3.3.2 Working memory tasks

One of the best ways to understand working memory is through a working memory task, such as the N-back task shown in Figure 3.6, one of the most common laboratory measures of working memory in humans. In a real test, these letter stimuli would be presented one at a time, either as visual images

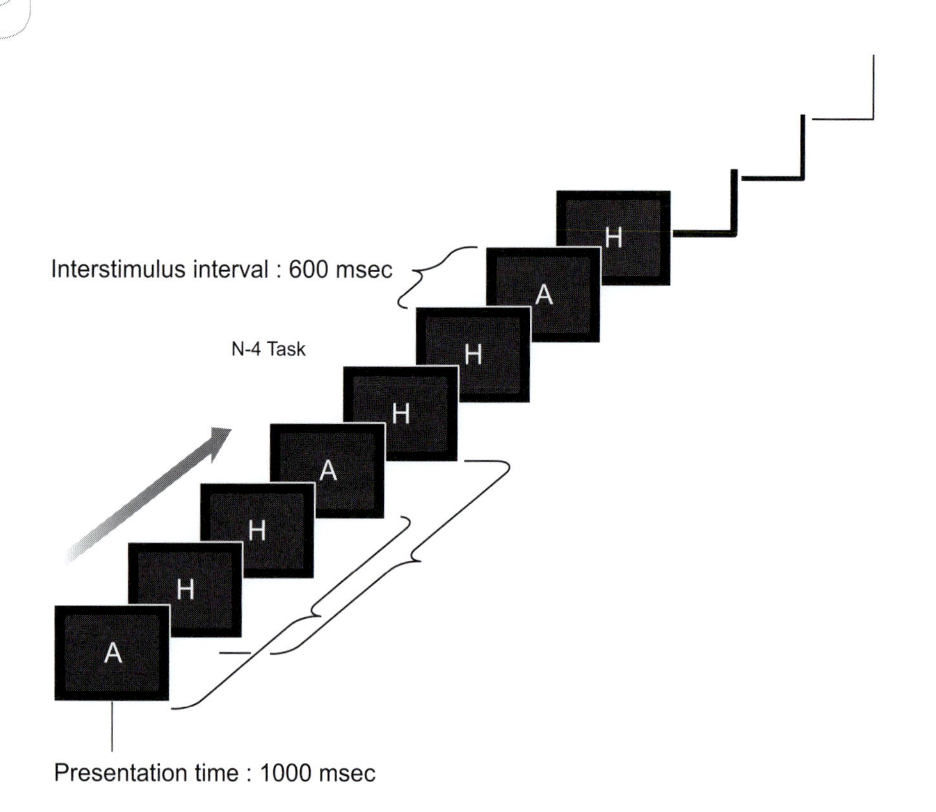

Figure 3.6 Experimental design of an N-back test. Stimuli are presented on a computer screen one at a time with a variable time delay between stimulus presentations. Subjects are instructed to respond (e.g. touch the screen or press a space bar) when the current stimulus matches the stimulus 'N' positions back.

or by having someone read them aloud. Participants are instructed to identify stimuli which match a stimulus presented N steps earlier in the sequence. Obviously, increasing the value of N increases the difficulty of the task. To perform efficiently, participants need to keep recently presented stimuli in mind *and* compare these to previously acquired information about the alphabet. Patients with neurodegenerative disorders, such as Alzheimer's disease, may have no problem naming letters (indeed many can still read) but their performance on the N-back task is dismal, even when N=2. This pattern of deficits is even more pronounced in patients with prefrontal cortex damage who show no amnesia for previous events but dramatic impairments in working memory (Jeneson *et al.*, 2010). This suggests that retrieval of information from long-term store is distinct from the ability to use this information in ongoing cognitive processing.

A critical feature of working memory tasks, including the N-back, is that the information which subjects must use to make a correct response varies from trial to trial. This introduces an element of unpredictability, both in terms of time and content. The principle has been applied to animal tests, allowing researchers to examine working memory in different species. Most of these tests take the form of a delayed matching task in which animals are shown a sample stimulus, such as a small plastic object, for a brief period of time. Following a delay in which the stimulus is no longer visible, subjects must identify which of two stimuli matches the original. Working memory is

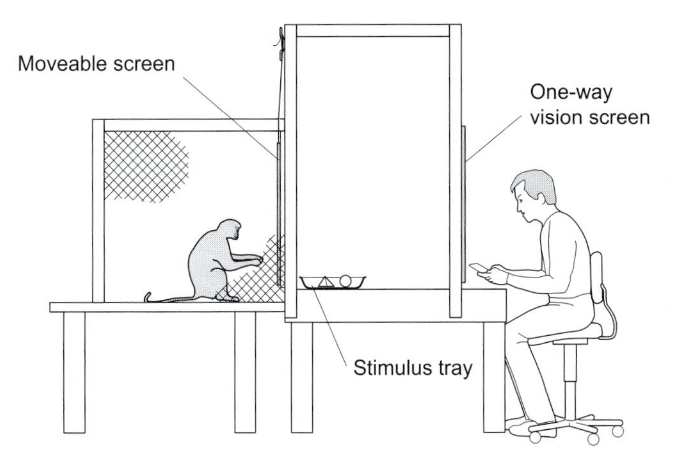

Figure 3.7 Wisconsin general test apparatus for assessing working memory in primates. An animal sits in front of a tray that contains recessed food wells at one end of the apparatus. An experimenter sitting at the other end of the apparatus places a sample stimulus over one of the food wells while the animal watches. A screen blocks the animal's view and the experimenter replaces the original stimulus with two stimuli: one that is the same as the original and one that is different. After a variable period, the screen is raised and animals are rewarded for choosing the matching stimulus. (In a modified version of the task, animals are rewarded for selecting the new or non-matching stimulus).

activated because subjects must keep information about the sample stimulus 'on-line' during the delay. The task was originally conducted with primates using an apparatus like the one shown in Figure 3.7; a modern version uses a computer screen and automatic reward dispenser. Not surprisingly, accuracy declines with increasing delay periods in a variety of species. Distracting animals during the delay period also reduces correct responses because they 'lose' information held in working memory.

Experiments using the delayed matching paradigm suggest that working memory is better (i.e. longer and larger capacity) in humans than in animals. Data from a single paradigm, however, are never sufficient to establish a 'truth.' Indeed, a challenge to the notion that working memory is limited in animals arose when the radial arm maze paradigm was developed for rats (Olton & Samuelson, 1976) (see Figure 3.8). On each trial, rats are placed in a central platform and are allowed to search the entire maze to find food rewards. These are placed in small wells at the end of each arm so they are not visible from the central platform. Rats quickly learn to retrieve all of the food without revisiting the same arms twice even in mazes with 24 arms (Roberts, 1979). To perform accurately, rats must be updating information constantly about which arms no longer contain food. The working memory component of the task can be increased by inserting a delay between arm visits. As an example, after a rat has retrieved 4 out of 8 food rewards and returned to the central platform, they are restricted from entering the alleys by doors that are inserted at each entranceway. In some experiments, rats are removed from the maze completely. After a delay, the doors are removed and rats are allowed to retrieve the remaining food. An error is scored if a rat enters an arm from which it already retrieved the food, either before or after the delay. Rats exhibit much better working memory in the radial arm maze than in typical delayed response tasks using visual stimuli and a computer screen. In some cases, rats can remember which arms they visited up to 24 hours later (Beatty & Shavalia, 1980).

3.3.3 Evolution of working memory

There are a number of reasons that working memory appears to be much better in the radial arm maze than in delayed response tasks (Shettleworth, 2009). One of the most obvious is that searching for food in a maze mimics foraging in the natural environment so rats have a natural tendency to perform in this task. Researchers have taken advantage of this, adapting the radial

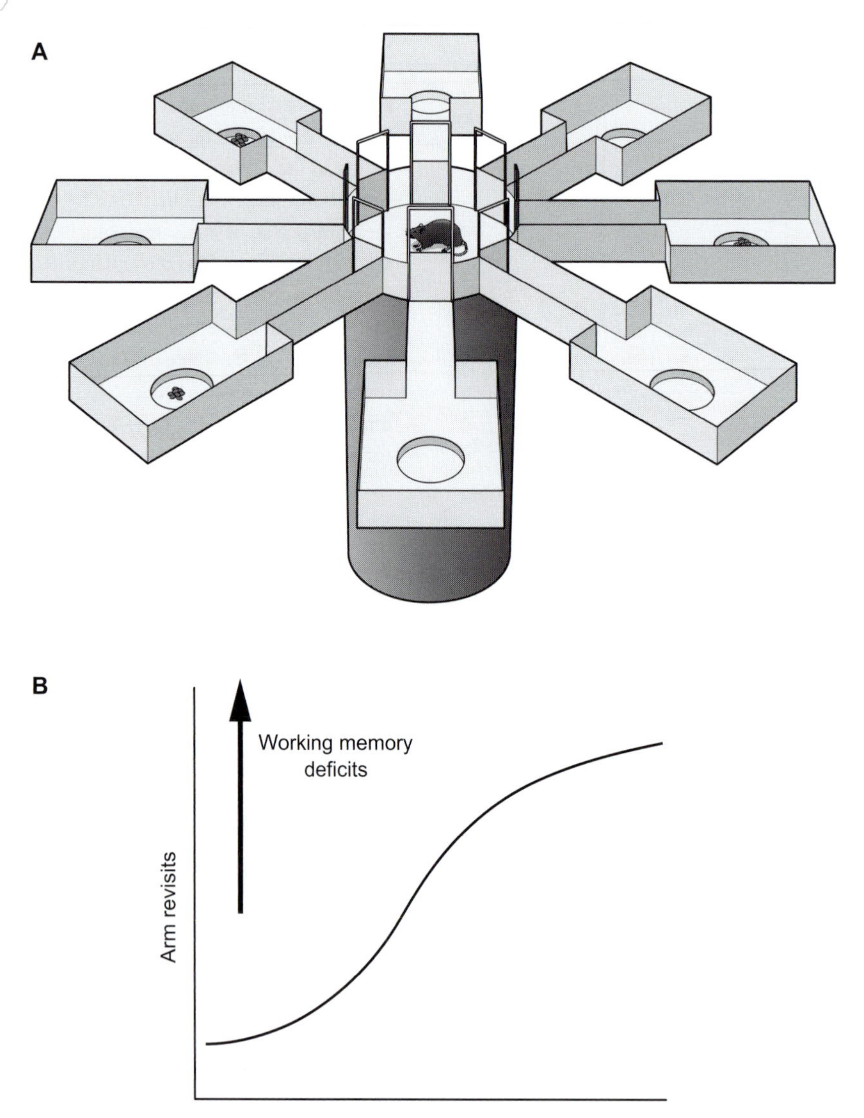

Figure 3.8 Radial arm maze paradigm for assessing working memory in rats. (A) Animals are placed in the center compartment of an 8-arm maze with doors to each alley closed. A food pellet is placed in four randomly selected arms (the same arms are continually baited for each rat). At the start of a trial, the doors are opened and rats are allowed to explore the entire maze. Rats cannot see the food until they reach the end of the arm but, over trials, they learn which arms contain food and visit all four arms in an orderly fashion without entering the non-baited arms. In a working memory trial, rats are allowed to retrieve two of the pellets and then are confined to the center compartment. After a delay period, the doors are opened so animals can access the entire maze. (B) As the time confined to the center increases, animals make more arm revisits, defined as returning to a previously baited arm where they have already retrieved the food.

arm maze to reveal working memory abilities in birds (Brodbeck, 1994), dogs (Fiset *et al.*, 2003), and insects (Brown & Demas, 1994). The existence of working memory in these different species should not be surprising as all animals face ecological challenges that require the updating of environmental information, such as the location of food sources. Even with exceptionally long-term memory storage, animals would have difficulty using this information effectively if they could not keep track of ongoing changes in their environment, such as when food in one location is depleted.

The main difference in working memory across species is how much information can be kept on-line, and how well it can be used in other cognitive tasks. Both honeybees (*Apis mellifera*) and fruit flies exhibit working memory in a delayed response task that lasts up to 5 seconds. Performance declines even at this short interval, however, if another stimulus (i.e. a distractor) is presented during the delay (Menzel, 2009). Insects have tiny brains (approximately one millionth the size of a human brain) and almost all of it is devoted to detecting and processing sensory input. Cognitive resources available for working memory, therefore, are restricted. It is too simplistic to conclude from these data that larger brains mean larger working memory capacity. At least in mammals, a critical component of working memory is neuronal activity in the frontal lobe, particularly the prefrontal cortex (see Box 3.3).

Of all brain regions, the prefrontal cortex has undergone the greatest expansion across human evolution (Rilling, 2006). Measured as a percentage of brain volume, the prefrontal cortex is larger in great apes (humans, chimpanzees, gorillas, and orangutans) than in any other animal. A primary difference between the brains of these close biological cousins is complexity: humans have much denser connections between prefrontal cortex neurons and other brain regions, allowing more efficient communication between neural sites that control cognitive processes. It seems probable, therefore, that working memory was augmented over human evolution. It is also likely that the process was shaped by natural selection because there is wide inter-individual variability in working memory abilities, which may have a genetic basis (Ando *et al.*, 2001). Interestingly, performance on working memory tasks correlates with standard intelligence tests but is influenced less by culture and previous learning than traditional measures such as the IQ test (Ackerman *et al.*, 2005).

Based on these two lines of evidence (prefrontal cortex expansion and heritability of working memory), some theorists argue that an increase in working memory capacity across evolution represents a *qualitative* shift in human cognition (Wynn & Coolidge, 2011). Put simply, working memory allows individuals to think about something that is not there. This symbolic representation provides a mental blackboard for constructing solutions to environmental problems. Designing tools and using language are prime examples of cognitive processes that rely on maintaining and updating information 'in the head.' This description can be extended to artistic works. The earliest known cave paintings were created approximately 30 000 years ago by artists who must have been imagining animals in their natural environment. These renditions could represent the beginnings of human culture, which began to flourish around the same period. Finally, because working memory involves the mental juggling of past and present events, it may form the basis of future planning, an enormous step in cognitive evolution. Although not everyone agrees, human enhancement in working memory capacity may be a "final piece in the evolution of modern cognition" (Wynn & Coolidge, 2011).

Box 3.3 How does the brain control working memory?

The contemporary definition of working memory was not formulated until the 1980s but, long before that time, clinicians and scientists speculated that the frontal lobes were critical for this process. One of the most famous patients in neurological history, Phineas Gage, was described as devising plans that he was unable or unwilling to execute after an accident destroyed a portion of his frontal lobe (Harlow, 1868). Close to 100 years later, surgical removal of one frontal lobe left a woman unable to carry out the sequence of tasks necessary to prepare dinner (Penfield & Evans, 1935). These case studies, replicated in later clinical reports (Stuss & Benson, 1986), identified a subregion of the frontal lobes, the prefrontal cortex, as important for keeping track of ongoing changes in environmental input (i.e. working memory) (Fuster, 1985). These human studies confirmed prior animal experiments showing that monkeys (*Cebus apella*) with prefrontal cortex lesions were unable to remember the location of a food reward for more than a few seconds (Jacobsen, 1935).

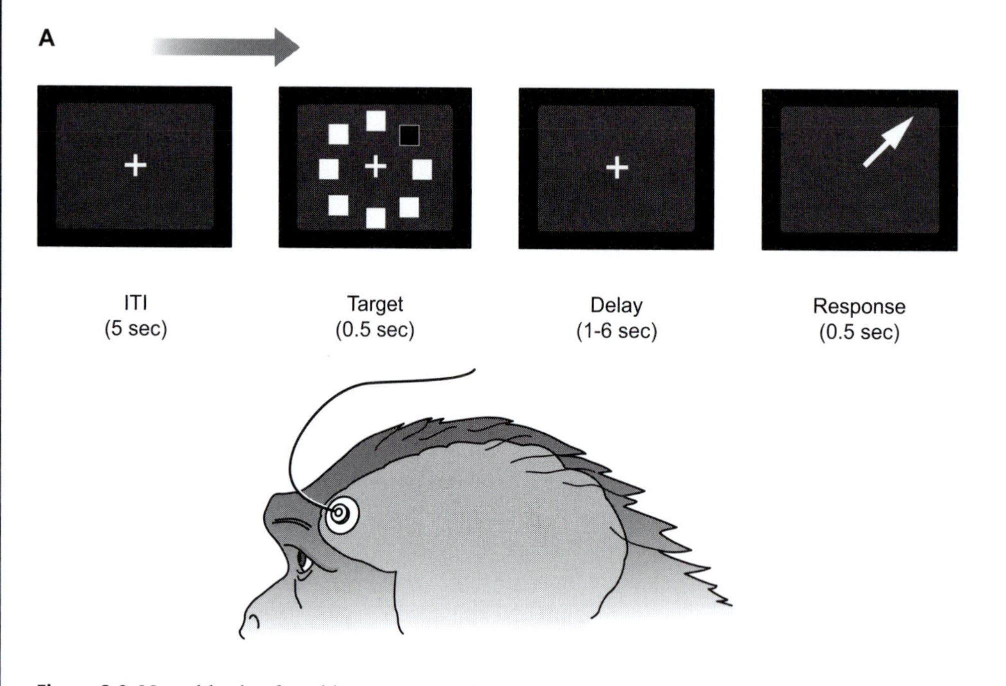

Figure 3.9 Neural basis of working memory. Electrophysiological recordings of prefrontal cortex (PFC) neurons are obtained while monkeys perform a working memory task on a computer screen. (A) During the initial inter-trial interval (ITI), monkeys fixate on a visual stimulus in the center of the screen. Five seconds later, an array of visual stimuli are presented (0.5 sec) around the periphery of the screen with one stimulus (target) marked in a different color. The stimuli disappear for a 1–6 sec delay period during which time animals must continue to fixate on the central stimulus. When the central stimulus disappears, animals are rewarded if they make an eye saccade to the previous location of the target stimulus. (B) PFC neurons fire during the delay period when animals must maintain the location of the target stimulus in working memory. The top panel shows firing patterns of individual neurons across time. The bottom panel shows the combined activity during ITI, target presentation, delay period, and when animals are making a response. Incorrect responses are characterized by decreased neuronal firing.

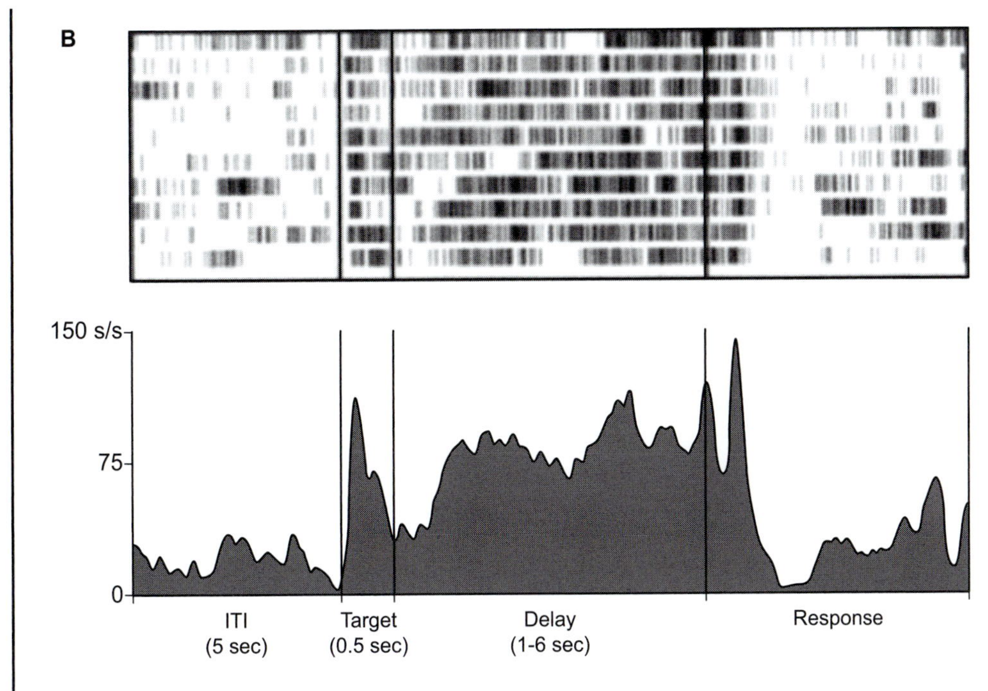

Figure 3.9 (*cont.*)

Researchers then proceeded to examine what happens in the prefrontal cortex when animals perform a delayed response task. Recording cellular activity within this region revealed that prefrontal neurons began to fire when the sample stimulus was presented and then continued to fire when it was removed (Fuster & Alexander, 1971). That is, neurons within the prefrontal cortex were active during the delay period when subjects must keep a record of the stimulus in working memory. More detailed anatomical work showed that neuronal activity increased in different areas of the prefrontal cortex when rhesus monkeys (*Macaca mulatta*) were required to remember a spatial location versus an object (Funahashi *et al.*, 1989). Most importantly, if the firing of a neuron was not maintained across a delay (either because the delay interval was too long or a distractor stimulus was presented), the animal was very likely to make an error during the choice phase (see Figure 3.9). This suggested that the active process of keeping information in short-term store is mediated by neuronal activity in the prefrontal cortex, a finding confirmed in brain imaging studies with humans (Courtney *et al.*, 1998).

Goldman-Rakic, who spent most of her career studying working memory and the prefrontal cortex, speculated that the capacity to hold information on-line evolved when neuronal firing persisted after the termination of a sensory stimulus. In other words, if a neuron continued to fire after an organism encountered a specific stimulus, a representation of that stimulus remained in working memory. This would provide a number of evolutionary advantages such as the ability to track prey or predators as they move in and out of hiding. Natural selection, therefore, may have played an important role in the evolution of working memory.

3.4 Reference memory

Just as discussions of short-term memory have been replaced by working memory, long-term memory has now been replaced by the term **reference memory**. Reference memory is a more accurate description of this type of memory because it implies an active process of referring to information in long-term store. Many researchers simply say 'memory' to distinguish reference memory from working memory, but this text will employ the more specific use of both terms. The study of reference memory changed dramatically in the 1950s with a very unassuming gentleman, who came to be known around the world by only his initials (Box 3.4).

Box 3.4 An unforgettable amnesic

He knew his name. That much he could remember. He knew that his father came from Louisiana and his mother from Ireland. He knew about the 1929 stock market crash and World War II and life in the 1940s. But he could remember almost nothing after that. In 1953, he underwent an experimental operation to reduce brain seizures, only to emerge fundamentally and irreparably changed. He developed profound amnesia, losing the ability to form new memories. For the next 55 years, each time he met a friend, each time he ate a meal, each time he walked in the woods, it was as if for the first time. He could navigate through a day attending to mundane details - fixing a lunch, making his bed - by drawing on what he could remember from his first 27 years. But he could not hold a job and lived, more so than any mystic, in the moment. For those five decades, he was recognized as the most important patient in the history of brain science. As a participant in hundreds of studies, he helped scientists understand the biology of memory as well as the fragile nature of human identity.

In the 1950s, many scientists believed that memory was distributed throughout the brain and not dependent on any one region. That began to change in 1962 when Brenda Milner presented a landmark study showing that a part of the patient's memory was fully intact. In a series of trials, she asked him to trace a line between two outlines of a five point star, one inside the other, while watching his hand and the star in a mirror. The task is difficult for anyone to master. Every time he performed the task, it struck him as an entirely new experience. Yet with practice he became proficient. At one point he said, "This was easier than I thought it would be." The implications were enormous. Scientists realized that there were at least two brain systems for creating memories, one for conscious and one for subconscious memories. This explains why people can ride a bike or play guitar after years of not touching either.

"He was a very gracious man, very patient, always willing to try these tasks I would give him" Dr. Milner reported. "And yet every time I walked in the room, it was like we'd never met." Nonetheless, he was as self-conscious as anyone. Once, a researcher visiting with Dr. Milner remarked how interesting a case this patient was. "He was standing right there," Dr. Milner said, "and he kind of colored - blushed, you know - and mumbled how he didn't think he was that interesting, and moved away."

On December 2, 2008, Henry Gustav Molaison - known worldwide only as H.M., to protect his privacy - died of respiratory failure in Connecticut, USA. He was 82. Hours after his death, scientists worked through the night taking exhaustive scans of his brain, data that will help tease apart precisely which areas of his brain were still intact and which were damaged, and how this pattern related to his memory. His brain has been preserved for future study, in the same spirit as Einstein's, as an irreplaceable artifact of scientific history. Henry Molaison left no survivors, but a legacy in science that cannot be erased.

Adapted from New York Times Obituary

An important point about H.M. is that his intellect, as well as his language and perceptual skills, remained intact following his surgery (Milner *et al.*, 1968). This helped to show that reference memory is independent of other cognitive processes. His ability to learn a new motor task, without remembering having practiced it, highlighted the distinction between 'knowing how' and 'knowing that.' Based on the thousands of research studies that H.M.'s case inspired, this distinction has been refined so that reference memory is now organized into two broad categories, declarative and non-declarative memory, each with its own subdivisions. These are defined by the *type* of information that is processed and by a specific brain region that mediates this processing. The processing occurs via the three stages described in Section 3.2. There is no reason to believe that encoding and consolidation differ across the subclassifications of reference memory, whereas retrieval of declarative versus non-declarative memories may be distinct. Intuitively, recalling what you had for dinner 2 weeks ago is different from remembering how to ride a bicycle. The former, but not the latter, involves an awareness of 'searching' for information in reference memory. For this reason, many people think that declarative memory is unique to humans, although (not surprisingly) not everyone agrees.

3.4.1 Non-declarative memory

A number of clinical case studies followed H.M., confirming that motor learning is preserved in amnesic patients. These studies also revealed other intact memories that were not measured in motor tasks such as mirror drawing. **Non-declarative memory** was introduced as an umbrella term to describe these different types of memory that do not depend on awareness or explicit knowledge to be expressed (Squire & Zola-Morgan, 1988).

Habituation and sensitization

The simplest type of non-declarative memory is modification of a response based on prior experience with a single stimulus. The response may decrease or increase with experience, changes described as **habituation** and **sensitization**, respectively. Habituation develops to stimuli with low motivational significance: sudden movement in the grass may elicit a startle response which then declines when it is not followed by the emergence of a predator (or anything else worth paying attention to). Habituation allows organisms to filter out irrelevant information so they do not waste time or energy responding to a stimulus each time it is encountered. The specificity of the phenomenon provides compelling evidence that animals remember the stimulus previously encountered. This principle is illustrated in Figure 3.10: habituation of aggressive responses in stickleback fish is most pronounced when the same intruder is encountered in the same environment (Peeke & Veno, 1973).

In contrast to habituation, sensitization develops to motivationally significant stimuli or events that an organism should pay attention to. Sensitization is studied less frequently than habituation for the practical reason that increases in a behavioral response may be more difficult to detect (i.e. if the response is already at a maximum). Moreover, depending on the paradigm, repeated presentation of a potentially harmful stimulus could be unethical. Recently the 'natural' experiment of warfare has provided insight into sensitization in humans. Loud and sudden noises elicit a startle response that can be measured through electrodes placed around the eye muscles. War veterans suffering from post-traumatic stress disorder (PTSD) show exaggerated startle responses in the lab, suggesting that they are sensitized to noise bursts that mimic gunfire (Morgan *et al.*, 1996). Interestingly, startle responses of veterans not diagnosed with PTSD are similar to a control group. Although there may

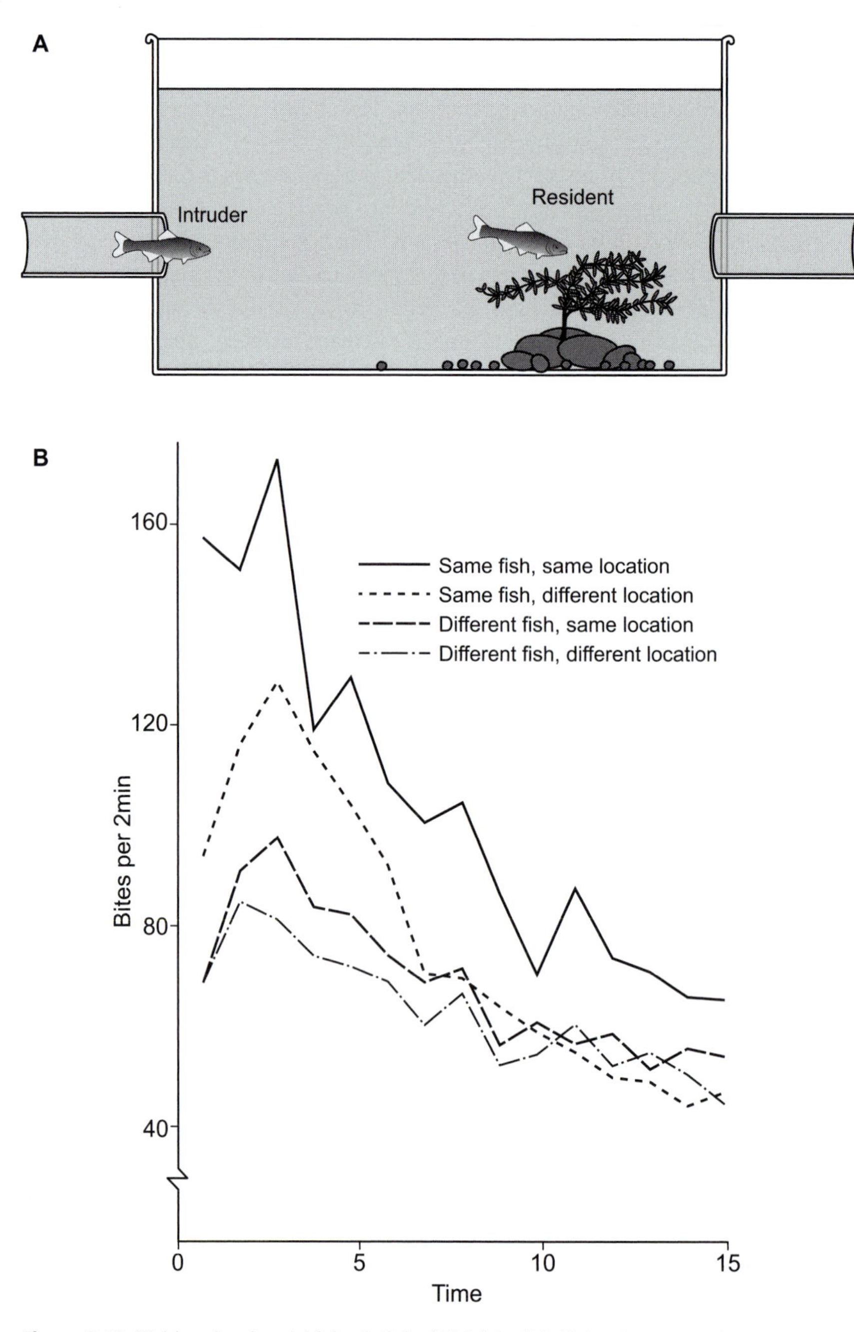

Figure 3.10 Habituation in stickleback fish. (A) Male fish living in one environment were exposed to an intruder that was introduced in one of two locations in the aquarium. Aggression of the resident fish toward the intruder declined across a 30-minute session. After a 15-minute break, aggression of the resident was tested in four different conditions: the same intruder introduced at the same location; the same intruder introduced at a different location; a different intruder introduced at the same location; or a different intruder introduced at a different location. (B) Aggression, measured as bites per 2-minute period, was lowest in the same/same condition and highest in the different/different condition.

Box 3.5 Habituation paradigm in developmental psychology

Young children love new toys but often become bored with the same old thing. Parents take advantage of this by pulling long forgotten items from the bottom of a toy chest. Scientist capitalize on the same effect using the habituation paradigm which compares the amount of time infants spend looking at familiar versus novel objects (Figure 3.11). In a typical setup, a large visual stimulus, such as a striped pattern, is repeatedly placed in front of an infant. For the first few minutes, subjects spend most of their time looking at the pattern but eventually ignore the stimulus and look elsewhere. At this point, the stimulus is removed and replaced with two stimuli, one is the previously viewed stimulus and one is new. Babies, chicks, and infant chimpanzees spend more time looking at the novel stimulus, revealing visual habituation to the familiar image (Colombo & Mitchell, 2009).

The habituation paradigm has been refined over the last 50 years, incorporating different stimuli across a range of ages. Indeed, hundreds of experiments in developmental psychology are conducted each year using a modified version of this original paradigm. These studies do not focus on habituation *per se*, but on the cognitive capacities of infants and young children. A primary goal in this research is to identify which cognitive abilities are present early in development and how these continue to develop in childhood.

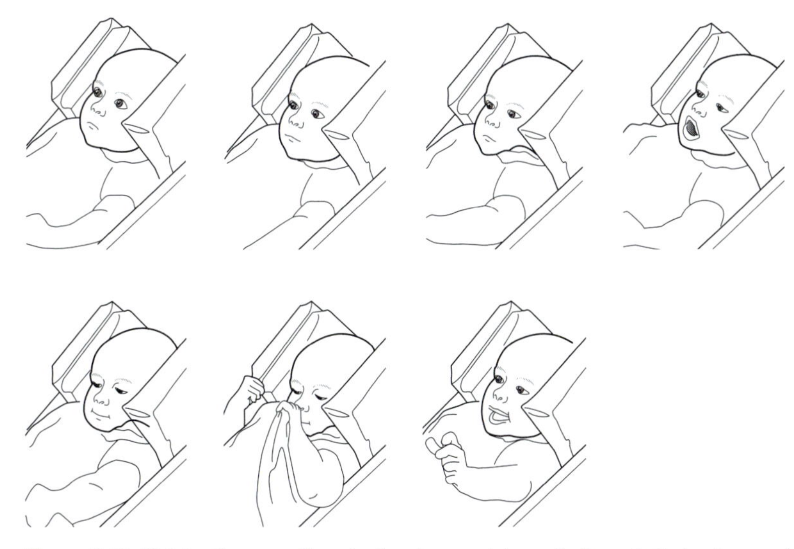

Figure 3.11 Habituation paradigm in developmental psychology. Infants view a visual stimulus presented on a computer screen. As time progresses, infants lose interest in the stimulus and spend less time looking directly at the stimulus. When a new stimulus is presented (final frame), interest is rekindled and infants spend most of their time looking directly at the stimulus.

An intriguing example is provided by a study examining what infants understand about the physical world (Spelke *et al.*, 1995). More specifically, do infants have a basic understanding of Newtonian principles such as continuity? Continuity describes how objects travel in a continuous path over space and time, with no gaps in either. In this study, infants were habituated to an object moving across a stage with two occluding screens. One group saw the object move behind the first screen, then "reappear" from behind the second screen without passing through the space separating the two screens. These infants subsequently looked longer at a test display of one object than a display of two identical objects, implying that the former was more novel than the latter. The authors proposed that the infants recognized discontinuity of motion in the habituation phase; since a single object cannot suddenly appear in another place, they inferred that there were two objects. Consistent with this interpretation, infants in a second group who were habituated to an object moving in a continuous manner (behind one screen, through the intermediary space, then behind the second screen and out again) looked longer at the test display of two objects (Spelke *et al.*, 1995). The authors concluded that a cognitive understanding of object continuity develops within the first year of life. The debate about when these abilities emerge and whether any individual test is a true measure of this cognitive ability is ongoing. Nonetheless, there is little, if any, disagreement that habituation is a powerful measure of memory for a previously experienced event.

have been pre-existing group differences in startle responses, the most likely explanation for the findings is that sensitization of a fear response emerges in individuals who are susceptible to stress-induced physiological changes. Current research is focusing on understanding the factors that predict this difference and whether it can be used to identify later mental health problems in high-stress occupations.

Given the evolutionary value of filtering out or paying attention to stimuli that recur in the environment, it is not surprising that both habituation and sensitization are ubiquitous throughout the animal kingdom (MacPhail, 1996). Both describe a change in a behavioral response so it should always be possible to identify sensitization or habituation of 'what' (e.g. escape or startle response). Habituation and sensitization are sometimes classified as 'non-associative' because they do not depend on learning a relationship between two or more stimuli. This position has been challenged by evidence that both habituation and sensitization are reduced, sometimes eliminated, in a new environment. This suggests that animals form associations between an environmental context and the motivational significance of a stimulus. Movement in the grass may be trivial in one location, but deadly in another.

Perceptual priming

Perceptual priming is the facilitated identification of a stimulus as a consequence of prior exposure to the stimulus (Tulving & Schacter, 1990). In a typical human priming experiment, participants are asked to read words as they are flashed on a screen. The rate of presentation is so fast that most people identify only a small proportion of the list. If words are repeated within a short interval, accuracy increases even though individuals report that they do not remember seeing the word before. A different task requires participants to name familiar objects from fragments of line drawings. Initially only a small portion of the image is presented so the pictures are difficult to identify. The image becomes recognizable as the lines are gradually filled in. On later trials, participants are able to identify the image at an earlier stage. Chapter 2 described perceptual priming in animals: birds are better at detecting prey in a visual search task if the item appeared in previous trials (Pietrewicz & Kamil, 1981). Priming is also observed in a bird version of the picture fragment task (Brodbeck, 1997).

Perceptual priming is categorized as non-declarative memory because it functions, primarily, at an unconscious level. Even patients with extreme deficits in declarative memory, like H.M., show intact perceptual priming (Hamann & Squire, 1995). The phenomenon has been observed in all sensory systems, with the strongest effects occurring when the prime and identifying stimuli are in the same modality. That is, an auditory stimulus may prime a visual stimulus (e.g. hearing a word may facilitate later identification of the written word) but the effect is reduced compared to two auditory or two visual stimuli. Moreover, it is not clear whether this cross modality enhancement is true perceptual priming; more likely, it reflects semantic priming in which a mental representation of the word meaning primes its later identification. Semantic priming is difficult to assess in animals because we have little understanding of the meaning that non-verbal subjects attach to different stimuli. If it does occur (which is probable), semantic priming would confer an evolutionary advantage much like perceptual priming. That is, enhanced recognition and identification of frequently occurring stimuli would facilitate a number of cognitive processes including categorization, decision making, and counting.

Classical conditioning

Classical conditioning, as one of the principal categories of associative processes, will be discussed in detail in Chapter 4. Briefly, classical conditioning is the process whereby a stimulus, through association with a motivationally significant event, acquires the ability to elicit a response. Pavlov's dogs salivated when they heard a bell because they associated the sound of the bell with food. Classical conditioning is often divided into behavioral responses, such as salivation, and emotional reactions, such as the uncomfortable feeling some people experience when they walk into a dentist's office. Even if the association that produces a specific behavior or emotional reaction can be identified, classical conditioning does not require awareness of the conditioning, which is the reason that it falls into the category of non-declarative memory.

Procedural memory

H.M.'s ability to learn the mirror drawing task (see Box 3.4) is a prime example of **procedural memory**, defined as a gradual change in behavior based on feedback. Although H.M. could not remember performing the mirror drawing task, he was able to observe his own mistakes on each trial, providing feedback on his performance. In everyday life, procedural memory is engaged when individuals tie their shoes, comb their hair, and control a cursor on the computer screen. More advanced skills like playing a musical instrument and performing an athletic feat are also examples of procedural memory. An important point about procedural memory is that these are not acquired through memorization of body movements involved in the activity. Admittedly visualization of an activity improves performance in many tasks, but it will not replace the hours of training that lie behind all high-level athletes and musicians.

Animals use procedural memory in many naturalistic behaviors in that experience often improves their ability to capture prey, court mates, or care for young. In the lab, procedural memory is assessed, most frequently, using operant conditioning tasks, summarized in Chapter 4. In most cases, performance gradually improves over a series of trials. Rats running through mazes or lever pressing for food get better with training. If this represents procedural memory, it should occur in the absence of conscious awareness (i.e. the definition of non-declarative memory). Obviously, it is difficult to know whether animals in these experiments are 'aware' of how and why their performance is improving, although scientists have devised clever ways to get around this issue (see Section 4.2.2 for details).

Studying procedural memory in humans comes with its own set of problems. It is one thing to show that amnesics can learn a motor skill task, but what does this tell us about memory in 'normal' individuals? Brain damage produces physiological changes beyond the immediate area of injury,

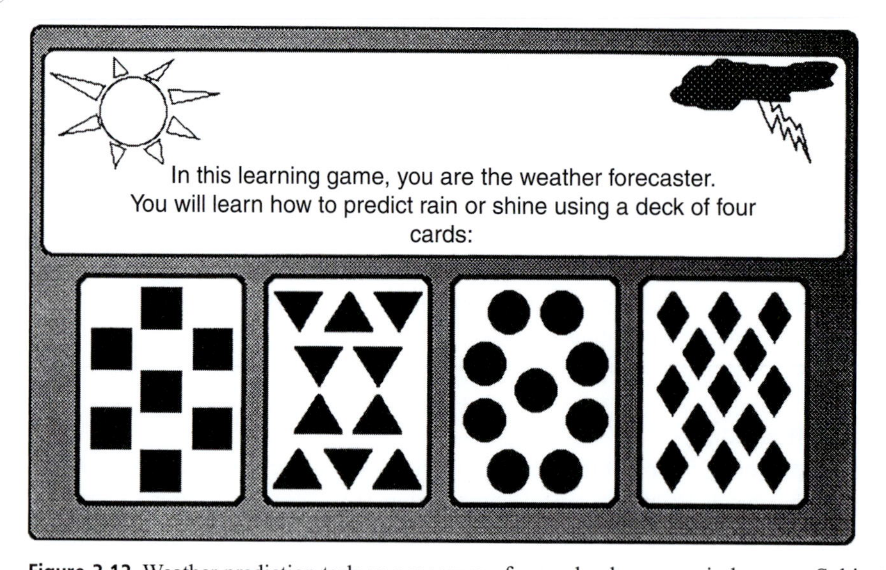

Figure 3.12 Weather prediction task as a measure of procedural memory in humans. Subjects decide on each trial which of two weather outcomes (rain or sunshine) will occur based on a set of one, two, or three cues (out of four possible cues) that appear on a computer screen. Each cue is independently associated to a weather outcome with a fixed probability, and the two outcomes occur equally often. There are four possible cue–outcome association strengths: throughout training, a cue is associated either 75%, 57%, 43%, or 25% (approximately) with sunshine. During each trial, one, two, or three of the four possible cues are presented. Subjects respond by pressing a key to predict the weather outcome, and feedback is given immediately after each choice (correct or incorrect).

making it difficult to extrapolate findings from an impaired to an intact brain. In H.M.'s case, the seizures that he experienced for years before his surgery could have altered brain systems that mediate reference memory. Knowlton *et al.* (1996) came up with a solution to this problem using a task that measures gradual changes in behavior based on feedback, with no conscious awareness of the learning. In this task, shown in Figure 3.12, participants are asked to predict weather patterns based on a set of one to four cards that are laid out in front of them. The actual outcome is determined by a probabilistic rule with each card being a partially accurate predictor of sunshine or rain. Feedback is provided on each trial but, with 14 possible combinations of cards, it is nearly impossible to identify which predict good or bad weather. Nonetheless, performance improves across trials even though subjects cannot explicitly name specific card-weather contingencies. Importantly, patients with brain damage in the same area as H.M.'s show normal learning on this task but no memory for the task itself. This is different from control subjects who remember doing the task; they just cannot explain why they are getting better at it.

Memory in the weather prediction task may seem categorically different from memory for a skilled activity such as dancing, but the two are grouped together because they are both examples of gradual feedback guided learning. In addition, the same brain structures are involved in the two types of learning, an important criterion for classifying subcategories of reference memory. Note also that the four categories of non-declarative memory described in this section are not the final word on this topic. Future research may lead to a refinement of these subcategories and/or further divisions of these memory systems.

3.4.2 Declarative memory

Humans cannot describe information that is retrieved in non-declarative memory and, in many cases, have no recollection of how previous experience altered their behavior. This makes it easy to compare non-declarative memory in humans and animals as many of the same tasks can be used with different species (e.g. classical conditioning to a fearful stimulus). This is in stark contrast to **declarative memory**, a knowledge-based system that is expressed through explicit statements (e.g. 'I remember when...'). Declarative memory is flexible in that it combines multiple pieces of information retrieved from long-term store. By definition it requires awareness: humans consciously reflect on the information they are retrieving, which is the reason that so many people question whether this process exists in animals.

Semantic memory

Tulving (1972) was the first to suggest that declarative memory should be divided into knowledge of facts and knowledge of episodes. He named the first **semantic memory** to describe general knowledge of the world that is not tagged to a particular event. This includes knowledge of words and their meanings as well as complex concepts such as gravity. The two types of declarative memory would be expressed when an individual remembers the definition of semantic memory (fact) and the occasion on which it was learned (episode). One case report as well as data from neuroimaging experiments suggest that these two processes involve dissociable brain systems (Tulving, 1989).

Episodic memory

Knowledge for events in a personal past is called **episodic memory**. This autobiographical information is unique to each individual as it is associated with a particular time, place, and event. If the debate about whether animals have declarative memory is heated, the issue of whether they have episodic memory is on fire! Researchers speculated on this possibility for many years, but it was not until a seminal paper showing that birds remember the 'what, where, and when' of an event (Clayton & Dickinson, 1989) that the scientific community was invigorated. In this experiment, western scrub-jays (*Aphelocoma californica*) cached peanuts or worms in separate trays (Figure 3.13) and

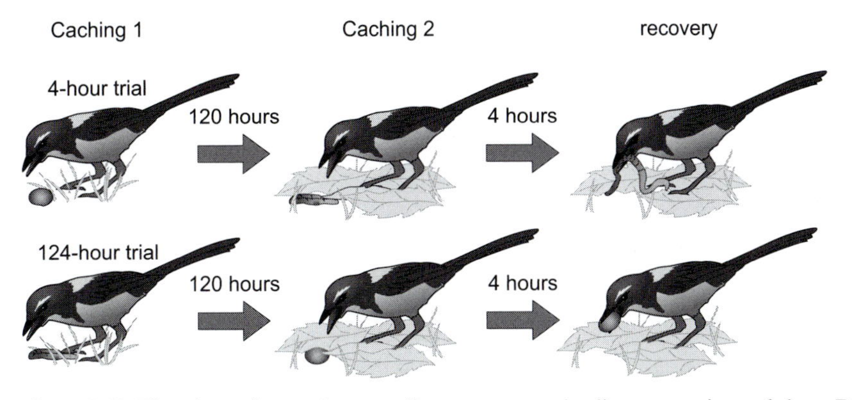

Figure 3.13 The what, where, when paradigm to assess episodic memory in scrub-jays. Birds were allowed to cache worms or peanuts in one of two locations. After a delay of 120 hours, they were allowed to cache the other item in a different location. Four hours later, birds were given the option of which of the two caches to recover. If worms were cached during the second caching (i.e. four hours before the test), birds recovered the worms first. If worms were cached at the first caching (i.e. 124 hours before the test), birds recovered the peanuts first (Clayton & Dickinson, 1989).

Researcher profile: **Professor Nicola Clayton**

Figure 3.14 Dr. Nicola Clayton.

Nicky Clayton has heard every possible joke about bird brains, but she doesn't mind. She is delighted to be associated with these animals from whom, she claims, we have learned so much. Clayton is responsible for much of this knowledge: her ground breaking work on cognition in corvids (a family of birds that includes jays, magpies, ravens, and crows) is recognized internationally by both the scientific community and the lay public. Clayton claims that her training in both zoology (BA Hons Oxford University) and Animal Behavior (PhD St. Andrew's University, Scotland) helped her to understand what scrub-jays do in the natural environment. She used this knowledge when she moved to the University of California at Davis where she designed experiments that capitalized on birds' natural tendencies, such as food caching. Clayton argued that birds must remember the 'what, where, and when' of their caching incidents, otherwise they would waste valuable time and energy flying back and forth to ensure that the food was still edible. In other words, there must be evolutionary pressure for episodic memory to emerge in these birds. Clayton pioneered a new procedure for testing this hypothesis (the what, where, when paradigm) revealing that scrub-jays could remember specific information about an incident in their personal past. "People seemed to think that if a monkey had done it, that would be one thing," said Clayton, "but a *bird*?"

Clayton returned to England in 2000 to take up a post at the University of Cambridge where she holds a distinguished chair and is a Fellow of the Royal Society. She continues to publish innovative and controversial studies, her latest showing that corvids have the ability to understand what other birds are thinking and to plan for the future. These cognitive capacities are generally reserved for humans and other apes so many people wonder how birds accomplish these tasks given that they have much smaller brains with different structural features. Clayton's theorizing in this area has led to an emerging theory that intelligence evolved independently in at least two disparate groups, the apes and the corvids. The impact of her work extends from animal to human cognition as well as to how and when these abilities develop in young children.

Clayton's expertise and passion extends to another field: dance. She is the first ever Scientist in Residence to Rambert, Britain's flagship contemporary dance company. She combines her love for these two parts of her life by teaching workshops on birds and tango. Clayton has also collaborated on new choreographic works inspired by science, including the Laurence Olivier award-winning *Comedy of Change*, based on Darwinian principles and commissioned in honor of Charles Darwin's bicentenary in 2009.

were allowed to retrieve the items at different times in the future. Scrub-jays prefer worms but, unlike peanuts, these decay over time. When the interval between caching and retrieval was short, birds searched first for the worms but this was reversed at long intervals when the worms would have decayed. Subsequent work showed that these birds could manipulate mental representations of these memories in that their choices changed when they received new information about the rate of decay (Clayton *et al.*, 2003).

Clayton and Dickinson's paper is a 'citation classic' with dozens of follow-up studies showing similar effects in other bird species, rodents, and non-human primates (Zhou & Crystal, 2011). As an apparent validation of this work, humans use episodic memory to solve a what, where, when task that models the scrub-jay paradigm (Holland & Smulders, 2011). Rather than settling the debate, these studies have only served to inflame it. A main point of contention is that animals can solve

Box 3.6 Hyperthymesia

Jill Price has a problem that many people would consider a blessing. She remembers the details of every day of her life since childhood. Given a particular date from the past 30 years, Price can immediately name the day of the week, the weather, and what she did. But Price is plagued by her remarkable memory which she describes as a running movie in her head that never stops. She cannot block out painful memories so is often overcome by anxiety from previous incidents that were highly emotional. After years of living in this state, Price decided to do something about it. She searched on the internet for descriptions of her condition, expecting to find other people who shared her abilities (and problems). There was nothing. She did, however, come across a webpage for James McGaugh, a professor at the University of California, Irvine, who is a world authority on flashbulb memories (see Box 3.1).

Price had no idea what to expect when she sent an email to McGaugh asking if he could help her. Many people would have ignored the message, assuming that it was a prank or a distorted exaggeration, but McGaugh responded in less than 2 hours. He admits that he was skeptical at first but was struck by Price's sincerity and intrigued by the possibility of learning something new about memory. Over the next few years, Price met regularly with McGaugh and his team who repeatedly verified her extraordinary claims (Parker *et al.*, 2006). She provided the scientists with journals that she has kept since childhood so they were able to confirm many details of the incidents she related. She never knew what she would be asked on any given session so she could not 'study' for these tests. Price can also recall the dates of specific TV episodes and important historical events (that she paid attention to). She has no idea how she accomplishes these feats, saying "I have always been about dates. I just know these things." (Parker *et al.*, 2006: page 41). She has a mental representation of a calendar that she refers to when accessing information about particular events, but cannot explain why it is organized as it is.

Surprisingly, Price says she has trouble with rote memorization. She is no better than the average person in remembering dates that have no special meaning for her and was not an exceptional student. She scores higher than average on some memory tests, but only excels on those that measure autobiographical information. McGaugh suggested that Price's profile of enhanced episodic memory warrants its own label, **hyperthymesia**, from the Greek thymesis for memory. Since the publication of Price's case, other hyperthymesic individuals have emerged. (After a news report on the study, McGaugh received dozens of emails from people claiming to possess enhanced autobiographical memory, but only a few met the criterion for hyperthymesia.) Like Price, they spend an inordinate amount of time ruminating on personal events and almost all keep detailed daily journals. These traits have led some researchers to suggest that the condition is a memory oriented form of obsessive compulsive disorder. This may not provide relief to Price and other patients but it could help researchers to understand this remarkable enhancement of episodic memory.

what, where, when tasks without using mental time travel, the subjective experience of going back to an incident that characterizes episodic memory in humans. Another criticism of the animal studies is that humans routinely retrieve episodic memories spontaneously, without prior knowledge that the information will be required at a later time. In contrast, scrub-jays underwent extensive training in the caching retrieval task prior to testing. An argument could be made that memory encoding is different when subjects are required to remember the details of an incident, particularly if they have repeated exposure to stimulus relationships (e.g. they could form mental representations of the conditions under which certain items will be encountered). Animal researchers persistently respond to criticism by modifying the what, where, when paradigm to tease apart these subtleties, but the issue is unlikely to be resolved in the near future. According to some proponents, skeptics simply raise the criterion bar for episodic memory in animals whenever new evidence contradicts their position. Clayton and Dickinson's experiment was designed, specifically, to model Tulving's original definition of episodic memory: the what, where, and when of an event. Having shown this ability in birds (no one doubts that animals can do the task, they argue about the cognitive process that explains it), stricter conditions were placed on the animal work. Whether these experiments provide evidence for a cognitive process in animals that is equivalent to episodic memory in humans may never be settled. The issue is complicated further by the fact that episodic memory in humans is still under investigation (see Box 3.6).

Research into declarative memory is ongoing: few scientists dispute the distinction between memory for facts and memory for episodes, but the line between the two is not always clear. For example, how should individuals classify knowledge about themselves? What about semantic information relating to a particular episode? Individuals who are deceived may later remember the other person as untrustworthy, but is this transfer of information from episodic to semantic the same process as children who encounter objects falling *down* and then develop a concept of gravity? These are only a subset of questions that scientists are pursing in the ongoing search to understand these processes.

3.4.3 Multiple memory systems model

Information in the preceding sections can be summarized by the multiple memory systems model, the idea that memory is divided into subcategories based on the type of information processed. The model also proposes that different subcategories of memory are associated with neural processing in distinct brain systems. The details, presented in Figure 3.15, are based on converging evidence from

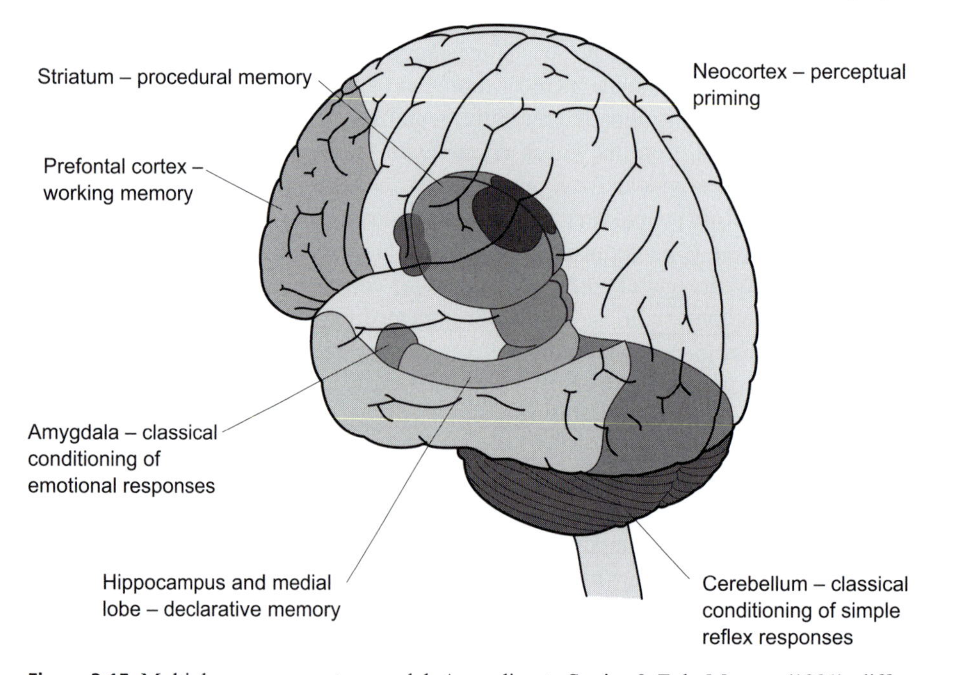

Figure 3.15 Multiple memory system model. According to Squire & Zola-Morgan (1991), different anatomical regions of the mammalian brain are associated with different subtypes of memory.

a variety of experiments including case studies of brain-damaged patients, neuroimaging, and experimental lesions in lab animals (primarily rats and monkeys). Although the general concept of multiple memory systems is accepted, the details of specific subcategories and the functional properties of each continue to be debated. A primary point of contention is what constitutes a memory system and how independent these are from other memory systems. Another potential problem is that some of the brain regions shown in Figure 3.15 have only been identified in mammals so it is not clear how the types of memory associated with these systems apply to other animal classes. A parsimonious solution to this problem is to adopt the *principle*, but not the specifics, of the multiple memory systems model. That is, memory is organized by biological systems that support the processing of different types of information, even if the details of these systems vary across species.

If this model of memory function is true, it raises the question of why evolution would favor the emergence of qualitatively distinct memory systems, rather than a single system that could adjust to different environmental constraints. One possibility is that specialization arose in memory processing because cognitive systems that store one type of information are better at solving particular cognitive problems. For example, episodic memory may have evolved when humans started living in groups and had to keep track of 'who did what to whom'; memory for these personal events is unlikely to facilitate memory for a motor skill. Indeed, cognitive processes supporting procedural and episodic memory may be incompatible in that skill-based learning depends on extrapolating commonalities across repeated occurrences of an action while ignoring the what, where, and when of each incident. Episodic memory is the reverse, defined by information tagged to specific events, rather than generalities across events. This distinction would produce a functional incompatibility that is the basis of the multiple memory system model (Sherry & Schacter, 1987). Finally, a memory system that appears to serve more than one function may arise if a system that evolved to serve one function has been co-opted by another (see discussion of adaptations in Section 1.5.2).

3.5 Neuroscience of memory

In the multiple memory systems model, the acquisition, storage, and retrieval of different types of information are associated with distinct neural substrates. But, what happens *within* these systems? Are there general properties of memory processing across systems? Although the pieces of the puzzle are not yet complete, current evidence points to commonalities in the cellular, molecular, and synaptic mechanisms that underlie different types of memory.

3.5.1 Cellular mechanisms

Most of what we know regarding the cellular basis of memory comes from the Nobel Prize winning work of Eric Kandel and his colleagues using the marine sea snail (*Aplysia californica*). The animal has a relatively simple nervous system making it a prime model for studying neural circuits that control behavior, as well as how these are altered during learning. A common behavioral measure in these studies is a reflex withdrawal response when the animal is touched. *Aplysia* exhibit both habituation and sensitization of this reflex; as expected, habituation develops to a mild stimulus whereas sensitization develops to a stronger one. Over habituation trials, sensory neurons release less of the excitatory neurotransmitter glutamate, causing a reduction in the firing of motor neurons controlling the response (Hawkins *et al.*, 2006) (Figure 3.16). These changes occur at several sites in

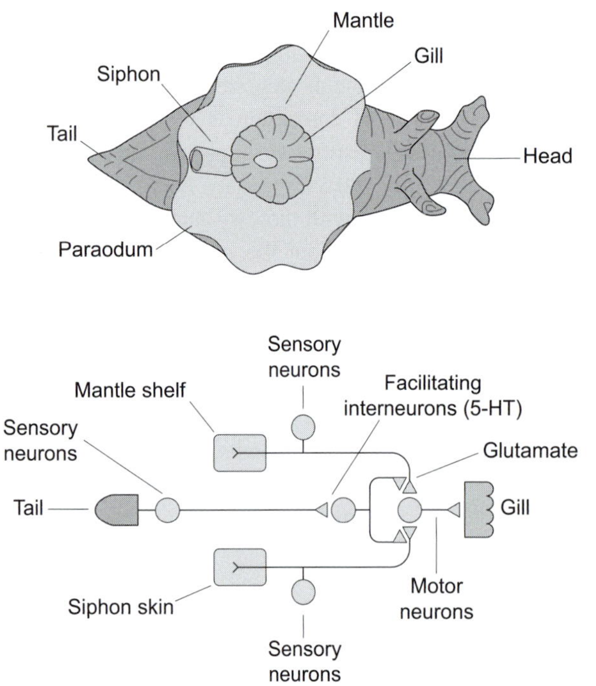

Figure 3.16 Cellular mechanisms of sensitization in *Aplysia*. A strong, potentially aversive, touch to the *Aplysia* body (e.g. the tail) induces a withdrawal response, mediated by glutamate release from the sensory to motor neurons. If the sensory stimulation is repeated, withdrawal responses become stronger (sensitization). This is mediated by increased release of serotonin (5-HT) from interneurons connecting sensory and motor neurons.

the reflex circuit, suggesting that memory is not confined to a specific location in the circuit. The same synapses that are depressed during habituation are enhanced during sensitization. In this case, the circuit is modulated by an interneuron that releases increasing amounts of serotonin (5-hydroxytryptamine, 5-HT) with each stimulation. 5-HT activates sensory neurons, causing intracellular changes that make it easier for the neuron to fire in the future. This potentiated response in sensory neurons increases the firing of motor neurons and, by consequence, the magnitude of the reflex withdrawal response. The same process has been described in other invertebrates and in some vertebrates, suggesting a common mechanism for reflex modification across species.

Changes in neurotransmitter release are short term so they cannot explain long-term changes in habituation or sensitization that *Aplysia* (or other animals) display. With sensitization, repeated 5-HT release causes intracellular changes that activate genes, thereby producing new proteins. These proteins alter neuronal function and lead to structural changes such as the growth of new synaptic connections. In contrast, long-term habituation leads to a loss of synaptic connections, although the mechanism that produces these changes is not as well understood. The structural changes associated with sensitization involve a long cascade of molecular changes, one of the most important being activation of the protein CREB. Research on *Aplysia* as well as the common fruit fly (*Drosophila melanogaster*) implicates CREB function in very simple forms of learning; genetic and pharmacological manipulations in rodents demonstrate that CREB is also required for more complex forms of memory, including spatial and social learning (Hawkins *et al.*, 2006). In addition, some human disorders characterized by severe deficits in learning, such as Alzheimer's disease and Coffin–Lowry syndrome, are associated with altered CREB function (Saura & Valero, 2011). Thus, molecular mechanisms of memory may be conserved across species with CREB-mediated gene expression acting as a universal modulator of processes required for memory formation.

Keep in mind, however, that there is no single genetic marker for memory. Even apparently simple forms of learning recruit multiple genes and proteins which act synergistically to modulate neuronal function and communication. These intracellular changes may span an extended timeframe, with some phases occurring immediately and others up to 24 hours following an experience. 5-HT is not the only neurotransmitter to be implicated in memory. Decreased acetylcholine (ACh) levels in Alzheimer's disease have linked this neurotransmitter to memory, although ACh-related deficits in these patients may reflect a disruption in attentional processes (Graef *et al.*, 2011). This raises an important point: given that memory has a role in most cognitive function, it is often difficult to determine whether a particular biological substrate has a role *only* in memory or in memory as it interacts with other cognitive processes.

3.5.2 Synaptic mechanisms

The molecular and cellular changes described in the previous section change the way that neurons communicate with each other. The Spanish neuroanatomist and Nobel laureate, Santiago Ramon y Cajal, was one of the first to suggest that memory involves improved neuronal communication (Kandel, 2001), but more than a half century passed before the idea caught on. Most people attribute this hypothesis to Donald Hebb (1949) and his cell assembly theory. With no experimental evidence (and no way to test it), Hebb proposed that connections between two neurons are strengthened if they fire at the same time. As a corollary, it becomes easier for one neuron to excite the other in the future (i.e. for a presynaptic neuron to cause an action potential in a postsynaptic neuron). According to Hebb, this mechanism of enhanced synaptic connectivity is the physiological basis of memory.

It was close to 25 years before Hebb's theory was confirmed experimentally using brain slices extracted from a rabbit hippocampus (Bliss & Lomo, 1973). In this preparation, which has now been used in thousands of studies (see Figure 3.17), electrodes are placed in the bundle of axons entering the hippocampus and in the group of neurons to which these connect. First, the axons are stimulated with a single electrical pulse and action potentials are recorded in the postsynaptic neurons: this provides a baseline measure of synaptic strength. Second, the axons are stimulated with repeated bursts of high-frequency pulses. Third, a single pulse is applied to the axons and postsynaptic responses are recorded. Postsynaptic responses are enhanced in the third phase, compared to the first phase, of the experiment and the effect can last for weeks. This persistent increase in synaptic strength following high-frequency stimulation of presynaptic neurons is called **long-term potentiation** (LTP). LTP results from increased responsivity of postsynaptic neurons, not increased neurotransmitter release from presynaptic neurons. Interestingly, when trains of pulses are presented at a slow rate, the strength of the connection is decreased, a phenomenon known as long-term depression (LTD).

LTP has a number of properties that make it a plausible neurobiological substrate of memory. For one, it is synapse specific: postsynaptic neurons may have hundreds of inputs but only those that are stimulated with high-frequency pulses are enhanced at a later time. Additionally, inputs will become associated if they are activated together: a weak input that does not elicit an action potential in postsynaptic neurons will acquire this ability if it is stimulated with a strong input. These two properties relate to memory in that organisms remember only a small amount of specific information from previous experiences and two events that occur together may become associated (classical conditioning is one of the best examples). Further evidence for an LTP memory connection is that manipulations which block LTP also block learning (Morris *et al.*, 1986; but see Bannerman *et al.*, 1995; Saucier & Cain, 1995) and LTP is mediated by a cascade of intracellular changes similar to those that mediate memory (e.g. protein synthesis) (Malenka & Bear, 2004). Perhaps the strongest evidence that LTP is a cellular form of memory comes from an occlusion study (Whitlock *et al.*,

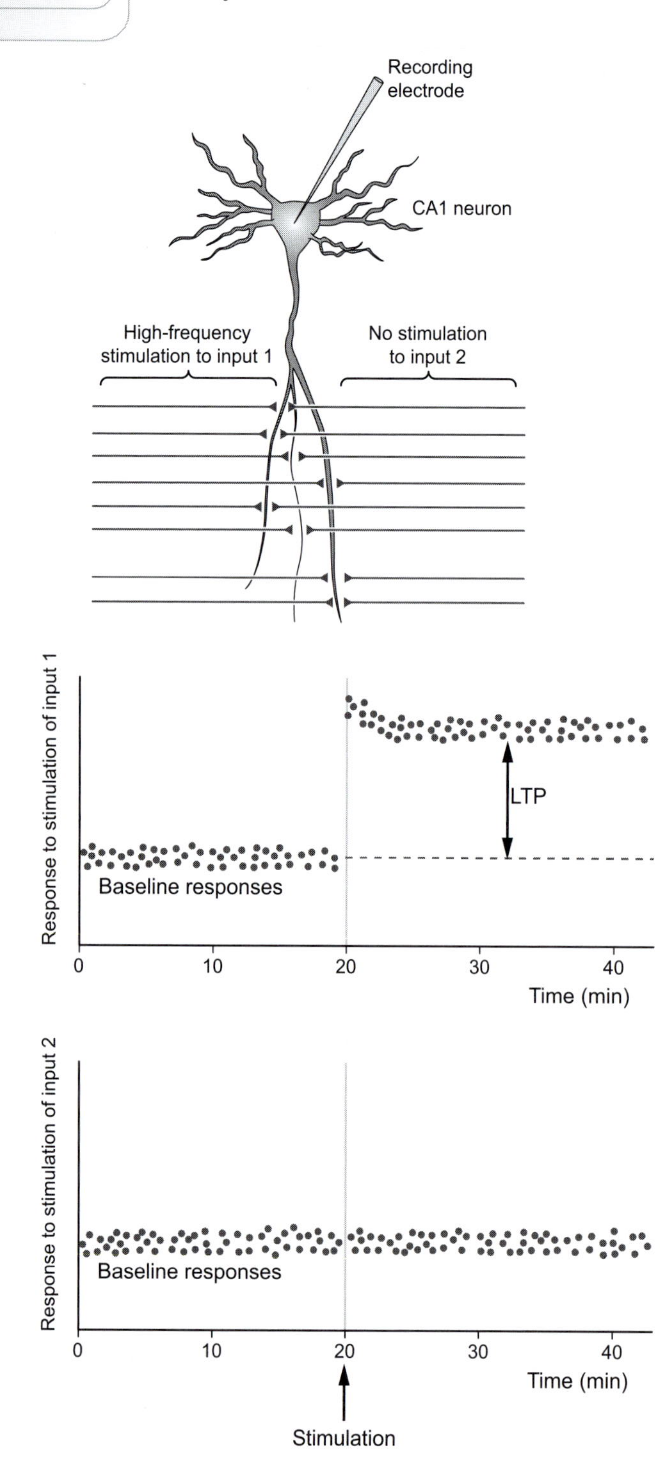

Figure 3.17 Long-term potentiation (LTP) in the hippocampus. Responses of a group of neurons in the CA1 region of the hippocampus are recorded when two inputs are stimulated alternately. Following this baseline testing, a string of high-frequency stimulation is applied to input 1. Responses are then measured to stimulation of each input: LTP is manifested as an increased response to stimulation of input 1, but not input 2, following the high-frequency stimulation.

2006). In this experiment, memory for an aversive event induced changes in brain chemistry that mimic those occurring during LTP. Most importantly, this training blocked the ability of a later experimental manipulation to produce LTP, presumably because the synapses were already potentiated.

LTP has been studied most thoroughly in the hippocampus, but it occurs in many other brain regions as well as the spinal cord (Malenka & Bear, 2004). The ubiquitous nature of this phenomenon has led to the proposal that LTP is a synaptic record of past experiences. This position may be too strong as synaptic plasticity could occur through more than one mechanism (i.e. LTP may not be the only one). Nonetheless, although many details remain to be established, most scientists agree that LTP is a key mechanism by which experience modifies behavior.

Chapter Summary

- Memory, the cognitive process of acquiring and retaining information for later retrieval, is manifested as experience-dependent modification of behavior. Memory is a fundamental component of all cognition, in most cases facilitating other cognitive processes. The ability to modify behavior based on experience is heritable and provides evolutionary advantages to species who inhabit unpredictable environments.
- Memory progresses through three phases: encoding converts sensory input to neural signals; consolidation modifies encoded representations to increase their stability; and retrieval accesses stored information. Memory may be modulated at any one of these stages by factors such as emotional state, pre-existing knowledge, or other ongoing cognitive processing. Given this ongoing processing, it is unlikely that retrieved memories are a snapshot of past events.
- Working memory is the active process of maintaining and updating information in short-term store. It is measured in both humans and animals by testing the ability to use recently acquired knowledge following a delay. These tests reveal inter-species differences in working memory capacity that are tightly linked to neural complexity in the prefrontal cortex. This has led to the theory that increased working memory is a qualitative shift in the evolution of human cognition.
- Reference, or long-term, memory may be conceptualized within a multiple memory systems model which associates the processing of different types of information with particular neural substrates in the mammalian brain. The largest division is between non-declarative and declarative memory, distinguished by conscious awareness in the latter. Non-declarative memory includes modification of a response to a single stimulus presentation (habituation and sensitization), perceptual priming, classical conditioning, and procedural memory. Declarative memory includes knowledge of facts (semantic memory) and personal events (episodic memory). A heated current debate in comparative cognition is whether an animal's ability to remember the what, where, and when of an incident is equivalent to episodic memory in humans.
- Memory is mediated by cellular and molecular changes, many of which are conserved across evolution. One example is CREB: activation of this intra-neuronal protein plays a critical role in several types of memory across many species. Intracellular changes, such as in CREB, produce synaptic plasticity that alters neuronal communication. These long-lasting effects may be mediated by LTP, the persistent increase in synaptic strength as a result of high-frequency stimulation. The properties and underlying mechanism of LTP make it an ideal (although not exclusive) model of memory.

Questions

1. If memory evolved to support other cognitive processes, such as decision making or social competence, why are there so many examples of memory failure?
2. Aging and age-related disorders, such as Alzheimer's disease, are characterized by disrupted memory processing, particularly of recent events. Could there be an adaptive value to these declining abilities across aging? Would these be different for 'normal' and pathological conditions?
3. Describe the processes that facilitate memory encoding and the evolutionary significance of each.
4. Design a memory experiment to distinguish deficits in consolidation from deficits in retrieval.

5. Working memory predicts performance on many neuropsychological measures including the IQ test. But unlike IQ, working memory tests are not culturally specific. With this in mind, should simple tests such as the N-back and picture completion replace more standard metrics of academic achievement and performance? What would be the advantages and disadvantages of doing so?
6. Drug tolerance, defined as decreased effectiveness of the drug with repeated use, is sometimes explained as habituated responding. Provide arguments for and against this idea.
7. In the ancient Greek epic poem *The Odyssey*, Odysseus' dog, Argos, is the only one to recognize his master after a 20-year absence. Although fictionalized, Homer's account suggests that humans from 2000 years ago believed dogs have reference memory, perhaps even episodic memory. If so, has modern science misrepresented animal cognition? How would a contemporary memory researcher explain this incident?
8. Explain the statement "Episodic memories become semanticized." What are the implications of this statement for the study of declarative memory?
9. Karl Lashley conducted a systematic search for the engram of memory (i.e. where memory is located in the brain) in the 1920s. He lesioned different parts of the rat cortex and then trained the animals to run through a maze. Regardless of where the lesion was placed or how large it was (in some cases he sliced through the entire cortex in back and forth motions with a wire), rats still relearned the maze. Some of them took longer than others and performance was not always perfect, but the data were consistent enough for Lashley to conclude that 'memory is not localized.' His findings were influential: most scientists believed that memory was a general property that emerged from overall brain function. Few people today agree with this conclusion, so how would they reinterpret Lashley's findings? What were the significant scientific advances that led to contemporary theories of how memory is represented in the brain?
10. What evidence supports the idea that LTP is a cellular model of memory? What evidence could be used against this idea?

FURTHER READING

Memory research is a rapidly advancing field so up-to-date books and articles are published regularly. These include edited volumes by different authors that focus on the following topics:

Human memory from a cognitive science perspective:

Baddley, A. D., Eysenck, M. W., & Anderson, M. C. (eds.) (2009). *Memory.* Hove: Psychology Press.

Animal cognition in the natural environment with specific chapters devoted to memory:

Dukas, R., & Ratcliffe, J. M. (2009). *Cognitive Ecology II.* Chicago, IL: University of Chicago Press.

A combination of human and animal research with extensive discussion of the neurobiological basis of memory:

Roediger, H. L., Dudai, Y., & Fitzpatrick, S. M. (eds.) (2007). *The Science of Memory: Concepts*. New York, NY: W. H. Freeman and Company.

Books by a single author are generally more cohesive, although they also tend to present a limited perspective on each topic. An entertaining explanation of how and why human memory may fail is presented in:

Schacter, D. L. (2002). *The Seven Sins of Memory.* Boston, MA: Houghton Mifflin Harcourt.

Kandel's autobiographical account of his life and career provides a captivating description of the social and scientific influences behind his Nobel Prize winning work. This is a 'must read' for anyone interested in how experience can alter ideas and how scientific theories develop over a lifetime.

Kandel, E. R. (2007). *In Search of Memory: The Emergence of the New Science of Mind*. New York, NY: W. W. Norton & Company.

A basic overview of how memory can be examined from an evolutionary perspective using cross-species comparisons may be found in:

Balda, R., & Kamil, A. (1989). A comparative study of cache recovery by three corvid species. *Anim Behav*, **38**, 486–495.

The same topic is developed extensively in a recent review that combines psychology, anthropology, and biology to present a new theory on the evolution of cognition:

Wynn, T., & Coolidge, F. L. (2011). The implications of working memory for the evolution of modern cognition. *Int J Evol Biol*, **2011**, 741357.

An easy to read summary of a controversial and engaging topic by an expert in the field is presented in:

Loftus, E. F. (2003). Make-believe memories. *Am Psychol*, **48**, 867–873.

There are many excellent texts that cover recent developments in the neuroscience of memory. These include:

Eichenbaum, H. (2011). *The Cognitive Neuroscience of Memory*. New York, NY: Oxford University Press.
Squire, L. R., & Kandel, E. R. (2008). *Memory: From Mind to Molecules*. Greenwood Village: Roberts & Company Publishers.

More succinct descriptions are provided in specific chapters of most behavioral neuroscience texts, such as:

Bear, M. F., Connor, B. W., & Paradiso, M. A. (2007). *Neuroscience: Exploring the Brain*. Baltimore: Lippincott, Williams & Wilkins.
Kolb, B., & Whisaw, I. Q. (2011). *An Introduction to Brain and Behaviour*. New York, NY: Worth Publishers.
Pinel, J. P. (2010). *Biopsychology*. Old Tappan: Pearson Education Inc.

4 Associative processes

Background

Like many animals, Japanese quail (*Coturnix japonica*) display stereotypical behaviors during courting. Males puff up their feathers, stretch their necks while they strut around, and emit a vibrating vocalization that lasts for several seconds. In some cases, a male will hold a worm in his beak to entice a female closer; when she approaches, he mounts and attempts to copulate with her. Males exhibit these courtship behaviors when females are present, but also when they are exposed to stimuli previously associated with receptive females. Domjan and his colleagues have studied these conditioned mating behaviors in great detail, showing that male quail prefer females that are the same color as those that they copulated with in the past (Nash & Domjan, 1991) and that they are quicker to mount females if they first encounter stimuli that predict her arrival (Gutierrez & Domjan, 1996). These cues may also affect physiological responses: testosterone levels are elevated (Graham & Desjardins, 1980) and more sperm is released during copulation (Domjan *et al*., 1998) in the presence of sex-related stimuli.

In writing about his own work, Domjan (2002) notes that it is not surprising that sexual behavior is influenced by so many different factors. Without it, no species would survive. So it makes sense that the behavior is flexible enough to be elicited by more than one set of stimuli. The importance of his work is in showing how sexual responses are altered by experience and how this experience can have a positive effect on reproduction. In other words, forming an association between environmental stimuli and sexual experience facilitates short-term behaviors (i.e. courting and copulation), but also increases long-term consequences of these behaviors (i.e. reproductive success). Associative learning may also influence sexual responses in humans, although most of the work on this topic relates to deviant behaviors (Akins, 2004).

Domjan's work builds on a long philosophical tradition. The idea that learned associations affect behavior dates back to Aristotle (384–322 BCE), who formulated a set of principles describing how ideas or thoughts are connected. These were expanded in the seventeenth and eighteenth centuries by the British Associationists, expounding the view that experiences are held together by associative connections. Uncovering the properties of these associations would explain how knowledge is acquired. As noted in the previous chapter, Ebbinghaus tested the laws put forth by the British Associationists, using himself as a subject. Based on these findings, he presented a series of associative laws explaining how memories are formed (Ebbinghaus,1885). With the turn of the twentieth century and the rise of behaviorism, the study of associative mechanisms took hold in North America. Many scientists were also influenced by the pioneering studies of Ivan Pavlov (1849–1936), a Russian physiologist who won a Nobel Prize for his work on the digestive system of dogs but is far better known today for his work on conditioning, a form of associative learning.

Chapter plan

This chapter begins where Pavlov's work took off, summarizing the last century of scientific research on associative processes. The chapter focuses on classical and operant conditioning, which are the two most commonly studied forms of associative learning. In order to understand how research in this area is conducted, the primary experimental paradigms for operant and classical conditioning are described. This is followed by a discussion of how an animal's evolutionary history impacts its ability to form associations. The mechanisms of classical and operant conditioning, as well as the phenomenon of extinction, are outlined as a means to understand how associative learning operates. This leads to an overview of the theories of associative learning. Finally, as with all cognitive processes, associative learning is represented by changes in the central nervous system; there is extensive evidence pointing to specific brain systems that control both classical and operant conditioning. Although the evidence is not complete, these provide working models for future investigations of associative mechanisms.

The majority of research on associative processes has been conducted by experimental psychologists working in the lab, although there are compelling and important studies of both classical and operant conditioning in the natural environment. These tightly controlled lab experiments allowed researchers to identify mechanisms of associative processes, including which environmental conditions produce the strongest conditioning. Ethologists and behavioral ecologists studied some of the same phenomena as a means to understand the evolution of naturalistic behaviors, although they seldom discussed their findings in terms of associative processes. By the 1960s, there was a growing realization that these two approaches are complementary (Hollis, 1997). Today, many researchers consider both proximate and ultimate explanations of associative learning, even if they continue to work only in one tradition.

4.1 Terminology

Every organism lives in an environment surrounded by hundreds, or even thousands, of stimuli. Those with motivational significance will elicit a behavioral response or, more accurately, a combination of responses.

4.1.1 Classical conditioning

Stimuli that do not initially elicit a response (at least not one that can be measured) may acquire the ability to do so through association with motivationally significant events, a process known as classical conditioning. The systematic study of classical conditioning began with Pavlov's work on conditioned reflexes. Indeed, the two are so closely connected that the terms classical and Pavlovian conditioning are used interchangeably. In the simplest version of Pavlov's experiments, a researcher rings a bell and then presents a dog with food (see Figure 4.1). This sequence is repeated and,

Figure 4.1 Ivan Pavlov and his fellow lab mates from the Physiology department at the Military Medical Academy in Petrograd, 1914.

eventually, the dog salivates when it hears the bell, presumably in anticipation of the food. Pavlov identified four components in these experiments that are now common terminology in classical conditioning: the **unconditioned stimulus (US)** food, the **unconditioned response (UR)** salivation, the **conditioned stimulus (CS)** bell, and the **conditioned response (CR)** salivation. Prior to conditioning, the US elicits the UR; this is sometimes called the unconditioned reflex. Following conditioning trials in which the CS is paired with the US, the CR is elicited by the CS, sometimes called a conditioned reflex. In the case of Pavlov's experiments, a dog salivating to the sound of a bell would be a conditioned reflex. Observation of this very simple response was the basis for an entirely new discipline (see Box 4.1).

4.1.2 Operant conditioning

In classical conditioning, a US is presented following the CS, regardless of how the animal responds to the CS. Pavlov's dogs did not need to do anything in order to receive the food after a bell rang. In contrast, an outcome, such as food, is presented in **operant conditioning** *only* if the animal performs a response. In that sense, the definition of operant conditioning is a change in behavior that occurs because the behavior produces some consequence (positive or negative). The prototypical example of operant conditioning is a rat pressing a lever to receive a sugar pellet. But any behavior that produces an outcome, either in the lab or the natural environment, can be described as an operant. This includes a cat pulling a rope to escape from a puzzle box, a rodent swimming to a hidden platform in a water maze, a bird pecking at tree bark to obtain food, or an animal avoiding a location where a predator was encountered in the past.

Thorndike, who pioneered the study of operant conditioning in the lab, noted that animals tend to repeat behaviors that produce satisfying effects and refrain from repeating those that lead to unsatisfying effects (the **Law of Effect**). A premise of this idea is that behavior is controlled by its consequences. This relationship between a behavior and its consequences can be expressed in one of

Box 4.1 Psychoneuroimmunology

As with many scientific 'breakthroughs,' the demonstration that immune responses can be classically conditioned was a chance discovery. In the early 1970s, Ader and Cohen (1975) were working with the immunosuppressant, cyclophosphamide. The drug is used to inhibit natural immune responses in a wide range of clinical conditions (organ transplants, autoimmune disorders), but it produces terrible side effects including nausea. Ader wanted to find a solution to this problem so he allowed mice to drink a flavored saccharin solution prior to an injection of cyclophosphamide. As expected, the mice developed an aversion to the taste of the solution. In order to test the effects of various anti-nausea agents, Alder force fed the saccharin solution to these mice. Unexpectedly, the animals kept dying. At the time, the explanation was not obvious but it is now clear that the saccharin solution was suppressing immune responses in animals that had experienced the saccharin–cyclophosphamide association. Ader demonstrated this experimentally by showing that mice exposed to the flavored saccharin water (i.e. the CS) prior to an injection of foreign cells developed fewer antibodies or a weaker immune response than did mice exposed to water (Ader *et al.*, 1990).

Subsequent work showed that classical conditioning can also *increase* immune system activity. In these experiments, an odor was paired with an injection of the drug interferon. Interferon increases the activity of natural killer cells in the bloodstream, which help the body to fight off infections, viruses, and foreign cells. In both mice (Alvarez-Borda *et al.*, 1995) and humans (Buske-Kirschbaum *et al.*, 1994), an odor CS paired with an interferon injection increased natural killer cell activity, even in the absence of the drug.

From a rather serendipitous finding that classical conditioning affects immune responses, an entirely new discipline emerged. Psychoneuroimmunology examines the interaction between psychological processes, the nervous system, and immune responses. Another way to think about this is how the immune system relates to brain function and how these alter cognition and behavior. An obvious application of this work is the study of immune system disorders and how these impact mental health.

four ways (see Figure 4.2): 1. **Positive reinforcement** describes a contingency between a response and an outcome that increases the probability of the response occurring. Giving a dog a treat when it comes to a call, praising children for doing their chores, or presenting a sugar pellet to a rat when it presses a lever are all examples of positive reinforcement. 2. **Negative reinforcement** also increases a behavior, in this case because it removes an aversive stimulus. Taking medication to relieve a headache or dressing warmly to avoid the cold are behaviors supported by negative reinforcement. Negative reinforcement can be tested in the lab by training animals to escape to another compartment when an aversive stimulus, such as a shock, is turned on. 3. **Punishment** decreases behavior through a contingency between a response and an aversive event. It includes slapping a dog for chewing slippers, scolding a child for making a mess, or fining a driver for parking in a reserved spot. In the lab, shocking an animal for lever pressing is an example of punishment. 4. **Omission** training, sometimes called negative punishment, decreases the probability of responding by withholding an outcome when the response occurs. Timeouts for young children work on an omission schedule: if the child behaves inappropriately, playtime and toys are removed for a short period. In the lab, *not* presenting a food pellet when a rat presses a lever is an example of omission training (in these cases, the animals are initially trained to lever press for food).

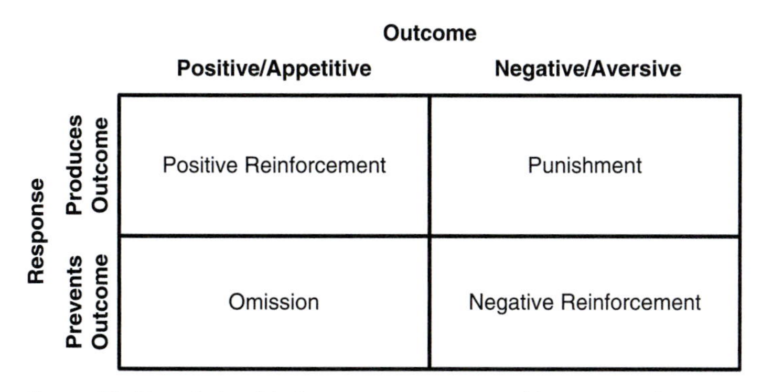

Figure 4.2 The relationship between a response and its outcome in operant conditioning. An operant response may produce or prevent an outcome that is either aversive or appetitive, leading to four different reinforcement relationships.

4.2 Experimental paradigms

Over the last century, most studies of classical and operant conditioning were conducted in the lab using rats or pigeons as subjects. This has changed in the last two or three decades as behavioral paradigms have been developed for other animals, including birds, fish, and invertebrates. Both forms of associative learning have been examined in the natural environment in a variety of species, although these are less common than lab-based studies. This is particularly true of operant conditioning which is almost always studied in the lab.

4.2.1 Classical conditioning

Most research into classical conditioning uses some variation of the paradigms described below.

Conditioned preference

Many USs elicit positive emotional reactions, such as happiness or excitement, that can be conditioned to stimuli predicting their occurrence. Common examples are pleasurable sensations evoked by the scent of a sexual partner or feelings of comfort evoked by the smell of a familiar holiday meal. With few exceptions, organisms will approach stimuli that elicit positive emotional reactions so classical conditioning is often manifested as approach responses to stimuli associated with rewards, such as food. This phenomenon was documented in carrion crows (*Corvus corone*), which approached and explored mussel shells after a small piece of beef was placed under the shell (Croze, 1970). The dependent measure, approach behavior, was assessed by examining crow footprints in the sand (see Figure 4.3). Initially, crows ignored the shells but after they (serendipitously and repeatedly) uncovered the beef, birds began to head directly to the shells. Insects, including many species of bumblebees, adjust their foraging behavior in the same way, approaching flowers of a particular color or shape that have been associated with higher nectar content in the past (Laverty, 1980).

In these field studies, crows and bumblebees developed conditioned approach responses to classically conditioned stimuli. The same phenomenon is measured in the lab using a conditioned place preference task in which animals prefer to spend time in an environment where they

Figure 4.3 Crow footprints in the sand.

experienced natural rewards, such as food, water, or sexual interactions (non-natural rewards including drugs and artificial sweeteners produce the same effect). Importantly, conditioned preference tests are conducted in the absence of the US. Pavlov's dogs salivated when they heard a bell, even if food was no longer presented. Similarly, rats return to a place where they encountered a sexually receptive partner that is no longer present. In conditioned preference tests, the latency to approach the CS, the number of CS approaches within a given time period, and the time spent in close proximity to the CS are all indices of classical conditioning. Conditioned taste preference is a variation on this methodology, measuring increased consumption of a flavored food that was paired with a natural reward (e.g. sucrose). Using these paradigms, preference conditioning has been documented in a wide range of mammalian species as well as invertebrates such as California sea snails (*Aplysia californica*), nematodes (*Caenorhabditis elegans*), and common fruit flies (*Drosophila melanogaster*).

Conditioned emotional reactions

Negative emotional reactions, such as anxiety or fear, may also be conditioned to stimuli predicting a US. Some people feel nervous or uneasy when they enter a dentist's office or an examination hall, based on their previous experience in these environments. Animals display similar responses to stimuli associated with aversive events. For example, many emit alarm calls and/or escape responses when they encounter a predator; the sight, sound, or smell of these predators will, over time, elicit similar responses. In addition, if rodents or other small mammals mistakenly touch a sharp object or eat something that makes them ill, they respond to these stimuli by vigorously digging around the object and covering it with dirt or bedding. Importantly, this response occurs when the US is no longer present (i.e. they do not touch the object or taste the food), confirming that the CS is eliciting the CR.

In the lab, conditioned emotional reactions are commonly measured using the conditioned fear paradigm. These tests capitalize on the fact that frightened rodents often freeze (Blanchard &

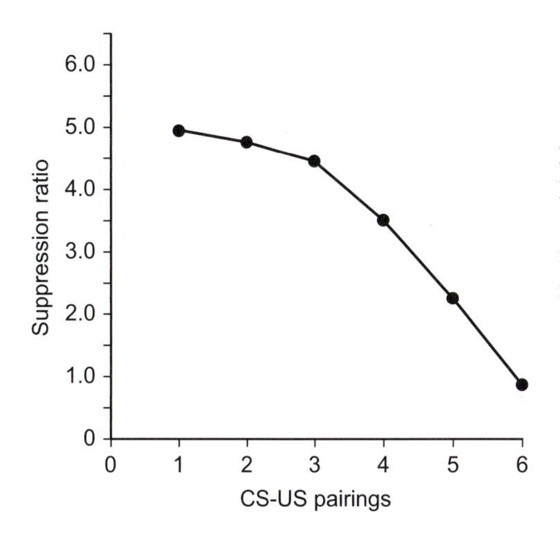

Figure 4.4 Conditioned suppression data. The conditioned suppression ratio is a common laboratory measure of classical conditioning in rodents. Lever pressing rates are recorded during the presentation of a conditioned stimulus (CS), such as a tone, that was previously paired with an aversive unconditioned stimulus (US), such as a shock. A suppression ratio of 0.5 indicates no change in responding when the CS is presented, thus no conditioning. With repeated CS–US pairings, the conditioned suppression ratio declines indicating that lever pressing is reduced during the CS.

Blanchard, 1969), particularly when escape is unlikely (Eilam, 2005). This may be an adaptive trait as mice that freeze when they detect a predator (before they are detected themselves) are less likely to be attacked and killed (Hirsch & Bolles, 1980). In the lab, freezing as a conditioned response is quantified by measuring the period of immobility (i.e. the CR) following the CS presentation. The conditioned suppression paradigm is a variation on this measure. First, rodents are trained to lever press for food; once lever pressing is stable, classical conditioning trials are instituted in which a CS (usually a tone or light) is paired with a shock. When animals freeze, they stop lever pressing so the number of lever presses that occur during the CS, versus the number of lever presses that occur during non-CS periods, is a measure of conditioned fear. This **suppression ratio**, as shown in Figure 4.4, is typically calculated as:

LP during CS/(LP during CS + LP during an equal period of time preceding CS)
LP = lever presses

Thus, a suppression ratio of 0.5 would indicate no reduction in responding and no conditioning, whereas a suppression ratio of 0 would reflect complete freezing during the CS and maximal conditioning.

Conditioned fear is typically studied in humans by pairing a neutral stimulus, such as a tone, with a sudden, loud burst of noise. With repeated pairings, the tone elicits a startle response, which is measured by electrodes placed on contracting muscles over the body. Conditioned emotional responses in this paradigm can also be measured as changes in heart rate, blood pressure, and galvanic skin response. The same responses can be measured in humans navigating through a virtual environment containing snakes, spiders, or other aversive stimuli (see Figure 4.5).

Conditioned taste aversion

Many people feel nauseated when they smell or taste a particular food. This often reflects a **conditioned taste aversion (CTA)** that developed when consumption of the food was associated with gastric illness, usually vomiting. It does not matter if the food itself *caused* the sickness; a coincidental food–nausea association produces a CTA. For example, an individual who eats a banana when they have a stomach virus may later have an aversive reaction to the smell and taste of this fruit, even if it was not the banana that made them sick. CTAs are easily observed in lab animals: after consuming a flavored substance, animals are made sick by injecting them with a

Figure 4.5 Schematic of a virtual reality setup to test conditioned fear in humans.

nausea-producing agent, such as lithium chloride, or exposing them to low-level gamma radiation. When they recover from the illness, animals are presented with the flavored food. If animals eat less of this food than another food that was not associated with illness, they have developed a CTA.

CTAs can be very powerful: they often develop with a single CS–US pairing and are sustained for long periods of time. This should not be surprising as poison avoidance learning is critical for survival. Animals can become very sick or die if they consume inedible items, so they must learn to avoid poisons and other toxins by associating the smell and taste of the food with illness. This is one reason that it is so difficult to develop effective poisons for rodents. After sampling a very small amount of the novel food, rats and mice feel ill and avoid this food in the future.

Note that the measure of classical conditioning (the CR) varies across these three types of paradigms. Given that there are dozens, or more accurately hundreds, of other classical conditioning tests, the number of ways in which scientists can measure this process is almost endless. In addition, for each CR, a researcher may choose to measure how quickly it is acquired, how large the response is once conditioning has occurred, and how long it lasts when the US is no longer present. This sometimes makes it tricky to compare the magnitude of conditioning across studies. Researchers (and students reading these studies) must pay close attention to the dependent measure in each study because discrepancies in research findings can sometimes be explained by differences in how the CR is assessed.

4.2.2 Operant conditioning

As noted previously, contemporary studies of operant conditioning can be traced back to Thorndike's experiments with cats. These animals were required to pull a rope, move a stick, or push a board to escape from a puzzle box. Operant conditioning (or "trial-and-error learning" as Thorndike called it) was evidenced by a decrease in the escape time over trials.

Discrete trials

This discrete trials setup, in which subjects have the opportunity to make one correct response for each time they are placed in a testing apparatus, is still common in lab-based research. The water maze, T-maze, and straight alleyway are all discrete trials operant conditioning paradigms, although most of these tests are used to assess other cognitive processes, such as spatial learning or decision making. A simple example of operant conditioning using discrete trials is the **conditioned avoidance** paradigm. In one version of this test, a rat is shocked for stepping off a platform; longer latencies to step from the platform on subsequent trials indicate better conditioning. Unlike the conditioned freezing paradigm described above, the presentation of the shock in the conditioned avoidance paradigm *depends* on the animal's behavior: if it does not step off the platform, it will not receive a shock. In classical conditioning tests such as conditioned fear, the US follows the CS regardless of what the animal does.

Free operant

When most students and researchers think about operant conditioning, they imagine a rat pressing a lever for food in a small chamber. This setup was designed by Skinner in the 1940s as a means to evaluate the ongoing responses of his subjects (usually pigeons or rats). Animals are placed in a chamber containing some manipulandum, usually a lever, which can deliver a food or sugar pellet. In contrast to the discrete trials paradigms, free operant methods allow animals to respond repeatedly (freely) once they are placed in the experimental chamber. One advantage of this method is that the experimenter can observe variations in responding across time. This information is represented on a cumulative response record such as those shown in Figure 4.6. The y axis of the graph represents the

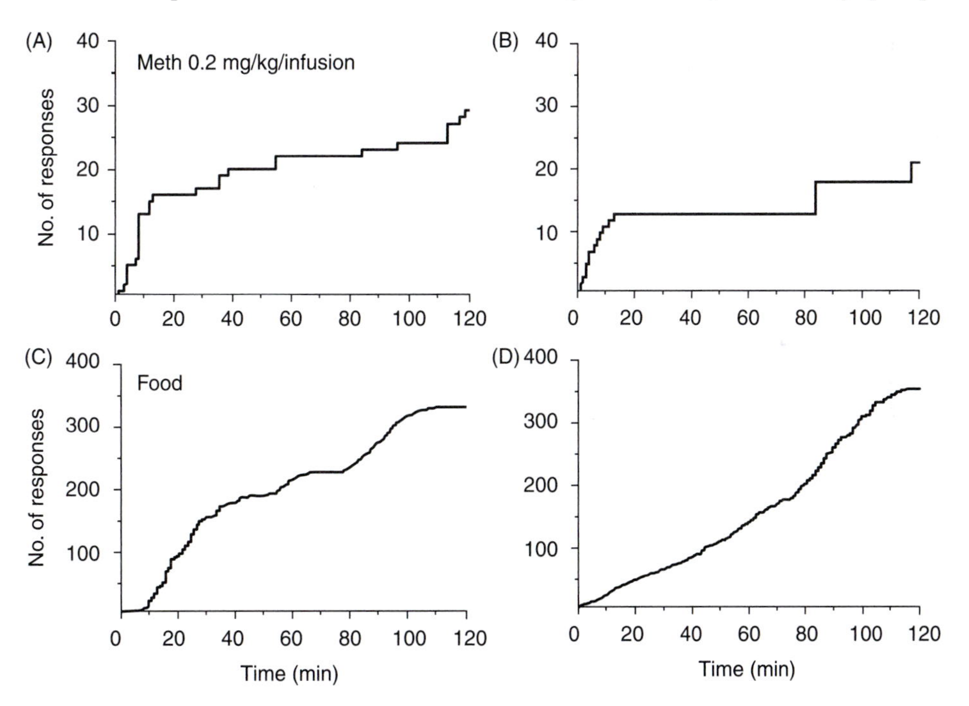

Figure 4.6 Cumulative response records for rats responding for methamphetamine (A and B) or food (C and D). Food reinforcement produced a steady rate of responding over time whereas methamphetamine produced a step-like pattern in which a drug infusion (i.e. presentation of the reinforcer) was followed by a pause in responding (A). When animals are responding for drug, they load up at the beginning of the session, presumably to attain an optimal level of drug in their system (A). This effect is not observed when animals are responding for food (C). Administration of the glutamate antagonist MTEP (B and D) led to a near-complete cessation of methamphetamine-reinforced responding (B) but had no effect on responding for food (D). The rapid lever presses that occur at the beginning of the methamphetamine session (B) indicate that animals are 'expecting' a drug infusion and may be frustrated by the lack of drug effect. Animals stopped responding for over an hour, tried the lever a few times, and stopped responding for another 40 minutes. This pattern is consistent with the idea that MTEP is blocking the reinforcing effect of methamphetamine.

total number of responses in the session and the x axis indicates time. Thus, the cumulative record provides a visual representation of when and how frequently the animal responds during a session.

In free operant paradigms, the relationship between responding and reinforcement is described by the **reinforcement schedule**. This is a rule (set by the experimenter) that determines how and when a response will be followed by a reinforcer. When every response produces a reinforcer, the schedule is called **continuous reinforcement** or CRF. More commonly, responding is reinforced on a partial or intermittent schedule. One of the simplest ways to produce partial reinforcement is to require animals to make a certain number of responses for each reinforcer. If the number is set, the schedule is called **fixed ratio (FR)**; if the number of required responses varies about a mean value, the schedule is called **variable ratio (VR)**. Thus, FR5, FR10, and FR50 schedules require animals to make exactly 5, 10, and 50 responses before the reinforcer is presented whereas VR5, VR10, and VR50 schedules require animals to make an *average* of 5, 10, and 50 responses. A progressive ratio (PR) schedule is a variation of an FR schedule in which animals must make an increasing number of responses for successive presentations of the reinforcer (Hodos, 1961). Typically, the schedule is set up so that animals make one, then two, then four, then eight responses for the first four reinforcers. The PR values continue to increase until animals stop responding. This break point is a measure of motivation or how hard animals will work for a single presentation of the reinforcer.

In contrast to ratio schedules, interval schedules provide reinforcement if a response occurs after a certain period of time. Under **fixed interval (FI)** schedules, the time from the presentation of one reinforcer to the possibility of receiving the next is constant. Under **variable interval (VI)** schedules, responding is reinforced after an average time interval has passed. For example, animals responding under an FI-15s schedule would be reinforced for the first response they make 15 seconds after the last reinforcer. Under a VI-15s schedule, reinforcement would be available *on average* 15 seconds after the delivery of the last reinforcer. Note that animals must still respond under interval schedules, but they are only reinforced for responses that occur after the interval has elapsed. Situations in which a reinforcer is presented at specified intervals, regardless of the animals' behavior, are called time schedules.

Schedules of reinforcement induce different patterns of responding, suggesting that the animals' behavior is determined (at least partially) by its payoffs. For example, acquisition is usually more rapid and declines more quickly under CRF than partial reinforcement schedules. This makes sense in that it is easier to form a response–reinforcer association if these always occur together and the behavior is more likely to cease when the reinforcer is removed if animals were reinforced for every response in the past. Even under partial reinforcement schedules, different patterns of responding emerge (see Figure 4.7 for details of each schedule). FR schedules are characterized by rapid responding up to the presentation of the reinforcer. Responding ceases after reinforcement is delivered and this post reinforcement pause increases with the size of the ratio. With very high FR ratios, animals may stop responding altogether, a phenomenon known as **ratio strain**.

FI schedules also induce post reinforcement pauses that vary directly with the length of the interval, presumably because animals learn that the reinforcer will not be available for a certain period of time. Responding recovers close to the completion of the interval to produce a 'scallop' in responding that peaks when the next reinforcer is obtained (see Figure 4.7). In general, response rates are steady under variable schedules, reflecting the fact that animals do not know when the next reinforcer will be delivered. This pattern can be interrupted by rapid bursts of responding that often occur after animals receive several reinforcers in a row. When the reinforcement payoff is low (i.e. high VR or VI schedules), response rates are reduced, responding becomes more sporadic, and pauses occur at irregular intervals. This makes sense in terms of what the animal has learned: reinforcement is infrequent and unpredictable. In other cases, an animal's behavior appears to be influenced by incorrect assumptions about response–reinforcer payoffs (Box 4.2).

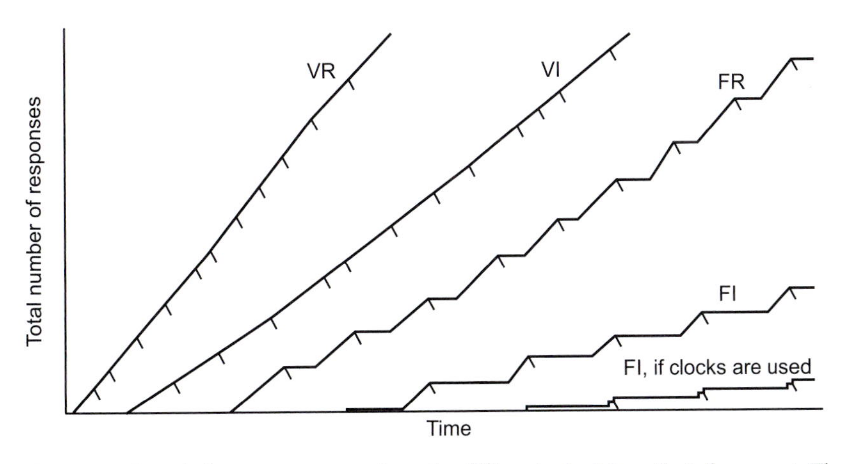

Figure 4.7 Cumulative response records under different schedules of reinforcement. The delivery of a reinforcer is marked by a vertical slash. Different schedules of reinforcement elicit different patterns of responding. FI responding in humans is very accurate if they can use clocks to time the intervals (bottom line to right of figure).VR = variable ratio; VI = variable interval; FR = fixed ratio; FI = fixed interval.

Box 4.2 Superstitious behavior

The defining feature of operant conditioning is that the presentation of the reinforcer depends on the organisms's response. In some cases, however, accidental pairings of a response with a reinforcer will increase the probability that the response will occur in the future, even if there is no causal connection between the two. Skinner (1948) observed this phenomenon in pigeons by presenting food at brief and regular intervals as animals were moving around a cage. Over trials, the birds developed responses that preceded the food delivery, but these differed across birds: one bird turned in counterclockwise circles, one made pecking motions at the floor, two swung their heads back and forth, one raised its head to one corner of the cage, and another bobbed its head up and down. Skinner surmised that each of these animals had been engaged in a particular behavior (e.g. head-bobbing) when the food was delivered, which reinforced the response that preceded it. As trials progressed, the birds were more likely to perform that particular response, which further increased the probability that it preceded the food presentation. And the cycle continued.

Skinner described behaviors that were maintained in the absence of a response–reinforcer contingency as superstitious. His results were replicated, but then reinterpreted based on a more detailed analysis of pigeon behavior (Staddon & Simmelhag, 1971). Specifically, it became apparent that pigeons were able to calculate the time between food presentations, engaging in some extraneous behavior during this interval, and then moving to the food hopper and pecking close to the time of food presentation. Behaviors that occurred during the food–food interval were often specific to individual birds, which is probably what Skinner was observing.

Skinner's findings aside, there is no doubt that many humans exhibit superstitious behaviors from walking around ladders to searching for four leaf clovers to wearing a particular piece of clothing at a sports competition. Clearly, not all coincidental occurrences of a response and a reinforcer lead to superstitious behavior, so researchers began to investigate the factors that produce this false belief in humans (Gmelch & Felson, 1980; Vyse, 1997). These studies concluded that superstitious behaviors are maximized when an infrequently performed response precedes the presentation of a highly valued outcome and the cost of engaging in the behavior is very low, but the payoff is potentially very high. Even under these conditions, the time and energy devoted to inconsequential responses make superstitious behaviors an evolutionary paradox. One explanation is that causality judgments in humans evolved to take into account incomplete information and superstitious behaviors are a side effect of this adaptation (Foster & Kokko, 2009). Moreover, superstitions are often acquired through social learning which influences how individuals respond in particular situations (Abbott & Sherratt, 2011). If a young boy *never* steps on a sidewalk crack and his mother *never* breaks her back, it helps to confirm his prior belief about this association. Regardless of how they appear, superstitious behaviors in humans are not likely to disappear. After all, why change something that is working?

Outside of the laboratory, different reinforcement schedules can be used to shape human behavior. For example, many sales reps receive a bonus for the number of items they sell (e.g. cars or insurance policies) which serves to increase the rate and number of sales. Lottery terminals in casinos are set up on variable schedules to induce high rates of responding with few breaks. Gamblers often note that they are reluctant to abandon a terminal after a series of losses because they are convinced that it is just a matter of time until their machine pays off. Unfortunately, sudden high wins help to reinforce this behavior, thereby promoting further gambling.

4.3 Associative processes as adaptations

Associative processes have been documented in almost every species studied, suggesting that they have been conserved across evolution. Nonetheless, there are many cases in which classical or operant conditioning does not occur (or is not observed according to some predefined measure). Not *all* of Pavlov's dogs developed CRs and many researchers are frustrated by the inability to train lab animals in simple operant tasks. In some cases, these discrepancies reflect methodological issues, such as an inability to identify factors in the lab that may influence responding. In many cases, associative learning is slow to develop or difficult to measure because the stimuli and/or responses used in an experiment are not meaningful in terms of the subjects' evolutionary history. To appreciate this point, it is important to conceptualize both classical and operant conditioning as adaptive traits (i.e. the outcome of natural selection).

4.3.1 Classical conditioning

The adaptive value of classical conditioning relates to the ability to predict significant events. If one stimulus reliably precedes another, organisms can use this information to their advantage. An obvious example is the ability to predict where and when food will be available. This was demonstrated in an experiment with American grasshoppers (*Schistocerca americana*) (Dukas & Bernays, 2000) given access to two types of food, one more nutritious than the other. For one group, the more nutritious food was associated with a particular spatial location, taste, and color (the other group had these changed every day). Both groups were able to feed on the two types of food, but insects that experienced the environmental cues paired with the better food group grew more

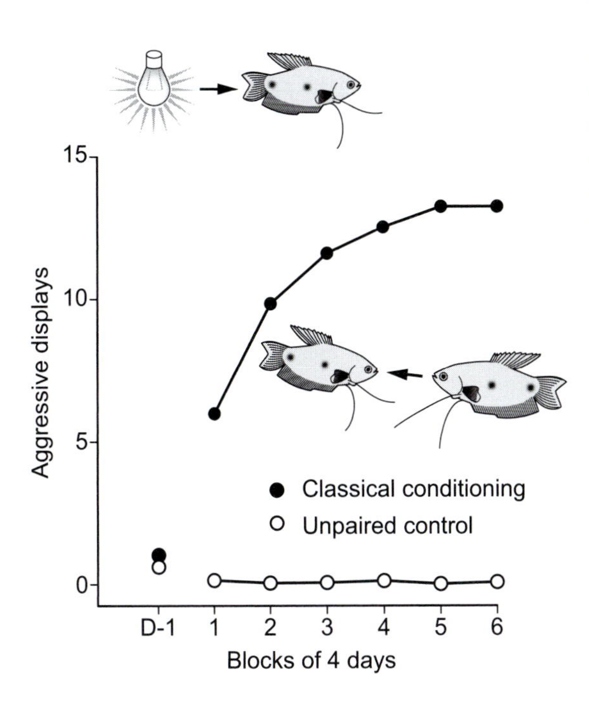

Figure 4.8 Classical conditioning in blue gourami fish. One group of male fish (classical conditioning) was presented with a light prior to the entry of a rival male into their territory. An unpaired control group experienced the same number of light presentations and rival entries, but these were not paired. On a subsequent test, the classical conditioning group showed more aggressive displays than did the control group (Hollis, 1984).

quickly. Thus, the ability to form an association between stimuli and motivationally significant events (i.e. classical conditioning) led to larger, probably healthier animals, which likely provided fitness advantages in the future.

Similar results have been obtained with more complex behaviors. For example, Hollis (1984) demonstrated that classical conditioning improves territorial aggression in male blue gourami fish (*Trichopodus trichopterus*). In this experiment (see Figure 4.8), one group of fish learned that a red light predicted the entry of a rival male. In contrast to fish that did not learn this association, the conditioned animals exhibited more bites and tail beating, almost always winning the fight and retaining their territory. When the same stimulus predicts the arrival of a female, male gourami fish exhibit courtship behavior. They also spend more time nest building, spawn more quickly, and produce more young than males who are exposed to a female with no prior signal (Hollis et al. 1997). All of these examples provide ultimate explanations for classical conditioning by illustrating how this cognitive process increases survival and reproduction.

4.3.2 Operant conditioning

Operant conditioning provides a different kind of fitness advantage, one that allows organisms to adjust their behavior based on its consequences. In many cases, operant conditioning serves to shape or modify pre-existing tendencies. The fixed action patterns described in Chapter 1 (Section 1.4) are behaviors that are triggered by stimuli in the environment (e.g. a red dot on a parent's beak elicits pecking in European herring gull (*Larus argentatus*) chicks), but these responses change with experience as chicks become more efficient at pecking (Hailman, 1967). The idea that cognitive processes supporting operant conditioning are inherited is supported by studies with human infants: babies as young as 2 months of age learn to kick vigorously to activate a mobile lying above their

Researcher profile: **Professor Karen Hollis**

Figure 4.9 Dr. Karen Hollis.

Karen Hollis' scientific studies are a model of comparative cognition research. Her initial training in psychology (BA) at Slippery Rock State College, Pennsylvania, was combined with a biological perspective during her PhD at the University of Minnesota. These two influences helped to shape her scientific perspective, which incorporates experimental manipulations into the systematic study of naturalistic behaviors.

The focus of Hollis' work is understanding how animals learn to predict biologically relevant events such as food, mates, or predators, in order to interact effectively with their environment. In Hollis' words: "The point of my research is to see how what psychologists say about learning can be brought to bear on what zoologists study in the field." This groundbreaking work is a regular feature on science shows in the media, including television, radio, and webcasts. Importantly, Hollis' initial studies identifying how classical conditioning may increase the survival and reproduction of fish has been extended to a variety of species. In that respect, Hollis' research program epitomizes the comparative approach: she and her colleagues have published experimental papers using ants, fish, dogs, or people as subjects.

Karen Hollis is currently a professor in the Interdisciplinary Program in Neuroscience & Behavior at Mount Holyoke College South Hadley. She also spends time as a visiting scientist in European labs (including Oxford, Cambridge, and the University of Paris) and is a regular speaker at conferences and symposia throughout North America.

head (Rovee-Collier, 1999). As with birds, parent–child interactions are also facilitated by operant conditioning. Babies provide social feedback, such as smiling and cooing, increasing the nurturing and attention that parents direct toward them. The advantages of this exchange are mutual; children experience better care and parents benefit from the increased health and potential fitness of their offspring.

4.3.3 Adaptive specializations

The adaptive significance of associative processes became apparent in experimental lab work in the second half of the twentieth century. Prior to that time, many psychologists adhered to the principle of **equipotentiality**, the idea that associations between different stimuli, responses, and reinforcers could be formed with equal ease. For example, Pavlov proposed that any stimulus could be used as a CS in his experiments and Skinner claimed that animals could learn any operant response (as long as it was physically possible) for any reinforcer. According to these scientists, the results of their experiments could be generalized to all instances of associative learning (and many believed to all cases of learning). This is clearly not true: some associations are easy to learn and others are not. A CTA often develops following a single CS–US pairing and may last for months or even years. In contrast, some stimuli never elicit a response, even if they are paired with a US on hundreds or thousands of occasions. The relative ease with which animals acquire certain associations is referred to as **adaptive specializations**, reflecting the idea that learning an association between two stimuli has conferred some evolutionary advantage on the species.

One of the most elegant examples of adaptive specialization comes from an experiment by Garcia and Koelling (1966), illustrated in Figure 4.10. In the first part of the experiment, rats were

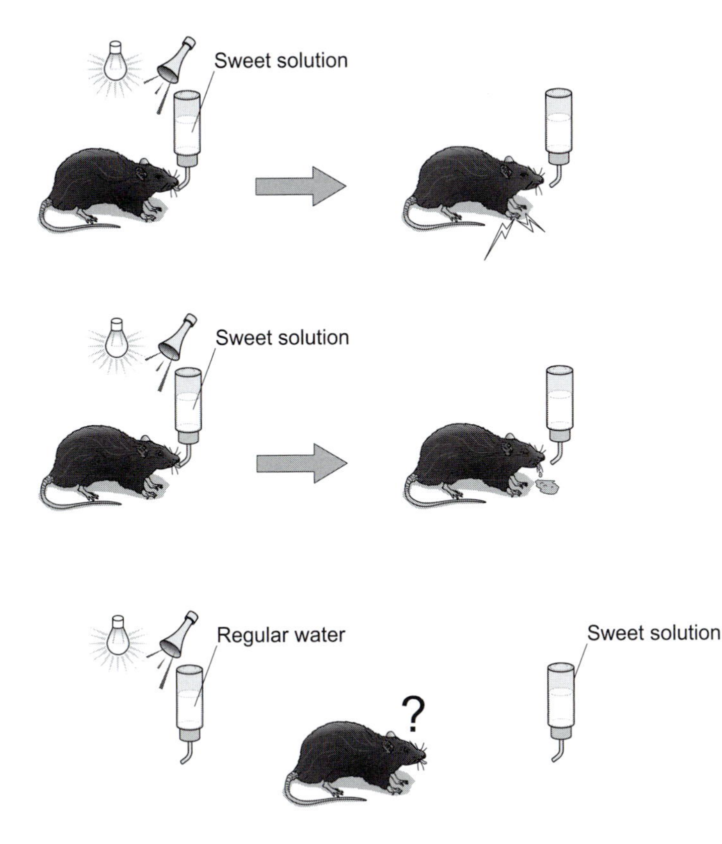

Figure 4.10 Experimental design for Garcia and Koelling's study demonstrating adaptive specializations in rats. All animals underwent classical conditioning trials in which a compound conditioned stimulus (CS) was presented while they drank sweetened water. The two CSs were the taste of the water and an audiovisual stimulus (light plus tone). Half of the animals received a shock (unconditioned stimulus; US) as they were drinking (top panel); the other half were made ill hours after drinking (middle panel). During testing (bottom panel), animals were presented with two drinking bottles: one containing a flavored water solution and one containing plain water linked to an audiovisual stimulus.

presented with a drinking tube containing flavored water. Every time the rats licked the tube, a brief audiovisual stimulus was presented that consisted of a clicking sound and a flash of light. Animals were then either shocked or made ill by x-ray treatment. Thus, all rats experienced the same CS (flavor plus light–sound combination) with half receiving a shock US and half receiving the x-ray treatment that made them ill. In the subsequent test, animals were presented with two licking tubes, one that contained flavored water and one that contained plain water linked to the light–sound cue.

The results of this experiment are shown in Figure 4.11. The group that was made ill avoided the flavored water and drank from the plain water tube that activated the audiovisual cue. In contrast, the shocked group avoided the tube linked to the audiovisual cue and drank the flavored water. The interpretation of these results is that rats have a tendency to associate a flavor with illness and a light–sound stimulus with shock. The same effect is observed in 1-day-old rat pups (Gemberling & Domjam, 1982), suggesting that the disposition is inherited, rather than acquired through experience. In a rat's natural environment, gustatory cues are a good predictor of food that may be poisonous, whereas audiovisual cues are a better predictor of environmental danger, such as predation. Animals that learned the predictive significance of these stimuli would be more likely to survive and reproduce, explaining why contemporary rodents acquire these associations so easily (Rozin & Kalat, 1971). Adaptive specializations are apparent in other species: pigeons rely heavily on vision for foraging, so they are more likely to associate food with visual, than auditory, stimuli (Lolordo *et al.*, 1982).

Thorndike was ahead of his time in recognizing the importance of adaptive specializations in operant conditioning. He tested whether cats could be trained to yawn or scratch themselves in order to escape from a puzzle box, concluding that these associations did not 'belong' together in the

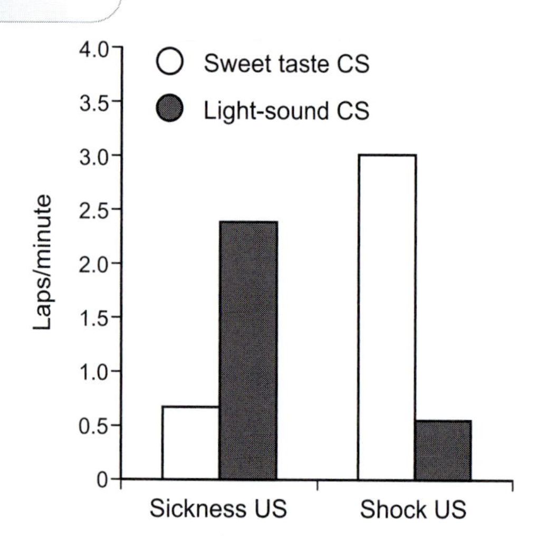

Figure 4.11 Results of Garcia and Koelling's experiment demonstrating adaptive specializations in rats. Bars indicate how much rats licked drinking tubes that contained a flavored solution (taste conditioned stimulus; CS) or produced an audiovisual CS. Rats that experienced sickness as an unconditioned stimulus (US) licked the tube containing the flavored (sweet taste) CS less frequently, whereas those that experienced the shock US showed fewer licks of the tube that produced the audiovisual (light–sound) CS.

animal's evolutionary history. This foreshadowed many unsuccessful attempts to train animals in operant paradigms, such as Hershberger's attempt to train chicks to run *away* from a food bowl (Hershberger, 1986), Bolles' attempts to train rats to stand on their hind legs to avoid a shock (Bolles, 1970), and the Brelands' difficulty in training animals to perform circus tricks if the trained response was incompatible with the animal's natural behavior (Breland & Breland, 1961). The phenomenon was investigated more thoroughly in a series of studies with golden hamsters (*Mesocricetus auratus*) (Shettleworth, 1975, 1978; Shettleworth & Juergensen, 1980). In these experiments, hamsters moved freely in a familiar environment and observers recorded when they engaged in naturalistic behaviors such as rearing, digging, or face washing. Different animals were presented with a food reward immediately after performing one of these behaviors. Despite consistency in reinforcement presentation, only some of the behaviors increased over training. That is, scratching against the walls or digging in the flooring sawdust increased when followed by food, but face washing and self grooming did not. Punishment (a mild electric shock) also had differential effects on behavior, some responses declined with punishment but some did not change.

All of these researchers discussed their findings in the context of evolution: operant responses that do not serve an adaptive purpose are difficult to acquire whereas those that enhance species' survival are easily acquired. Sevenster (1973) demonstrated this principle in male three-spined stickleback fish (*Gasterosteus aculeatus*) that were trained to bite a rod or swim through a ring to gain access to another fish (see Figure 4.12). When males could gain access to another male, the biting increased but swimming through a ring did not. The opposite occurred with access to a female: swimming through a ring increased, but biting the rod did not. The finding that access to another male fish is an effective reinforcer for biting, and that access to a female is an effective reinforcer for ring swimming, fits with the animal's evolutionary history. Biting is a component of aggressive behavior that occurs when a resident male encounters an intruder whereas swimming through a ring is more characteristic of fish courting behavior, characterized by males rapidly chasing females as they dart through the water.

Adaptive specializations may also explain some irrational fears in humans. People readily acquire exaggerated fear responses to stimuli associated with threats in the natural environment, and many phobias likely stem from an evolved tendency to fear objects or situations that were dangerous to human ancestors. This explains why humans and monkeys associate pictures of snakes or spiders

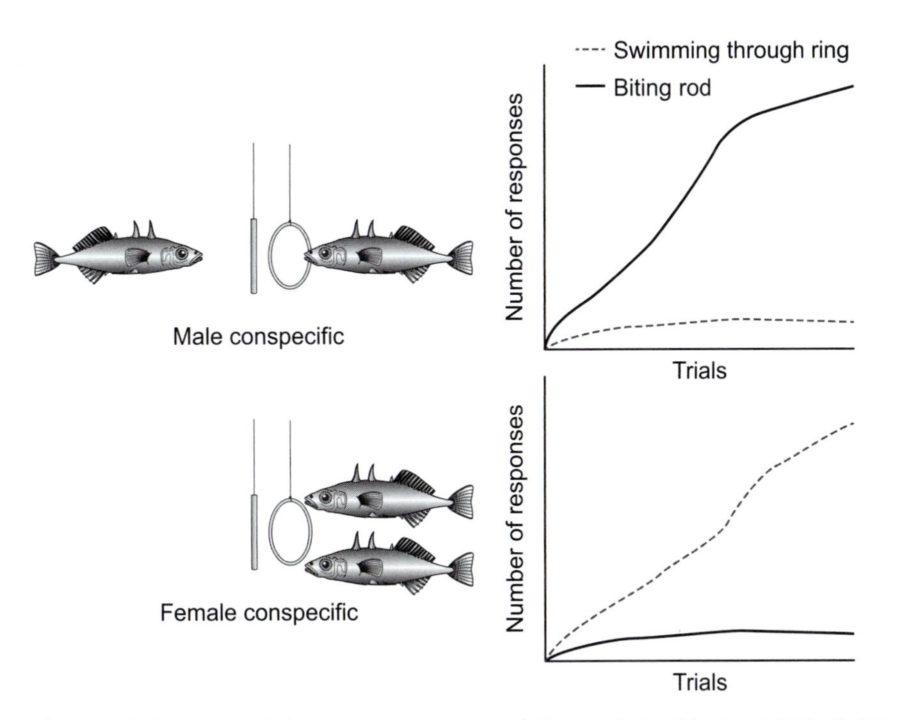

Figure 4.12 Experimental design to test response–reinforcer relationships in stickleback fish. Male fish were trained to either bite a rod or swim through a ring to gain access to male or female conspecifics.

with shock more readily than pictures of flowers or houses (Öhman *et al.*, 1985). As noted in Chapter 3, this effect is not a simple distinction between threatening and non-threatening stimuli: humans are more likely to develop conditioned responses to stimuli that predict snakes or spiders than those that predict guns or electric outlets (Öhman & Mineka, 2001).

By the end of the twentieth century, most scientists accepted that adaptive specializations impact associative learning. Many researchers still focus exclusively on proximate causes of classical or operant conditioning, but their research questions are often framed within an evolutionary perspective. For example, a scientist may investigate which brain regions mediate associative learning, and then ask why these neural systems are preserved across evolution. If nothing else, researchers can capitalize on the fact that it is much easier to train animals in classical or operant tasks when one works with, rather than against, adaptive specializations.

4.4 Mechanisms

Adaptive specializations provide an ultimate explanation of associative learning in that being able to predict when one stimulus follows another or which outcome will follow a response should help animals to survive and reproduce. In contrast, proximate explanations describe associative learning in terms of causal factors that produce optimal conditioning. These include whether one stimulus is expected to follow another, how closely the two occur in time, and how easy they are to detect. Proximate mechanisms of associative learning also include physiological processes, a topic that will be discussed in Section 4.6.

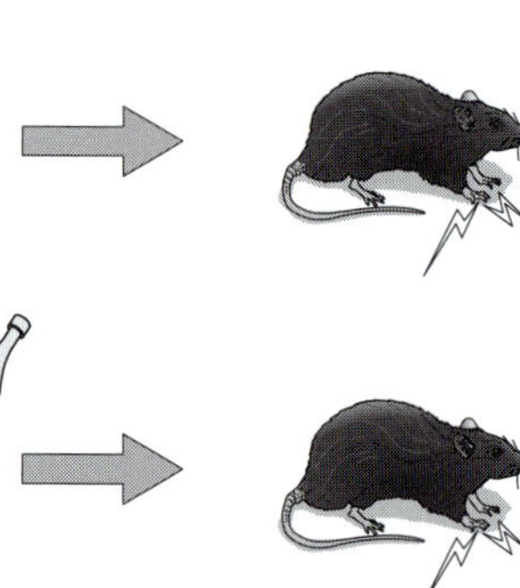

Figure 4.13 Experimental design for Kamin's (1969) blocking experiment. The blocking group underwent simple conditioning trials in which a single conditioned stimulus (CS) was followed by an unconditioned stimulus (US). For example, a light preceded a shock. These animals then experienced compound conditioning trials (light plus tone) followed by a shock. Responses to the tone were measured during testing. The control group underwent the same procedure with the exception that the first phase consisted of CS and US presentations that were not paired.

4.4.1 Informativeness

It seems intuitive that the more frequently two stimuli are paired, the stronger the association between them will be. But contiguity alone cannot explain associative learning, a realization that came about following a classic experiment by Kamin (1969). In this study (see Figure 4.13), one group of rats underwent a classical conditioning protocol in which a CS (tone) preceded a shock (US). As expected, a freezing response (CR) developed to the tone. Following these trials, a light was presented at the same time as the tone and both were followed by the US. Stimuli presented simultaneously are called compound stimuli and are labeled individually as CS1 (tone), CS2 (light), etc. A separate group of control rats experienced CS–US pairings with the compound stimulus (tone and light combined) but had no prior exposure to either stimulus alone.

In the test session, Kamin compared CRs to the light in the experimental and control groups. Note that all animals experienced exactly the same number of trials in which the light preceded the shock.

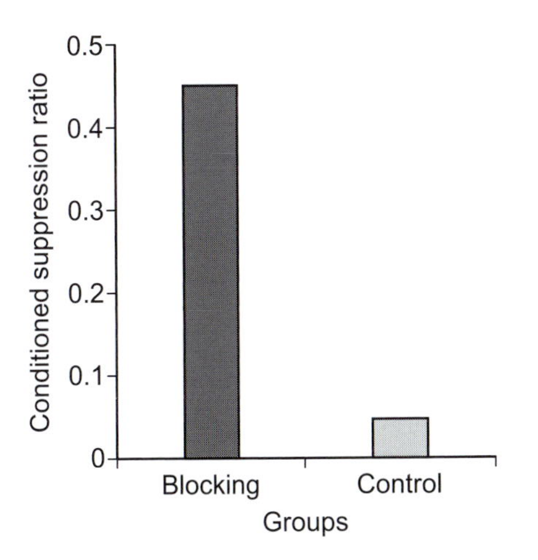

Figure 4.14 Results of Kamin's (1969) blocking experiment. Conditioning to the stimulus added during compound conditioning trials (CS2) was measured in a conditioned suppression experiment with the blocking group showing almost no suppression of responding (i.e. no freezing to the light). In contrast, the control group showed marked conditioned suppression indicating that, unlike the blocking group, they had formed an association between CS2 and the unconditioned stimulus (US).

Thus, if the frequency of CS–US pairings alone determines classical conditioning, the CR should be the same in control and experimental groups. Of course, this is not what happened. Figure 4.14 shows that the control group exhibited robust freezing to the light, whereas the experimental group did not. In Kamin's terms, the initial tone–shock pairings *blocked* subsequent conditioning to the light. He discussed his findings in terms of 'surprise,' noting that the animals' previous experience with the tone made the light irrelevant as a predictor of the US. Thus, in a **blocking** experiment, CS2 conveys no new information about the occurrence of the US so conditioning does not occur.

Kamin's experiment has been replicated thousands of times with different stimuli and different organisms, confirming that the frequency of CS–US pairings cannot, in itself, explain classical conditioning. Originally it appeared that blocking was limited to vertebrates, as insects, at least western honeybees (*Apis mellifera*), did not show the effect (Bitterman, 1996). Subsequent studies modified the paradigm so that the two CSs were in the same modality (either odor–odor or color–color) or in different modalities with the first (blocking) CS being more salient. Under these conditions, honeybees exhibit blocking (Couvillon *et al.*, 1997, 2001) although it is still not clear whether this is mediated by the same process in different species.

Another way to reduce the predictive value of a CS is to present it alone, prior to any CS–US pairings. This describes the phenomenon of **latent inhibition** in which previous exposure to the CS, in the absence of the US, retards subsequent conditioning to the CS. (The term CS pre-exposure is often preferred to latent inhibition because it does not infer that inhibitory processes underlie the behavioral effect.) Organisms first learn that the CS has no motivational significance; they must then inhibit or overcome this information when the CS is presented with the US at a later time. Both blocking and latent inhibition fulfill an important biological function in that they limit cognitive processing of stimuli that are meaningless to the organism.

4.4.2 Temporal contiguity

It seems intuitive that it would be easier to form an association between two stimuli if they occur close together in time. Many laboratory experiments confirmed this assumption: a US that immediately follows a CS and a response that immediately produces a reinforcer induce robust conditioning. Researchers went on to demonstrate that classical conditioning is reduced when the US is delayed because stimuli present during the intervening interval become better predictors of the US. The same appears to be true in operant conditioning. Animals can learn an operant response for a delayed reinforcer but the longer the interval to the reinforcer, the more likely it is that animals will form associations between extraneous stimuli and the reinforcer (Dickinson, 1980), thereby reducing the strength of the response–reinforcer association.

The problem with this general principle of temporal contiguity is that it does not apply to all cases of associative learning. For example, CTAs develop with very long CS–US (taste–nausea) intervals, up to a few hours in rodents and even longer in humans. Any ill effects related to eating would occur after the food is digested, absorbed in the bloodstream, and distributed to body tissues. Organisms must be able to retain information about the stimulus properties of food over a long interval if they are to later avoid food that made them sick. This is a prime example of how ultimate explanations (adaptive specializations) influence proximate explanations (temporal contiguity) of associative learning.

Even within the same paradigm, changing the CS–US interval alters how animals respond. Timberlake (1984) demonstrated this effect in a classical conditioning experiment with rats. When a light CS predicted food at very short intervals (less than 2 seconds), rats developed a CR of handling and gnawing the CS. When the US occurred more than 5 seconds after the light CS, rats developed a conditioned foraging response as if they were searching for the food (Timberlake, 1984). The fact that a different CR developed under these two conditions is evidence that rats were learning the predictive temporal relationship between the CS and the US.

The relationship between temporal contiguity and the development of conditioned responding provides further evidence for the biological relevance of associative learning. If animals use associations to make predictive inferences about their world, then the temporal intervals that conform to CS–US or response–reinforcer relationships in the natural environment should produce the best conditioning.

4.4.3 Stimulus salience

Even if one controls for temporal contiguity and informativeness, the rate and magnitude of conditioning to different stimuli may vary. This is true in both the lab and the natural environment, but is illustrated most effectively by thinking about a typical classical conditioning experiment. The arrival of the experimenter (often at the same time each day), the placement of the animal in a testing apparatus, the sound of automated equipment starting an experiment, as well as other extraneous cues may all become effective CSs because they ultimately predict the presentation of the US. Researchers attempt to control for these confounds but, even in tightly controlled lab studies, it is difficult to eliminate cues that are not explicitly part of the experimental design. The issue is even more complicated in the natural environment where almost every US is preceded by a cluster of CSs. The technical term for this phenomenon is **overshadowing**: one stimulus acquires better conditioning than other stimuli in the environment, even if they are equal predictors of the US. In blocking, a stronger association develops to one CS because it was presented with the US first (i.e. before the other CS), whereas in overshadowing a stronger association develops because the CS is more salient.

The most obvious explanation for overshadowing is that animals notice or pay attention to one stimulus at the expense of the others. This is often described as stimulus salience, which is the likelihood that a stimulus will be attended to. Salience is often equated with the conspicuousness of a stimulus, or how well it stands out from the other background stimuli. In general, salience increases with the intensity of the stimulus: brighter lights, louder noises, or stronger smells attract

Figure 4.15 Naturalistic and artificial stimuli in sexual conditioning experiments. The stimulus on the left is made of terrycloth and only resembles the general shape of a female quail. The stimulus on the right was prepared with head and neck feathers from a taxidermically prepared bird.

more attention. It is important to remember, however, that salience is not a fixed property of the stimulus in that the ability of an organism to attend to a particular stimulus depends on the evolution of its sensory system (see Chapter 2, Section 2.1 for details).

The salience of a stimulus can also be increased by altering the motivational state of the animal. Food-related cues, such as cooking aromas, are far more salient when individuals are hungry than when they are sated. Not surprisingly, one of the best ways to increase the salience of a stimulus is to make it similar to cues that animals encounter in their natural environment. Male quail will develop a conditioned sexual response to a CS, such as a light or terrycloth object, that predicts access to a female quail (US). If this arbitrary cue is made more realistic by having it resemble a female quail (see Figure 4.15), more vigorous responding develops to the CS (Cusato & Domjan, 1998).

Stimulus salience impacts operant conditioning in the same way that it affects classical conditioning. Animals are much quicker to acquire responses to stimuli that are salient, and responding is increased when the stimuli have biological significance to the animal.

4.4.4 Extinction

In both classical and operant conditioning, responses will decline if the US or reinforcer is no longer presented. For example, birds will eventually stop approaching shells that covered a piece of food if the shells are now empty, and rats will stop lever pressing for food if the food is withheld. This gradual reduction in responding is called **extinction**. It may seem intuitive that animals forget or erase an association that was acquired during conditioning, but this is not how extinction occurs. For one thing, a CR will reappear if a delay period follows extinction, even when the organism has no further experience with the US. This **spontaneous recovery** indicates that the original association is still available to the animal. Second, if a novel stimulus is presented during extinction, the response rapidly recovers. For example, a dog will develop a conditioned salivation response to a bell that predicts food; when the bell is presented without the food, salivation declines. If a new stimulus, such as a light, is then presented with the bell, the dog will salivate again which seems counter-intuitive given that extinction has already occurred. The most common explanation is that animals are distracted by the new stimulus and forget that the CS no longer signals the outcome. The same phenomenon occurs in operant conditioning when rats reinitiate lever pressing in the presence of a loud noise. In some cases, the renewed response may be as robust as the pre-extinction response.

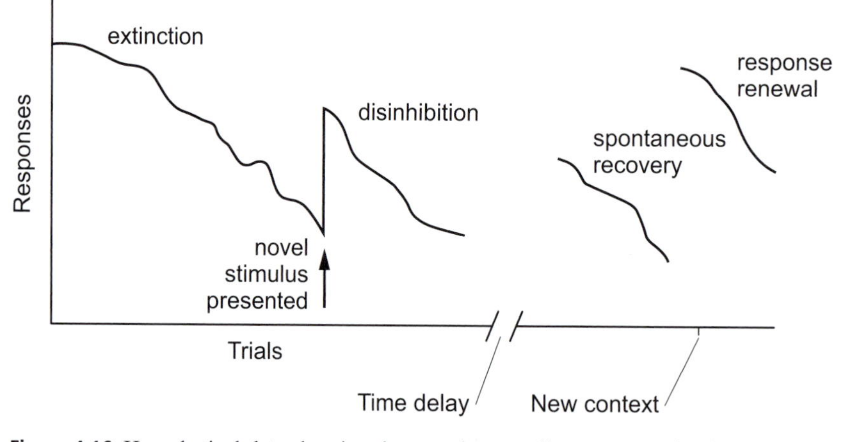

Figure 4.16 Hypothetical data showing decreased responding across extinction sessions. Responding may recover, at least partially, under different conditions. These are described as disinhibition, spontaneous recovery and response renewal (see text for definitions). Note that these changes could equally apply to operant and classical conditioning.

This phenomenon is called **disinhibition** to reflect the idea that the novel stimulus is disrupting a process that actively inhibits the original association (CS–US in classical conditioning and response–outcome in operant conditioning). Third, extinction is context specific in that inhibition of responding is stronger in the environment in which extinction trials took place. In the laboratory, the environmental context can be altered by changing the flooring, the wall patterns, adding odors, etc. If extinction trials are conducted in a new context, the response declines but then re-emerges when the CS is presented in the original context (Bouton, 1994). This **response renewal** is another piece of evidence that CS–US associations are inhibited, not eliminated, during extinction. In other words, extinction is not simply forgetting. Hypothetical changes in responding under these different conditions are shown in Figure 4.16.

The mechanisms of associative processes, described above, are general principles that apply to many (but not all) instances of classical and operant conditioning. Some exceptions were already noted, including the fact that CTAs emerge over long CS–US intervals. Nausea from ingestion of a toxic food takes time to develop, so the ability to form these associations over long intervals is adaptive. Similarly, the ability of one stimulus to 'block' another depends on its adaptive significance: in pigeons, visual stimuli predicting food cannot be blocked by auditory stimuli and auditory stimuli predicting a shock cannot be blocked by visual stimuli (Lolordo *et al.*, 1982). These examples emphasize that the mechanisms of associative processes must always be evaluated within the context of adaptive specializations.

4.5 Theories

Classical and operant conditioning are clearly governed by associative processes, but there was considerable debate in the first half of the twentieth century over *what* associations control these behaviors. Some researchers argued that animals form stimulus–stimulus (S-S) associations during conditioning, others that they learn stimulus–response (S-R) associations, and still others that

Figure 4.17 Pigeons pecking a key that predicted a reward. The pigeon on the left was trained with a water US and the pigeon on the right with a food US. The water-trained pigeon pecks at a slow rate with its beak closed and swallows frequently, responses that mimic drinking. The food-trained pigeon pecks with an open beak at a more rapid rate as if it is attempting to eat the light.

response–outcome (R-O) associations control responding. A number of very clever experiments were designed to tease apart these hypotheses but, as frequently occurs with such issues, no one theory can explain all of the data.

4.5.1 What associations are formed in classical conditioning?

Pavlov was the first to argue that classical conditioning involves S-S associations, an effect he called stimulus substitution. According to this position, a connection is formed between the CS and the US such that the CS becomes a functional substitute for the US. Compelling evidence for this position was provided by Jenkins and Moore (1973) who trained pigeons to associate a light with the presentation of either food or water. Pigeons developed a classically conditioned response of pecking the light but the characteristics of the peck varied with the US. As shown in Figure 4.17, a water US produced a CR that was a slow pecking with the beak closed, often accompanied by swallowing. In contrast, a food US produced a CR that was a sharp, vigorous peck with the beak open at the moment of contact, as if the pigeons were pecking at grains of food. It appeared that the animals were attempting to 'drink' or 'eat' the CS, seemingly confirming the idea that the CS was a substitute for the US.

Further evidence for S-S associations in classical conditioning comes from **sensory preconditioning** and **devaluation** experiments. In the first, two stimuli (CS1 and CS2) are presented together with no US; both stimuli are motivationally neutral so no observable response (CR or UR) is

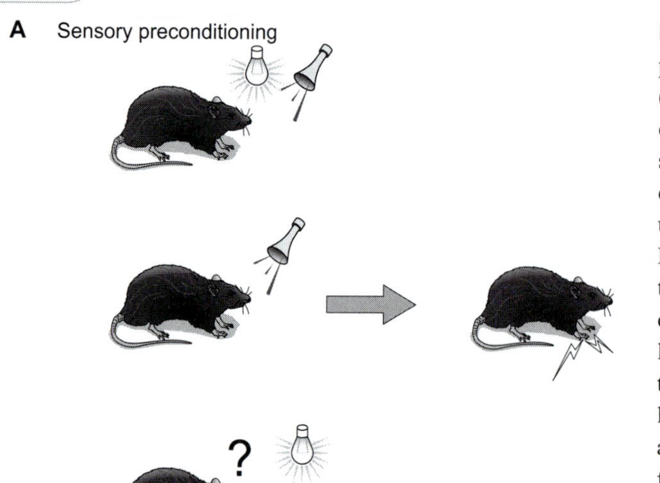

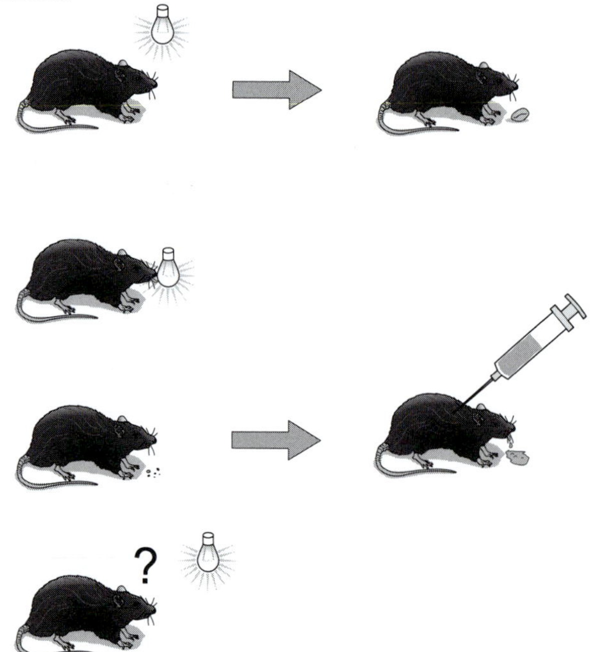

Figure 4.18 Experimental design for sensory preconditioning (A) and devaluation (B) studies. (A) In the first phase of a sensory preconditioning experiment, rats are presented with two neutral stimuli (light and tone). In the second phase, a conditioned stimulus (CS) predicts an unconditioned stimulus (US) (tone–shock). Responses to the light (alone) are measured during testing. (B) The first phase of a devaluation experiment involves classical conditioning (e.g. light–food). When conditioned responses (CR) to the CS have developed (e.g. the rat approaches and licks the light), the US is presented on its own and associated with an aversive outcome (e.g. food followed by sickness). Responses to the light (alone) are measured during testing.

elicited. Then, CS1 is followed by a US in standard classical conditioning trials. After a CR is established, CS2 is presented alone. If CS2 elicits a CR, animals must have formed an S-S association (CS1–CS2) prior to conditioning (see Figure 4.18A). Devaluation experiments follow a different logic: if the CS is associated with the US, changing the value of the US should change responding to the CS. To test this idea, animals undergo CS–US pairings until a CR develops. Then, the US is made aversive (devalued) in the absence of the CS. Finally, the CS is presented in the absence of the US. Note that animals have never experienced the CS and the aversive US. If the CR is reduced, compared to a group that did not undergo the devaluation experience, one can conclude that animals learned an S-S association during conditioning trials (see Figure 4.18B).

Sensory preconditioning and devaluation effects have been demonstrated in hundreds of experiments using dozens of different species. While not discounting these data, some researchers point out that S-S theories have limited utility because they make no prediction about what CR will develop across conditioning. That is, most USs elicit a variety of hormonal, emotional, and behavioral responses (i.e. URs); when these are paired with neutral stimuli, the CRs that develop may mimic the UR or be in the opposite direction. For example, a mild shock increases heart rate, whereas a CS that predicts the shock *decreases* heart rate (Hilgard, 1936). Preparatory response theories account for this discrepancy by suggesting that classical conditioning prepares organisms for the appearance of motivationally significant events (Kimble, 1967). The CR mimics the UR when this is the most adaptive response to the US (e.g. eyeblink to a puff of air or salivation to food), but is in the opposite direction to the UR when countering the US is more adaptive (e.g. decreased heart rate to impending shock). The idea that CRs prepare individuals for the imminent arrival of the US has been used to explain more complex behavioral responses, such as drug tolerance (see Box 4.3).

Preparatory response theories, including those that explain drug tolerance, are a form of S-S theories, even though the CR is predicted to be different (in fact, opposite) to the UR. That is, the CS and US may be connected even if the first is not a substitute for the second. Animals may also form S-R associations during classical conditioning (Rizley & Rescorla, 1972), although these are probably less common than S-S associations. Nonetheless, it is certainly the case that associations are formed between multiple stimuli and events in the natural environment and that these may form hierarchical relationships with each other (Rescorla, 1988). In other words, classical conditioning does not consist of learning a simple relationship between two stimuli, although it is typically measured this way in the lab. One of the greatest challenges in this research is to dissociate the formation of associations from their behavioral expression (Wasserman & Miller, 1997). This learning–performance distinction is a perennial problem in any behavioral study. Regardless, classical conditioning remains one of the most powerful models for studying experience-dependent changes in behavior.

4.5.2 What associations are formed in operant conditioning?

There are three fundamental elements in operant conditioning: the response (R), the reinforcer or outcome (O), and the stimulus (S) which signals when the outcome will be available. Thorndike proposed an S-R account of operant conditioning, arguing that cats in his puzzle boxes formed associations between stimuli in the box and the operant response that led to the escape. The outcome (in this case escape) increased subsequent responding because it strengthened the S-R association. Thorndike formulated this principle into a Law of Effect which stated that "if a response in the presence of a stimulus is followed by a satisfying event, the association between the stimulus and the response is strengthened" (Thorndike, 1911). Thorndike also suggested that responses followed by negative consequences are 'stamped out,' although he later abandoned this idea. Other researchers, most notably Hull (1930), used the term **habit learning** to describe the tendency to perform a particular response in the presence of a particular stimulus. According to him, the strength of the habit was a function of the number of times that the S-R sequence was followed by a reinforcer. As

Box 4.3 Preparatory responses and drug tolerance

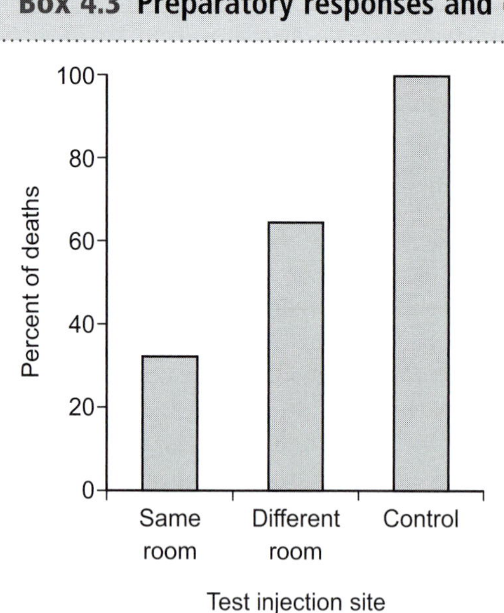

Figure 4.19 Development of conditioned tolerance to morphine in rats (Siegel *et al.*, 1982). On conditioning sessions, rats were injected daily with morphine in the same room. On test day, they were injected with a higher dose of the drug, either in the same room or in a different room. Control animals received a morphine injection for the first time on test day. Twice as many rats overdosed when they were injected in a different room on conditioning and test days. Almost all of the control rats were killed by the injection. Thus, rats that received conditioning and testing injections in the same environment showed the strongest drug tolerance.

Drug tolerance occurs when higher and higher doses of a drug are required to get the same effect. Tolerance develops rapidly to opiates so many drug addicts regularly inject a dose of heroin that is 30–40 times higher than a dose that would kill most people. Drug tolerance cannot be explained, entirely, by the pharmacological properties of the drug because animals and humans with the same level of drug exposure exhibit very different levels of tolerance. One suggestion is that cues associated with the drug act as a CS that elicits preparatory responses to the drug (Siegel, 1983). Tolerance develops because an association between these stimuli and the drug injection becomes stronger such that compensatory mechanisms become better at countering the drug effects. One prediction of this theory is that tolerance should be stronger when the drug injection always occurs in the same environment (i.e. the same cues reliably predict the injection).

Siegel and colleagues tested this hypothesis in rats that were repeatedly injected with heroin in one of two distinct environments. A control group of rats received sucrose/water injections. The following day, the dose of heroin was doubled for all animals. One group was injected in the same environment that they received the original injections and one group in a different environment. The dependent measure was the number of overdoses in each group. As shown in Figure 4.19, animals that received the test injection in a new environment were much more likely to overdose.

Siegel (1984) related his findings to human addiction, pointing out that accidental overdoses often occur when individuals inject a dose that, given their drug history, should not be fatal. He questioned former addicts who had almost died and found that many of their near overdoses were preceded by unusual circumstances. Addicts often administer drugs in a particular setting so environmental cues (including other people) become CSs that act as preparatory signals. These counter the upcoming drug effects but, in their absence, the full effects of the drug are experienced, often with disastrous consequences.

Figure 4.20 Experimental design for an operant devaluation study. Rats are trained to lever press for a food reward, such as sucrose pellets. After the response is acquired, the outcome is devalued independent of the operant contingency. For example, rats are allowed to eat sucrose pellets and then are made sick with a lithium chloride injection. At a later time, rats are returned to the operant chamber and lever pressing responses are measured in extinction (i.e. no sucrose pellets are presented). Responses in the devaluation group are compared to a control group that was trained to lever press but did not undergo devaluation sessions.

habit strength increased, the probability that a subject would perform the given response in the presence of the appropriate stimulus also increased.

Tolman was one of the strongest critics of Hull's work, arguing that S-R theories turned animals (including humans) into automata, with no understanding of how their behavior changed the environment. His view of operant conditioning was that animals form associations between their response and the outcome that it produces. This is nothing more complicated than saying that organisms understand what will happen when they do something. The problem for scientists during Tolman's time was that R-O theories like his require animals to have mental representations, both of their response and of the goal that they wish to achieve, a position that many were reluctant to adopt.

A partial resolution to the S-R versus R-O debate in operant conditioning comes from devaluation experiments. A devaluation protocol for classical conditioning is described in Section 4.5.1. The same procedure is used in operant conditioning with the outcome being devalued independently of the operant contingency (see Figure 4.20): animals learn to lever press for food, the food is associated with illness, and lever pressing responses are tested at a later time. Adams and Dickinson (1981b) were the first to show that lever pressing by rats is reduced following devaluation, indicating that animals formed R-O associations during training. In other words, rats connected what they did with what happened.

An interesting twist on the finding is that devaluation is ineffective if animals are well trained. Using Hull's terminology, the response becomes habitual, governed by S-R associations. This matches human habits, described as automatic responses to environmental stimuli. Habits often develop when the same action is performed repeatedly, such as travelling the same route to work or school; most people do this with little conscious awareness of their actions. Even if they intend to deviate from the usual route (i.e. to stop by the post office one day), they may end up at their final destination before they realize it.

The transition from R-O to S-R systems in operant conditioning is an example of how cognitive processing helps organisms adjust to changing environments. When humans or animals initially learn a task, they must attend to the consequences of their actions so that they can modify their responses accordingly. If the environment remains relatively stable and responses consistently produce the same outcome, habitual responding takes over. In this way, animals can cope with both predicted and unpredicted events in their environment.

Devaluation experiments helped researchers formulate new theories of operant conditioning, but many pointed out that the effects in these studies are small and often transient (Holland, 2008). Perhaps the most important outcome of these studies is that they emphasized a cognitive view of operant responding. If animals form R-O associations (even if these do not last), they must have a cognitive representation of their actions as well as the consequence of those actions. The association that is formed will interact with other learned associations, including those that control classical conditioning. In other words, even if animals form S-R and/or R-O associations, these connections are unlikely to explain all instances of operant conditioning.

4.5.3 Rescorla–Wagner model

Rather than focusing on which associations are acquired during classical or operant conditioning, a number of researchers examined associative learning by asking how animals code the logical relationship between events in their environment. One of the most influential of these theories is the Rescorla–Wagner model (Rescorla & Wagner, 1972), formulated to explain classical conditioning and the phenomenon of blocking. In a blocking experiment, the stimulus added during compound conditioning trials (CS2) does not elicit a CR because it does not provide any new information about the arrival of the US (see Section 4.4.1 above). CS1 already predicts the US, so adding CS2 is redundant. As Kamin noted in his original experiment, if the US presentation is not surprising, then no new learning will take place. Based on this premise, Rescorla and Wagner formulated a mathematical equation that presents classical conditioning as an adjustment between expectations and occurrences of the US (see Box 4.4).

There are four important components to the Rescorla–Wagner model that help to explain different aspects of classical conditioning. First, the model describes changes in conditioning on a trial by trial basis. If the US is stronger than expected (e.g. more food pellets or a stronger shock), the strength of the CS will increase, meaning that it will elicit stronger CRs in the future. The larger the difference between the expected and actual strength of the US, the larger the increase in CS strength will be (i.e. more conditioning). The first trial produces the greatest change in conditioning because the US is completely unexpected. In contrast, if the US strength is less than expected, CS strength will decrease and CRs will decline on the next trial. Second, increases in CS salience increase conditioning. In essence, stimuli that are more noticeable (e.g. brighter lights or louder tones) produce better conditioning. This principle explains overshadowing. When more than one CS is present on a trial, conditioning to the weakest or least noticeable stimulus is minimized even if that stimulus can elicit a CR when it is paired with the US on its own. Third, the magnitude of the US defines the maximum level of conditioning that may occur. This asymptotic level produces a ceiling effect in that further conditioning trials do not produce any change in behavior. In other words, a small amount of unpalatable food will never elicit the same number of saliva drops as a large and succulent reward, no matter how frequently it is paired with a CS. Finally, the strength of the US expectancy will be equal to the combined strength of all CSs present on that trial. This is true even if the CSs never occurred together on previous trials. Animals may undergo light–shock and tone–shock conditioning at different times so that each CS acquires its own conditioning strength. When these are combined on a compound conditioning trial, the expected US strength will be the sum of the two. If this combined value is at the maximum US strength, no new conditioning will occur. This is what happens in blocking: one CS has already acquired an associative strength that is at (or close to) the maximum US strength so adding a new CS in conditioning trials does not produce any conditioning. In other words, the conditioning strength has been used up by CS1 so there is none left for CS2.

Despite its general utility and appeal, the Rescorla–Wagner model cannot explain all aspects of classical conditioning. The most obvious are latent inhibition and sensory preconditioning. According to the model, a stimulus should not acquire (or lose) any associative strength when the

Box 4.4 The Rescorla–Wagner model

According to the Rescorla–Wagner model, changes in the associative strength of a CS on each conditioning trial can be calculated as follows:

$$\Delta V = \alpha\beta(\lambda - \Sigma V)$$

V the associative strength of a CS on a given trial.
ΔV the change in associative strength (V) on that trial.
α a learning rate parameter determined by the salience of the CS. **α** ranges from 0 to 1 and remains constant if the CS does not change. Bright lights and loud tones will have a higher **α** value than will dim lights and soft tones.
β a learning rate parameter indicating preparedness of the US to be associated with the CS. **β** ranges from 0 to 1 and remains constant if the US does not change. Strong shocks and large amounts of food will have a higher **β** value than will weak shocks and small amounts of food.
λ the maximum amount of conditioning or associative strength that a US can support. It has a positive value when the US is presented and is 0 when no US is presented.
ΣV the sum of the associative strength of all CSs present on that trial.
λ – SV an error term that indicates the discrepancy between what is expected (SV) and what is experienced (λ). When (λ – SV) is zero, the outcome is fully predicted and there are no changes in associative strength (ΔV).

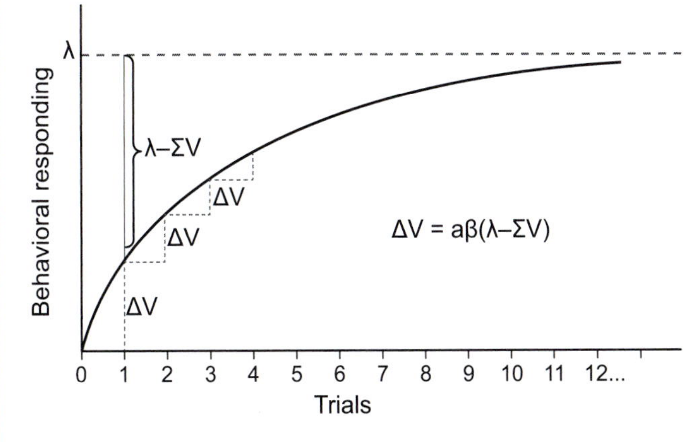

Figure 4.21 Hypothetical data illustrating the Rescorla–Wagner model of conditioning. The model predicts the negatively accelerating function that characterizes changes in responding across classical conditioning trials. Units on the y axis represent hypothetical associative strength (V) of a conditioned stimulus (CS). The change in associative strength on each trial is represented by ΔV. **α** represents the salience of the conditioned stimulus (CS); **β** represents the preparedness of the unconditioned stimulus (US) to be associated with the CS. The error term (λ – SV) represents the discrepancy between what is expected and what occurs on a given trial. Asymptotic responding, or the maximum amount of associative strength that a US can support, is represented by λ.

This formula explains a typical **learning curve** (see Figure 4.21) in which each trial produces a smaller change in behavior. Initially, the US is surprising so the difference between what is expected following the CS (nothing) and what occurs (the US) is large. This difference gradually decreases as the CS is associated more frequently with the US. The formula also explains extinction following conditioning: with no US, **λ** = 0, and the error term will be negative (ΣV is positive because conditioning has already occurred). Under these conditions, the associative strength of the CS is reduced and responding declines.

Any model is hypothetical so many researchers were excited by the idea that neural activity could be mapped onto the Rescorla–Wagner equation (Hollerman & Schultz, 1998), linking psychological theory with neuroscience mechanisms. In studies with non-human primates, Schultz and his colleagues showed that midbrain dopamine neurons, projecting to the striatum and frontal cortex, fire when a US is unexpected, but not when it is expected. This "prediction error" is based on previous CS–US associations, matching the (λ – SV) error term in the Rescorla–Wagner model. Neurons in other regions, such as the striatum, orbitofrontal cortex, and amygdala, code the quality, quantity, and preference for rewards, possibly reflecting other parameters in the model (i.e. α, β, and λ). To link these events biologically, Schultz proposes that the dopamine error signal communicates with reward perception signals to influence learning about motivationally significant stimuli.

US is not present. Thus, there is no way to account for a CR developing to a stimulus that was never paired with a US (sensory preconditioning) or the reduction in conditioning that follows CS pre-exposure (latent inhibition). Moreover, even though the model predicts a decline in responding during extinction, it cannot account for spontaneous recovery, disinhibition, or response renewal, suggesting that the model is not an accurate description of the mechanisms by which extinction occurs. Subsequent modifications to the model were able to deal with these problems, although inconsistencies between what the model predicted and what happened in the lab were still evident. Because of this, a number of other theories developed that were purported to be better explanations of classical conditioning. Some focused on the attention that organisms direct toward the CS (Mackintosh, 1975; Pearce & Hall, 1980), whereas others focused on comparing the likelihood that the US will occur in the presence and absence of a CS (Gibbon & Balsam, 1981). Each theory has its strengths, but no single model was able to account for all aspects of classical conditioning. Nonetheless, if a theory is to be judged on the research and discussions that it generates, the Rescorla–Wagner model is one of the most successful in psychology.

4.5.4 Associative Cybernetic model

There are fewer formal theories of operant conditioning than of classical conditioning for three primary reasons. First, classical conditioning is easier to study in the lab because animals do not need to be trained in an operant task. Second, the experimenter controls US (reinforcer) presentations in classical, but not operant, conditioning, making the connection between learned associations and behavioral changes more straightforward. Third, and perhaps most important, many scientists assumed that classical and operant conditioning were mediated by the same processes, so a theory of one should explain the other. In that respect, the Rescorla–Wagner model has been applied to operant conditioning by assuming that responses with surprising outcomes produce the greatest increments in learning.

The Associative Cybernetic model (de Wit & Dickinson, 2009; Dickinson & Balleine, 1994) is one of the only formal models of operant conditioning. Although it is not as well known as Rescorla–Wagner (and therefore has not generated as much research), the Associative Cybernetic model is an excellent example of how a behavioral phenomenon can be translated into theoretical descriptions. The model, shown in Figure 4.22, consists of four principal components: a habit memory system that represents S-R learning; an associative memory system that represents R-O associations; an incentive system that connects the representation of an outcome with a value (e.g. rewarding or punishing); and a motor system that controls responding. The motor system can be activated directly by the habit memory or via associative memory connections through the incentive system. The 'cybernetic' in the model refers to the fact that an internal representation of the value assigned to an outcome feeds back to modulate performance. (Cybernetics is the study of communication and control systems.) Thus, the Associative Cybernetic model explains the interaction between S-R and R-O learning systems, and describes how responding is modified by its consequences.

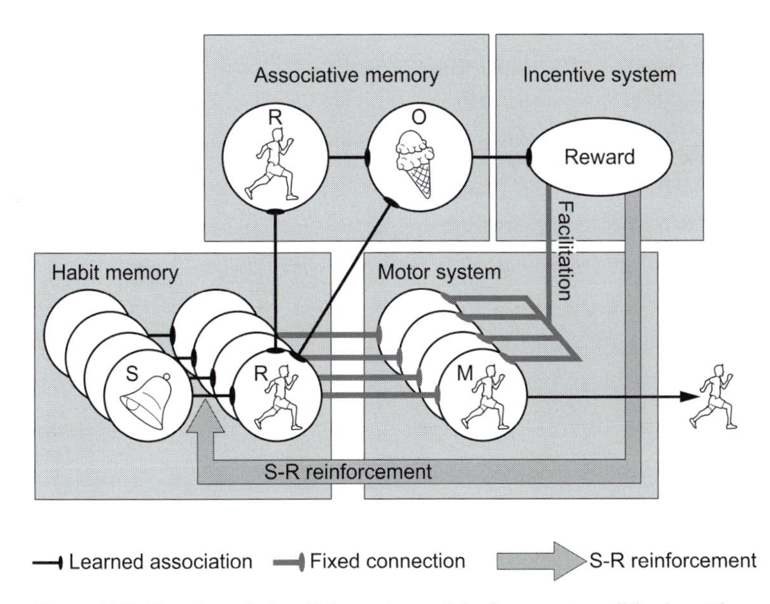

Figure 4.22 The Associative Cybernetic model of operant conditioning. The model consists of four principal components that interactively produce changes in behavioral responding. See text for details.

As with Rescorla–Wagener, some components of the Associative Cybernetic model have been applied to specific neural substrates (Balleine & Ostlund, 2007). The details are not complete, but any evidence that neural mechanisms can map onto parameters of a specific model adds credence to these theoretical accounts of learning.

4.6 Neuroscience of associative processes

The fact that associative learning is observed in almost every species studied *could* indicate that the process is conserved across evolution. If this is true, classical and operant conditioning may have descended from a common cognitive mechanism (i.e. homology). In contrast, different species may have converged on the same cognitive solution (i.e. forming associations) to learn about causal and predictive relationships in their environment (Shettleworth, 2010). One way to distinguish between these possibilities is to compare biological processes mediating classical and operant conditioning in different animals. Cross-species similarities would support (although not prove) the conservation hypothesis. Currently, there is not enough biological evidence to confirm one theory or the other, although research examining this issue is progressing rapidly. The following sections outline current evidence on brain systems that mediate associative learning, beginning at the level of neural circuits, moving to connections between neurons, and finishing with intracellular (molecular) mechanisms.

4.6.1 Neural circuits

Most of what is known regarding the neural circuits of associative learning comes from the conditioned eyeblink paradigm. All animals, including humans, blink when a puff of air hits their eye. This reflexive response is commonly studied in rabbits because the rate of spontaneous blinking is low in these animals. Thus, if a rabbit blinks when it hears a tone that was previously paired with

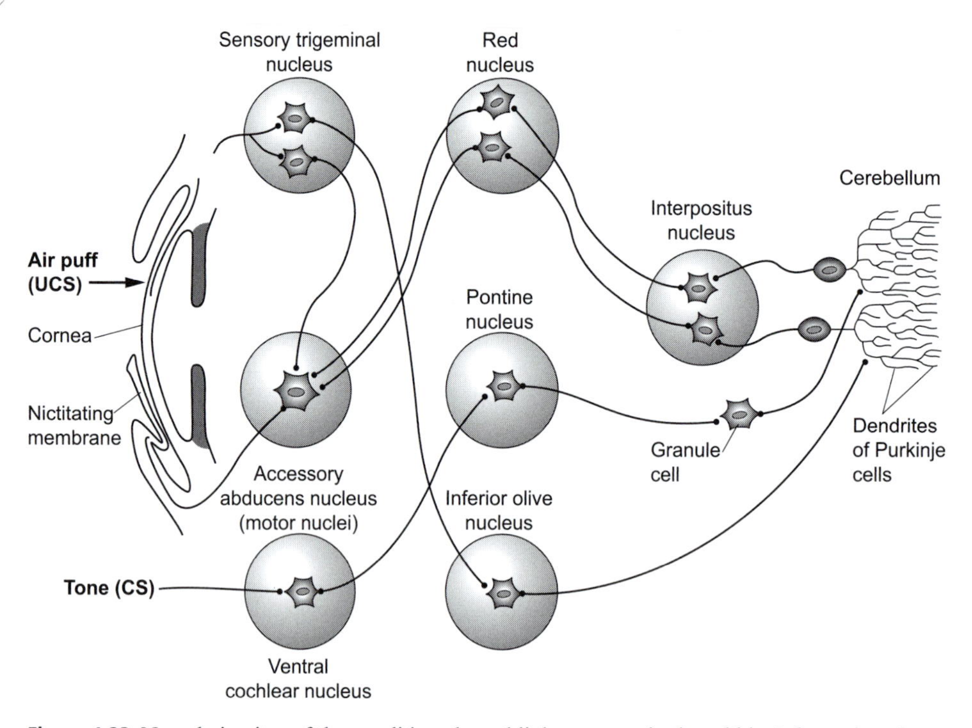

Figure 4.23 Neural circuitry of the conditioned eye blink response in the rabbit. Information about an auditory conditioned stimulus (CS) enters the central nervous system via the ventral cochlear nucleus and converges with signals transmitted from the airpuff unconditioned stimulus (US) at the level of the interpositus nucleus. Eye blink responses, both conditioned and unconditioned, are controlled by the accessory abducens nucleus (motor nuclei).

an air puff, it is likely due to the conditioning process. The majority of this work was conducted by Thompson and his colleagues who determined that the conditioned eyeblink response in rabbits is mediated in the cerebellum (Thompson & Krupa, 1994). Figure 4.23 presents a schematic of the classical conditioning circuits identified by these researchers. This includes the reflex circuit that causes an eyeblink after an air puff hits the eye (US–UR) and the pathway through which the auditory signal (tone CS) enters the brain. The critical question is how the CS accesses this US–UR circuit in order to produce a CR. In the eyeblink paradigm, the convergence of signals happens in the interpositus nucleus of the cerebellum. The same brain region is involved in other conditioned reflexive responses in a variety of mammals (e.g. leg flexion to a CS predicting shock) (Thompson & Steinmetz, 2009), supporting the idea of homology in associative mechanisms, at least in this class of animals.

Similar schematics have been devised for other classically conditioned responses, such as CTA, conditioned fear, and conditioned approach. Details of these circuits may be found in many neuroscience textbooks. There are relatively few cross-species comparisons of these mechanisms, but available evidence points to similarities in the neural systems that mediate conditioned emotional responses across mammals (Phelps & LeDoux, 2005). In all classical conditioning, the brain systems that carry signals in and out of the brain will vary depending on the sensory properties of the CS as well as the response that is used to measure conditioning. For example, an auditory CS will be transmitted through a different circuit than visual or tactile CSs, and an escape response will be mediated through a different output pathway than a salivary response. As with the conditioned eyeblink response, however, the important point about this mechanism is that the emergence of a CR coincides with changes in the neural circuit connecting CS and US signals.

The same principle holds true in operant conditioning: behavioral changes are reflected as alterations in neural connections. Even so, classical and operant conditioning in mammals are mediated through dissociable brain structures (Ostlund & Balleine, 2007), providing concrete evidence for a distinction between these two associative processes. In addition, the neural systems that control R-O and S-R associations are distinct. S-R responding, or habit learning, is often equated with procedural learning, discussed in Chapter 3, a process that is dependent on the dorsal striatum. In contrast, R-O contingencies are mediated though a network of brain regions that begins with signals generated in the medial prefrontal cortex (Tanaka *et al.*, 2008). This region computes response contingencies (i.e. R-O associations) and then sends this information to the orbitofrontal cortex. The orbitofrontal cortex codes the motivational significance of reinforcers (Rolls, 2004) so it is likely that associations between the outcome and its value are formed in this brain region. Signals from the orbitofrontal cortex are transmitted to the dorsal striatum, which controls behavioral responses. Like classical conditioning, the details of the circuit will vary depending on the response to be emitted and the stimuli that precede it. This hypothesized neural circuit may not mediate all instances of operant conditioning, but it provides an alternative framework for studying the biological basis of associative learning.

4.6.2 Cellular mechanisms

The changes in neural communication that underlie associative learning have been described at the cellular level using classical conditioning of the gill withdrawal reflex in *Aplysia*. Initially, a mild touch to an outer part of the animal's body, the mantle shelf, does not elicit a response; when this stimulus precedes a tail shock (US), however, a CR develops to the mantle touch (CS). A control condition is included in these studies in which a light touch is applied to the siphon (another part of the *Aplysia* body) that is not explicitly paired with the US. The body touches that are and are not associated with the US are referred to as CS^+ and CS^- respectively. As with sensitization of the gill withdrawal reflex (see Section 3.5.1 for details), the US causes facilitatory interneurons to release serotonin (5-HT) on the presynaptic terminals of incoming sensory neurons. Sensory neurons conveying CS^+ information fire when the mantle is touched and if this occurs in close temporal proximity to the US, the strength of the CS^+ sensory–motor neuron connection is increased, making it easier to elicit a CR in the future. On subsequent trials, the incoming CS^+ signal produces a greater postsynaptic potential in the motor neuron, causing a gill withdrawal response in the absence of the US (Hawkins *et al.*, 1983). Because the two neurons have not been active at the same time, synaptic strength is not altered in the CS^- sensory–motor neuron circuit (see Figure 4.24).

Serotonin is involved in other forms of classical conditioning, although none has been characterized as thoroughly as the gill withdrawal response in *Aplysia*. Neurotransmitters, such as glutamate and dopamine, are also implicated in classical conditioning so the cellular mechanisms of associative learning clearly involve an interaction between different neural systems. Many of these act by modifying pre-existing circuits (Johansen *et al.*, 2012), similar to the presynaptic modulation of sensory neurons described in *Aplysia* classical conditioning. There are relatively few studies examining the cellular mechanisms of operant conditioning; those that have been conducted suggest differences in how neurons code classical and operant conditioning, at least in invertebrates (Baxter & Byrne, 2006).

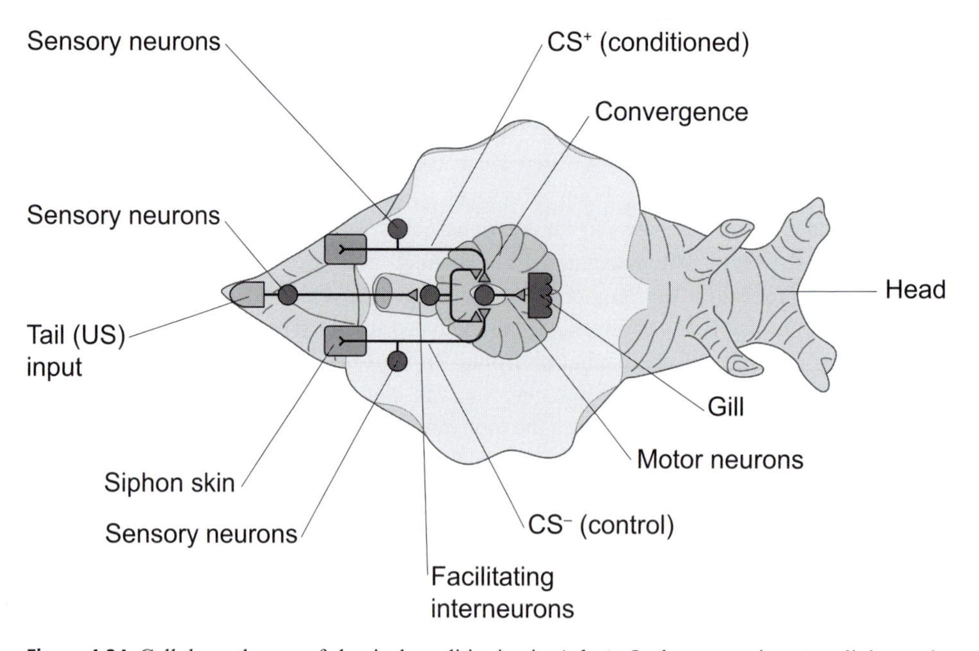

Figure 4.24 Cellular pathways of classical conditioning in *Aplysia*. In these experiments, a light touch to one part of the animal's body is the conditioned stimulus (CS^+) and a touch to another part of the body is the control stimulus (CS^-). During conditioning trials, the CS^+ is paired with a shock unconditioned stimulus (US) applied to the animal's tail. The CS^- is presented an equal number of times but not explicitly paired with the US. With training, the CS^+, but not the CS^-, elicits withdrawal. The cellular mechanisms of this CS^+–US association are as follows: the CS^+ (light touch) primes the sensory neuron by making it more excitable. The US excites facilitating interneurons that synapse on presynaptic terminals of sensory neurons. This combination produces a stronger response in the motor neurons. This strengthened connection between sensory and motor units is capable of eliciting a response in the future even when a US is not presented.

4.6.3 Molecular mechanisms

Finally, associative learning involves changes that occur within neurons. Benzer (1973) developed a technique to study these by exposing fruit flies to radiation or chemicals that produced genetic mutations. Different behavioral changes were observed with different mutations and some of these related to associative learning. For example, fruit flies will avoid an odor that was previously paired with a shock (classical conditioning); mutant flies (referred to as 'dunce' flies) fail to learn this association even though they show no deficit in responding to these stimuli on their own (Dudai *et al.*, 1976). Dunce flies have a mutation in the gene that codes for the enzyme cyclic AMP phosphodiesterase, which itself breaks down the intracellular messenger cyclic AMP (cAMP). Defects in the cAMP signaling pathway were later identified in two other mutants, named 'rutabaga' and 'amnesiac,' both of which show deficits in classical conditioning.

Research with *Aplysia* supports the idea that the cAMP pathway is critical for associative learning. During classical conditioning, a CS sets off action potentials in sensory neurons that trigger the opening of calcium channels at nerve terminals. The US produces action potentials in a different set of neurons that connect (or synapse) on to terminals of the CS neurons. When neurotransmitter is released from the US neurons, it binds to the CS neurons and activates adenylate

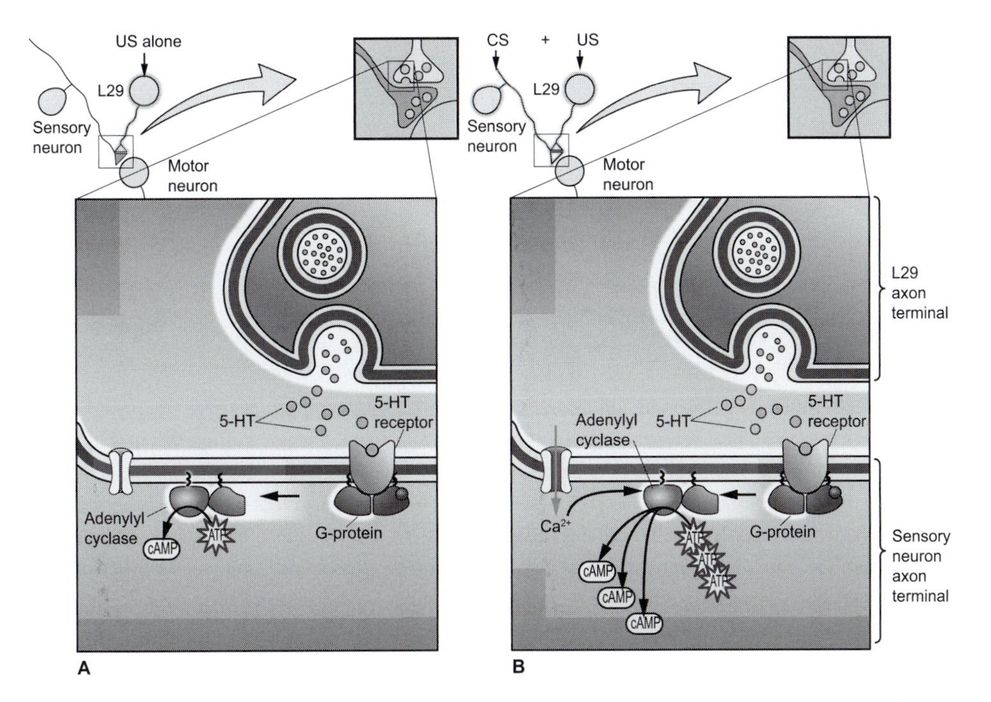

Figure 4.25 Molecular mechanisms of classical conditioning. (A) Presentation of the unconditioned stimulus (US) increases 5-HT release from the facilitating interneuron. This activates the postsynaptic neuron and increases adenylyl cyclase levels in sensory neurons. (B) When a conditioned stimulus (CS) precedes the US, it excites the sensory neuron, causing an opening of calcium channels on the postsynaptic cell. This leads to even higher levels of adenylyl cyclase following the US presentation because calcium increases the intracellular production of adenylyl cyclase.

cyclase within the cell. Adenylate cyclase generates cAMP and, in the presence of elevated calcium, adenlyate cyclase churns out more cAMP. Thus, if the US occurs shortly after the CS, intracellular second messenger systems are amplified within CS neurons. This causes conformational changes in proteins and enzymes within the cell that lead to enhanced neurotransmitter release from the CS neuron. The consequence is a bigger behavioral response, as shown in Figure 4.25. The same process may underlie the contribution of cAMP to associative learning in rodents (Josselyn *et al.*, 2004).

It should not be surprising that the details of this figure resemble those of Figure 3.17 describing the molecular mechanisms of long-term potentiation (LTP): LTP is a cellular model of memory and classical conditioning is one type of memory. Moreover, as with LTP, many of the molecular changes that occur during associative learning culminate in activation of the intracellular protein, CREB (Johansen *et al.*, 2012). Thus, the contribution of CREB to classical conditioning in both vertebrate and invertebrate models probably reflects its more general role in memory processing.

A recurrent theme in associative learning, touched on throughout this chapter, is the question of whether classical and operant conditioning are mediated by the same process. The issue was never resolved, completely, at a behavioral level but the two appear to be dissociable at a molecular level. For example, mutations that disrupt adenylate cyclase in fruit flies produce deficits in classical, but

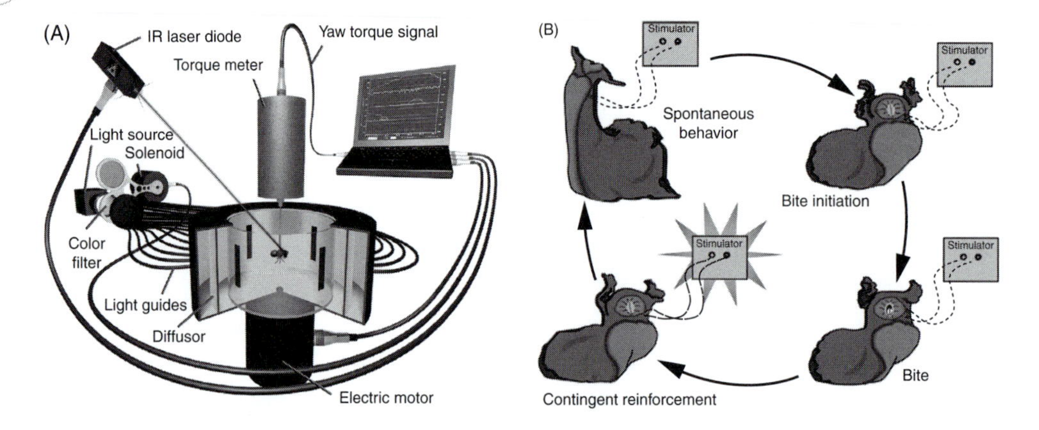

Figure 4.26 Invertebrate operant conditioning paradigms. (A) *Drosophila* learn to fly toward a particular visual stimulus to receive a heat reward. The fly is tethered and suspended inside a drum that acts as a flight-simulator. Four pairs of vertical bars on the outside of the drum change color when the animal flies toward them. During training, flying toward one color (e.g. blue) turned on a heat source. Over training, flies approach the blue color more frequently, regardless of where the vertical lines were located. Thus, even when they had to redirect their flight path or change directions, the flies approached the heat-associated color more frequently than the other colors. (B) *Aplysia* learn to perform a biting response to receive brief stimulation of the esophageal nerve. One day prior to training, animals are implanted with a stimulating electrode on the anterior branch of the left esophageal nerve. During training, the animal moves freely in a small aquarium and spontaneous behavior is monitored so that the initiation of a biting response can be noted. Stimulation is applied immediately following a bite (contingent reinforcement) or at random intervals unrelated to the biting response (not shown). Over trials, the rate of spontaneous biting increases, but only in animals that experienced the stimulation following the bite.

not operant, conditioning whereas mutations that disrupt protein kinase C (PKC) have the opposite effect (Brembs & Plendl, 2008). PKC is an enzyme that is activated by cAMP so this step is downstream to cAMP, as opposed to the upstream effect of adenylate cyclase. PKC also appears to be important in *Aplysia* operant conditioning (Lorenzetti *et al*., 2008). Figure 4.26 illustrates how operant conditioning is assessed in these invertebrate models.

Despite cross-species similarities in the neural mechanisms of classical and operant conditioning, there is no consensus that these processes are homologous. Given the adaptive value of being able to predict motivationally significant events, different animals may have converged on the same solution across evolution. Currently, there is no way to distinguish between these two possibilities.

Chapter Summary

- Organisms ranging from insects to humans are capable of forming associations between stimuli and events. The existence of this common trait across a variety of species suggests that animals are designed to detect and store information about relationships in their environment. The two most studied forms of associative learning are classical and operant conditioning. Classical conditioning describes a predictive relationship between a neutral and a motivationally significant stimulus; operant conditioning describes a causal relationship between a response and its consequences.
- Classical conditioning is measured, most commonly, using some variation of the conditioned preference, conditioned emotional reaction, or conditioned taste aversion paradigms. Operant conditioning can be assessed using discrete trials or free operant measures. In the latter, different schedules of reinforcement (i.e. the relationship between responding and reinforcer presentation) produce different patterns of responding, suggesting that animals have some knowledge of the payoff provided by each schedule.
- The adaptive value of associative learning may reflect the ability to predict biologically relevant events (classical conditioning) and to associate responses with their consequences (operant conditioning). This idea is supported by adaptive specializations, the relative ease with which some associations are formed, based on an animal's evolutionary history.
- Proximate mechanisms of associative processes can be studied by examining factors that produce the most rapid and robust responding. In both classical and operant conditioning, these include informativeness, temporal contiguity, and stimulus salience. There are exceptions to these general rules, such as CTA learning which is maximized with long CS–US intervals. Although responding declines when the reinforcer is removed (extinction), it is unlikely that animals have forgotten the associations that were acquired during classical or operant conditioning.
- Early theories of associative learning attempted to identify which relationships explain the emergence of classical and operant responses. After long debates and hundreds of experiments, it became clear that animals can form stimulus–stimulus, stimulus–response, or response–outcome associations depending on the conditions in an experimental paradigm. A more productive approach to theorizing about associative learning is to examine how animals code the relationships between environmental events. One of the most influential of these theories, the Rescorla–Wagner model, explains changes in associative strength as a difference between what an organism expects (based on previous experience) and what occurs on any given trial.
- Information on the biological basis of associative processes has increased exponentially over the last two decades. Scientists now have a detailed understanding of the molecular, cellular, and neural circuit changes that underlie both classical and operant conditioning. A large portion of this work has been conducted with invertebrates; evidence, to date, suggests that the same underlying principles govern associative learning in mammals, including humans.

Questions

1. Approach responses often lead to a reward, such as food or access to a sexual partner. Thus, animals may approach stimuli associated with these rewards because the consequence of this action is associated with a positive outcome. If this is true, conditioned approach responses would be

maintained by classical, not operant, conditioning. How could one tell the difference between the two?

2. In a standard CTA experiment, rats are exposed to a novel flavor then made ill with an injection of lithium chloride 12–24 hours later. What are the control conditions that should be included in this experiment and what confound would each rule out?
3. Name and explain the four different reinforcement relationships in operant conditioning.
4. Design a behavior modification program to change a child's nail biting behavior. How would this be adjusted if the behavior only occurred under certain situations (e.g. when the child is stressed)?
5. If a variety of species exhibit adaptive specializations in the same classical and operant conditioning experiments, does this prove that associative processes are conserved through evolution? Why or why not?
6. A rat is trained to press a lever for food and then the food is removed until the rat stops pressing the lever. Describe three ways in which the response could be renewed.
7. What is the difference between blocking and overshadowing?
8. Why do drug addicts frequently overdose when they inject the drug in a new environment? Is this an example of classical or operant conditioning?
9. The Rescorla–Wagner model uses a mathematical equation to describe changes in associative learning. How does this equation change when a US is removed on one conditioning trial? Assume that associative strength is at 100 and the combined value of alpha and beta is 2, chart the change in associative strength across five conditioning trials when the US is removed.
10. A mild touch to an *Aplysia*'s mantle shelf does not initially elicit a response; when this stimulus precedes a tail shock (US), however, a CR develops to the mantle touch (CS). Describe how this phenomenon occurs at the cellular level.
11. Based on what we know about classical conditioning, what physiological changes were probably occurring in Pavlov's dogs?

FURTHER READING

The early theories of associative mechanisms can be found in the following texts:

Cuny, H. (1962). *Ivan Pavlov: The Man and His Theories*. New York, NY: Fawcett.

Hull, C. L. (1943). *Principles of Behavior: An Introduction to Behavior Theory.* New York, NY: Appleton-Century-Crofts.

Pavlov, I. P. (1927). *Conditioned Reflexes*. Oxford, UK: Oxford University Press.

Thorndike, E. L. (1911). *Animal Intelligence: Experimental Studies*. New York, NY: Macmillan.

Tolman, E. C. (1932). *Purposive Behavior in Animals and Men*. New York, NY: Appleton-Century-Crofts.

Contemporary views of the same topic are presented in:

Bouton, M. E. (2007). *Learning and Behavior: A Contemporary Synthesis*. Sunderland, MA: Sinauer Associates Inc.

Dickinson, A. (2012). Associative learning and animal cognition. *Philos Trans R Soc Lond B*, **367**, 2733–2742.

Domjan, M. (2003). *The Principles of Learning and Behavior* (5th edn.). Belmont, CA: Wadsworth/Thomson Learning.

Holland, P. C. (2008). Cognitive versus stimulus-response theories of learning. *Learn Behav*, **36**, 227–241.

Pearce, J. M. (2008). *Animal Learning and Cognition: An Introduction* (3rd Edn.). New York, NY: Psychology Press.

Shanks, D. R. (2010). Learning: From association to cognition. *Ann Rev Psychol*, **61**, 273–301.

For more specific discussions of the relationship between proximate and ultimate mechanisms of associative learning, see:

Hollis, K. L. (1997). Contemporary research on Pavlovian conditioning: A 'new' functional analysis. *Am Psychol*, **52**, 956–965.
MacPhail, E. M. (1996). Cognitive function in mammals: The evolutionary perspective. *Cognitive Brain Research*, **3**, 279–290.

Excellent reviews of the neurobiological mechanisms of associative learning include:

Hawkins, R. D., Kandel, E. R., & Bailey, C. H. (2006). Molecular mechanisms of memory storage in *Aplysia*. *Biol Bull*, **210**, 174–191.
Suzuki, W. A. (2008). Associative learning signals in the brain. *Progr Brain Res*, **169**, 305–320.
Thompson, R. F., & Steinmetz, J. E. (2009). The role of the cerebellum in classical conditioning of discrete behavioral responses. *Neuroscience*, **162**, 732–755.

5 Orientation and navigation

Background

The Haida of the Pacific northwest in Canada recount a legend in which the daughter of a tribal chief dreams of a large, shining fish that leaps out of the water. She longs, desperately, to possess this fish that has never been seen, even by the elders in her village. She cries and cries until she grows sick with grief, so her father calls together his tribal council to ask their advice. A wise old medicine man brings forth a raven, who explains that he has seen such fish far, far away, at the mouth of the mighty river which leads to the village inlet. To show his loyalty to the tribe, the raven offers to bring one of these fish back to the chief's daughter. He flies directly to the river mouth, dives into a swarm of fish and rises with a young salmon in his talons. By chance, he has caught the son of the Salmon chief so, as he turns to fly back to the village, salmon scouts jump out of the water in great arcs to see the direction in which he is flying. Led by the chief, a horde of salmon swim up the river in pursuit. The raven, arriving far ahead of the salmon, places the fish at the feet of the tribal chief's daughter and then tells the village members to prepare nets across the river to catch the salmon that will soon arrive. When all of the fish are caught, they are strung between a rock and a large cedar which then becomes a totem pole. The totem is carved with both a raven and a salmon to honor the animals that have brought such abundance to the tribe. Since that time, the salmon return each year to the village, swimming rapidly and frantically against the river current in an attempt to rescue one of their sons.

This tale, the Coming of the Salmon, provided the Haida with an explanation for the collective and seasonal movement of one group of animals in their environment. Other people, in different parts of the world, have generated similar stories to explain why animals undertake long, and often arduous, journeys. In the nineteenth century, many Europeans believed that birds spent the winter hibernating in mud at the bottom of lakes. Although it now seems farfetched, it was probably more believable than the idea that small birds could travel from Europe to Africa and back again in one year without getting lost. At that time, most people travelled very little and they had few ways to access information from other geographical locations. Today, many people visit different parts of the globe and are able to observe animals in their natural habitat, either directly or indirectly through video recordings. It is also possible to follow animal movement, even across vast distances, using technologies such as global positioning systems (GPS) and radar tracking devices. This information has been invaluable in studies of how and why animals move through their environment, although many questions remain unanswered.

Chapter plan

This chapter focuses on spatial behaviors, examining how humans and animals are able to move through their environment without getting lost. Even very simple organisms are not stationary and many humans and animals spend a large part of their waking hours traveling from one place to another. The ability to do this effectively and efficiently is critical for a variety of activities. Foraging or hunting for food, searching for mates, caring for young, and returning home afterwards all depend on cognitive faculties that direct and control spatial movement within familiar and unfamiliar environments.

The chapter will discuss a range of spatial behaviors from basic orientation responses to complex navigational feats of migration. The first section outlines evolutionary explanations for these behaviors, which are typically divided into processes of orientation and navigation. Cognitive mechanisms of orientation point animals in the right direction; these are described in terms of kinesis and taxes, two of the simplest spatial responses in the animal kingdom. The next sections focus on navigation, which is divided into small-scale (short distance) and large-scale (long-distance) travel. Sections 5.3 and 5.4 consider issues such as how different species compute the distance and direction of travel, and what they understand about spatial relationships in their surroundings. The final section of the chapter describes the neuroscience of spatial behaviors with a focus on neural systems that mediate small-scale navigation.

5.1 Finding the way

Animals do not search randomly for food or mates, nor do they rely on chance to return home afterwards. Rather, natural selection has favored cognitive processes that facilitate the ability to find the way, both to and from specific locations. This first section outlines the adaptive significance of finding the way: ultimate explanations for 'why' animals are able to move through their environment without getting lost. 'How' this occurs is described in the subsequent three sections (5.2–5.4).

5.1.1 Getting there and knowing where

Finding the way can be divided into two distinct processes: getting there and knowing where (Benhamou, 2010). Getting there depends on an elementary system that directs animals toward different, often unknown, locations. Virtually all organisms display some form of this behavior when they orient to places where they may find food, mates, or new habitats. Even bacteria get somewhere new by moving from areas of low to high nutrient content. The getting there system operates by initiating motor patterns in response to internal signals, such as hormones, or to external stimuli, such as sensory cues. In contrast, the knowing where system allows animals to reach a particular destination, regardless of a current position. It relies primarily (although not exclusively) on previous experience to identify the goal location. Knowing where allows animals to locate hidden food stores, to find the way around familiar environments, and to travel back and forth between seasonal habitats.

Spatial behaviors guided by either 'getting there' or 'knowing where' often appear to be goal directed. This is particularly true of the latter as animals searching for food or returning home *seem* to have a predetermined destination in mind. However, even if animals consistently arrive at the same place, this does not prove that they had the intention of doing so. In many cases, the cognitive mechanisms that signal when to stop control behavior at the end of a journey. In other words, animals don't 'know' that they have arrived until they get there. This issue is discussed in more detail throughout the chapter.

Although they will be discussed separately, the processes of getting there and knowing where often operate together. One example is **dispersal**, the movement away from a parent or conspecific population due to declining resources, overcrowding, or mate competition. Dispersal generally involves movement to an unfamiliar environment, but animals seldom wander aimlessly as they find their way to this new home. Dispersing animals must possess cognitive mechanisms that direct their travel (i.e. getting there) and help them to identify suitable ecological conditions when they arrive (i.e. knowing where). Without these in place, the advantages of dispersal (i.e. finding a better habitat) may not outweigh the disadvantages (e.g. energy resources required for a long journey).

5.1.2 Adaptive value

The adaptive value of finding the way is examined, most frequently, in terms of two behaviors: caching and migration. Caching allows animals to hide a valuable resource, usually food, when it is plentiful, to be retrieved at a later time, when it is not. Not all animals cache food, although the behavior is displayed by a variety of birds, mammals, and insects. One of the most impressive is the caching ability of Clark's nutcracker (*Nucifraga columbiana*), an avian species that hides seeds in up to 10 000 separate locations, which it then recovers a year later. If evolutionary pressures have shaped the behavior, caching should be more pronounced in harsh environments where food is less plentiful at certain times of the year. This hypothesis was verified in a comparison of four corvid species that inhabit different ecological zones of the San Francisco peaks in Arizona (Balda & Kamil, 2006). Clark's nutcracker and pinyon jays (*Gynmorhinus cyanocephalus*), which live at the highest elevation with the most severe winters, cache a larger number of seeds than Mexican jays (*Aphelocoma wollweberi*) and western scrub-jays (*Aphelecoma californica*), which live at the lowest elevation.

Caching is a labor-intensive activity so it would not be adaptive if animals were unable to locate their hidden food in the future. Stated as a scientific hypothesis, species that depend on cached food for survival should display superior memory for spatial locations (Balda & Kamil, 1989). Figure 5.1

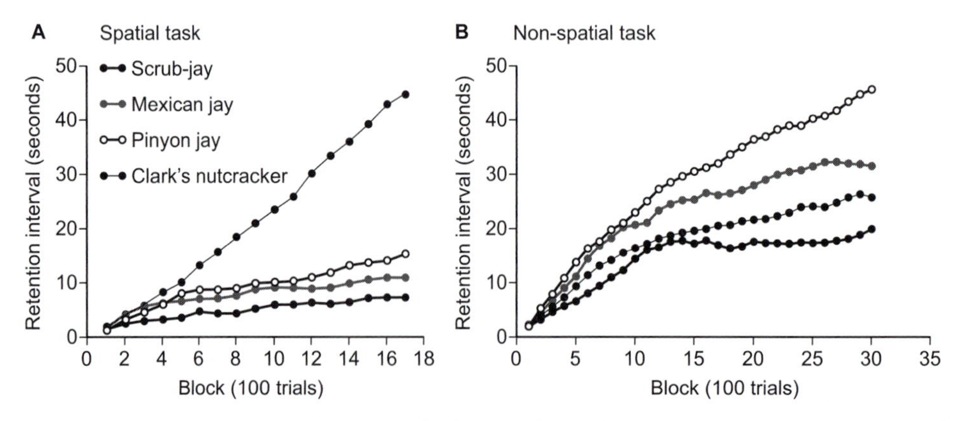

Figure 5.1 Spatial and non-spatial memory in four corvid species. (A) When tested in the lab, Clark's nutcrackers outperformed three other species in the crow family on a spatial memory task. (B) In contrast, this species did not exhibit superior abilities in a non-spatial memory task (remembering the color of a circle). Because they live in a harsher environment than the other bird species, Clarke's nutcrackers must rely more heavily on cached food sources to survive.

shows the results of an experiment testing this idea in which performance in a spatial memory task increased across the four corvid species that inhabit progressively harsher environments (Olson *et al.*, 1995). The findings could not be explained by differences in general cognitive abilities because the same species/performance relationship was not apparent in a non-spatial memory task. Compared to the other two species, Clark's nutcrackers and pinyon jays also have larger wings and bigger bills, both of which facilitate the transport of seeds over long distances. It seems likely, therefore, that both cognitive and morphological traits evolved in these species in response to the ecological challenge of reduced food availability during particular seasons (Balda & Kamil, 2006).

Differences in spatial memory related to ecological pressures have also been observed between subpopulations of the same species that inhabit different environments. For example, black-capped chickadees (*Poecile atricapilla*) from Alaska exhibit higher rates of successful cache recovery in the lab than do chickadees from Colorado where the winters are shorter and less severe (Pravosudov & Clayton, 2002). Not all studies, however, report better spatial memory in caching than in non-caching animals. Indeed, data summarized across a large number of these studies indicate that the findings should be described as 'supporting trends,' rather than 'conclusive results' (Smulders *et al.*, 2010). Smulders and colleagues go on to suggest that the adaptive significance of spatial memory should be examined by considering which components of this cognitive process reflect the natural history of a species. That is, the duration, capacity, and specificity of spatial memory should vary with the ecological constraints of an animal's natural environment. This appears to be the case in that animals retrieving caches after extended periods (e.g. up to 1 year or later) show the longest memory retention for stored locations in the lab, and those with the greatest number of caches in the natural environment can remember the most locations of hidden food (Smulders *et al.*, 2010). This approach, identifying which components of spatial memory should be affected by specific ecological pressures, has helped to uncover evolutionary adaptations of spatial memory, although many questions remain. One of the most controversial is whether evolutionary pressures have led to sex differences in this cognitive process (see Box 5.1).

Box 5.1 Sex differences in spatial memory

If spatial memory is shaped by natural selection, there may be sex differences in this ability because evolutionary pressures to find the way are different in the two sexes. For example, males often spend more time and travel a greater distance than females when they are looking for food or potential mates. This is reflected in animal experiments that consistently report a male advantage on tasks, such as the radial arm maze, that rely on memory for the spatial layout of an environment. Interestingly, there is no sex difference on this measure for species in which males and females show similar patterns of territory exploration. Male meadow voles (*Microtus pennsylvanicus*) have a home range up to four times larger than females whereas the home range size is similar for male and female pine voles (*Microtus pinetorum*). Male meadow voles, but not male pine voles, outperform female conspecifics in lab tests of spatial memory (Gaulin & Fitzgerald, 1986).

The male advantage in spatial memory is frequently offset by a female advantage in object recognition. In tests of this process, animals spend a few minutes exploring an environment that contains two objects (e.g. small Lego constructions). After a delay period, they are returned to the environment where one of the objects has been replaced with a new one. Object recognition is measured as increased time exploring the novel, compared to the familiar, object. Female rats (Saucier *et al.*, 2007) and mice

Figure 5.2 There is a common assumption that boys and girls use different strategies to solve spatial tasks.

(Bettis & Jacobs, 2009) outperform males on this task, suggesting that they have better memory for the unique features of cues in their environment. This would explain why females, but not males, make more mistakes on a radial arm maze task when objects within the maze are randomly moved (Williams *et al.*, 1990). In contrast, performance declines in males, but not females, when the objects stay in the same place but the overall geometry of the testing environment is altered.

Similar patterns of sex-related differences have been observed in other animals and in humans (Sandstrom *et al.*, 1998), leading to the idea that males rely more heavily on cues that provide global directional information, whereas females are more sensitive to information about the position of uniquely identified cues (Healy, 2006). These differences reflect averages with large overlap in the distribution of scores for each sex; thus, one cannot assume that any one woman will have better cue- and worse directional-based spatial memory than any one man, and vice versa. Nonetheless, the overall pattern of results suggests that sex differences in finding the way, originally described as a male superiority in spatial memory tasks (see Figure 5.2), are more accurately interpreted as a preferential reliance on directional cues in males and positional cues in females (Chai & Jacobs, 2010). An ultimate explanation for this distinction is based on a hunter gatherer division of labor: males inherit cognitive processes that promote hunting, whereas females inherit those that promote gathering, including the ability to identify and remember the location of edible plants (Silverman *et al.*, 2007).

As with caching, ecological pressures have shaped how animals are able to find the way during **migration**, the seasonal movement between spatially distinct habitats. Migration is a timely and physically challenging endeavor: the cost is enormous in terms of additional energy required for the journey as well as potential problems along the route. These disadvantages must be offset by fitness advantages in the new habitat. A case in point is that many birds spend summer months in northern Canada where they can feed on immense populations of protein rich insects over long daylight hours, but then travel to more southerly locations when these conditions decline. As with this example, selection tends to favor migration when there are predictable variations in environmental conditions across seasons. In all likelihood, short-distance migration preceded long-distance migration in that specific groups of animals began to travel further over successive generations as their environments, and the resources within them, changed.

During this evolutionary process, physiological traits that facilitate long-distance travel emerged in some species. These include a larger wing span in migrating birds and an increased capacity to store fat in most migrating mammals. No doubt, cognitive processes supporting migration also evolved in these species. Consistent with this idea, memory for the spatial location of a feeding site persists for at least 12 months in garden warblers (*Slyvia borin*), whereas the closely related, but non-migratory, Sardinian warblers (*Slyvia melanocephala*) retain the same information for only 2 weeks (Mettke-Hofmann & Gwinner, 2003). Further comparisons of migratory and non-migratory species are likely to reveal other differences in cognitive processes related to migration.

5.1.3 Flexibility

Finding the way, both getting there and knowing where, involves some level of flexibility. Caching is an obvious example as animals must be able to hide and retrieve food in different locations and at different times, depending on food availability and weather conditions. The same is true of migration: journey onset is shifted if weather conditions are unfavorable and animals may detour around storms or alter stopover locations if feeding grounds are overcrowded (Morell, 2006). Over repeated journeys, migrating western marsh harriers (*Circus aeruginosus*) exhibit less variability in departure times than in routes taken, suggesting that cognitive mechanisms controlling the direction of travel are more flexible than those controlling its timing (Vardanis *et al.*, 2011). Migration itself is flexible in that closely related species, as well as subpopulations of the same species, frequently display contrasting migratory behaviors. Within a single colony of European blackbirds (*Turdus merula*), the distinction between individuals that migrate and those that do not is related to social dominance (Lundberg, 1988). Given the increased competition for limited resources during winter months, subordinates benefit by moving to more fertile areas whereas dominants remain in place and avoid the energy expenditure of long-distance travel. Importantly, a single bird can switch from migrating to remaining resident across seasons, confirming individual flexibility of this behavior.

Much of the flexibility that animals exhibit in finding the way is explained by developmental processes. As outlined in Chapter 2 (Section 2.2.1), environmental input during a sensitive period alters sensory functions in adulthood. Early experience, therefore, could modify how organisms use environmental cues to find the way. This appears to be the case with convict cichlids (*Archocentrus nigrofasciatus*): fish of this species are far less likely to navigate using geometric cues, such as the corner of a tank, when they are raised in a circular tank than when they are raised in a rectangular one (Brown *et al.*, 2007). Similarly, pigeons locate their home loft using sensory cues, but whether they rely on olfactory, auditory, or geomagnetic cues depends on their experience during development (Wiltschko *et al.*, 1987). Olfactory cues that exert the most power in this situation are those that birds experienced early in life, confirming that a sensitive period controls this process (Mehlhorn & Rehkamper, 2009).

All animal species (including humans) are faced with the ecological challenge of finding the way. Retrieving food, returning home, or moving to a new habitat all depend on cognitive mechanisms that support getting there and knowing where. The variability in how animals solve these spatial problems is often explained by the evolutionary history of a particular species. For example, how an organism moves, where and when its food is available, and how quickly it travels through the environment will favor the emergence of one spatial strategy over another. Developmental experience interacts with these inherited traits, allowing animals to respond flexibly to current ecological demands. In other words, finding the way, either for caching, migration, or other spatial behaviors, is best served by a combination of inherited and acquired mechanisms.

5.2 Orientation

The primary mechanism of getting there is **orientation**, defined as an angle measured with respect to a reference. The reference may be a home location, a previous direction, or some external force such as gravity or a geomagnetic field. In simplistic terms, orientation is the process of taking up a particular bearing, regardless of the ultimate destination. If an animal is orienting in a certain direction but moved laterally, it will continue on the same path even though it will now reach a different endpoint. Movement in space that is controlled by orientation, therefore, sets an organism in a particular direction, not toward a particular goal.

5.2.1 Kinesis

Kinesis, possibly the simplest form of orientation, is non-directional movement in response to a stimulus. This definition appears to be contradictory in that kinesis is a form of orientation, which is directional. If kinesis is described in terms of its function, however, it fits within the larger topic of orientation because kinetic movements guide animals to a preferred location. Although this guidance may be indirect (i.e. without reference to a directional stimulus), it is not random. Kinesis is dependent on the intensity of the stimulus so movement increases in the vicinity of a stimulus, and decreases as animals move away. A classic example of this phenomenon occurs in common rough woodlice (*Porcellio scaber*), who exhibit increased locomotion in environments that are dry and well lit. The brighter and drier the conditions, the more the woodlice move. They slow down as they reach darker and moister areas, eventually stopping in a location that provides an appropriate habitat (Fraenkel & Gunn, 1961).

Larvae of the brook lamprey (*Lampetra planeri*) display a similar behavior, moving in the light until they are buried in mud at the bottom of lakes or streams (see Figure 5.3). Jones (1955) conducted a simple experiment to show that this response was non-directional (i.e. kinesis). He placed larvae in an aquarium in which they could not burrow at the bottom. A light at one end of the tank gradually diminished to darkness at the other end. Activity levels were highest at the lit end and were elevated even further when the light intensity was increased. Larvae stopped moving in the far end of the tank. In both the natural environment and artificial lab setting, larvae end up in dark environments NOT because they are avoiding the light, but because they stop moving in its absence.

5.2.2 Taxis

Often contrasted with kinesis, taxis is directional movement in response to a stimulus. The movement may be toward (positive) or away from (negative) the stimulus. Taxes may also be defined in terms of the stimulus eliciting the response: phototaxis is movement in response to light;

Figure 5.3 European brook lamprey (*Lampetra planeri*), spawning.

chemotaxis is movement in response to a chemical. Taxis results in behaviors that are narrowly focused on a specific stimulus. One of the best examples occurs in male moths that fly perpendicular to the wind direction until they detect the scent of female pheromones: they immediately turn upwind to follow the chemical gradient of the trail (Vickers, 2006). The same chemotaxic response is observed in male marine planktonic copepods (*Temora longicornis*), a small ocean crustacean that navigates vast areas of three-dimensional aquatic space to locate mates (Weissburg *et al.*, 1998). Many terrestrial insects as well as some birds and fish use a similar process of trail following, suggesting that chemotaxis may be a common mechanism of orientation across a variety of animals and habitats.

Negative taxis (i.e. a response that moves an animal away from a stimulus) occurs in many insects, including larvae of the housefly (*Musca domestica*). When photoreceptors are stimulated on either side of the head, flies move in the opposite direction. They swing their head from side-to-side as they fly, alternately activating left and right photoreceptors. This taxis response produces a zigzag flight pattern away from the light source (Barnard, 2004). Other animals orient away from particular environments, such as closed spaces of forests or open spaces of fields. This negative taxis may contribute to success in finding new habitats during dispersal or other large-scale movement as it directs animals away from unfavorable ecological conditions.

Kinesis and taxis are discussed, most frequently, as fundamental reactions in relatively simple organisms. They are included in this chapter because even seemingly straightforward responses to environmental input involve some level of information processing (i.e. cognition). Neither kinesis nor taxis can fully explain how animals move through their environment, but the fact that a specific sensory stimulus sets a specific motor pattern in action could be one of the building blocks for orientation in more complex species, including humans.

5.2.3 Setting the course

Beyond kinesis and taxis, other cognitive processes may facilitate orientation. Collectively, these are referred to as setting the course because they point organisms in a particular direction, using one or more cues as an orientation reference. These compass-like mechanisms have a flexible component in that many animals can orient to more than one type of cue. For example, wood mice (*Adopemus sylvaticus*) deprived of visual or olfactory input can still orient within their home range as well as non-deprived controls (Benhamou, 2001), suggesting that they are compensating for the lack of one sensory cue by relying on another. The ability to orient to different sensory cues facilitates behaviors as fundamental as foraging and mate choice because the mechanism is not restricted to current environmental conditions (e.g. if scent marks are obliterated, animals can use other signals to orient in the correct direction). These findings also highlight that the cognitive process of orientation does not depend on specific sensory input, at least in some animals.

Setting the course is easy to conceptualize as the initial stage of any spatial behavior, but this cognitive process may also continue to operate once animals are in motion. For example, if obstacles or detours are encountered along a route, animals will need to reorient using a new, possibly different, reference cue. Even if they continue on a direct path, orientation mechanisms may be initiated when animals encounter olfactory, visual, or other sensory cues along a journey. The manner in which orientation interacts with mechanisms that control ongoing spatial movement will be discussed in the next two sections.

5.3 Small-scale navigation

Navigation is knowing where, defined more precisely as the process of identifying a specific location, regardless of one's current position. Navigation is apparent as animals move through their environment to retrieve hidden food, revisit feeding sites, or return home. In contrast to orientation, therefore, navigation is determined by a destination or goal. This section and the next (Section 5.4) discuss the cognitive representation of these goal locations, as well as how animals use this information to guide behavior. The material is divided into small- and large-scale navigation, although the distinction between the two is not absolute. In general, small-scale navigation refers to spatial movement within a few kilometers, often familiar territory of the home range. Adaptive behaviors that are facilitated by small-scale navigation include foraging, caching, and territorial defense.

5.3.1 Landmarks

Small-scale navigation is often accomplished using **landmarks**, stimuli that have a fixed spatial relationship to a goal location. Because these cues are stationary, they provide both distance and directional information to the goal. Tinbergen (1951) presented one of the first demonstrations of landmark use by a type of digger wasp, European beewolf (*Philanthus triangulum*). He noticed that wasps hover above their nest before a foraging expedition and he speculated that this behavior allowed the insects to recognize cues marking the nest entrance. To test his idea, Tinbergen placed rings of pine cones around some of the nests and then moved these to a nearby location after the wasps had departed. When they returned, wasps searched unsuccessfully for the nest in the middle of the displaced cones (see Figure 5.4).

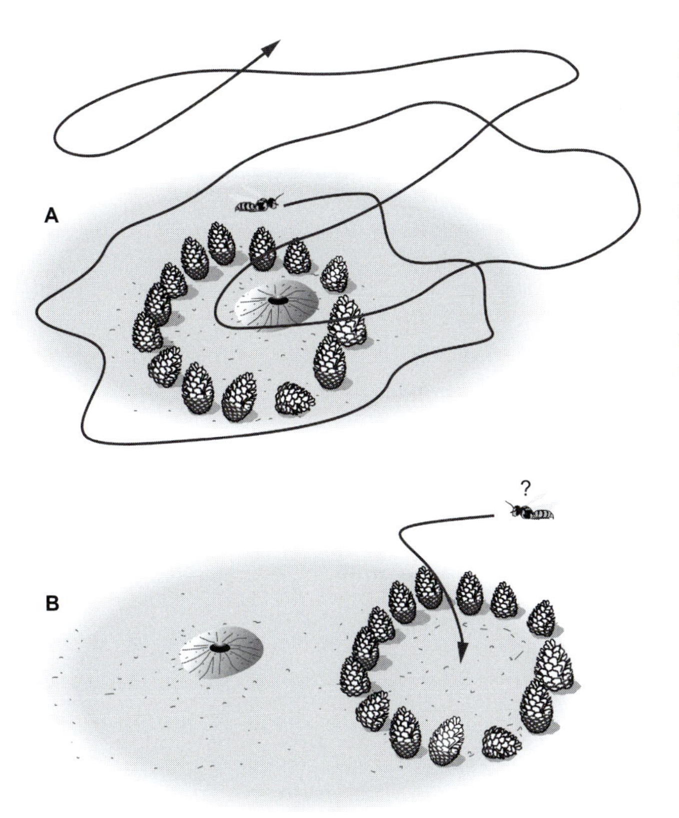

Figure 5.4 Landmark use by digger wasps. Tinbergen (1951) placed pine cones around the periphery of a wasp nest. After the wasps left the nest to forage, pine cones were moved to a nearby location. When they returned, wasps searched for the nest in the middle of the pine cones, indicating that they were using these cues during spatial navigation. Lines indicate the typical flight pattern of a wasp when it leaves (A) and returns to (B) the nest under these conditions.

Since Tinbergen's publication, navigation by landmark use has been observed in honeybees (Chittka *et al.*, 1995), rats (Roberts & Pearce, 1998), and turtles (Lopez *et al.*, 2000), as well as several species of birds (Jones *et al.*, 2002) and non-human primates (Deipolyi *et al.*, 2001; MacDonald *et al.*, 2004; Sutton *et al.*, 2000). Most of these animals appear to be judging the goal location based on its absolute distance from a landmark. This is demonstrated in experiments using an expansion test in which subjects are trained to locate a goal within an array of landmarks. During testing, the landmarks are all moved further away from the goal so that the absolute distance between the goal and the landmarks increases, but the geometric relationship between them remains the same. Under these conditions, only human adults search for the goal in the location that bears the same spatial relationship to the entire array (Marsh *et al.*, 2011). Young children initially search in a region that is the same distance from one of the landmarks as the original goal, but outgrow this tendency as they mature. Even at a young age, however, children who search unsuccessfully using an absolute distance strategy may shift to using a relational one. This contrasts with search patterns of non-human primates, including Sumatran orangutans (*Pongo abelii*) and common marmosets (*Callithrix jacchus*), who stick with an ineffective (absolute distance) search pattern (MacDonald *et al.*, 2004; Marsh *et al.*, 2011). Many animals can be trained to use relational strategies to locate a goal, but (other than humans) the only species that appears to do this spontaneously is the honeybee (Cartwright & Collett, 1982). This likely reflects some specialized aspects of the bee visual system that is still not clearly understood.

Landmark navigation is closely related to navigation using **beacons**, which are stimuli that mark the goal location (as opposed to having a fixed spatial relationship to the goal). Beacon

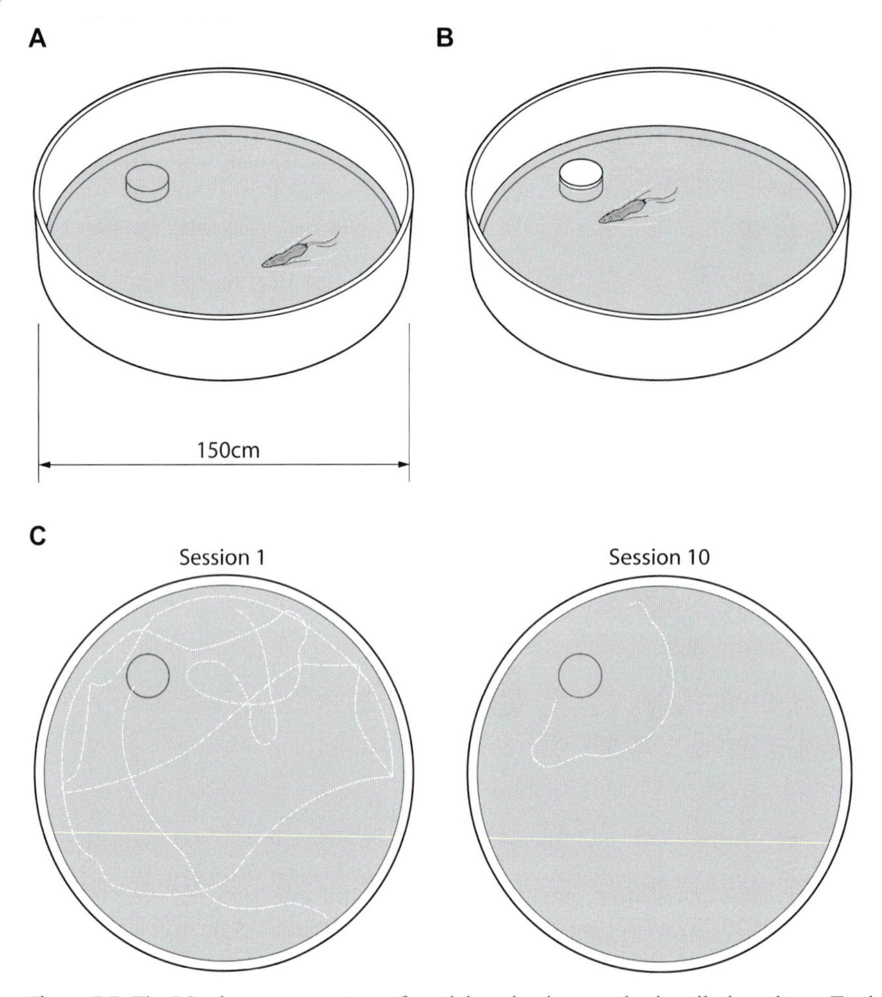

Figure 5.5 The Morris water maze test of spatial navigation, used primarily in rodents. Testing is conducted in a circular pool (diameter approximately 150 cm) filled with opaque water. (A) In the hidden platform version of the test, a platform is placed in the same location of the pool slightly below the water surface. (B) In the visible platform version, it is placed slightly above the water surface and is moved on each trial. For both tests, animals are placed in the water at a different location on each trial. Initially, animals swim randomly around the pool until they bump into the platform and can climb on top to escape. Over trials, they learn to swim directly to the platform. Swim patterns early and late in training for a typical rat in the hidden platform version of the Morris water maze are shown in C.

navigation is characterized by direct approach to the goal location, analogous to conditioned approach responses described in the previous chapter (Section 4.2.1). Navigation using landmarks versus navigation using beacons can be examined in two different versions of the Morris (1981) water maze task (see Figure 5.5). In the first, animals are lowered into a small pool where they must swim around until they locate a hidden platform. The water is made opaque by adding non-toxic paint so animals cannot see the platform from the water surface. A number of large objects are placed on the walls around the pool, which serve as visual cues during navigation. Because they are released from a different starting point on each trial, animals cannot travel directly toward one visual cue or another; they must learn to locate the platform based on the spatial relationship between the cues and the platform (i.e. landmark use). In the second version of the task, the platform is visible but moves on each trial; animals learn to swim directly to the platform (i.e. beacon use) because visual landmarks do not provide reliable information about the goal location.

The fact that rodents learn both hidden and visible platform versions of the water maze task suggests that they can use either landmarks or beacons in small-scale navigation. The same is true of many insects, birds, fish, turtles, and other mammals. A fundamental question in comparative cognitive research, therefore, is whether a particular species is more *likely* to employ one strategy over the other. This can be tested using a cue competition task in which animals are trained to locate a target (usually a food reward) that has a fixed relationship to environmental cues (i.e. landmarks) and is also marked by distinctive cues at the goal location (i.e. beacons). During testing, beacons, but not landmarks, are moved: if animals search in the original location that still bears the same spatial relationship to environmental cues, rather than going directly to the displaced cues, they are navigating by landmarks, not beacons and vice versa. An experiment with this setup showed that food caching birds are more likely to use landmarks to locate food than are non-caching birds (Clayton & Krebs, 1994). This difference reflects ecological circumstances in that food caching birds must rely on prominent features of the landscape (e.g. trees or protruding rocks) to locate hidden stores, particularly over changing seasons when the appearance of cache sites may be altered.

Dozens of papers have been published using paradigms such as the cue competition task to tease apart which species use landmarks versus beacons in small-scale navigation. A recent review of these studies confirms that food caching animals rely more heavily on landmarks when they are further away, but switch to beacons in close proximity to the goal (Gould *et al.*, 2010). This strategy works well in the natural environment because beacons may be more difficult to detect (either to see or to smell) at a greater distance. The same principle directs navigation in honeybees (Collett & Baron, 1994) and social wasps (Collett, 1995) that fly toward different cues depending on how close they are to a feeding station. Many other animals, including rufous hummingbirds (*Selasphorus rufus*), use both beacons and landmarks in small-scale navigation (Healy & Hurly, 1995). If these birds are trained to find sucrose in artificial flowers of different colors, they return to the location where they found the food, even when the colors of the flowers are switched. In the natural environment, flowers seldom move so using landmarks (rather than a color beacon) to return to the food is an adaptive strategy. Landmarks are not as useful when animals are foraging in unfamiliar territory; in these situations, hummingbirds rely on color to locate food-containing flowers.

The ability to switch from landmarks to beacons, and back again, may explain some of the controversies regarding small-scale navigation because specific training or testing conditions may increase the likelihood that a certain species will navigate using one strategy or the other (Healy, 2006). For example, even if an experiment is set up with large visual landmarks around a maze (i.e. landmarks), rats may find the food reward by approaching olfactory cues near the goal location (i.e. beacons). If researchers fail to identify how animals solve a spatial task, they may reach erroneous conclusions regarding navigational mechanisms in that species. Thus, the most successful studies of spatial navigation acknowledge inherited tendencies to use one type of cue or another, while recognizing that there is a great deal of flexibility in how these cues are used by both humans and animals.

Researcher profile: **Dr. Sue Healy**

Figure 5.6 Dr. Sue Healy.

A major challenge in comparative cognition research is to integrate studies in evolutionary biology with those in experimental psychology. This has never deterred Professor Sue Healy whose research on spatial navigation combines theories and techniques from the two traditions. Healy's original plan to be a medical doctor was derailed when she began a biology degree at the University of Otaga in New Zealand. Her encounters with a stimulating group of friends (ready to argue about science for hours, particularly over gin and tonic) and inspiring female lecturers cemented her decision to pursue zoology and physiology. After completing her undergraduate degree, Healy opted for 2 years of 'overseas experience' that quickly extended into a long-term commitment when she landed a job as a research assistant to John Krebs at Oxford University. Krebs was beginning his research into the relationship between food storing, spatial memory, and the hippocampus (see Section 5.5.1) and Healy took up the topic with such passion that she then transferred into a PhD. She claims that she has never lost her enthusiasm for the topic and has used this work as a foundation to investigate the cognitive abilities of animals in the 'real' world. Her work on small-scale navigation is a prime example of how observational data inform studies with experimental manipulations, and vice versa. In Healy's works "I try to add neural and/or endocrinological data, to build a comprehensive picture of the role that cognition plays in what, how, and why the animal is behaving as it is."

According to Healy, one of the best parts of her work is that it provides the opportunity to travel to many gorgeous places and spend time watching all kinds of cool animals going about the business of their lives. This includes studies of vocal mimicry in Australian spotted bowerbirds (*Ptilonorhynchus maculatus*), visual cue navigation in rufous hummingbirds in the Canadian Rockies, and nest building of African masked weaver birds (*Ploceus velatus*). But above all, Healy raves about her interactions with other scientists, including her partner: "His enthusiasm for invertebrates more than matches my own for cognition." She goes on to say that the people around her make the intellectual challenge of doing science so much more enjoyable. "I have been enormously lucky in my students, postdocs, and collaborators – they have been, and are, a bunch of smart but also, importantly, fun people."

5.3.2 Path integration

Rather than relying on landmarks, **path integration** allows animals to return to a starting location (usually their home) by keeping track of the combined distance and direction travelled on a single journey. Path integration, also called dead reckoning, was revealed in a study with foraging desert ants (*Cataglyphis bicolor*) that leave their nest and take a circuitous route searching for food; as soon as food is found, they turn and run, more or less directly, back to the nest (see Figure 5.7). If the nest is moved, ants exhibit a back and forth search pattern centered on the previous nest location. If ants are picked up and moved at the beginning of a return trip, they behave as if they were not displaced, running the computed distance and direction that would bring them to the goal, then frantically searching where the goal would have been located (Wehner & Srinivasan, 1981). Path integration is studied most commonly in insects, but it has also been observed in birds and a number of mammalian species (Gallistel, 1989). This includes mother Mongolian gerbils (*Meriones unguiculatrus*), which exhibit path integration during pup retrieval: although their search involves

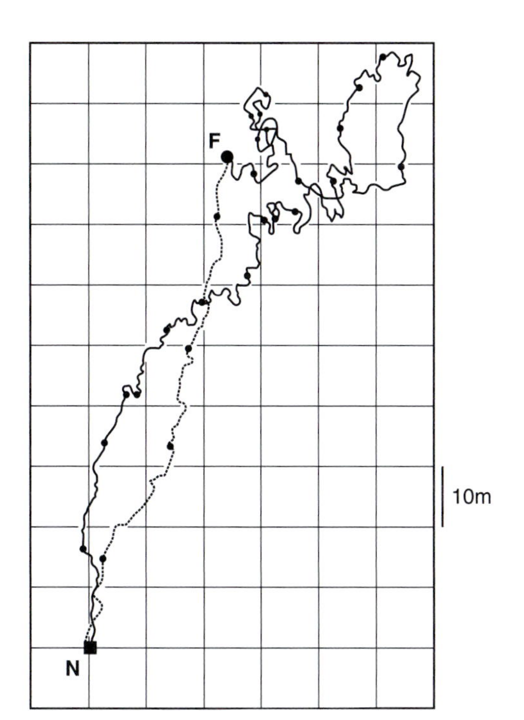

Figure 5.7 Path integration in desert ants during foraging. The solid line marks the travel path of an ant as it leaves its nest (N) and locates food (F). The dotted line shows the ant's travel route back to the nest. The outbound path is 354.5 m; the maximal distance from the nest is 113.2 m. Small filled circles represent 60-second intervals along each path.

numerous twists and turns, when the mothers find a pup, they carry it directly back to the nest even in complete darkness (Mittelstaedt & Mittelstaedt, 1980).

Path integration necessitates continuous monitoring of one's present location in relation to a starting point. In other words, an organism must receive ongoing feedback on its own movement through time and space in order to calculate a direct path home. Despite demonstrations of path integration in different species, there is no common formula for how this is accomplished. Even insects use different mechanisms to keep track of distance and direction on each journey. Honeybees use visual flow information, the rate at which images cross the eye surface, to calculate return distance to a food source (Srinivasan *et al.*, 1996) whereas desert ants use a pedometer, keeping track of the number of steps they have taken from a starting point (Wittlinger *et al.*, 2006). These differences make sense as bees fly through their environment whereas ants walk; spiders use a similar mechanism to ants (Seyfarth *et al.*, 1982). At the same time as they are calculating distance travelled, insects monitor the direction of their travel using the sun as a reference point. Amazingly, they are able to compensate for the movement of the sun across the sky, suggesting that some internal clock is marking where the sun should be at a particular time (details of this cognitive process will be covered in Chapter 6). This is true even when insects are reared under conditions in which they only see the sun for part of the day (Dyer, 1998).

Unlike insects, path integration in mammals appears to involve the **vestibular system**, which is composed of sense organs in the inner ear that monitor body rotation and movement in relation to gravity. For example, rats with vestibular lesions display path integration when the environment is lit (i.e. they travel directly back to a nest after finding food) but appear lost in the dark, searching randomly until they stumble across their home (Wallace *et al.*, 2002). Unlesioned rats travel directly to their nest in both conditions, although they probably rely on different cues to do so (i.e. visual in the light and vestibular in the dark). Disrupting vestibular signals in other ways, such as slowly rotating an animal's environment, has similar effects suggesting that this is a primary mechanism of path integration in mammals (Knierim *et al.*, 1996).

A major problem with path integration is that minor errors in calculation can have profound effects on navigation. If an ant begins a homeward journey 10 meters from the nest and is mistakenly oriented a few degrees to the side, they may miss their home completely. The problem is magnified the further the animal is from their final destination. For this reason, many researchers believe that path integration is used during the initial stages of a return trip, but that animals rely on other cues, such as landmarks, to navigate when they are closer to a goal. They may also use landmarks along the way to reset the path integration calculator. This helps to minimize errors that would accumulate along a lengthy route (Knierim *et al.*, 1996).

5.3.3 Environmental geometry

Many ideas about small-scale navigation, particularly how animals locate food in a familiar environment, were altered by a classic experiment with rats (Cheng, 1986). The study design, shown in Figure 5.8, involved a rectangular arena in which the corners were marked by distinct olfactory and tactile cues. In the exposure phase, a food reward was placed at a random location in the arena; rats were allowed to consume part of the reward and then removed from the arena for a short period. When they were returned, the remainder of the reward was hidden in exactly the same place under the sawdust flooring. Rats quickly learned to return to the location and uncover the reward. On test trials, the arena was rotated after the exposure phase but the cues stayed in the same location. The results were telling: half of the rats searched in the correct location but the other half dug through the sawdust in the diagonally opposite corner. This effect, dubbed **rotational bias**, indicates that these animals were using information about the geometric relationships in their environment, rather than cues such as smell, to locate the food reward.

Subsequent studies tested other animals in similar paradigms, often with mixed results. For example, chicks use feature cues, such as wall color, to locate food in a rectangular room that has been rotated, but if these cues are removed, the birds exhibit a standard rotational bias effect

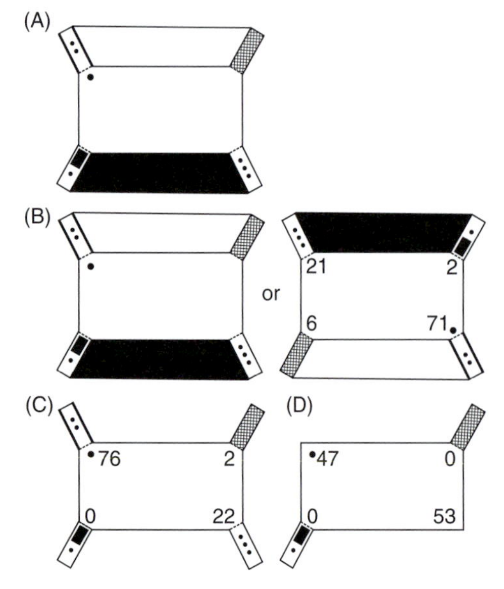

Figure 5.8 Experimental setup and results of a study testing the use of geometric cues in rats. In all panels, the view of the rectangular testing box is shown from above. Filled circles represent the location of hidden food. (A) The testing environment consists of distinct geometric cues in that three walls are painted black and one is painted white; the corners each contain different visual, tactile, and olfactory cues. Numbers indicate the percentage of trials on which rats search in each corner for the hidden food during rotational transfer tests (B). (C) All four walls are painted black so geometric cues do not provide unambiguous information about where the food is located. The corners contain distinct cues that are used to locate the hidden food. (D) All four walls are painted black but the corner containing the hidden food and the diagonally opposite corner contain no cues. The two other corner panels contain feature cues so the location of the food cannot be ascertained by geometric or feature cues. Numbers in C and D indicate the percentage of trials on which animals search in each corner for the hidden food under these conditions.

(Vallortigara *et al.*, 1990). The same is true of pigeons: they navigate using environmental geometry when features in the environment become more difficult to discern (Kelly & Spetch, 2001). Eventually it became clear that a range of species from monkeys to birds to fish exhibit a rotational bias effect under certain experimental conditions (Cheng *et al.*, 2006). The effects are not limited to rectangular corners; animals can locate a food reward using different types of angles, the length of major landmarks such as walls, as well as the spatial relationship among objects in the environment. As with beacons and landmarks, some species exhibit a tendency to use one type of geometric cue over another, which probably reflects aspects of their natural environment. For example, pigeons rely on the amplitude of an angle more than the length of a wall to locate food in a parallelogram-shaped room; these may mimic images of cliff crevices where pigeons nest (Lubyk & Spetch, 2012).

In humans, the tendency to use geometric versus feature cues changes over development. Toddlers exhibit a rotational bias effect (i.e. they use geometric cues) when they are searching for a hidden toy, ignoring non-geometric cues such as the color of a wall. Six-year-old children performing the same task rely more heavily on feature cues; adults do the same, although they revert to geometric cues if the feature cues are not salient or reliable (Hermer & Spelke, 1996; Wang & Spelke, 2002). Spelke and colleagues argue that only humans show this developmental shift in cue use, although they acknowledge that many animals *can* use either feature or geometric cues, depending on the experimental setup. The same researchers go on to note that geometric cues may allow both humans and animals to establish a new path direction. In other words, the ability to use environmental geometry may be more important for orientation than for navigation. This makes sense in that angles and other geometric cues are view dependent, so they are only reliable when animals are in the same location. They also change as organisms move through their environment so they are likely to be used at critical points along a journey, such as ‘this is where I should turn.’ Environmental geometry would also facilitate ongoing navigational processes, such as path integration, because errors along a route could be corrected if animals use this information to orient in a new (correct) direction.

Despite widespread evidence that animals and humans can use environmental geometry in orientation and navigation, there is considerable debate about what this reflects at the cognitive level. Some argue that many of the experimental findings cited above could be explained by simpler processes, such as associative learning (Pearce *et al.*, 2004), whereas others propose that the processing of geometric relationships occurs independently of other cognitive functions (Cheng *et al.*, 2006). No one doubts that meaningful aspects of the natural environment are defined by specific geometric shapes, but how this information is used to guide movement through space is still not clear.

5.3.4 Cognitive maps

A classic image in psychology is of a rat running through a maze like the one shown in Figure 5.9. Behavioral psychologists in the early twentieth century proposed that animals reached the goal box in these mazes, successfully avoiding dead ends, by learning to turn down alleyways that led to the

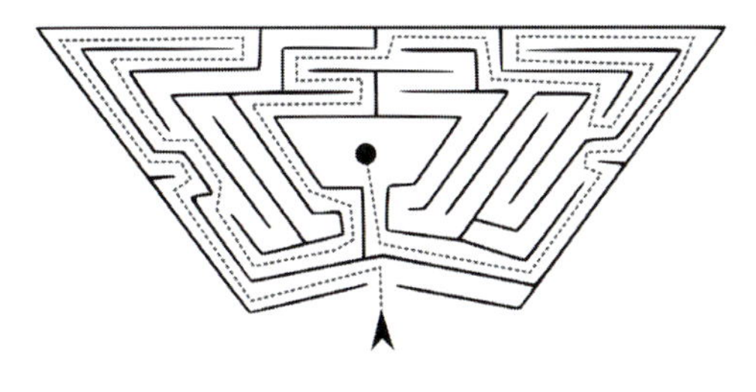

Figure 5.9 Hampton court maze that was used as a model for spatial learning experiments with rats in the early twentieth century.

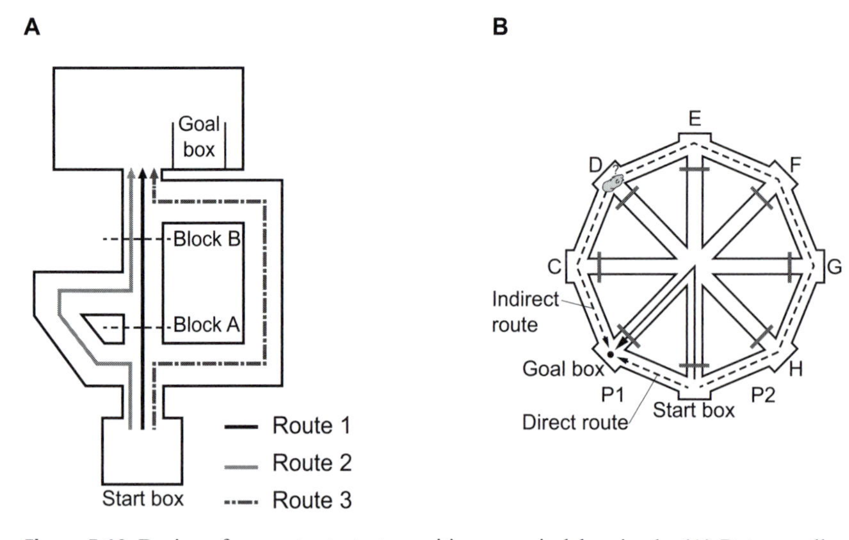

Figure 5.10 Design of apparatus to test cognitive maps in lab animals. (A) Rats are allowed to freely explore a maze that has different routes from a start box to a goal (food) box. Over training, they select the most direct route (1) between the two. When a block is placed at A, they select the next most direct route to the food: Route 2. When a block is placed at B, rats select Route 3, suggesting that they have a cognitive map of the entire maze, understanding that Block B obstructs Route 1 as well as Route 2 (Tolman & Honzik, 1930). (B) Hamsters are trained to navigate a circular maze from a start box (A) to a goal box (e.g. B) where food is located. During training, the arms leading to the center of the maze are blocked so animals must traverse the circumference of the maze to reach the goal box. During testing, the blocks to the center arms are removed; animals are placed in A with the path around the outer maze blocked so that they must travel to the center of the maze. From the center point, hamsters then take the most direct route to the goal box (Chapius & Scardigli, 1993).

food. Tolman (1948) argued against this interpretation: he hypothesized that animals find their way by forming a cognitive map of the environment as they travel through the maze. According to Tolman, this mental representation included the spatial relationship among landmarks and beacons, as well as the animal's own position with respect to these cues. The concept of cognitive maps fits an evolutionary perspective in that a mental image of the animals' surroundings could facilitate adaptive behaviors such as foraging, nest building, and territorial defense.

One of the strongest arguments for cognitive maps is evidence that animals can plan novel routes to a familiar location. Tolman and Honzik (1930) demonstrated this phenomenon in the lab using a maze with three different routes to the food reward (see Figure 5.10A). During the initial training, all of the alleys were opened and rats learned to find the most direct route (alley 1) between the start and goal boxes. Then, a barrier was placed at A. When rats encountered the barrier, they turned around and took alley 2, the next most direct route. Finally, a barrier was placed at B. This time when rats turned around, they took alley 3 to the goal box indicating that they 'understood' barrier B was blocking both of the first two alleys. Tolman concluded that rats had formed a spatial map of how the three alleys fit together. Behavioral psychologists, such as Hull, explained the findings in terms of stimulus–response learning, pointing out that rats could solve the maze using cues in the room (e.g. lights or pieces of furniture) as landmarks or beacons. Dozens of papers were published over the next 50 years, purportedly supporting or contradicting the existence of cognitive maps. Figure 5.10B shows an updated version of one of these studies using hamsters (Chapius & Scardigli,

1993). Animals were trained to take one route (e.g. A – F – E) but then barred from this route during testing. Importantly, the ability to use landmarks was minimized by rotating the maze and starting the animals from different locations on each trial. Even with these additional controls, hamsters were more likely to use the most direct route (e.g. A – G – E) during testing, suggesting that they had formed a mental representation of the maze design.

Other findings support the idea of cognitive maps in different species. For example, if chimpanzees (*Pan troglodytes*) watch an experimenter hiding food at various locations in their enclosure, they retrieve the food in a different order from which it was hidden (Menzel, 1973). That is, they appear to have a spatial map of each location and take the most direct route from one hiding site to the next. Vervet monkeys (*Chlorocebus pygerythrus*) do the same thing (Gallistel & Cramer, 1996), although they cannot remember as many locations as the chimps. The water maze task (successfully navigated by rats, mice, guinea pigs, and other animals) provides indirect evidence for cognitive maps in that animals go directly to the hidden platform from new start locations. Certain behaviors in the natural environment can also be explained by cognitive maps. Meerkats (*Suricata suricatta*) escape directly to the nearest hole when they hear an alarm call (Manser & Bell, 2004) and honeybees take novel shortcuts between hives and feeding stations (Menzel *et al.*, 2005), suggesting that both species know the spatial layout of their environment. Reminiscent of the chimps in Menzel's study, marsh tits (*Poecile palustris*) do not follow the same path when they store and retrieve seeds (Shettleworth & Krebs, 1982), indicating that they know where the seeds are stored, not just how they got to the storage sites in the past. Finally, jumping spiders (*Portia labiata*) select the most efficient route to a food source when obstacles, such as broken branches, block the shortest path (Tarsitano & Andrew, 1999).

The concept of cognitive maps is appealing because many people conjure up a mental image of their spatial environment when they give or receive directions. If animals could do the same thing, it would provide further evidence for cognitive maps in non-human species. The process of giving and receiving directions has been studied extensively in honeybees who communicate the location of a food source through an elaborate dance (Chapter 12 discusses honeybee communication in detail). Gould and Gould (1988) capitalized on this knowledge by placing a sugar source on the shore of a lake within the foraging area of a nearby hive. Bees who flew to the feeding site transmitted information about the sugar location to the rest of the hive. The sugar source was gradually moved to the center of the lake and some of the bees transported with it. When foraging bees returned to the hive, again communicating the location of the food source, nest mates did not fly to the new site. According to the authors, honeybees 'knew' that a sugar source was unlikely to be situated in the middle of a lake because they possessed a spatial map of their environment. The study was often cited as evidence for sophisticated cognitive abilities in insects, but the findings were never replicated (a key component of scientific research). In fact, the Lake study, as it became known, was never published itself, just reported in other review articles by the authors. Most importantly, when the experiment was repeated, honeybees followed directions to a sugar source in the middle of a lake just as frequently as to a land location the same distance away (Wray *et al.*, 2008). At least in this study, honeybees were not discriminating between food located in probable and improbable locations. It is unlikely, therefore, that these insects were navigating using a cognitive map.

The insect studies aside, many people are willing to believe that mammals or birds possess cognitive maps, but the evidence supporting this claim is not irrefutable. As with Tolman's initial experiments, critics point out that rats and other animals can navigate mazes using some combination of environmental cues or a sophisticated form of path integration, even if they have not formed a mental map of their environment. Similarly, birds and primates that retrieve food in a different order from which it was hidden may simply go to the closest cue, retrieve the food, and then proceed to the next cue. Even if researchers agree that a particular species can form a cognitive map, they may disagree on the nature of this representation. A cognitive map may be a precise record of how spatial elements (e.g. cues and geometry) are connected in the environment (Alvernhe *et al.*, 2012), a more general association between major landmarks and the organisms' position within this space (Gallistel, 1990), or something in between.

Even in humans, cognitive mapping is a hotly debated topic. The main point of contention is how mental representations of spatial information are organized. One proposal is that humans form cognitive maps of different environments, but have difficulty combining these even if they are linked in the real world. Wang & Brockmole (2003) examined this by asking individuals to imagine navigating between small and large spaces in their environment (e.g. within a room in a particular building versus that building within a larger complex of buildings). They concluded that individuals switch between spatial representations as they move through an environment, mentally situating themselves within the small or large space, but not both. For example, finding a room within a building, then choosing which door of the building provides the most direct access to another building involved shifting from one cognitive map to another. According to the authors, this shift is mediated, or at least facilitated, by language, suggesting that the ability may be unique to humans. A complementary line of research focuses on understanding how other cognitive processes, such as memory, relate to spatial mapping (Burgess, 2006). These processes must interact because cognitive maps will be modified and updated as organisms move through their environment. There is little doubt that humans and animals 'know where' they are in relationship to spatial cues in their environment; whether they do so by referring to a cognitive map is unresolved.

5.3.5 Interactive systems

The preceding sections emphasize that most species can use different cues during small-scale navigation, even if they have a tendency to employ one strategy over another. This predisposition to rely on a particular cognitive process often reflects the ecological constraints of a species. For example, food caching animals may use some form of a cognitive map, rather than environmental geometry, to locate food because they regularly approach hidden stores from different directions. In addition, exposure to specific cues during development (Brown *et al.*, 2007), as well as training in a particular paradigm (Cheng *et al.*, 2006), can influence which cognitive mechanism animals use during navigation. If animals learn that one set of cues (e.g. landmarks) are not reliable predictors of a goal location, they may revert to an alternative strategy (e.g. environmental geometry) even if they were unlikely to do so in the natural environment. The fact that many species can use spatial information in flexible ways allows them to adapt to changing environmental conditions (Gould *et al.*, 2010), some of which involve endogenous physiological signals (see Box 5.2).

Once it became clear that animals have different mechanisms of small-scale navigation at their disposal, researchers began to investigate how these function together. One suggestion is that navigation is guided by *either* landmarks or environmental geometry: animals may switch between the two as they move through the environment, but only one of the processes is directing behavior at a given point in time. Evidence supporting this claim comes from experiments using the cue competition paradigm described in Section 5.3.1. In general, learning to navigate using landmarks makes it difficult to find a goal using geometric cues and vice versa. As with landmarks versus beacons (discussed previously), animals appear to rely on one strategy or the other, but do not combine information from the two (Shettleworth, 2010). It is unlikely that this distinction reflects an inability to use multiple sources of spatial information because food caching birds (at least Clark's nutcrackers) are more accurate in retrieving food as the number of landmarks increases (Kamil &

Box 5.2 Hormonal modulation of spatial navigation

Box 5.1 outlined evidence for sex differences in spatial navigation, more accurately described as a preferential reliance on directional cues in males and positional cues in females. But this distinction is not absolute: in many species, males outperform females on spatial memory tasks *only* at certain times of the year. This observation led Galea and colleagues to speculate that changes in circulating levels of sex hormones underlie a male superiority on laboratory tests of spatial memory (Galea *et al.*, 1996). To test this idea, they compared male and female performance in the water maze across reproductive cycles. Female deer mice (*Peromyscus maniculatus*) were worse at finding the hidden platform during the breeding season whereas males of this species were better. In both males and females, performance was inversely related to circulating levels of the female sex hormone, estradiol. In contrast, higher exposure to testosterone (the primary male sex hormone) *in utero* was associated with improved performance. Importantly, sex differences were not apparent prior to puberty, suggesting that differences in spatial memory emerge when sex hormones are activated.

An important aspect of Galea's study is that she tested spatial memory under conditions of naturally fluctuating hormones. Manipulations that artificially block or enhance these levels often produce contradictory effects because they push hormones beyond a normal physiological range. Although they were specifically interested in proximate mechanisms of spatial navigation, Galea and colleagues provide an ultimate explanation for their findings that increasing levels of male and female sex hormones have opposite effects on spatial navigation (improving and disrupting this ability respectively). They note that male reproductive behaviors would be facilitated by increased spatial exploration as animals search for mates, but female reproductive strategies would favor staying close to home. This is particularly true when females are close to giving birth, which may explain why women show deficits in spatial memory tasks at the later stages of pregnancy (Buckwalter *et al.*, 1999). Indeed, Liisa Galea often describes her findings in terms of her own experience, desperately searching for her car in the parking lot when she was 9 months pregnant (Galea, personal communication).

Cheng, 2001). In other words, cognitive load is not a problem in these tasks; it is the cognitive process of using landmarks to locate a goal that operates independently of navigation by environmental geometry.

In contrast, path integration and cognitive mapping may work in parallel to landmark or environmental geometry navigation. It makes sense that these processes operate simultaneously as cognitive maps (if they exist) are a combined representation of landmarks and environmental geometry. Path integration may work in a different way, acting as a backup system to landmark navigation when animals get off track (Cheng *et al.*, 2007). This implies that computations that would allow animals to return home by a direct route are ongoing, but only come into play if cues in the environment are unreliable or unstable. Even desert ants appear to use landmark cues preferentially, only resorting to path integration if they are displaced from their travelling route.

In sum, small-scale navigation is accomplished in a variety of ways: animals may locate a goal using landmarks, environmental geometry, path integration, or cognitive maps. Although a particular species may exhibit a tendency to rely on one process or another, most animals can navigate using more than one of these mechanisms. Frequently, animals switch between strategies, depending on how close they are to the goal and what they have learned about spatial navigation in the past. When cognitive processes controlling navigation are activated simultaneously, they may compete to control behavior (e.g. landmarks and environmental geometry), interact cooperatively (e.g. landmarks and environmental geometry facilitate cognitive maps), or function in a hierarchy (e.g. landmarks override path integration). These interactions highlight the flexibility of small-scale navigation in that the ability to rely on different processes allows animals to adjust their behavior to the ever-changing world around them.

5.4 Large-scale navigation

Large-scale navigation is defined as long-distance travel to a goal, often through unfamiliar territory. Distinguishing large- from small- scale navigation is not always easy as it can be difficult to know whether a particular environment is unfamiliar to animals not reared in captivity. Moreover, the difference between 'short' and 'long' distances will depend on the species under study (e.g. ants versus birds). Nonetheless, small- and large-scale navigation are usually examined independently, even if researchers agree that the two may involve overlapping mechanisms (Frost & Mouritsen, 2006).

5.4.1 Homing versus migration

One of the amazing feats of large-scale navigation is **homing**, the ability to return to a nest or burrow after displacement to a distant, often unfamiliar, site. Humans exploited the homing abilities of birds during the nineteenth and early twentieth centuries by sending messages over long distances attached to the legs of homing pigeons (*Columba livia*), appropriately named carrier or messenger pigeons. These displaced birds were able to return, more or less directly, to a start location, suggesting that cognitive processes supporting this behavior provided some fitness advantage in the past. One of these relates to breeding, which is expressed as **natal homing** (returning to one's birthplace to reproduce). Natal homing occurs most commonly in aquatic animals such as green sea turtles (*Chelonia mydas*). These animals hatch at night on a sandy beach and head directly to the water where they spend most of their life swimming in open seas and feeding at distant locations (sometimes at the other side of the Pacific Ocean from where they were born). The beaches that are ideal for feeding are not the same as those needed for nesting so three or four decades later, when females are ready to reproduce, they return to their natal site to lay eggs. In all likelihood, this long-distance travel between feeding and birthing sites did not arise spontaneously: although the two locations are now separated by an ocean, they may have been closer together when turtles evolved more than 100 million years ago. Natal homing of Pacific salmon (genus *Oncorhynchus* that includes eight different species) is just as amazing. They spawn in freshwater streams that feed into the ocean, spend 4 or 5 years swimming in the Pacific, then return to their birthplace when they reach sexual maturity (Dittman & Quinn, 1996). These journeys can be arduous, covering up to a few thousands kilometers, much of it upstream (see Figure 5.11).

Migration (large-scale, seasonal movement between spatially distinct habitats) can be even more physically demanding. Migration of the bar-tailed godwit (*Limosa lapponica*) includes an 11 000 km non-stop flight, from Alaska to New Zealand, with no pause for feeding (Gill *et al.*, 2005). Humpback whales (*Megaptera novaeangliae*) feed in cooler, northern oceans and travel up to 25 000 km each winter to reach warmer waters where they breed and give birth (Rasmussen *et al.*, 2007). The longest known migratory route of a land mammal is undertaken by tundra caribou (*Rangifer tarandus*) in northern Canada, which travel approximately 5000 km between summer and winter habitats, covering 25–50 km per day (Bergerud *et al.*, 2012). Finally, insects such as monarch

Figure 5.11 Salmon jumping up a waterfall to spawn near Katmai National Park, Alaska.

butterflies (*Danaus plexippus*) display an amazing transgenerational migration in which a single butterfly partakes in only one segment of the Canada to Mexico journey. The insects reproduce along the way and successive generations travel to the next migration stop (Frost & Mouritsen, 2006).

Despite the obvious similarities, homing and migration do not always occur in the same groups of animals: some animals that migrate do not exhibit homing, and vice versa. One suggestion is that homing and migration may have evolved from a common process with differential evolutionary pressures leading to homing in some species and migration in others. Indeed, many researchers believe that homing and migration involve similar navigational mechanisms, particularly in avian species (Mehlhorn & Rehkamper, 2009). For this reason large-scale navigation is discussed in this chapter as one topic, which includes both homing and migration. The next section examines cognitive mechanisms that underlie both behaviors, with an emphasis on similarities and differences between the two.

5.4.2 Mechanisms

The cognitive mechanisms mediating large-scale navigation must provide animals with information on when to leave, which direction to go, and where to stop. These processes are frequently studied in birds using an Emlen funnel (Emlen & Emlen, 1966), a small circular bird cage with a funnel-like top that is covered by glass or a wire screen (see Figure 5.12). An ink pad is placed at the bottom of the cage so that foot traces are left on the sloping sides of the cage as birds hop about or try to escape. Using this setup, Gwinner (1977) showed that the night-time activity of migratory species reared in captivity increases around their usual time of migration, suggesting that these birds possess an endogenous (i.e. genetic) mechanism that keeps track of yearly cycles. This migratory restlessness, usually referred to by the German term **zugunruhe**, is synchronized with external cues

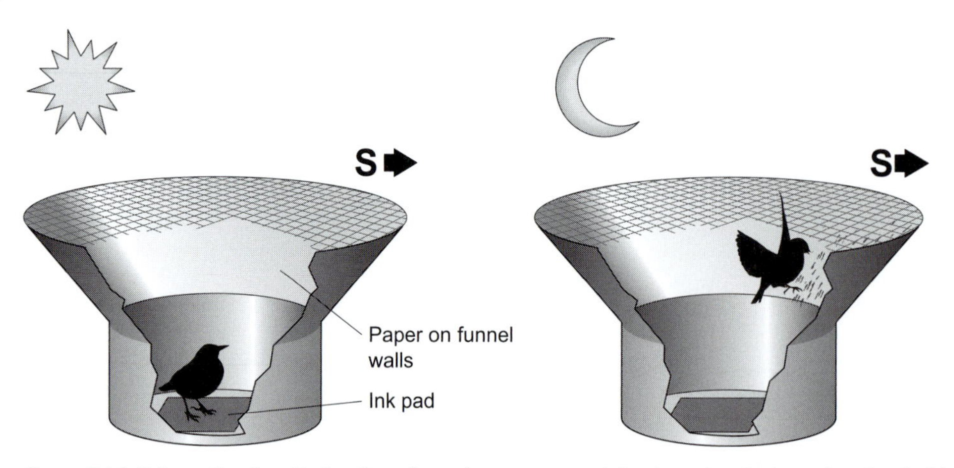

Figure 5.12 Schematic of an Emlen funnel, used to assess spatial orientation during migration in birds. Birds are placed in an inverted cone with an ink pad lining the bottom of the cage. When the bird hops or flutters up the sloping walls, traces of ink are left on the funnel sides. The top of the cage is clear so that the bird's image of the night sky (e.g. star patterns) can be manipulated.

marking seasonality (primarily changes in daylight), although many species display the behavior even when these cues are absent in the lab. Migratory onset in nonavian species is mediated by a similar mechanism: an endogenous annual clock that interacts with environmental changes such as a reduction in the length and intensity of daylight (Gwinner, 2003). The initiation of natal homing depends on the same signals, combined with changes in circulating levels of reproductive hormones.

As with timing, the direction of migratory travel appears to be under endogenous control. This is manifested in the natural environment when first time migrants successfully travel to their winter destination on their own (this does not occur in all migratory species, but the fact that it occurs at all is impressive). Helbig (1991) tested inheritance of migratory direction in two populations of European blackcaps (*Sylvia atricapilla*) that migrate to either Spain or Africa. As expected, young birds who had never migrated exhibited zugunruhe toward the winter destination of their parent population. Cross-breeding the animals resulted in a direction of travel almost exactly midway between the two. Subsequent research, using a variety of techniques, confirmed a genetic contribution to migratory direction in birds (Berthold, 1998), a principle that likely applies to other animals. From a cognitive perspective, the question of interest is how these endogenous signals point animals in the right direction. The answer, on a general level, is that migrating and homing animals are equipped with a global reference system that allows them to determine their own position in relationship to some external referent. These 'referents' can be divided into four separate topics.

Sun compass

Kramer (1952) was the first to suggest that birds (specifically European starlings, *Sturnus vulgaris*) use the sun's position to maintain a constant direction during migration. This 'sun compass' hypothesis was supported by a series of studies examining flight orientation of captive birds under different solar conditions (Wiltschko & Wiltschko, 2003). The majority of these studies used a phase shift manipulation in which the light–dark cycle in the colony room was out of synch with the natural environment (e.g. rather than daylight beginning at 6 am and ending at 6 pm, lights in the

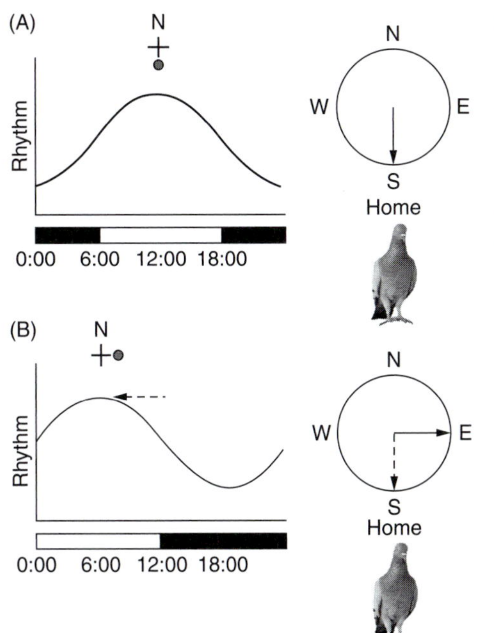

Figure 5.13 Effects of a phase-shift manipulation on sun compass orientation of homing pigeons. (A) Under natural conditions, circadian rhythms peak at midday when the sun is in the south. Pigeons use this information during migration to orient in a southerly direction. (B) If pigeons are housed on a light–dark cycle that is advanced 6 hours, the circadian rhythm is shifted so the midday peak now occurs during the natural morning. When released, birds orient toward the sun, which is now in the east.

animal house turned on at 2 pm and off at 2 am). If this cycle continues for more than a few days, an animal's circadian rhythm will shift (as will a human's). Under these conditions, animals made errors in migration orientation that were predicted by the difference between where the sun was and where the sun should be, given the animal's circadian cycle (see Figure 5.13). Thus, cognitive processing that allows birds to use the sun for navigation relies on an internal clock that keeps track of the sun's position over a 24-hour period. Amazingly, the timing mechanism also takes into account seasonal shifts in the sun's position.

Insects are also capable of using a time compensated sun compass in large-scale navigation (Frost & Mouritsen, 2006). For example, monarch butterflies will orient in the wrong direction during migration if they have undergone a phase shift manipulation (Froy *et al.*, 2003). Although they do not migrate, ants (Wehner & Muller, 2006) and honeybees (Dyer & Dickinson, 1994) keep track of the sun's position across the day and use this information to return home from foraging trips. Thus, there is at least one common mechanism of small- and large-scale navigation in insects. Some food caching birds appear to use a sun compass to locate hidden stores, although the evidence for this claim is still preliminary (Gould *et al.*, 2010). Similarly, there are relatively few studies examining whether mammals use the sun as a directional cue during navigation.

The sun may be less important for navigation in mammalian species because it is not a reliable cue in poor weather conditions or when animals are behind large landmarks (e.g. a mountain). This limitation is overcome in insects that have evolved a sensory system to detect polarized light. Polarized light is produced by the sun's rays hitting the earth's atmosphere; the resulting three-dimensional pattern varies as the sun moves across the sky, making it a reliable time of day cue even when the sun is obscured. Some vertebrates, including certain species of fish and amphibians, appear able to detect polarized light (Frost & Mouritsen, 2006) but, given the biological structure of the eye, it is unlikely that this ability is present in mammals.

Star compass

Even with the compensation for polarized light, a sun compass would be useless to animals who navigate during the night. An obvious way to deal with this is to use stellar constellations, the organization of stars, as directional cues. Emlen (1970) tested for a star compass in indigo buntings (*Passerina cyanea*) that were reared in captivity with a normal light–dark cycle, but different exposure to the night-time sky. Those that never saw the star constellations showed restlessness at the time of migration, but no preferred direction of travel. Two other groups were housed in a planetarium where they had exposure to star configurations that were artificially manipulated. One viewed the normal rotation of stars around the north–south axis whereas the second group viewed a configuration that rotated around the east–west axis (i.e. abnormal). When the migratory season arrived, the first group showed a southerly (i.e. correct) direction of travel but the second group did not. This star compass navigation appears to be limited to birds that use the axis of star rotation as a reference for north–south directions (Wiltschko *et al.*, 1987). Unlike a sun compass, orientation to stars is not time compensated and it may be important only in migration (i.e. not in small-scale navigation or homing) (Bingmam *et al.*, 2006). Humans can also orient using star constellations; in previous centuries this information was used to navigate across large bodies of open water. Unlike birds, however, star compass use in humans is acquired, primarily, through the transmission of knowledge within a community.

Magnetic cues

In addition to celestial cues, both the sun and stars, migrating and homing animals may use variations in the earth's magnetic field as a directional cue (Wiltschko & Wiltschko, 1995). This idea became popular when it was shown that pigeons carrying small coils that distort the earth's magnetic field had difficulty homing (Walcott & Green, 1974). Bird migration is also affected by magnetic fields: the orientation of migratory restlessness in lab-reared birds matches the earth's magnetic field (Freake *et al.*, 2006) and this preferred direction of travel can be altered by simulating a shift in the magnetic field surrounding a cage. Similar experiments revealed that honeybees (Hsu & Li, 1994), American cockroaches (*Periplaneta Americana*) (Vacha, 2006), California spiny lobsters (*Panulirus interruptus*) (Boles & Lohmann, 2003), common mole rats (*Cryptomys hottentotus*) (Marhold & Wiltschko, 1997), and big brown bats (*Eptesicus fuscus*) (Holland *et al.*, 2006) are able to detect changes in magnetic field direction, although these processes have been studied far less frequently than in birds.

Magnetic cues may be particularly important for marine animals because ocean environments are altered significantly across seasons and weather patterns. Moreover, celestial cues, including polarized light, are only available to animals that remain at or near the water surface (Lohmann *et al.*, 2008). The best evidence for magnetic field detection by migrating marine animals comes from an experiment with loggerhead sea turtles (*Caretta caretta*) (Lohmann *et al.*, 2004). Animals were tethered in a large pool filled with seawater so that their preferred direction of travel could be recorded. The underlying magnetic field was altered by a large magnetic coil surrounding the pool. Turtles swam in a direction predicted by the magnetic field: if the field indicated that they were north of their migration site, they oriented south but if the field was set up to mimic magnetic fields that would occur south of their destination, the turtles swam north. The importance of magnetic cues in the natural environment was confirmed in an experiment that placed disrupting coils on the heads of green sea turtles during homing (Benhamou *et al.*, 2011), and in an observational study that tracked humpback whales during migration (Horton *et al.*, 2011).

Many other reports provided evidence that animals (primarily birds and turtles) use magnetic field information during large-scale navigation, but the interpretation of these results is not straightforward. For example, although the initial orientation of homing pigeons is disrupted by magnetic coils, the birds still find their way home (Mehlhorn & Rehkamper, 2009). Recent technological advances, such as satellite tracking, showed that navigation sometimes remains intact in sea turtles and migrating birds fitted with disturbance magnets (Wiltschko & Wiltschko, 2012). A further paradox is that chum salmon (*Oncorhynchus keta*) and rainbow trout (*Oncorhynchus mykiss*) can detect magnetic fields, but orientation to these cues does not appear to influence ocean navigation (Lohmann *et al.*, 2008). Even simulated magnetic displacement does not produce consistent effects on navigation, or at least not effects that are predicted by the earth's magnetic field (Phillips *et al.*, 2006). A possible resolution to these contradictions is that magnetic maps, like a star compass, may be used to derive north–south coordinates during initial orientation but that other mechanisms take over once animals are heading in the correct direction.

Olfactory cues

The fourth and final mechanism of large-scale navigation, orientation to olfactory cues, is the most controversial. The idea, put forward to explain pigeon homing (Papi *et al.*, 1972), suggests that birds learn how different atmospheric gases are spatially distributed at their loft. When they are released from a novel location, they detect the proportion of different gases being carried on the wind and fly in a direction that maximizes the similarities between odor profiles at their current location and at their home. The olfactory hypothesis is supported by evidence that blocking olfactory input in pigeons disrupts homing (Papi *et al.*, 1980), although these results, and later ones reporting similar effects, were criticized because manipulations (e.g. lesions) often produce other deficits. Critics also pointed out that atmospheric odors are unstable, so could not be used as reliable navigational cues. In response, Wallraff (2004) showed that the relationship between atmospheric gases (although not their absolute levels) remains constant within an environment. Moreover, introducing 'false' odors at an unfamiliar release site fools birds into flying in the wrong direction (Bingman *et al.*, 2006). A consensus is now emerging that homing pigeons, as well as some other bird species, may use olfactory cues during navigation, although it is unlikely that this mechanism plays a major role in migration. The fact that disrupting olfactory input *only* affects homing when birds are released from unfamiliar sites suggests that they revert to using other cues (e.g. visual landmarks) when these are available.

The other well-documented instance of olfactory cue use during navigation is natal homing of Pacific coho salmon (*Oncorhynchus kisutch*). It is still not clear how these fish find their natal river from distant ocean sites; the best guess is that they use a sun and/or magnetic compass. Once they reach the river mouth, however, they swim upstream and locate the spawning ground using chemical cues (Scholz *et al.*, 1976). This process resembles imprinting (see Chapter 1, Section 1.4) in that salmon exposed to specific olfactory cues during development will return to breed in streams artificially scented with these chemicals (Dittman *et al.*, 1996). It is still not clear where the olfactory cues are coming from that salmon respond to. Aquatic plants or odors emitted from conspecifics are two of the most likely candidates.

The mechanisms that determine which direction to travel during large-scale navigation (e.g. sun, star, magnetic, and olfactory cues) are not the same as those that signal where to stop. In most migrating species, the termination of long-distance travel occurs through a genetic program that keeps track of travel time. This was demonstrated, initially, in a lab study using European blackcaps from different parts of Europe: the length of migratory restlessness mapped onto the migratory

distance of each population (Berthold & Querner, 1981). In other words, birds that had longer migratory routes spent more time in zugunruhe. The same phenomenon is present in other migratory species (Bingman *et al.*, 2006). It is still not clear whether this internal clock is signaling how long the animal has been travelling or when its energy reserves are depleted (or both). Regardless, if speed and direction of travel remain relatively constant across seasons, either calculation will place animals at their final destination, or at least close enough that they can then rely on familiar landmarks to locate their habitat.

5.4.3 Role of experience

The previous section provides compelling evidence for endogenous control of large-scale navigation, but experience also has a profound effect on this behavior. Indeed, the discussion of olfactory cue use emphasized that animals (at least pigeons and salmon) rely on previously acquired information to return home. Learned cues may be less important during the initial stages of homing, particularly if animals are released from an unfamiliar site. Once they reach the home vicinity, however, they appear to navigate using familiar sensory cues (Wiltschko *et al.*, 1987).

For many years migration was considered to be different from homing in that it was supposedly due, entirely, to inherited factors. Opinions began to change in the mid-twentieth century with a classic 'translocation' experiment in which thousands of European starlings were displaced from the Netherlands to Switzerland prior to migration (Perdeck, 1958). When released, first time migrants travelled in a direction and over a distance that would have taken them to their wintering home in Spain. That is, they made no compensation for their displacement (see Figure 5.14). In contrast, second-year migrants flew in the right direction to reach their southerly habitat, even though they had no experience with the Swiss landscape prior to departure.

The experiment has been repeated with different avian species in other parts of the world, often with mixed results. Some first time migrants compensate for displacement, both when they are moved by an experimenter and when they are set off course by natural elements such as wind patterns (Thorup *et al.*, 2011). The distinction between birds that do and do not adjust their migratory routes when they are moved may reflect species differences or specific environmental conditions at the capture and release sites (Thorup *et al.*, 2010). However, even if first time migrants perform better than originally suggested, they are consistently worse at finding their way than are older conspecifics. The current challenge is to identify why this occurs. Some suggest that migrating birds acquire information about magnetic fields, wind patterns, or landscape topography which they use as orientation cues on subsequent voyages but, to date, there is no convincing evidence for either hypothesis. Still, there is little disagreement that migratory experience facilitates later navigation. Conservationists capitalize on this fact in their efforts to protect endangered species (see Box 5.3).

5.4.4 Synthesis of mechanisms

In most species, large-scale navigation is controlled by more than one process. This provides greater precision in reaching the target destination and also allows animals to adjust to current conditions, such as weather or geographical location. A large-scale tracking study verified that humpback whales use a combination of sun, star, and magnetic cues during migration, and that none of these in isolation can explain the accuracy of navigation by these animals (Horton *et al.*, 2011). Importantly, experience may shift the control of navigation from one mechanism to another. For example, migrating birds orient using magnetic cues if they have not been exposed to the movement of the

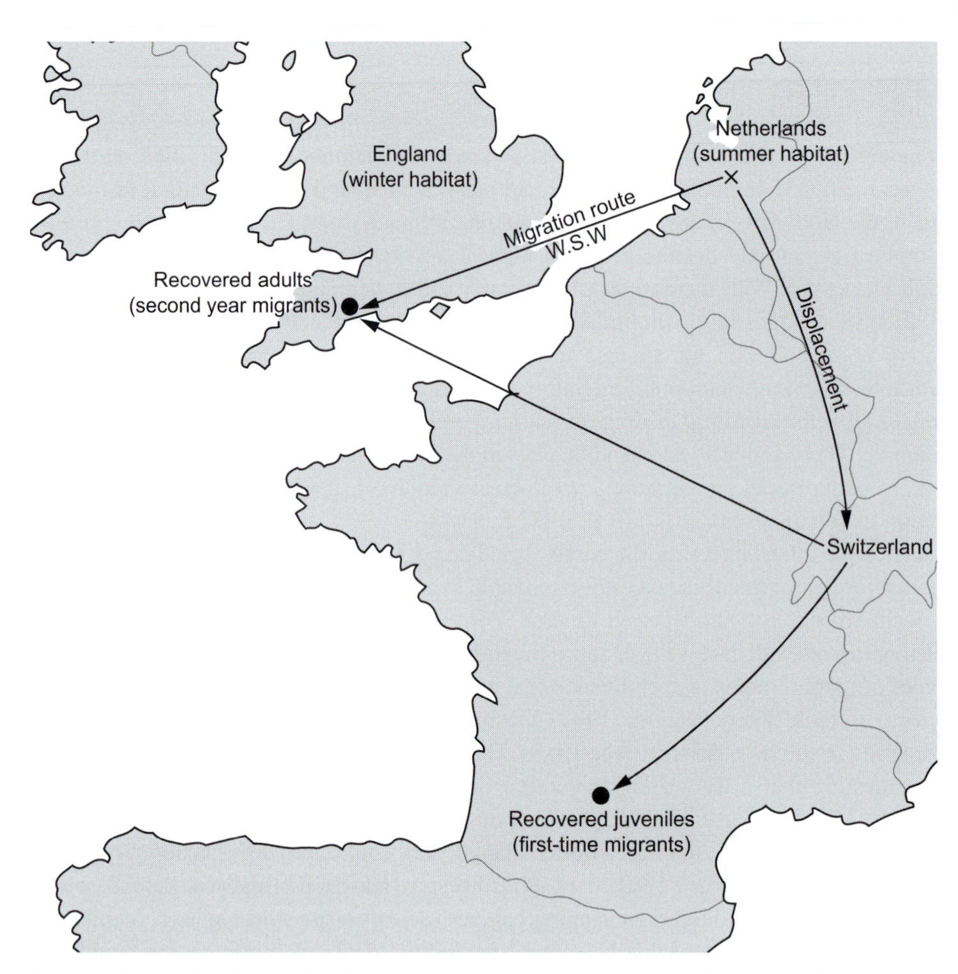

Figure 5.14 Results of a translocation experiment in which European starlings were displaced from the Netherlands to Switzerland prior to migration. Arrows indicate flight patterns of first-time and second-year migrants.

sun across the sky. With this experience, however, a sun compass becomes predominant and birds only revert to using magnetic cues if the sun is not visible (Bingman *et al.*, 2006). Similarly, homing pigeons with more experience navigating under cloud cover show better accuracy in returning home, probably because they have learned to rely on magnetic cues when necessary (Mehlhorn & Rehkamper, 2009).

Given that animals can use different cues during large-scale navigation, the current challenge is to understand when and how these work together. Some researchers propose a sequential model in which different mechanisms guide behavior depending on the stage of a journey. In support of this idea, magnetic coils attached to the heads of green turtles disrupted navigation *only* when animals were 4 to 50 km from their home (Benhamou *et al.*, 2011). Thus, although turtles can detect and respond to geomagnetic cues, they only use these during the middle stages of large-scale navigation, a finding that may explain contradictory evidence from previous experiments (Alerstam, 2006). Salmon homing appears to fit the same model: one process (that has yet to be identified) guides fish to the river mouth, the point at which olfactory cues from the spawning site take over. Similar

Box 5.3 Migration and conservation

Humans have been interested in animal migration since the beginning of civilization. In the past, they used this information to hunt effectively, to avoid being hunted, and to locate natural resources such as water and fertile land. Now, many people are studying migratory patterns in an effort to protect species whose numbers are rapidly declining due to human activity. One of the most devastating is habitat destruction that comes with increasing urban sprawl and industrialization. Even attempts at environmentally friendly changes (e.g. windmills) may impact certain species by disrupting migration routes and creating noise pollution for animals with highly developed auditory systems. Not surprisingly, global warming is severely disrupting migratory patterns as animals may return to seasonal habitats that are no longer ideal for breeding, feeding, or rearing young. Walrus (*Odobenus rosmarus*) migration is a prime example. These animals leave land habitats in the far north to breed on more southerly located ice floes. Due to rising ocean temperatures, ice floes are rapidly disappearing which is inhibiting walrus reproduction in certain areas.

Despite these pessimistic reports, the survival challenges facing contemporary species are not new: all living animals are proof that their ancestors were able to adapt to altered environmental conditions, even if these occurred within a short period of time. For example, migrating birds would have arrived in temperate regions after the last ice age, approximately 10 000 years ago. This suggests a relatively rapid evolution of migration routes, a phenomenon that has accelerated in many parts of the world during the last 50 years. Berthold and colleagues (1992) documented this change in blackcaps from the Arctic that began wintering in southern Britain in the 1950s. The numbers increased dramatically over the next half century, suggesting that selection had favored a novel migratory route that took birds to a habitat approximately 1500 km north of their previous destination. Selective breeding verified that genetic variation could account for this behavioral change and confirmed that phenotypic variability in migratory behaviors is inherited (Pulido *et al.*, 2001). At least for blackcaps, natural selection has helped these birds to adapt to rapidly changing habitats. Scientists are working hard to understand both the extensions and limitations of this behavioral flexibility; ideally this information could be used to facilitate shifts in bird migratory patterns in anticipation of changing environmental conditions.

These efforts have been facilitated by modern tracking technologies that allow researchers to record changes in migration routes with greater precision. Knowing where and when animals will travel can be critical for species' protection. As an example, the Institute for Environmental Monitoring and Research in Québec, Canada, has been tracking migratory caribou since 1991 and using this information to develop policy statements on the operation and construction of hydroelectric dams in northern Canada. Similar discussions of animal conservation often extend to the international level. Animal migration does not respect political boundaries so animals that are 'protected' in one part of the world may be actively hunted in others. In an attempt to reduce this practice, environmental groups frequently lobby governments or provide educational material on animal conservation to local agencies in the targeted area. In rare instances, some migrating bird species have been trained to follow light aircraft as a means to avoid inhospitable migration routes or destinations (www.operationmigration.org).

Notwithstanding these measures, it is clear that many species will not adapt to the changing environment. For those that do, it is likely to be a very different world.

models are being developed for migration although the huge variability in migratory patterns, both within and between species, makes it difficult to provide a comprehensive analysis of this behavior.

Finally, any model of large-scale navigation must describe how this process relates to small-scale navigation. According to some researchers, the two processes represent the same underlying phenomenon, even if they support different behaviors (e.g. caching versus migration). A similar idea was put forward by the renowned ornithologist Grinnell, who proposed that migratory adaptations are rather simple extensions of spatial behaviors that animals exhibit on a daily basis (Grinnell, 1931). Others note that the mechanisms mediating small-scale navigation may have evolved to support large-scale navigation and vice versa, but there is no way to distinguish between the two (Jacobs & Schenk, 2003). From a theoretical perspective, both functional and mechanistic explanations of small- and large-scale navigation are likely to overlap; the separation of the two probably relates to the difficulty of studying the two processes simultaneously (Frost & Mouritsen, 2006).

5.5 Neuroscience of spatial behaviors

The neuroscience of spatial behaviors can be divided into physiological mechanisms of orientation and navigation. Orientation is manifested as spatial responses to external stimuli (e.g. the sun), so the behavior must rely on neural processing of sensory cues. This process is well documented (see Chapter 2 for a detailed description) and scientists are now investigating how these neural signals are translated into orientation responses. One of the most promising lines of research examines how birds detect and respond to geomagnetic cues. One proposal is that cue detection occurs through iron mineral sensors in bird beaks that are activated by shifts in the earth's magnetic field (Falkenberg *et al.*, 2010); the problem with this idea is that these iron containing cells have no link to the brain (Treiber *et al.*, 2012). A more plausible hypothesis (Mouritsen & Hore, 2012) is that birds possess a light-dependent magnetic detection mechanism that operates through the visual system. This idea is based on evidence that magnetic orientation only functions under specific wavelengths of light (Wiltschko *et al.*, 2010). Later work identified proteins in the eye that are light-dependent magnetic sensory molecules; neurons containing these molecules project to higher brain regions involved in orientation (Zapka *et al.*, 2009). Although the work is far from complete, studies such as these are helping to reveal how orientation is mediated at a neural level. This information may be applicable to large-scale navigation (Mouritsen & Hore, 2012), which also depends on orientation responses to external cues. For example, birds may inherit a tendency to migrate south in the fall, but this behavior is modified by their ability to detect and respond to sensory cues, such as the sun's position. The current challenge is to understand the neural mechanisms that mediate this interaction.

5.5.1 Role of the hippocampus

By far, the majority of research into the neuroscience of spatial behavior has focused on small-scale navigation, particularly the idea that cognitive maps are formed in the hippocampus. Although it was Tolman (1948) who originally suggested that animals form cognitive maps as they move through their environment, the theory was not widely accepted until neurobiological studies showed a relationship between hippocampal function and cognitive mapping (O'Keefe & Nadel, 1978).

Place cells

Some of the strongest evidence relating cognitive maps to the hippocampus came from the discovery of **place cells** in rats (O'Keefe & Dostrovsky, 1971). Place cells are pyramidal neurons in the hippocampus that fire when animals are in a particular location, regardless of the direction that they are facing. One cell may fire when the animal is in one end of a maze and another may fire at the opposite end so different cells fire as animals move through their environment. The spatial area that produces the greatest response in a specific neuron is called a **place field** (see Figure 5.15). Place fields are formed within minutes of animals entering an environment and remain stable for weeks or months (O'Keefe & Speakman, 1987). Place fields may overlap in that more than one neuron may fire at a particular location and the same cell may fire in a different location of another environment. Cells with similar properties have been identified in the homing pigeon, although there is some debate as to whether these constitute place cells (Bingman *et al.*, 2006). Nonetheless, the general impression from these studies is that an animal's position in space is signaled by a unique pattern of firing neurons, pointing to an internal (i.e. cognitive) representation of the spatial environment.

Research on place cells exploded in the decades following O'Keefe's original report (for review, see Mizumori & Smith, 2006). Some studies suggested that place cells respond to local visual cues in that rotating a maze shifted the place fields of individual neurons (O'Keefe & Conway, 1978). Other studies contradicted these findings, suggesting that maze rotation does not disrupt place fields (Olton *et al.*, 1978). In addition, once place fields are formed, they remain stable in the dark (Quirk *et al.*, 1990) and place cells fire in both blind and deaf rats (Save *et al.*, 1998). Experiments such as these led to a reconceptualization of place cells, including the idea that they mediate path integration by updating an animal's current location with respect to a goal (McNaughton *et al.*, 1996). Eventually it became clear that place cells also respond to non-spatial stimuli, such as odor, as well as other task features such as reward availability (Eichenbaum *et al.*, 1999) and motivational state (Kennedy & Shapiro, 2009). This made it difficult to conclude that hippocampal pyramidal neurons are coding (only) the spatial environment.

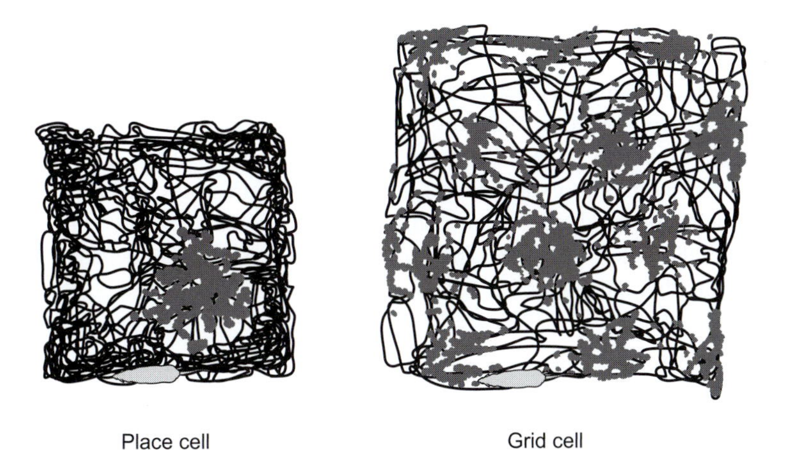

Figure 5.15 Firing patterns of place (left) and grid (right) cells in the entorhinal cortex of rats (Hafting *et al.*, 2005). Lines indicate the path of a rat as it explores a square-shaped environment. Gray areas indicate the locations at which the place or grids cells fire action potentials. Place cells are confined to a specific area of the environment whereas the grid cells fire in a pattern of regularly spaced clusters.

Current theories of how the brain represents space are built around the idea of grid cells, which are located in the entorhinal cortex (Fyhn *et al.*, 2004). Place fields of grid cells are small, equally spaced clusters; connecting the centers produces a grid pattern of equilateral triangles. This regular and systematic organization contrasts with the somewhat random arrangement of hippocampal place fields, suggesting that grid cells provide a more accurate representation of the local environment. The entorhinal cortex sends projections to place cells in the hippocampus, as well as to neurons in other brain regions that code changes in head direction (McNaughton *et al.*, 2006). These neural connections would allow information about an animal's current location to be combined with their direction of movement, an interaction that is critical for effective navigation. Current work is investigating how these processes relate to other aspects of spatial behavior (e.g. calculating the distance to a goal) and how these are mediated at a neural level. Although there is little doubt that the hippocampus and linked structures, such as the entorhinal cortex, are activated during navigation, whether this reflects coding of a cognitive map is still widely debated.

Hippocampal size and spatial navigation

Comparative studies of hippocampal size have also implicated this brain region in spatial navigation. For example, the volume of the hippocampus in food caching birds is nearly twice as large as that of birds that do not cache food (Sherry *et al.*, 1989). Moreover, within families of birds that cache, relative hippocampal size correlates with the amount of caching in the natural environment (Hampton *et al.*, 1995; Healy & Krebs, 1992). The same relationship is evident in mammals: Merriam's kangaroo rats (*Dipodomys merriami*) that cache food in various locations have larger hippocampi than banner-tailed kangaroo rats (*Dipodomys spectabilis*) that store food in one place (Jacobs & Spencer, 1994). Similarly, hippocampal size is reflected in the species-specific sex differences, outlined in Box 5.1. That is, male meadow voles have larger hippocampi than females of the same species, whereas male pine voles do not (Jacobs *et al.*, 1990).

Laboratory studies are not entirely consistent with the field work in that hippocampal volume does not always correlate with performance on spatial memory tasks, at least in some bird species (Balda & Kamil, 2006). Nor does the relationship between hippocampal size and navigation extend to all spatial behaviors because there is no difference in hippocampal volume of migratory and non-migratory birds (Sherry *et al.*, 1989). This latter finding can be reinterpreted in the context of learned cues in that the hippocampus is larger in first-year migrants compared to juveniles of the same species, a difference that is not apparent in non-migratory birds (Healy *et al.*, 1996). This suggests that increases in hippocampal volume may be related to navigational experience, a finding that has an interesting analog in humans (see Box 5.4).

Lesion and activation studies

Further support for the idea that the hippocampus has a role in spatial navigation comes from both lesion and activation studies. Rats with hippocampal lesions show deficits in the radial arm maze, but only when they must rely on spatial cues outside of the maze to locate the rewards (McDonald & White, 1993). Similar effects have been observed in other mammals (Squire, 2009). Surprisingly, displaced homing pigeons with lesions encompassing the avian homolog of the hippocampus are able to return directly home, but cannot locate their nesting loft once they arrive (Bingman *et al.*, 1984). The interpretation of this finding is that birds rely on other navigational mechanisms during the initial stages of homing but once they encounter familiar landmarks, a hippocampal-based system guides behavior. This fits with evidence that humans with hippocampal damage, such as H.M. (see Box 3.4), have difficulty finding the way through familiar environments. Of course, these

Box 5.4 The London taxi driver study

Birds and mammals that rely heavily on navigational skills often have larger hippocampi (relative to the rest of their brain) than closely related species that are not so dependent on this skill. Maguire and colleagues tested whether the same difference is apparent in humans by studying a group of individuals with extensive navigational experience and expertise: London taxi drivers. In order to be licensed as a taxi driver in the UK capital, individuals must memorize a map of the city center, including the names and locations of 25 000 different streets and 20 000 major landmarks. Studying for the 'The Knowledge of London Examination System' test can take 3 to 4 years. Even after several attempts, only half of the applicants pass. Once they are working in the profession, taxi drivers must engage these superior navigational skills on a regular basis. Maguire hypothesized that this experience would be reflected in structural brain changes, a prediction that was verified using non-invasive brain imaging: taxi drivers had larger posterior hippocampi compared to age-matched controls, and the more years of driving experience, the larger the size of this brain region (Maguire *et al.*, 2000).

One drawback to this study is that hippocampal volume was measured after individuals had learned to navigate through London streets. It may be that taxi drivers were able to take on their profession *because* they had larger hippocampi, although this would not explain the relationship between driving experience and hippocampal size. To verify their idea that brain differences develop with navigational training, Woollett and Maguire (2011) conducted another study in which they examined hippocampal volume before and after individuals studied for 'The Knowledge' test. Initially, the size of the hippocampus was similar in yet to be trained taxi drivers and a matched control group. Over the next four years, 39 of the 79 trainees passed the test. A post-training scan revealed an increase in hippocampal volume *only* in this group. Neither the failed trainees nor the control group showed these changes.

Prof. Ingrid Johnsrude who conducted the image analysis in the original study remarked that this finding was one of the 'cleanest' she had ever seen. It is not that the size of the effect was larger than she had observed in other imaging studies, but when she analyzed which brain regions differed, the only regions of significance were located within the hippocampus. Most imaging studies show differences in the predicted regions (e.g. in this study, the hippocampus) but also in many other brain sites, including extraneous effects that show up in the ventricles or in white matter (the axons rather than the cell bodies of neurons). "But the London cabbie study showed nothing else, just beautifully symmetrical bilateral activation in the posterior hippocampus." (Johnsrude 2011, personal communication).

patients have many other cognitive deficits (e.g. memory), which makes it difficult to isolate their impairments to spatial behaviors. The same is true of animal studies in that lesions frequently produce unintended deficits that may confound the behavioral measure of interest.

A complementary approach, therefore, is to examine which brain regions are active during spatial navigation. Place cell research, summarized above, verifies that neurons in the hippocampus and connected regions fire as animals move through familiar spaces. Neuronal activity can also be measured using immediate early gene expression in post-mortem brains. These studies verify that homing pigeon navigation using familiar landmarks (Shimizu *et al.*, 2004) and cache recovery in black-capped chickadees (Smulders & DeVoogd, 2000) both activate hippocampal neurons. In humans, neuronal activity is measured using non-invasive techniques, such as functional magnetic resonance imaging (fMRI). Using this technique, Maguire and colleagues (1998) examined brain systems mediating spatial navigation as participants 'moved' through a virtual environment. At the beginning of the test, participants explored a small virtual town that included a complex maze of streets. Once they were familiar with the environment, participants were asked to navigate from one point to another as quickly as possible. Individuals who were able to perform the task most accurately exhibited higher levels of activity in the hippocampus (an effect that was lateralized to the right side of the brain), suggesting that they were relying more heavily on this brain region to do so.

In sum, converging evidence from a variety of techniques points to an important role for the hippocampus in small-scale navigation across a variety of species. In some ways, it is amazing that this brain system has a near similar function in birds and mammals, despite close to 300 million years of independent evolution (Bingman & Sharp, 2006). It is still not clear how the hippocampus operates to guide spatial behaviors, but the process likely involves some map-like representation of familiar landmarks that *may* resemble a cognitive map, depending on how this is defined (see Section 5.3.4).

5.5.2 Non-hippocampal mechanisms

Other brain regions have been linked to small-scale navigation in both birds and mammals, although these control different aspects of spatial behavior than the hippocampus. For example, the striatum acts as a response reference system by signaling to the rest of the brain which spatial responses (e.g. turn right or left) are appropriate in a particular context (Mizumori & Smith, 2006). This hypothesis is based on evidence that striatal lesions disrupt the ability to learn response–reward associations (McDonald & White, 1993) and that neuronal firing in the striatum codes a rat's location, as well as the direction and timing of specific movements (Wiener, 1993). The difference between navigation using a hippocampal versus a striatal based system can be revealed in experiments in which animals are trained to locate food in one arm of a T-maze (see Figure 5.16). On test day, the maze is rotated so that the arms are now in a different spatial location. If the animal travels down the arm that is in the 'correct' location relative to the environmental cues, it is showing place learning but if it turns in the same direction as training trials (thereby ending up in a different spatial location), it is showing response learning. Blocking hippocampal activity during testing disrupts place learning but has no effect on response learning in rats; blocking striatal activity has the opposite effect (Packard & McGaugh, 1996). Humans also engage different neural systems when they are 'moving' through a virtual maze, and these reflect the use of one navigational strategy over the other. More specifically, brain imaging scans show that increased activation in the hippocampus is associated with place learning whereas heightened striatal activity is associated with response learning (Marchette *et al.*, 2011). Notably, activation in these different regions did not predict whether individuals were able to find their way through the maze, but revealed which strategy they were using when they did so.

Even if they make different contributions to spatial navigation, the hippocampus and striatum must interact as animals and humans move through their environment. This may occur through the retrosplenial cortex, which receives direct input from the hippocampus and sends connections to the striatum and other brain regions involved in motor control (for review see Mizumori & Smith, 2006). Neurons in the retrosplenial cortex (located in the posterior cingulate cortex) respond to spatial cues, such as head direction and egocentric movement, although lesions can have variable and inconsistent effects on navigation. Eventually it became clear that retrosplenial lesions disrupt the ability to switch from one spatial strategy to another (Pothuizen *et al.*, 2008), leading to the idea that this brain region integrates different modes of spatial navigation (Vann *et al.*, 2009).

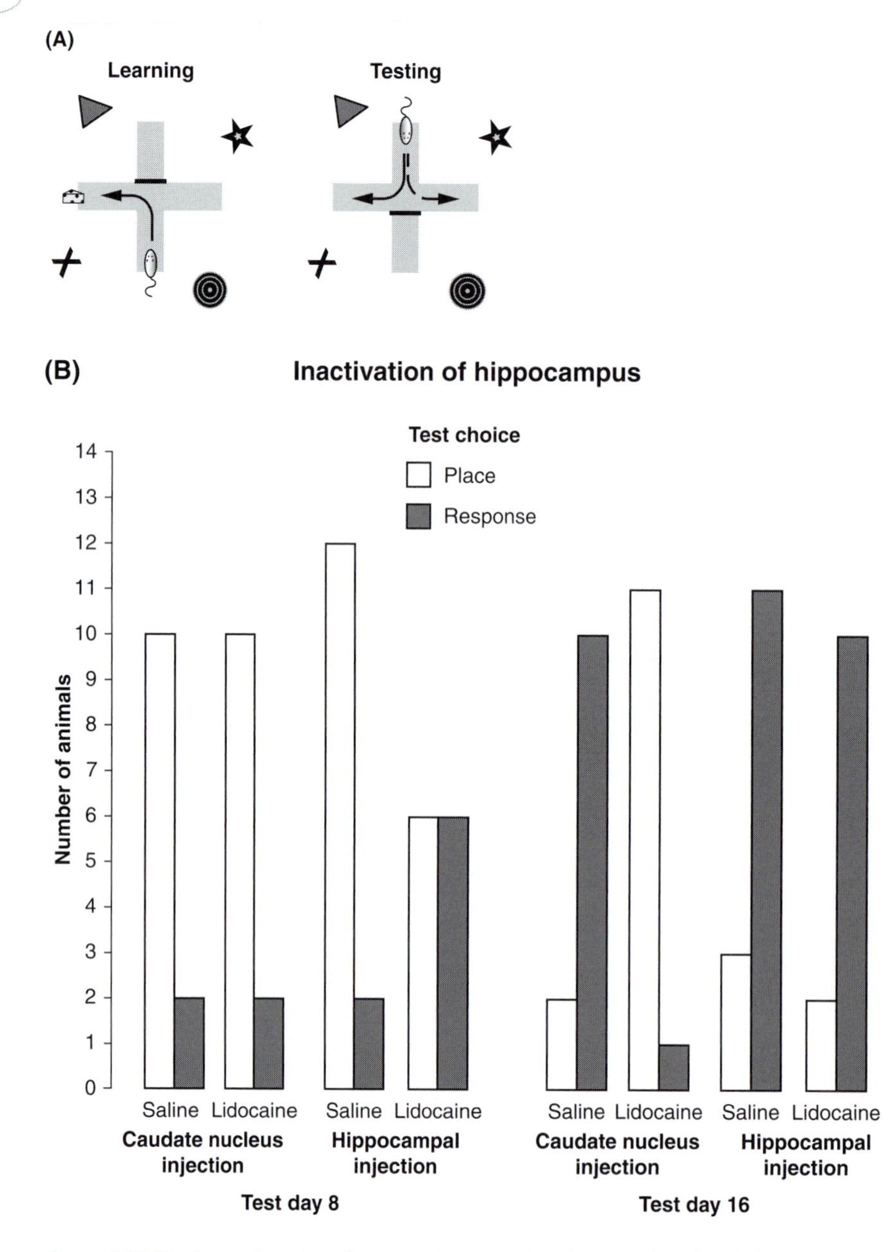

Figure 5.16 Design and results of an experiment testing the neural basis of place versus response learning in rats. (A) Rats were trained to turn in one direction of a plus maze to locate a food reward. Visual cues surrounding the maze were stationary and could be used as orientation cues. During testing, rats were released from a start point 180° from the original start point. (B) After 8 days of training, 10/12 rats used a place rather than a response strategy to locate food during testing (saline injections); this pattern was reversed after 16 days of training. Inactivation of the caudate nucleus disrupted the shift from place to response learners whereas inactivation of the hippocampus inhibited a place learning strategy on both days.

Research into the neurobiology of small-scale navigation is advancing rapidly; current theories emphasize an interaction between various brain systems, including (but not limited to) those outlined above. A guiding premise of this work is that specific cognitive systems process different types of information related to spatial movement. This may include an analysis of directional cues, egocentric orientation, or the relationship between local and distal cues in an environment. Many of the structures implicated in these mechanisms are present in birds, mammals, and reptiles (Jacobs and Schenk, 2003), suggesting a preservation of these important cognitive functions across evolution.

Chapter Summary

- Animals and humans find their way through the environment using cognitive mechanisms that can be described as getting there and knowing where. The getting there system is mediated by orientation and uses external stimuli, such as a light source, as a reference for spatial movement. The simplest orientation responses are kinesis and taxes, which are non-directional and directional forms of orientation respectively.
- The knowing where system mediates navigation, allowing animals and humans to identify a specific place from different spatial locations. In general, short distance (small-scale) navigation is examined separately from long-distance (large-scale) navigation, although it is very likely that the functions and mechanisms of these two probably overlap, at least in some circumstances.
- Small-scale navigation supports behaviors such as foraging, caching, and territorial defense, whereas large-scale navigation is used during homing and migration. The cognitive mechanisms that mediate small-scale navigation include landmark use, path integration, environmental geometry, and cognitive maps. The latter is the most controversial, with no current consensus on what constitutes a cognitive map and which animals (if any) possess these. Regardless of how a cognitive map is conceptualized, it is clear that most animals can use more than one spatial strategy to locate a goal. This flexibility allows different species to adapt to changing environmental conditions, both within and across generations.
- Animals determine the direction of travel in large-scale navigation using a global referencing system. This may involve a sun compass, a star compass, magnetic cues, or some combination of all three. Certain animals, particularly pigeons and salmon, use olfactory cues during large-scale navigation, but it is not clear whether this is limited to homing or whether other species rely on the same cues. In addition to the direction of travel, migrating animals possess a genetic program that signals when to leave and where to stop. These inherited traits are modified by experience, producing relatively rapid shifts in migratory behavior across generations.
- Historically, behaviors reflecting small-scale navigation, such as foraging, were explained by learned cues (e.g. landmarks) whereas large-scale navigation was described in terms of inherited traits (e.g. orientation to geomagnetic cues). The distinction between the two is not absolute: migration and homing are facilitated by experience with environmental cues and effective caching and foraging depend on cognitive mechanisms that have evolved to support both short- and long-distance travel.
- Research into the neural mechanisms of spatial behaviors has focused on brain systems that control small-scale navigation. Evidence that the hippocampus is integral to this process includes findings that hippocampal neurons fire in response to a variety of spatial cues, the size of the hippocampus is positively correlated with navigation experience in both the lab and natural environment, and lesions of the hippocampus produce deficits in spatial learning tasks across a variety of species. Structures connected to the hippocampus, including the entorhinal cortex, striatum, and retrosplenial cortex, make separate contributions to small-scale navigation, although it is still not clear how these regions interact as animals and humans move through their environment.

Questions

1. Given that mnemo is the Greek suffix for memory, how would you define mnemotaxis? How would this process relate to the processes discussed in this chapter?

2. Describe the proximate and ultimate mechanisms of beacon versus landmark use. In other words, what are the mechanistic and functional explanations of navigation using beacons and landmarks?
3. The water maze and the radial arm maze are both used to measure spatial memory (refer to Chapter 3 for descriptions of each paradigm). What are the advantages and disadvantages of each task? Suggest both practical and theoretical reasons for using one task or the other.
4. There is some evidence that the use of environmental geometry in small-scale navigation depends on visual experience during development. How would a research program examining spatial learning need to be modified to take this into account?
5. Design an experiment to test the hypothesis that pigeon homing depends on a sun compass. If the hypothesis were confirmed, but birds still found their way home on a cloud covered day, what would this mean?
6. In many mammals, males disperse but females do not, whereas the opposite pattern occurs in many avian species. Why would this sex difference in spatial behavior be reversed in these two classes of animals?
7. Given that many species use compass-like mechanisms to navigate, how could these be calibrated within an individual?
8. Migratory behaviors, including the distance and direction of travel, are shifting more rapidly than would be expected by random selection. How could this be explained? Are there likely to be any constraints on the flexibility of migration patterns across different species? If so, what would these be?
9. The magnetic map hypothesis of large-scale navigation is controversial. Many animals appear capable of responding to shifts in magnetic fields but this is not proof that these cues are important for orientation or navigation. What would be the evolutionary advantage of the ability to detect changes in the earth's magnetic field if these cues are not used during spatial behavior? What other function could magnetic field detection serve?
10. In Perdeck's (1958) translocation experiment, second-year migrants were able to reach their winter habitat following displacement. Given that they had no experience with the Swiss landscape, how could this have happened? What cues were they using? If they started their journey with a random search flight pattern and then set off on the correct course, what would this indicate about underlying mechanisms in these birds compared to birds that oriented correctly at the beginning of the flight?
11. A primary criticism of place cell research is that the vast majority of work has been conducted with rodents, specifically lab-reared rats or mice. Given that the hippocampus is preserved across many species, is this a drawback to theories of place cell function or is it simply the result of research practicalities (i.e. lab rats and mice are readily available and easy to implant with electrodes)? Why or why not?

FURTHER READING

Advanced discussions of the philosophical and psychological notions of space in different animals can be found in:

Gallistel, C. R. (1990). *The Organization of Learning*. Cambridge, MA: MIT Press (focus on the first six chapters).

O'Keefe, J. & Nadel, L. (1978). *The Hippocampus as a Cognitive Map*. London, UK: Oxford University Press.

A general overview of topics related to spatial navigation is discussed in separate chapters of this very informative on-line text:

Brown, M. F., & Cook, R. G. (eds.) (2006). *Animal Spatial Cognition: Comparative, Neural and Computational Approaches*. [On-line]. Available at: www.pigeon.psy.tufts.edu/asc/

A summary of the evidence supporting an evolutionary theory of sex differences in spatial memory, with a focus on humans, can be found in:

Silverman, I., Choi, J., & Peters, M. (2007). The hunter-gatherer theory of sex differences in spatial abilities: Data from 40 countries. *Arch Sex Behav*, **36**, 261–268.

To discover amazing feats of animal migration with beautiful accompanying illustrations:

Hoare, B. (2009). *Animal Migration*. Berkeley, CA: University of California Press.

This review paper summarizes the mechanisms of both small- and large-scale navigation in aquatic animals, and compares these to navigation in terrestrial species:

Johmann, K. J., Lohmann, C. M. F., & Endres, C. S. (2008). The sensory ecology of ocean navigation. *J Exp Biol*, **211**, 1719–1728.

Advanced students, particularly those with an interest in the behavioral ecology of navigation, will be engaged by this comprehensive text. The material includes numerous examples of how animal behavior, in this case migration, is modeled in theoretical and mathematical formulas:

Milner-Guiland, E. J., Fryxell, J. M., & Sinclair, A. R. E. (eds.) (2011). *Animal Migration: A Synthesis*. Oxford, UK: Oxford University Press.

6 Timing and number

Background

A popular tale recounts the story of a hunter who wished to shoot a crow, but the bird repeatedly eluded him by flying just out of range whenever he appeared. Hoping to deceive the crow, the hunter asked a friend to go with him behind a hunting blind and then leave after a short time. Surely, the hunter thought, if one man left, the crow would not realize that one remained and would return to the area. Yet, the bird was not fooled by this ruse, or another attempt in which three men entered and two emerged. Only if five men entered the blind was the crow's ability to keep track of the hunters exceeded.

Credit for the original report of the hunter's story is often given to Sir John Lubbock, though Ayn Rand's mention of the story in *Introduction to Objectivist Epistemology* (Rand & Peikoff, 1979) helped to spread the tale. If true, the story would suggest that even without a formalized symbol system for counting (e.g. the use of Arabic numerals), a bird species can track at least one to four potential predators, represent this quantity while the predators are out of sight, and then update this representation if the predators are seen to leave. It would also suggest, though, that there is a limit to how many hiding individuals can be precisely represented.

Anecdotes such as this, of course, do not provide the basis for good scientific study. Yet, whether or not the story is true, it raises questions that comparative cognition researchers, often alongside developmental and cognitive psychologists, have been working to answer for many years. Along the way, many commonalities have been found between how the animal mind represents number and time, and studies of one have often shed light on the other. Numerical cognition and timing both involve discriminating magnitude, and there is now good reason to think that some neural systems may be shared by these two processes. Additionally, both seem to be present across many animal species and emerge very early in human development. Indeed, given what is now known, the hunter's story is not farfetched.

Chapter plan

This chapter begins with a description of animals' ability to keep track of time, both for long durations such as days and short intervals on the order of minutes. The study of interval timing connects closely with the study of numerical processing; research has repeatedly pointed to similar or shared cognitive systems that allow for interval timing and approximating number. After discussing the approximate number system, a second system for small quantities will be detailed, followed by a consideration of operations such as addition, subtraction, and ordering. Sensitivity to time and number is likely adaptive for many behaviors, including foraging; timing may provide information as to the availability of food sources that replenish over delays, and numerical discrimination can provide a basis for choosing foraging sites. The importance of these abilities suggests that one might expect to see evidence for them across a wide range of species, and as will be seen, the study of timing and number has benefitted greatly from a comparative approach.

6.1 Periodic timing

It is relatively easy to observe instances in which animal behavior is influenced by time. Many pet owners know the joy of being awakened by a restless dog that wants to walk early on a Sunday morning, and while on that walk, it is common to witness that the early bird truly does get the worm. Two primary types of timing have been characterized, one for cyclical activity that occurs during a 24-hour period, which will be referred to as **periodic timing**, and one timing system for much shorter durations on the order of seconds or minutes (**interval timing**). The latter will be presented in Section 6.2.

Almost every organism that has been studied shows a rhythm of activity or physiological state that matches the day/night cycle of earth. Importantly, though, this rhythm can persist even when organisms are moved to consistent darkness or light. The **circadian rhythm** appears to be the result of an internal clock with an approximately 24-hour cycle that can undergo **entrainment** by, or fall into synchrony with, the external day/night cycle. For example, in an experiment with cockroaches (*Leucophaea maderae*) that were normally active at dusk, covering the insects' eyes did not stop the rhythm of daily activity (Roberts, 1965). However, without the light/dark input from the environment, the rhythm followed a slightly shorter period. Activity rose every 23.5 hours, meaning that over time, activity occurred earlier and earlier before dusk. This **free-running rhythm** is thus only close to 24 hours; it is *circa* (near) a day. When sight was restored by uncovering the eyes, the cockroaches' internal rhythm slowly became synched with (entrained to) the natural day/night cycle again. Across species, there appears to be an independent internal clock that can be modified in potentially adaptive ways by the environment to influence behavior.

6.1.1 Time and place learning

The internal circadian rhythm is typically discussed in terms of physiology, not cognition. However, animals may learn about events that occur at certain times of day, linking the event to the internal clock. For some carnivores, for example, the emergence of prey may influence daily activity routines, thus optimizing hunting efforts (e.g. Rijnsdorp *et al.*, 1981). Many flowers produce nectar at certain times of day, and thus bees and hummingbirds may effectively feed through time and place learning. Inspired by these observations in the wild, researchers have systematically examined time and place learning in the laboratory.

In one classic example, Biebach *et al.* (1989) housed garden warblers (*Sylvia borin*) in an enclosure consisting of a central room surrounded by four feeding rooms (Figure 6.1). Unlocked

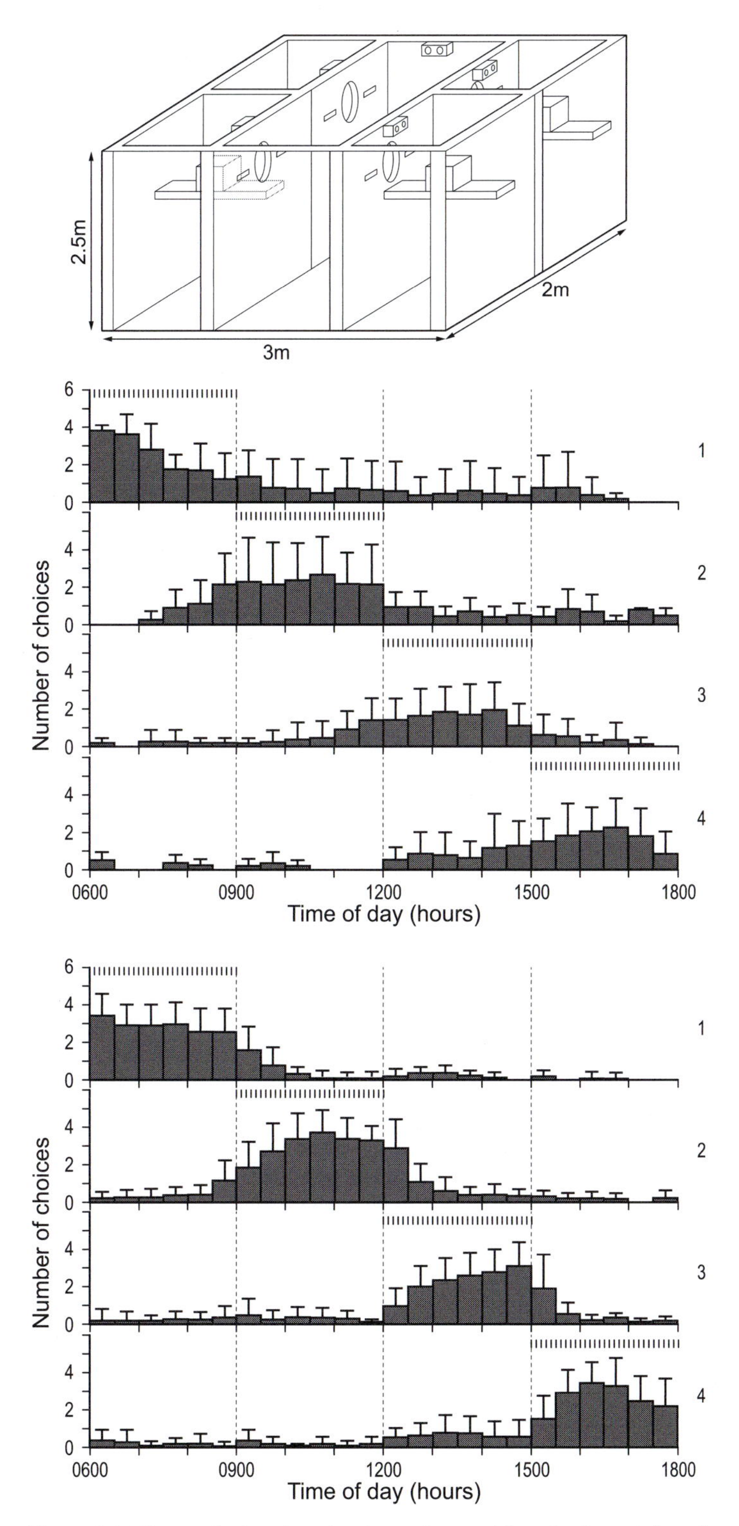

Figure 6.1 Time and place learning in garden warblers. In the training phase (middle graph), food was only accessible in a feeder in one room for every 3-hour block. During the test phase (bottom graph), the warblers continued to search for food in relation to the time of day, even though all rooms were accessible.

food containers were only available in one feeding room at a given time, corresponding to each 3-hour block of a 12-hour light cycle (i.e. Room 1 had food available from 0900 to 1200 hours, Room 2 from 1200 to 1500 hours, etc.). If a bird entered a room at the correct feeding time, it could eat from the unlocked container for 20 seconds. Over 10 days, the birds learned to enter rooms during the appropriate 3-hour block. In subsequent testing, each of the four rooms had food containers that were unlocked throughout the 12-hour light cycle. Although birds could now eat in any of the rooms, at any time of day, they continued to visit the rooms based on the original schedule. In a second experiment, birds were prohibited entry into any of the rooms during one 3-hour block. When entry was again permitted during another block, the birds reliably entered the appropriate room for that block, suggesting that performance was not simply based on a learned pattern of activity alone (e.g. *enter Room 1*, *then Room 2*, etc.), but was due to the linking of the food locations with the circadian rhythm (Krebs & Biebach, 1989; see also Saksida & Wilkie, 1994).

6.1.2 Mechanisms of time and place learning

For the garden warblers in the previous example it is likely that the state of the internal circadian rhythm acted as a contextual stimulus to which the correct food location was associated. Additionally, the circadian clock itself was likely entrained to the light/dark cycle. Evidence for this comes from an experiment in which the lights remained on for 24 hours; birds still visited the rooms during their specific 3-hour block, at least during the first day of testing. After that, the visits to feeding rooms began to diverge from the 24-hour clock, instead appearing to follow the free-running rhythm (Biebach *et al.*, 1991).

In mammals, the light-entrained circadian clock is thought to be located in the suprachiasmatic nucleus (SCN) of the hypothalamus. The eyes provide information to the SCN about illumination through the photopigment melanopsin, which is sent from retinal ganglion cells to the SCN. The SCN can then pass on information to the pineal gland, which, in turn, secretes the hormone melatonin. In humans, the circadian rhythm can be entrained (by manipulating the ambient light) to shorter and longer periods than the earth's 24 hours, though the differences were slight: 23.5 hours and 24.65 hours (Scheer *et al.*, 2007).

6.2 Timing intervals

In periodic timing, an internal clock can be entrained to important environmental stimuli such as the day/night cycle, in turn influencing behavior. In interval timing, however, behavior can be influenced by relatively arbitrary, short durations that are signaled by relatively arbitrary stimuli that can occur at any time of day (Shettleworth, 2010). The importance of short durations on learning was discussed in Chapter 4 in relation to temporal contiguity of stimuli; associations between two stimuli are formed more easily if they occur close together in time. The work presented here focuses instead on the ability to time short durations and discriminate between them. Much of this work has been done in the laboratory where experimenters create the relevant intervals, though timing of naturally occurring intervals likely influences behavior in the wild.

6.2.1 Laboratory procedures

In 1981, Roberts developed a new procedure for examining timing in animals. The **peak procedure**, which was actually based on previous work by Catania (1970), involved exposing rats (*Rattus norvegicus*) to many trials in which food could be accessed a fixed time after the onset of a stimulus

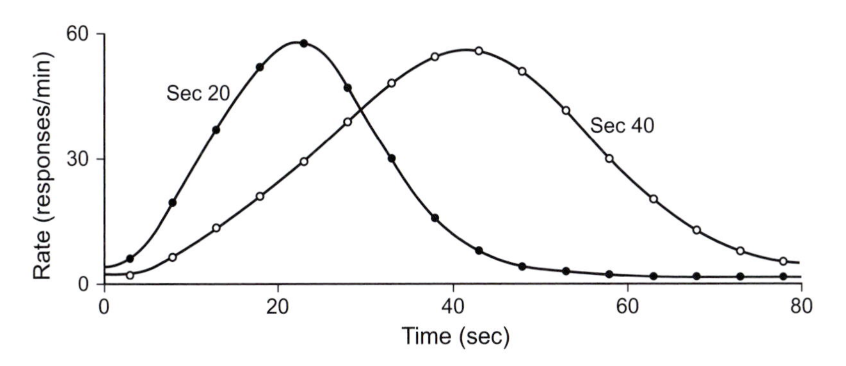

Figure 6.2 Example peak procedure data from Roberts (1981) for rats trained with reward for lever press at 20 seconds ("Sec 20") and 40 seconds ("Sec 40"). Roberts (1981) details the statistical methods of fitting the curves in detail.

(e.g. a light). When a light came on, it signaled that food could be obtained by pressing a lever 20 seconds later. After food was obtained, the light turned off in preparation of the next trial. Occasionally, 'empty trials' were included in which food was *not* delivered upon the first lever press at 20 seconds, and instead, the trial ended when the light turned off much later. Recording the rate of lever presses on these trials reveals a rather unsurprising finding: the rats do not press the lever only and exactly at 20 seconds. They are hungry and tend to press multiple times, both before and after 20 seconds, on these empty trials. Indeed, without consciously counting seconds, adult humans would have difficulty being exact in their responses. But, two aspects of the behavior are particularly noteworthy. First, the rats' responses peak very close to 20 seconds and follow a relatively normal distribution; the timing is quite accurate (Figure 6.2). Second, if a stimulus signaled that food could be obtained after 40 seconds, responses again peaked at 40 seconds and followed a normal distribution, but now, the spread of the curve (the variability) was much greater (Roberts, 1981). In fact, the variability of the curves is proportional to the length of the interval being timed, which has been termed the **scalar property** (Gibbon, 1977).

The general finding from the peak procedure that responses to longer time intervals occur with greater variability fits with Weber's Law, which was introduced in Chapter 2. Other experimental procedures produce similar results. Church and Gibbon, for example, presented rats with certain intervals that would be reinforced, while other durations yielded no reinforcement. The rats were placed in a lighted chamber, but the lights would turn off for intervals that ranged from 0.8 to 7.2 seconds. When the lights were turned back on, lever pressing would be rewarded if the darkness interval had been 4 seconds. After extensive experience, rats' lever-pressing behavior was highest if the darkness lasted exactly 4 seconds, and the behavior decreased as the interval duration differed from 4 seconds. That is, **temporal generalization** occurred; the responses continued for durations that were different yet close to the target duration. Further, analogous to the results from the peak procedure, the generalization gradients for target durations longer than 4 seconds had greater variability (Church & Gibbon, 1982).

Although the majority of experimental evidence comes from rats and pigeons, interval timing appears to be present across many vertebrates and invertebrates (Lejeune & Wearden, 1991). Bumble bees (*Bombus impatiens*), for example, have been tested using fixed-interval procedures in which a reward is received on the first response that occurs after a fixed interval of time has elapsed. The response in question was the extension of the proboscis, which naturally occurs during feeding. A bee could obtain sucrose by extending its proboscis through a small hole in the experimental chamber. During a trial, a chamber light was illuminated, and the interval (e.g. 6 seconds) began. The first proboscis extension that occurred after the interval elapsed yielded a sucrose reward. Typically, vertebrates tested using these procedures withhold responses until later in the duration, and the highest probability of responding occurs at the end of the interval. Similarly, bumble bees responded maximally at the end of the interval (Boisvert & Sherry, 2006).

6.2.2 Timing in the wild

It is possible that, in the wild, the timing ability of bumble bees supports their feeding behavior. Pollinating species often show 'resource fidelity'; they repeatedly visit the same nectar sources. Food sources such as flowers do replenish their nectar after a pollinator has fed, but this process takes time. To feed efficiently, then, pollinators should schedule visits to food sources according to temporal schedules (e.g. Boisvert & Sherry, 2006). What this would require is remembering not only where the feeding site was but also how much time has passed since the last visit.

As mentioned in Chapter 5, rufous hummingbirds (*Selasphorus rufus*) can use landmarks to remember locations and avoid recently depleted flowers. Similar experimental techniques can be used to examine whether and how these birds keep track of the time interval between visits. In one such study (Henderson *et al.*, 2006), wild, territorial, male rufous hummingbirds living in the Rocky Mountains of Alberta, Canada, were tested using arrays of eight artificial flowers that could contain a sucrose solution. In each bird's territory, experimenters refilled four flowers 10 minutes after the male fed and emptied them. The other four flowers were refilled 20 minutes after the bird fed. (Both of these durations are shorter than the typical nectar refilling rate of flowers but allowed the experimenters to observe the hummingbirds' behavior.) After experience with these refill rates over many days, the birds were revisiting the flowers that replenished after 10 minutes sooner than they revisited those that replenished after 20 minutes. Further, the visits appeared to match the refill schedules (Figure 6.3). Similar to the laboratory studies of interval timing, the visits peaked at the appropriate times, but there was variance. While it was not the case that this variance was truly proportional to the refill durations, there was greater variance for the larger, 20-minute duration.

The rufous hummingbirds seem to have learned that the time until refilling differed among the flowers, and more specifically, they learned that the refill times were 10 and 20 minutes. Another way to consider this finding is the following: the birds remembered the refilling times for each of the eight flowers and then updated this information across the many visits to the flowers throughout the day. Although there are not many studies of interval timing in the wild, it is possible that procedures such as these may complement and confirm results from the laboratory. The natural feeding behavior of pollinators is a likely situation in which interval timing may be observed, yet other situations in which animals utilize food sources that can be depleted and then renewed might also indicate the timing of intervals (e.g. the feeding behavior of 'cleaner' fish that remove ectoparasites from other fish species; Salwiczek & Bshary, 2011).

6.2.3 Models of interval timing

The system that underlies interval timing is typically assumed to be different than the system that allows for circadian timing. That said, different models of interval timing exist, three types of which do well to describe the current data and will be described here. Although there is no consensus regarding which model is best, each has inspired new research and paradigms. Ultimately, it may be that advances in the neuroscience of timing will help to decide among these models, a topic that will be considered in Box 6.1.

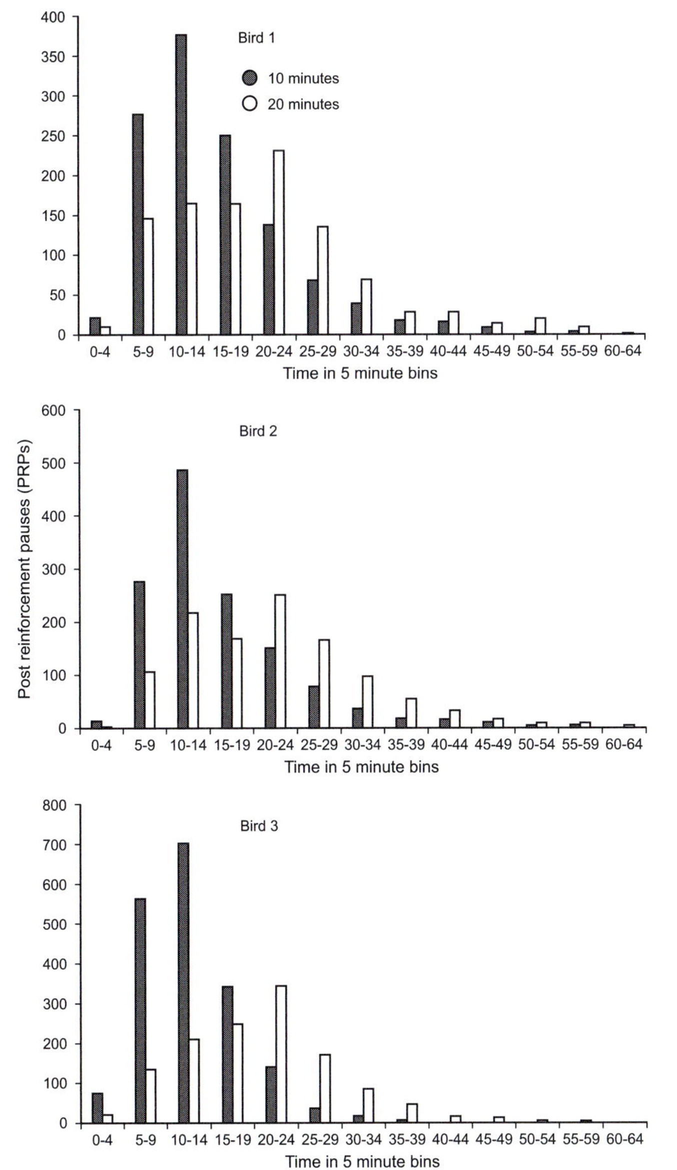

Figure 6.3 Example artificial flower and data from Henderson *et al.* (2006). The graphs show that the birds visited the artificial flowers at intervals consistent with the refill rate (10 or 20 minutes), with greater variance for the 20-minute duration.

Box 6.1 Interval timing in the brain

In humans, the neuronal bases of interval timing have been hypothesized to comprise a broad system spanning the thalamus and striatum, as well as the frontal and parietal cortices (Buhusi & Meck, 2005). The details of the system, including the individual functions of different brain areas, have yet to be determined. One goal of the ongoing research is to utilize neuroscience findings to elucidate the cognitive processes of timing by distinguishing among the models presented above, but along the way, the research has also improved our knowledge of some neurological disorders.

Early work found initial support for the Information Processing Model from pharmacological manipulation involving dopaminergic drugs. Data consistent with a separate clock component came from studies with rats in which methamphetamine (a dopamine agonist) led to peak responses that were shifted to the left of peaks in control conditions. That is, the increase in dopamine seemed to act in such a way as to speed up a clock. In contrast, haloperidol, a dopamine antagonist, caused a rightward shift or deceleration of the clock. However, subsequent data have been inconsistent, with some studies showing mixed results or suggesting that the clock was actually slowed with dopamine increase.

A direct relationship between dopamine and an interval timing clock has also been challenged by studies of neurological disorders. Patients with Parkinson's disease, in which nigrostriatal dopaminergic projections degenerate, show worse interval discrimination when they are not taking dopaminergic medication (L-dopa) than when they are taking the drug (Malapani *et al.*, 1998). This finding may appear to be in support of the pharmacological studies just described, but actually, for these patients, timing did not generally appear to 'slow down' as would be predicted with a decrease in dopamine. Instead, two comparison times were judged more similarly, as if one interval was sped up and the other one slowed down.

Currently, none of the existing cognitive models has complete support from neuroscientific studies, though as research begins to define the roles of the various areas within the timing network (thalamus, striatum, frontal and parietal cortices), certain models may gain support. In turn, the cognitive effects of neurological disorders may be better understood. Additionally, as knowledge is gained regarding how humans and animals process other magnitudes such as number, the understanding of timing processes in the brain may become clearer (see Section 6.3.3).

The Information Processing Model of Timing

Also known as the Scalar Expectancy Theory or the Pacemaker/Accumulator Model, the Information Processing Model by Gibbon, Church, and Meck (e.g Gibbon *et al.*, 1984) is possibly the most well known of the three theories. In general, it is a model of the manner in which a current event may be compared to a representation, stored in memory, of a past event (Figure 6.4). More specifically, the model is based on a hypothetical **pacemaker** that emits pulses at a constant rate. When a to-be-timed event starts, a switch is activated, allowing pulses to be accumulated until the event ends. At the end of the event, the switch is deactivated, and the information in this accumulator is held briefly in working memory. The contents of working memory are transferred to reference memory, which also contains information as to whether a response (e.g. lever press) was rewarded. For subsequent events (e.g. trials in an experiment), the value in working memory will be compared to the contents of reference memory and a decision will be made to respond or not to respond.

The comparison process works by computing a ratio of the value in working memory to the rewarded value that is stored in reference memory. It is commonly assumed that when reference

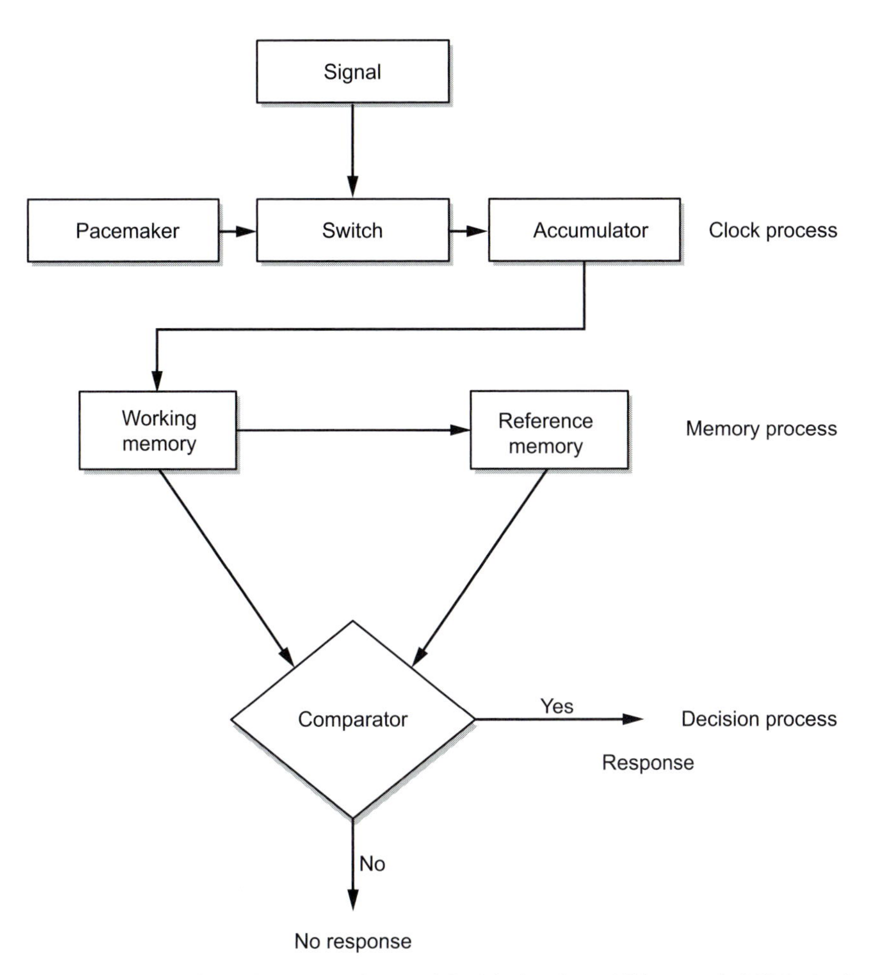

Figure 6.4 The Information Processing Model of timing from Gibbon *et al.* (1984). At the start of an event, a switch is closed, allowing signals from a pacemaker to accumulate. Working memory of the magnitude can then be compared to reference memory to guide a decision process.

memory is accessed during the comparison phase, a single criterion value is obtained. Yet, it is proposed that for each decision (trial) this value can come from a distribution of different criterion times stored in reference memory. This variation among criterion times is the result of variability in the speed of the pacemaker between trials, and if the pacemaker varies, more error will naturally accrue for longer intervals than shorter ones. Thus, by incorporating a source of variance, the Information Processing Model accounts for the fact that timing is scalar in that the variance increases with the length of the timing interval.

The Oscillator Model of Timing

The Information Processing Model has proven seminal in the study of interval timing, but it may not fit the actual neurobiology of timing organisms. One model that can potentially be a better fit to the manner in which the brain works is the Oscillator Model (e.g. Buhusi & Meck, 2005; Church & Broadbent, 1990). Although the details of the model go beyond the scope of this book, several aspects of this connectionist model can be explained by building on the Information Processing Model.

First, using a suggestion by Gallistel (1990), this model replaces the pacemaker with multiple **oscillators**. That is, this model no longer posits a pacemaker that emits pulses, but instead considers the presence of oscillators, mechanisms that fluctuate in a rhythmic, sinusoidal manner. Importantly, each of these oscillators is proposed to have a different period. For a timed event, the duration is coded not by an accumulation of pulses but by vectors that represent the phase of each of the oscillators. Further, for this model, the reference memory is stored differently than it was in the Information Processing Model. In the Oscillator Model, memory for intervals is stored within a matrix (familiar to those with a background in linear algebra) rather than as a distribution of values with a mean close to the actual interval.

Though mathematically complex, the Oscillator Model has the advantage of potentially being a more accurate fit to the underlying biology. Since single neurons and groups of neurons can generate oscillatory activity, the model might better guide the neuroscientific study of timing. However, in its present form, it is meant to be a cognitive model, presenting only analogs to brain structures. Additionally, there are limitations to the model, including assumptions that must be made to fit Weber's Law that still require verification (Staddon & Higa, 2006).

Models explaining timing without internal clocks

Some explanations of interval timing do not incorporate a pacemaker, oscillators, or other clock-like mechanisms and instead account for timing in other ways. The Behavioral Theory, for example, describes timing in a more traditional behavioristic manner rather than through a cognitive model like those described above. Because animals often engage in relatively stereotyped behavior starting at the beginning of a to-be-timed event, the behavioral state at the time of reinforcement can come to be associated with the reward. An experiment by Machado and Keen (1999) illustrates this process. Pigeons observed a light turn on for 4 or 16 seconds, and depending on the duration, pecking a blue or red key would be rewarded. After some experience with these contingencies, the light started to elicit a series of responses, starting with pecking the light and then moving around the box. On most 4-second trials, the birds were still pecking at the light when it turned off, and at this point, they went to the corresponding key. On the 16 second trials, they had begun to move around by the time the light turned off and then went to the appropriate key. It is thus possible that the correct responses were not due to an internal timing system, but instead may have been the outcome of the formation of associations between certain behaviors and a key that yielded reward.

Another explanation for how animals may come to respond after short intervals without a specialized internal clock comes from a memory model by John Staddon and colleagues (e.g. Staddon, 2005; Staddon & Higa, 1999). In laboratory situations in which animals receive food at regular intervals, it is possible that the decay of memory from one reward to the next can itself be associated with food. If, say, a pigeon receives reward for pecking every 30 seconds, any one event of reward will create a memory trace. This memory will become weaker over time and its strength at 30 seconds may come to be associated with the reappearance of food if the rate of decay is similar trial after trial.

Conclusions

Despite clever, well-controlled research and models such as these, it has been difficult to define the process of interval timing. Models must fit the behavioral data, but also should find some analogy with neurobiological data, the latter of which is relatively new and improving as technologies advance. Modifications of old models or creation of new models will likely be necessary. It might also be the case that the processes outlined in the models presented above are not mutually exclusive. For example, an interval timing system may serve to encode durations of signals, but then certain behaviors could become associated with post-interval responses (Shettleworth, 2010). Additionally, as the next sections will detail, similarities with the process of counting may ultimately improve our understanding of timing.

6.3 The approximate number system

Discriminating between two durations and discriminating between two arrays of objects share a common feature: both are instances of discriminating magnitude. Indeed, many have hypothesized that timing and numerical processing originate from a similar, if not common, representational system. One piece of evidence for this hypothesis is that numerosity discrimination also fits Weber's Law. Other evidence comes from studies showing that animals keep track of number and time simultaneously. In 1983, for example, Meck and Church presented rats with either two or eight bursts of noise that were 1 second each. Reward could be obtained if the rats pressed one lever if two bursts had occurred and another lever if eight bursts occurred. In this situation, number and time covary: two bursts took 2 seconds, and eight bursts took 8 (Figure 6.5). After experience, trials were presented that tested either for control by time or control by number. On trials testing number, either two or eight bursts were played, but each occurred over 4 seconds. On trials testing timing, four bursts were played either over 2 or 8 seconds. The rats remained accurate regardless of which dimension they were tested on; they responded with the 'long' lever either for 8 seconds or 8 bursts, and the 'short' lever for 2 seconds or 2 bursts.

Similar results were found for pigeons (Roberts & Mitchell, 1994), suggesting that across animals, the duration and numerosity features of stimuli are processed simultaneously. This hypothesis will be revisited throughout this section, though the primary focus will be on how animals and humans represent number. Then, in Section 6.4, a second proposal will be explored, one that considers a separate system for representing small numbers. Importantly, though, this chapter will follow the recent traditions of this field of study by limiting the term 'counting' to situations in which quantities are precisely labeled with a symbol system (e.g. Arabic numerals). In fact, it is likely that the numerical processing that is detailed here forms the foundation for the mathematical, counting skills that humans learn (see Box 6.2).

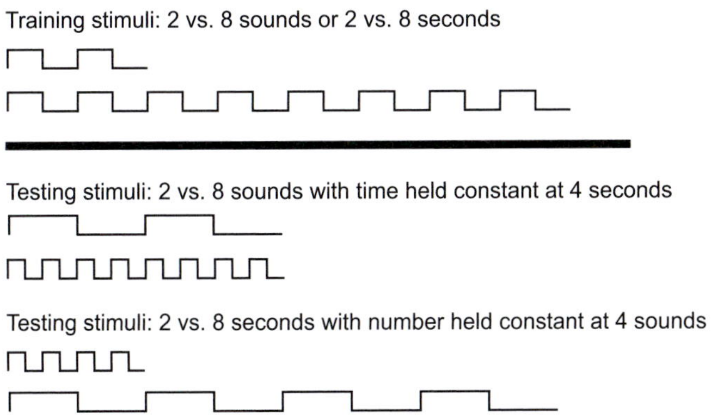

Figure 6.5 Example training and testing stimuli from Meck and Church (1983). Each 'hill' represents a stimulus (sound burst) and its duration.

6.3.1 Discriminating the number of responses

When human adults are told to press a key a certain number of times, as fast as possible, they can be reasonably accurate, even if counting is made unlikely by asking the participants to say the word "the" with every key press (Cordes *et al.*, 2001). That is, humans have a sense of how many times they have responded, even without engaging the verbal counting system. Rats appear to be able to keep track of responses as well. In one study (Platt & Johnson, 1971), each rat had to press a lever a certain number of times (ranging from 4 to 24 across different groups) and then place its head into a food tray to receive reward. If a rat moved to the food tray before having completed enough presses, the trial was ended and a new one began. The peak response at the food tray occurred at or near the correct number of presses.

For both the humans and rats, though, there was variability; key or lever presses were not always exact. Similar to the studies of timing, the variance increased as the number increased, following Weber's Law. Thus, not only do tasks requiring subjects to keep track of the number of responses show some of the same variability signatures as timing, these signatures are present in both rats and humans. That is, although humans have an ability to learn symbols for numbers and engage in counting, they also have a system for non-verbally estimating numerical value that is shared across species.

6.3.2 Comparing numerosities

It is relatively easy to look at the third array in Figure 6.6 and quickly determine that there are more 'O' signs than 'X' signs. It is not immediately apparent that the ratio is 70:30, but there is clearly a difference in **relative numerosity**. The ability to gauge which of two arrays has more (or less) has obvious practical advantages. It could, for example, be a cue to ideal feeding locations. Outside of the visual domain, recognizing the number of individuals in a rival group by using their vocalizations – relative to the size of one's own group – might impact decisions regarding aggressive responding (e.g. McComb *et al.*, 1994).

In a classic example of relative numerosity discrimination, Honig and Stewart (1989) began by training pigeons that pecking at an array of all red dots would yield reward, but responding to an

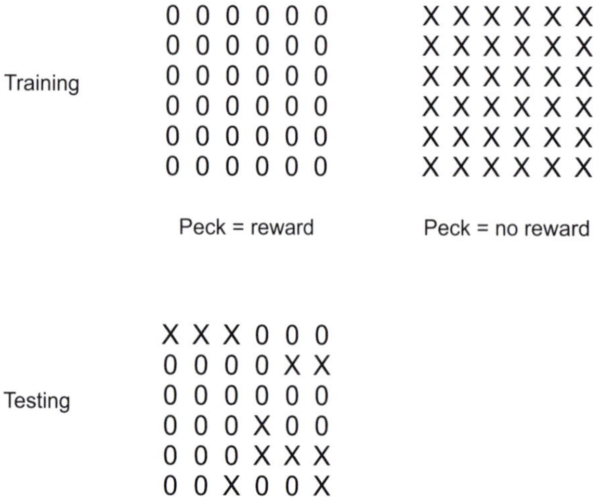

Training

O O O O O O
O O O O O O
O O O O O O
O O O O O O
O O O O O O
O O O O O O

Peck = reward

X X X X X X
X X X X X X
X X X X X X
X X X X X X
X X X X X X
X X X X X X

Peck = no reward

Testing

X X X O O O
O O O O X X
O O O O O O
O O O X O O
O O O X X X
O O X O O X

Figure 6.6. Example training and testing stimuli from Honig and Stewart (1989). In this example, the test array has a greater number of the previously rewarded O's than the X's, and the rate of pecking responses was directly related to the proportion depicted in the array.

array of all blue dots would not yield reward. Once pigeons were responding accordingly, test trials were included that had red and blue dots intermixed (analogous to the O and X signs in Figure 6.6). On these test trials, pigeons' rate of responding was directly related to the proportion of red dots. That is, pigeons pecked more frequently if there were more red dots than blue dots, and the responding increased as the number of red dots increased.

The results could mean that pigeons are sensitive to relative numerosity, but an alternative explanation exists. In these arrays of dots, number is confounded with surface area. An array of all red dots may have, say, 36 dots, but it also has a certain surface area of 'redness.' An array of 20 red dots to 16 blue dots will have more red color than blue color. Thus, correct responding in Honig and Stewart's experiment may not have been due to comparing numerosities but comparing surface areas. In the natural world, these two qualities often covary, yet if it is the ability to enumerate that is the focus of a research question, then an experimental design would have to differentiate between them. Follow-up studies to Honig and Stewart (1989) did just this by controlling for surface area in various ways (e.g. altering the size of the dots, using X's and O's instead of dots, etc.), and pigeons still showed the same pattern of results, providing evidence that they were indeed comparing numerosities.

Number discrimination without training

The studies just mentioned have all relied, to some extent, on training. While these studies do suggest sensitivity to number, evidence for *spontaneous* numerosity discrimination would mean that this ability is not simply the result of long laboratory experience but something akin to what Gallistel (1993) referred to as a 'mental primitive,' or natural ability. That is, spontaneous use of number would suggest that it is salient to animals, not secondary to other characteristics like duration, size, or intensity. There is reason to think that this is the case. In a study by Cantlon and Brannon (2007), rhesus monkeys (*Macaca mulatta*) were given a delayed match-to-sample task in which matches could be made based on number or on one other dimension (i.e. color, shape, or surface area; Figure 6.7). Each of the monkeys had previous experience with delayed match-to-sample tasks using photographs of scenes, but not all had previous experience in tasks that used number as a key feature. Cantlon and Brannon found that regardless of previous number experience, the monkeys primarily based their matching on number even though they could have responded based on another dimension.

Human infants also show spontaneous encoding of number. After seeing arrays of eight dots over and over until habituation, 6-month-old infants dishabituated (looked longer) to an array of 16 dots than another array of 8 dots (Xu & Spelke, 2000). However, infants of this age did not discriminate if the dot arrays differed by a smaller ratio (e.g. 8 vs. 12 dots, or 2:3 ratio). Similar results come from studies using non-visual stimuli; infants will discriminate 8 from 16 but not 12 sounds (Lipton & Spelke, 2003). Thus, it appears that when infants are discriminating between two numerosities, the ratio difference is important: 1:2 ratios are discriminable, but 2:3 ratios prove difficult. In fact, the ratio difference appears to be more relevant than the absolute number difference, a feature of Weber's Law. Infants will discriminate 4 from 8 sounds (Lipton & Spelke, 2004) even though the absolute number difference (4) is the same as it is for a comparison they have difficulty with (8 versus 12).

Again, though, particularly with visual stimuli, it is important to consider whether infants are truly representing number or whether their behavior is primarily influenced by another magnitude dimension such as surface area. Some researchers have suggested that for infants, features such as surface area may be preferentially encoded over number, particularly for small arrays of items (e.g.

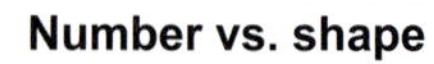

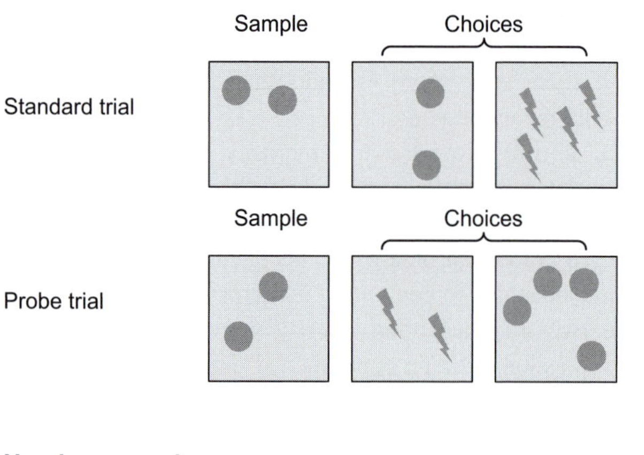

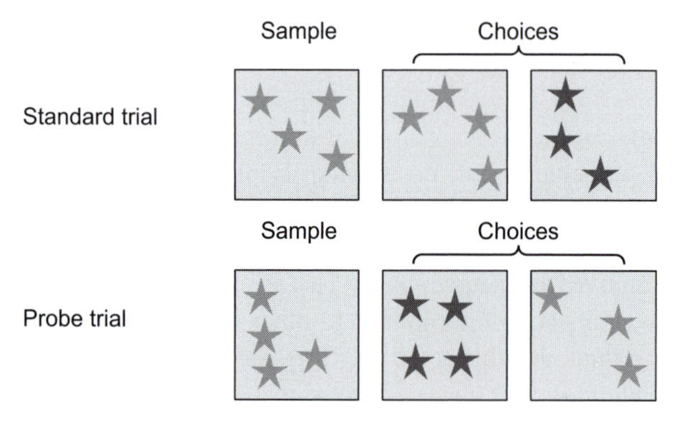

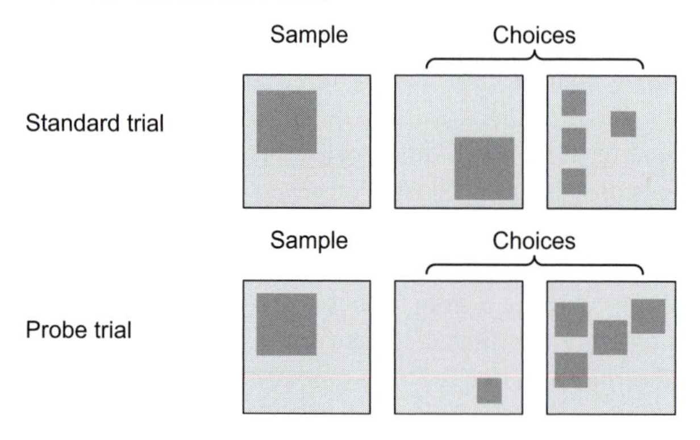

Figure 6.7 Example stimuli used in Cantlon and Brannon (2007). In the delayed match-to-sample task, rhesus monkeys primarily chose matches based on number.

Clearfield & Mix, 1999; Feigenson *et al.*, 2002). Yet, similar to findings from studies with monkeys, it appears that infants will often ignore surface area and attend to number. Relatedly, in situations in which number changes but surface area remains constant, encoding surface area is difficult for infants (Brannon *et al.*, 2004; Cordes & Brannon, 2008, 2009a). Together, the (sometimes conflicting) results thus far suggest that infants and animals can encode number *and* variables like surface

Box 6.2 The approximate number system and math skills

Although it would appear that infants and animals share a similar approximate number system, humans in most cultures go on to learn a representational system for formal mathematics and counting. Whereas the approximate number system is imprecise and comparisons between numerosities are dependent on ratio differences, the language-based representations of number are exact (Feigenson *et al.*, 2013). With the former, one can quickly determine that an array of 16 is different than an array of 32, but with the latter, one can know that there is exactly a 16-unit difference between these arrays (and that 16 indeed also differs from 17).

Humans use the approximate number system through adulthood, and the speed at which adults determine which of two quantities is greater continues to depend on the ratio difference (Moyer & Landauer, 1967). Several researchers have proposed that this approximate number system forms the foundation for formal mathematical computation and counting (e.g. Butterworth, 2005), and research has begun to support this claim. In a study with 3- and 5-year-old children using a task similar to the relative numerosity experiments by Honig and Stewart (1989) described above, performance correlated with scores on a standardized mathematics test even after children's age and verbal skills were controlled (Libertus *et al.*, 2013). Similarly, at 14 years, variation in children's performance on an approximate number task correlated with variation in mathematical ability after controlling for more general cognitive abilities (Halberda *et al.*, 2008). In adulthood, the correlation continues to hold (e.g. DeWind & Brannon, 2012; Libertus *et al.*, 2012). As strong as these findings are, there are also studies that suggest that more research is needed. The relationship is not always found in studies with children, or if found, a similar relationship is not always found for adults (e.g. Holloway & Ansari, 2009). At the very least, it may be that the measurements used matter; some mathematical abilities may be more related to the approximate number system than others.

Given the correlational nature of the research, it is logical to ask whether there is truly a causal relationship such that enhanced approximate number ability makes some people better at math. Although this area of study is still new, Feigenson and colleagues (2013) have proposed a causal relationship based on three sets of findings: (1) individual differences in the approximate number system exist even in infancy (Libertus & Brannon, 2010), (2) the precision of this system predicts math scores later in development (e.g. Mazzocco *et al.*, 2011), and (3) training the approximate number system may improve formal math ability (Park & Brannon, 2013). There is still much to be examined, of course, but the comparative study of numerical competence may ultimately play a role in our understanding of formal math and perhaps advance mathematics education.

area, but the priority given to one feature over another may depend on the situation. It is easy to imagine, for example, that a salient feature of food may often be its continuous extent rather than the number of individual elements, yet for objects (e.g. number of conspecifics), a salient feature may be number.

6.3.3 Mechanisms of the approximate number system

Cognitive models of the approximate number system posit that a real-world value is represented in the mind as an analog of that value. This **analog magnitude representation** is a 'noisy' representation of the set of items, yet it is proportional to the number of items being represented. Models start to differ, however, when considering how this analog magnitude representation is formed. The Information Processing Model, presented above to describe interval timing, can also work with number after slight modification. Meck and Church (1983) posited that if a switch connecting a pacemaker to an accumulator could open and close for each object or event (instead of staying closed for the entire event as when timing), the contents of the accumulator would increase incrementally, ultimately resulting in an analog representation of the number. In contrast, an alternative model proposes that events or objects are detected in parallel (not incrementally as with the pacemaker/accumulator) and then summed to form an analog representation (e.g. Dehaene, 2009).

Whichever model ultimately forms the best description of the process of representing number, one idea appears uncontroversial: the performance of both human and animal participants is predicted by Weber's Law. To review, for animals and humans, the sensitivity to differences between numerosities does not appear to depend on the absolute difference (e.g. that 8 and 16 differ by 8 units) but on the ratio (e.g. that 8 and 16 have a 1:2 ratio). It mathematically follows, then, that for a certain absolute difference in numerosity, discrimination is easier if the arrays are smaller (e.g. 8 versus 16 is easier than 16 versus 24 even though both comparisons differ by 8 units). This is the basis of Weber's Law, and experiments across species find these hallmarks repeatedly.

As noted earlier, behavior based on Weber's Law also is seen in interval timing tasks. As a further example, 6-month-old infants discriminate among intervals that differ by a 1:2 ratio, but not a 2:3 ratio, over a range of durations; the absolute difference between the durations did not modulate performance (vanMarle & Wynn, 2006). In fact, as infants develop and become better at discriminating closer ratios for number (e.g. 2:3 ratios), they also become better at discriminating the same ratios for duration (Brannon *et al.*, 2007). Thus, two magnitudes, number and timing, are seemingly processed in a similar way, showing similar behavioral signatures of Weber's Law.

Timing and counting also appear to show similar neural processing. The brain systems that are involved may process many types of magnitude judgments, including loudness, size, and brightness, among others. In a review of relevant literature, Cantlon *et al.* (2009) note that the parietal cortex, specifically the intraparietal sulcus, plays an important role in the processing of magnitude (Figure 6.8). The existing neuropsychological research with humans suggests that deficits in time,

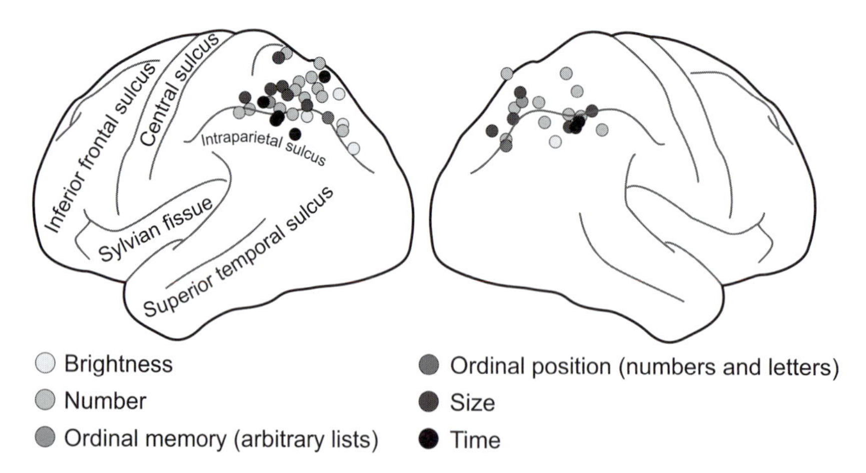

Figure 6.8 A depiction of findings, across many studies, implicating the parietal cortex in magnitude processing. Cantlon *et al.* (2009) created this figure – an inflated human cortical template – based on a review of the existing literature.

Researcher profile: **Dr. Elizabeth Brannon**

Figure 6.9 Dr. Elizabeth Brannon.

As an undergraduate majoring in physical anthropology at the University of Pennsylvania, Elizabeth Brannon was on the path to become a field primatologist. Through the School for Field Studies, she had studied cotton-top tamarins (*Saguinus oedipus*) in their natural habitats in Panama. While working with Drs. Dorothy Cheney and Robert Seyfarth, Brannon was exposed to research approaches that blended fieldwork with the study of cognitive processes. She quickly became interested in the animal mind, and after receiving a Masters degree in anthropology at Columbia University, transitioned to the field of psychology where she earned her Ph.D in Dr. Herb Terrace's laboratory.

During graduate school, Brannon developed new studies of primate numerical processing that have greatly enhanced the understanding of the evolutionary bases of human numerical cognition. Now, as a professor at Duke University, Brannon's work has become an example of the effectiveness of the comparative approach. She examines numerical cognition by examining its development in human infants as well as its evolution within the primate order. To do so, she must maintain two laboratories, one for testing human participants and one for monkeys and lemurs. The research utilizes a range of methods including behavioral measurement and neuroscientific approaches such as fMRI, EEG/ERP, and single unit recording in monkeys.

The comparative approach is key to Brannon's research program: "By studying cognitive abilities in non-human animals and human infancy we can understand both the evolution and the ontogeny of cognition and thereby define the primitives upon which adult human cognition is built." More specifically, studying the approximate number system has become an ideal way for Brannon to examine cognitive processes that occur in the absence of language. Her work demonstrates how abstract numerical concepts are possible independently of language. Brannon's work also suggests that these non-verbal aspects of numerical cognition may, in turn, causally influence symbolic math abilities: "the internal representation of approximate numbers and their operations provides a critical foundation for uniquely human symbolic mathematics and may indicate new interventions for math educators" (Park & Brannon, 2013).

size, and number comparisons can co-occur due to parietal lesions (e.g. Walsh, 2003), and in patients without lesions, fMRI studies suggest that there is overlap in the parietal cortex for processing these magnitudes (Fias *et al.*, 2003; Pinel *et al.*, 2004). In rhesus monkeys, the parietal cortex is also implicated in magnitude tasks; ventral intraparietal neurons respond either to numerosity or length, but a subset respond to both (Tudusciuc & Nieder, 2007).

The research is ongoing, but there are signs that the approximate number system is part of a system that is shared across species, and at least in humans, emerges early in development. Neuroscience and behavioral studies are beginning to shed light on whether number processing is part of a more general magnitude processing system, but there is still the possibility that some aspects of numerical ability are cognitively unique. Consider, for example, that in the laboratory tasks that form the basis of the research, participants must compare and discriminate magnitudes. Some have suggested that perhaps the shared aspects of reasoning about magnitudes occur late in the processing when these comparisons are made. Earlier, when objects and events are individuated, numerical representations may be formed within their own unique neural systems (e.g. Cantlon *et al.*, 2009). As neuroscience techniques advance and cognitive theories are elaborated, the details of the evolutionarily primitive system or systems that allow for numerical competence will likely be elaborated.

6.4 Representing small numbers

Although the approximate number system allows for adaptive processing of magnitude, it still has limitations, most notably the lack of precision in discriminating represented quantities when they differ by smaller ratios. Indeed, adherence to Weber's Law is considered to be the signature property of this number system. It is all the more surprising, then, to find that in certain experimental tasks in which small quantities are presented, a different sort of limitation, or signature property, is often observed for both humans and animals. In this section, research examining the representation of small numbers will be presented, along with proposals regarding the cognitive systems that may underlie this representation.

6.4.1 Set size signature

On the island of Cayo Santiago off the coast of Puerto Rico, a large group of free-ranging rhesus macaques have been studied for many years. The species is not native to the island; the monkeys are descendants of a group that was brought to the island from India in the 1930s for research purposes. More recently, the monkeys have been participants in studies of cognition, including spontaneous numerical discrimination. In one such experiment (Barner *et al.*, 2008, based on Hauser *et al.*, 2000), a researcher approached lone monkeys and placed two boxes on the ground about 2 to 5 meters from the monkey. The researcher placed a different number of food items (e.g. apples) one at a time into each box and then walked away, allowing the monkey to approach and choose one of the boxes (Figure 6.10). After a box was chosen, the researcher came back, typically causing the monkey to scurry away.

Given what has been previously discussed in terms of numerical discrimination, one might expect that monkeys' ability to reliably choose the box with the greater number of apples would fit Weber's Law such that success would be dependent on the ratio difference. However, in this task, monkeys' behavior depended on the size of the sets. That is, they chose the box hiding the larger amount of apples when the boxes held 1 versus 2, 2 versus 3, or 2 versus 4 apples, but not when they contained 2 versus 5 or 3 versus 5. The 2:5 ratio (e.g. 2 versus 5 apples) should be more discriminable than 1:2 (e.g. 2 versus 4 apples), yet the monkeys succeeded in the latter and not the former. The results suggest that when discriminating between small quantities, performance is determined by the size of the arrays (the **set size signature**); comparisons in which one of the sets is greater than 4 are difficult for the monkeys to make.

Testing human infants on an almost identical task has yielded similar findings. Ten- and 12-month old infants crawled toward the larger quantity when comparisons were 1 versus 2 and 2 versus 3, but not when one of the sets was 4 or greater (Feigenson *et al.*, 2002). Infants even chose randomly when the comparison was 1 versus 4. Thus, again, it was the size of the sets that determined the success (though here, the infants seemed more limited than the mature monkeys and had difficulty with arrays greater than 3). The same pattern of findings was also found using a second method in which infants were timed as they searched a box for previously hidden treats: an experimenter placed items into an opaque box, though in some cases, items were surreptitiously removed before the infant began to search. In this way, one can compare, for example, an infant's search time after one object is placed and he or she retrieves it to an instance in which two objects are placed yet one has been secretly removed. Infants reliably searched longer after retrieving one object when they had observed two objects being placed in the box than when they had only seen one object placed. The longer search times occurred for arrays up

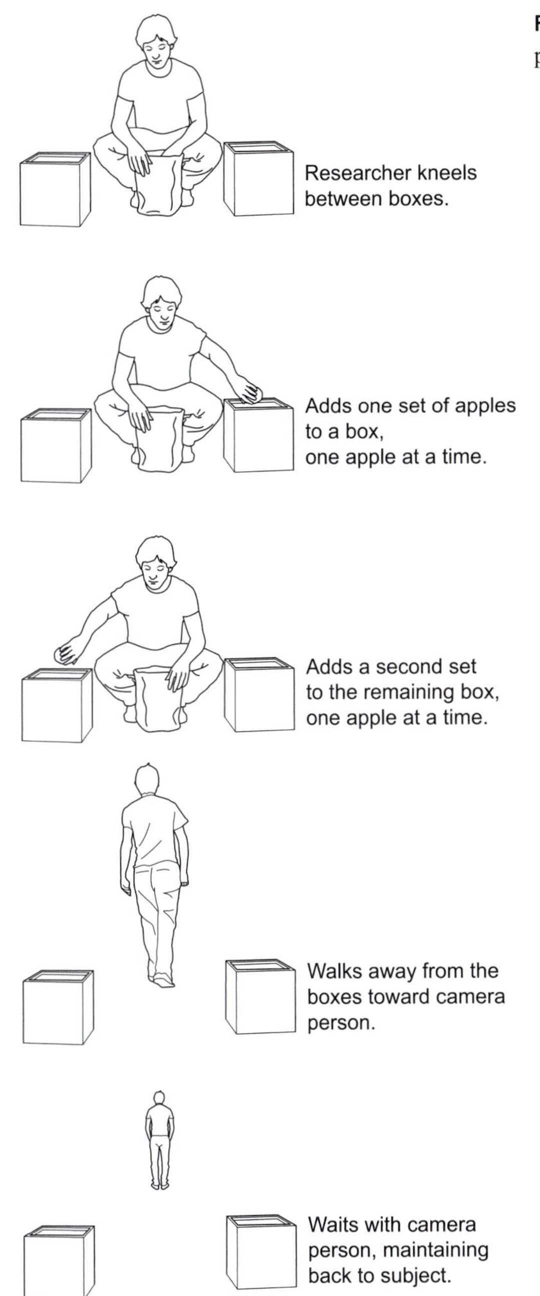

Figure 6.10 Test events in Barner *et al.* (2008), from the perspective of a rhesus monkey participant.

to three items, but there was no indication from the search behavior that the infants precisely represented sets larger than three (Feigenson & Carey, 2003).

Outside of the primate order, other species have shown behavior that also might be indicative of a set size signature with small numbers. Uller *et al.* (2003), for example, found that red-backed salamanders (*Plethodon cinereus*) would approach a clear tube containing 3 fruit flies (*Drosophila virilis*) over a tube containing 2 flies, but not 6 over 4 flies. Young domestic chicks (*Gallus gallus*) similarly discriminated arrays of dots as long as sets remained 3 or fewer (Rugani *et al.*, 2008). In both of these cases, though, the arrays were either completely visible, or partially occluded, which would likely not pose the representational demands that monkeys and infants faced on tasks in which the contents of opaque containers had to be stored in memory.

In contrast to the above studies, a study with mongoose lemurs (*Eulemur mongoz*) found that discrimination between small numbers *was* dependent on Weber's Law and not a set size limit. Lemurs are **prosimians** (primates that split from the common ancestor to monkeys, apes, and humans) and part of the motivation for the study was to examine whether numerical sensitivity was likely present in the common ancestor of all primates. In a search time task similar to what has been used with human infants (e.g. Feigenson & Carey, 2003, described above), the lemurs' discrimination was determined by the ratio difference, not the absolute value. Specifically, 1:2 ratio differences were discriminated, even when quantities exceeded 4 (e.g. 2 versus 4 and 4 versus 8), but 2:3 ratios were not (Lewis *et al.*, 2005). Thus, in this study, the lemurs appeared to be relying on analog magnitude representations of number. The reason for the differences in results in this study and those with human infants and rhesus macaques is still under consideration. One possibility that must be ruled out pertains to the task design itself. The lemurs were tested many times whereas the wild rhesus macaques each only received one test trial. It is possible that repeated exposure encouraged the use of analog magnitudes for the lemurs. Indeed, the mechanisms underlying the representation of small numbers and the situations in which they are deployed by an animal are the topic of current discussion.

6.4.2 Mechanisms for representing small numbers

Based on the set size signature found across various species in tasks assessing small number discrimination, many researchers have proposed that there are at least two distinct mechanisms for processing number: a precise small number system and the approximate number system (e.g. Carey, 2004; Feigenson *et al.*, 2004; Xu, 2003). One proposed mechanism that likely underlies small number discrimination in some situations is an **object-tracking system**. This system was described originally by researchers examining human visual attention in order to explain the ability to track objects in the world (e.g. Kahneman *et al.*, 1992; Scholl, 2001). The mechanism entails a small number of mental 'indexes' that 'point' to individual objects and enable people to keep track of them as they move in and out of view. Across many different types of tasks (e.g. tracking the movement of four specific objects among many other moving objects, Pylyshyn & Storm, 1988), adult humans appear to be limited to tracking about four objects on average. Thus, the set size limit seen in the aforementioned studies of small number discrimination fits the general model of the object-tracking system. Importantly, though, the object-tracking system is not meant to be a model of a *number* system *per se*. Instead, the output of a system such as this might be to track, for example, that 'Object, Object, Object' exists in one container while 'Object, Object' exists in another container, rather than outputting an analog magnitude for the numerical value of the two sets. Yet, the output of the object-tracking system still provides a basis for comparison between numerosities.

Additional evidence for the existence of a separate system that may be utilized when representing small sets comes from studies in which human infants and animals must discriminate between two sets of objects in which one set is small and the other large (e.g. 2 versus 8). Human infants as well as a species of fish (guppies, *Poecilia reticulata*) can discriminate sets in which both numbers are small or both numbers are large but have difficulty when one number is small and the other is large (e.g. Piffer *et al.*, 2012; vanMarle, 2013; vanMarle & Wynn, 2011; Xu, 2003). It is thus possible that there are two systems governing discrimination in these cases, and further, the output of one system cannot be compared to the output of the other (e.g. Gallistel, 2007; vanMarle, 2013). Specifically, analog magnitude representations will provide information about the total value of the set, but object-tracking representations will not.

In contrast, some researchers have proposed that both small number and large number discrimination can be processed by a single system of analog number representation. The system, though, will be primarily effective in situations in which the ratio difference is large, and for smaller ratios, the output of an object-tracking system will dominate (e.g. Cordes & Brannon, 2009b; Xu, 2003). In sum, though it is clear that many species can process and discriminate both small and large quantities, researchers are still actively examining the role of these two potential cognitive systems. Outstanding questions remain regarding what situations might elicit different systems, the development of these systems in humans and other species, and their evolution.

6.5 Operations

In their seminal work on numerical cognition, Gallistel and Gelman have detailed many aspects of non-verbal number representation, including the features that indicate a true concept of number. The ability to perform **operations**, or procedures on numbers, is a proposed hallmark of this concept (e.g. Gallistel & Gelman, 1992, 2000). For example, a true concept of number should include an understanding that number is an ordered system. That is, not only are 4 and 8 different quantities, but also 8 is larger than 4 (Cordes *et al.*, 2007). Evidence for operations in animal species and humans has been growing over the past two decades, with special focus placed on addition and subtraction, as well as an understanding of ordinality. More broadly, operations also could underlie foraging behavior, particularly when foraging decisions are based on sensitivity to the rate at which a site provides food (e.g. Cantlon & Brannon, 2010), which will be discussed in Chapter 7.

6.5.1 Addition and subtraction

It is reasonable to think that addition and subtraction are operations that are specific to symbolic representations of number like Arabic numerals. From early in schooling, humans are taught to write out numbers, inserting '+' and '−' signs to denote adding and subtracting. Yet, it is also not a stretch to recognize that the non-verbal numeric representations discussed thus far in this chapter can also serve as the operands for these procedures.

A study by Wynn (1992) with human infants provided a non-verbal experimental procedure with which to examine spontaneous addition and subtraction of small quantities. Five-month-old infants were shown one doll being placed onto a stage, and an occluding screen was placed in front of the doll. A second doll was then placed on the stage, also behind the screen. If infants can represent the first doll after it was occluded, and then add the second doll to this representation, they should respond with longer looking if the screen is dropped to reveal only one doll ('1 + 1 = 1') than if it drops to reveal two dolls ('1 + 1 = 2'). Infants did look longer at the impossible outcome in the addition event, and they also looked longer at an impossible subtraction event ('2 – 1 = 2') than a possible event ('2 – 1 = 1'). An additional experiment found that infants looked longer at events depicting '1 + 1 = 3' than '1 + 1 = 2,' suggesting that the representation of objects behind the screen was more precise than simply 1 + 1 = *more than 1*.'

In Wynn's original study, however, other magnitudes such as surface area were not controlled. Because all of the dolls were the same size, it is possible that infants were not responding to number *per se* but to an impossible amount of 'doll stuff' (see Mix *et al.*, 2002). Subsequent studies controlled for other variables by either using alternative objects (e.g. Simon *et al.*, 1995) or by showing that addition can occur across modalities (Kobayashi *et al.*, 2004). For the latter study, 5-month-old infants first learned to associate the arrival of animated dolls with tones. In test trials, one doll was moved behind a screen and then a single tone or multiple tones occurred indicating the arrival of additional unseen dolls behind the screen. Infants looked longer to events in which an impossible outcome occurred (e.g. '1 doll + 2 tones = 2 dolls', or '1 doll + 1 tone = 3 dolls') than events in which a possible outcome occurred (e.g. '1 doll + 1 tone = 2 dolls', or '1 doll + 2 tones = 3 dolls').

In many ways, the addition abilities seen in these studies fit the findings from the choice tasks detailed in Section 6.4.1. In the choice tasks, which occurred years after the looking-time studies, observers saw objects added one by one into an opaque container, and approach responses appeared to be based on representing all of the objects. Animals have also shown the ability to add and subtract in studies in which looking time to possible and impossible events was measured. Rhesus macaques look longer when impossible addition and subtraction outcomes are revealed, both when the quantities are small ('1 + 1 = 3'; Hauser & Carey, 2003) and large ('4 + 4 = 4'; Flombaum *et al.*, 2005). Four lemur species (*Eulemur fulvus*, *E. mongoz*, *Lemur catta*, and *Varecia rubra*) were also found to look longer at impossible outcomes of '1 + 1' events depicted with lemons, even when surface area was controlled (Santos *et al.*, 2005).

The ability to represent simple addition and subtraction operations has also been seen in bird species. In these studies, looking time is not used as the dependent measure, and instead, behavior such as pecking keys (pigeons: Brannon *et al.*, 2001) or searching (newly hatched chicks: Rugani *et al.*, 2009) is examined. Thus, there is evidence across a variety of experimental paradigms that, similar to the ability to discriminate quantities, the representation of arithmetic operations is found beyond the primate order (see also Box 6.3). As will be seen below, though, the underlying cognitive mechanisms for these operations remain under consideration.

Box 6.3 Training animals to use number symbols

In the study of numerical cognition, some research programs have focused on the ability of animals to learn associations between number words or symbols (e.g. Arabic numerals) with quantities. For example, Alex, an African gray parrot (*Psittacus erithacus*), was trained to vocalize English number words and associate these words with Arabic numerals as part of a larger training process that included learning to respond to various spoken questions. When Alex was shown two numerals (e.g. '2' and '4') of the same size, and asked "Which bigger?", he was able to respond vocally with "Four" (see Pepperberg, 2006, for review). To do so, Alex likely was linking the vocalizations to the visual symbols and the corresponding numerosities.

Boysen and colleagues taught chimpanzees to use Arabic numerals through a process that began with initial training of one-to-one correspondence between food pieces and non-edible small objects (e.g. 2 food pieces on a plate would be matched to a plate that contained 2 small cloth balls). Over time, the object plates were replaced with Arabic numerals, and ultimately the chimpanzees could touch a corresponding numeral for arrays of 0–8 objects (Boysen, 1993). This research program also provided one of the first examinations of arithmetic in an animal species (Figure 6.11). When one of the chimpanzees, Sheba, was allowed to search a room for hidden food items or numerals in various

Figure 6.11 Testing room for examining a chimpanzee's ability to add. The circles depict food items, although on some trials the food was replaced with Arabic numeral placards.

locations, she was able to return to the starting location and report, by touching a numeral, the summed amount (Boysen & Berntson, 1989). The ability to sum symbolic representations of quantity has subsequently been found in other studies with chimpanzees (e.g. Beran, 2001; see also Matsuzawa, 1985) and with Alex the parrot (Pepperberg, 2006). Additionally, Olthof *et al.* (1997) found that squirrel monkeys (*Saimiri sciureus*) that had been trained with Arabic numerals could compare pairs of numerals so as to pick the pair that summed to the greater amount.

The process of training associations between numerals and quantities is time-consuming, and it has been proposed that these studies do not reveal much about numerical cognition beyond that of studies examining spontaneous numerical discrimination and operations (e.g. Shettleworth, 2010). One particularly seminal finding, however, has been the role of numeral use in a reversed contingency task. Boysen and Berntson (1995) created a task in which numeral-trained chimpanzees had to choose the smaller of two amounts of candies to receive the larger amount. In this situation, though, the chimpanzees seemingly could not inhibit choosing the larger amount of candy, even though that choice led to a smaller reward. In contrast, when the Arabic numerals were used in place of the candy arrays, chimpanzees readily chose the smaller number and subsequently gained the larger amount. Thus, the numerals seemed to represent magnitude symbolically and aid the inhibition of prepotent responses to actual food quantities. This work has proven informative to the study of inhibitory control in young children (Carlson *et al.*, 2005) as well as the different ape species (e.g. Vlamings *et al.*, 2006).

Cognitive mechanisms of addition and subtraction

In its traditional definition, addition involves combining two or more number representations to form a new representation, the sum. The approximate number system would thus account for addition because an analog magnitude representation could be combined with another analog magnitude in such a way as to yield a total, or sum. It gets trickier when one considers the object-tracking system, however, because as discussed above, that system does not create a representation of a set; only the set's individual components are represented. For example, in an addition task with small numbers, one object is tracked as it goes behind a screen, and then another object is tracked as it too goes behind a screen. Under object-tracking models, the representation is then 'Object, Object', which could be compared to the impossible outcome of 'Object' that is seen when the screen drops on test trials.

Thus, in addition and subtraction tasks using small quantities, it is possible that observers are not creating a magnitude sum after all, but instead are relying on the object-tracking system. This system, however, cannot account for tasks in which operations are performed on large quantities due to the set size limit. One such study has already been mentioned (i.e. '4 + 4' events, Flombaum *et al.*, 2005), and others include studies with human infants ('5 + 5' and '10 – 5' events, McCrink & Wynn, 2004) and chimpanzees ('6 + 10' events, Beran & Beran, 2004). In these studies, it is proposed that magnitudes – and sums – are being represented by the approximate number system.

6.5.2 Ordinality

At a basic level, a number can simply serve to differentiate items and be used as a label (like Wayne Gretzky's jersey number '99'). This type of representation of number is on a **nominal scale**, and it does not allow for inferences to be made about relative magnitudes (e.g. 99 is greater than 50), though basic equality/inequality determinations can be made (e.g. 99 does not equal 50). In contrast, when numbers are represented on the **ordinal scale**, they are ordered in a rank based on their relative magnitudes. Tests of relative numerosity like those described in Section 6.3.2 and the addition and subtraction studies described above suggest that animals do represent number on an ordinal scale, but it is important to note that behavior on some tasks can be interpreted as the result of representation on a nominal scale. If an animal is trained to choose the larger of two numerosities, for example, it is possible that what is ultimately learned is a set of pairwise contingencies (e.g. *pick 7 when paired with 6, but pick 6 when paired with 5*, etc.) instead of an ordinal rule (e.g. *pick the greater magnitude*).

A more direct test of knowledge of numerical order would entail testing animals on a range of numbers outside of the range they were trained in. Brannon and Terrace (1998, 2000) presented rhesus macaques with a touch-sensitive screen that displayed the arrays of size 1–4 simultaneously; the elements of the arrays differed in size and shape and they appeared in various locations from trial to trial (Figure 6.12). After the monkeys learned to touch the arrays in increasing order, new test trials were inserted that contained arrays of size 5–9, but only two arrays were presented at a time. Even with these novel pairs (e.g. 7 and 9), the monkeys responded correctly by touching them in increasing order, though as predicted, response times and accuracy followed Weber's Law.

A question remains as to whether the monkeys learned the concept of ordinality from the training trials or if they already had this concept prior to the study. A clue comes from one monkey that was unsuccessfully trained to respond to arrays of 1–4 in an arbitrary order that was not based on increasing or decreasing magnitude. The fact that this monkey did not learn the order, but later was able to learn an increasing order, suggests that monkeys

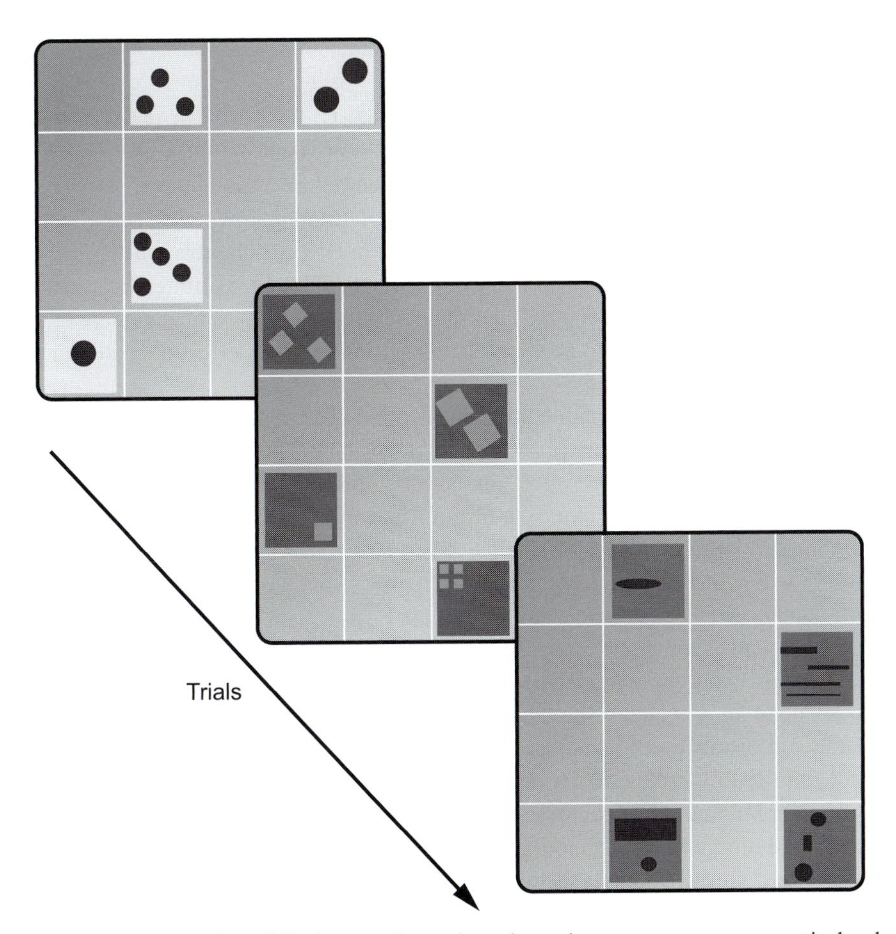

Figure 6.12 Examples of displays used to train and test rhesus macaques on numerical order in Brannon & Terrace (2000). The elements of the arrays varied in size, shape, and location from trial to trial.

may spontaneously order magnitudes (Brannon & Terrace, 2000). The arbitrary order would thus have been in conflict with this existent representation, and formed a more difficult order to learn.

In sum, a wide range of animal species can not only discriminate between numerosities but can also add, subtract, and order. Though the cognitive and neural mechanisms supporting a concept of number are still under investigation, the comparative study of numerical processing, as well as timing, has led to great advances in our knowledge. These findings may ultimately influence mathematics education as well as the understanding of certain neurological disorders. This chapter will end, however, by presenting in Box 6.4 an ongoing question for research into numerical cognition, one that is specific to human cognition, but in turn helps to complete the comparative framework. If the cognitive systems for number that humans share with other animals form the foundation for our human-unique mathematics, what may be the intervening role of language, particularly the role of words for precise quantities like 'eight' or 'ten'?

Box 6.4 Language and numerical cognition

Humans share with other animals an ability to precisely represent small numbers and approximately represent large numbers. But, human adults are also able to manipulate exact numbers in ways that go beyond the abilities of animals. For example, we can quickly determine how many items will remain after 3 are removed from 6, likely by using our counting skills. In Box 6.2, the relationship between the approximate number system and formal math skills was discussed, but here, a different relationship will be considered: the relationship between language and number cognition. Specifically, there has been a resurgence of interest regarding the extent to which language might affect or even create our number concepts.

Western language systems allow for infinite enumeration, but the languages of some other cultures have a very narrow set of number words (e.g. 'one, two, many'). These languages offer a unique opportunity to examine the relationship between number cognition and language (e.g. Gordon, 2004; Everett, 2005). For example, the Mundurukú tribe of the Amazon in Brazil uses count words for 1, 2, and 3 consistently, but not for larger quantities. When asked how many items are in an array of five or more, they may reply with words that indicate 'some' or 'many.' Another tribe, the Pirahã, do not even use words for 1 and 2 consistently.

Number discrimination and simple operations have been examined in these groups. For example, Pica *et al.* (2004) found that Mundurukú participants were able to discriminate and add large approximate numbers (up to 80 dots) in a manner similar to a French-speaking group. Further, both groups' performance on the estimation tasks varied according to Weber's Law; the judgments were more accurate for larger ratio discrepancies. However, in *exact* arithmetic tasks with numbers larger than 4 or 5 (e.g. '8 – 5'), the Mundurukú participants had lower performance than the French participants. Similarly, the Pirahã had difficulty representing medium-sized sets of 4 or 5 (Gordon, 2004).

This cross-cultural work further supports the claim that there is an approximate number system that is shared across humans and many animal species that does not rely on language, yet it also suggests that for humans, number language allows for precise arithmetic beyond small numbers. Which came first, though, the language or the concept of exact quantity? One particularly controversial proposal is that number language itself actually creates the precise number concept (e.g. Gordon, 2004). An alternative proposal is that number language allows for a precise representation of quantities by perhaps reducing memory demands and enabling more efficient coding, but that the initial development of a number concept is independent of language (e.g. Frank *et al.*, 2008; Gelman & Butterworth, 2005). Both proposals thus implicate language as a factor in connecting the abilities we share with animals to our unique precise number skills. Determining the exact role of language, however, is a challenge for developmental psychologists and linguists, but in concert with comparative work with animals, will help complete our understanding of numerical cognition.

Chapter Summary

- Two primary types of timing have been characterized, one for cyclical activity that occurs during a 24-hour period (periodic timing) and one timing system for much shorter durations on the order of seconds or minutes (interval timing). The internal circadian rhythm for periodic timing is typically discussed in terms of physiology, not cognition. However, animals may learn about events that occur at certain times of day, linking the event to the internal clock.
- Laboratory experiments examining interval timing, including those using the peak procedure, have found that for animals and humans the ability to discriminate among short durations depends on Weber's Law. Greater variability in responses occurs as the to-be-timed durations get longer.
- Different models of interval timing exist, three types of which do well to describe the current data. Although there is no consensus regarding which model is best, each has inspired new research and paradigms.
- Discriminating between two durations and discriminating between two arrays of objects share a common feature: both are instances of discriminating magnitude. Indeed, many have hypothesized that timing and numerical processing originate from a similar, if not common, representational system.
- For human infants and animals, when discriminating approximate numerosities, the ratio difference is important: 1:2 ratios are discriminable, but 2:3 ratios prove difficult. In fact, the ratio difference appears to be more relevant than the absolute number difference, a feature of Weber's Law.
- Cognitive models of the approximate number system posit that a real-world value is represented in the mind as an analog of that value. This analog magnitude representation is a 'noisy' representation of the set of items, yet it is proportional to the number of items being represented.
- Based on the set size signature found across various species in tasks assessing small number discrimination, many researchers have proposed that there are at least two distinct mechanisms for processing number: a precise small number system and the approximate number system. One proposed mechanism that likely underlies small number discrimination in some situations is an object-tracking system.
- Some have proposed that a true concept of number should include an understanding that number is an ordered system and the ability to perform simple addition and subtraction. Each of these numerical operations has been observed across a wide range of species.
- Though the cognitive and neural mechanisms supporting a concept of number are still under investigation, the comparative study of numerical processing, as well as timing, has led to great advances in our knowledge.

Questions

1. By what mechanism might an animal come to learn that a prey species will appear at a certain time of day in a certain location?
2. Describe the peak procedure for interval timing, including likely results for intervals of 20 seconds versus 40 seconds.

3. How does the Information Processing Model account for the fact that timing is scalar in that the variance increases with the length of the timing interval?
4. Describe evidence for the claim that rats can process number and short time intervals simultaneously.
5. Which discrimination would be easier for human infants to make, 8 versus 16 dots or 16 versus 24 dots? Why?
6. What brain areas are likely involved in the processing of magnitude, including timing and number?
7. What are the signature properties of the approximate number system and the object-tracking system?
8. How might the approximate number system and the object-tracking system each solve a simple addition problem (e.g. '1 + 2')?
9. Describe a test procedure to examine the ability of rhesus macaques to represent number on an ordinal scale.
10. What similarities and differences exist between humans and other animals in relation to numerical cognition?

FURTHER READING

Recent review papers have discussed the processing of magnitude, including time and number, noting the commonalities and possible shared mechanisms.

Cantlon, J. F., Platt, M. L., & Brannon, E. M. (2009). Beyond the number domain. *Trends in Cogn Sci*, **13**, 83–91.

Cordes, S., Williams, C. L., & Meck, W. H. (2007). Common representations of abstract quantities. *Current Directions in Psychological Science*, **16**, 156–161.

Dehaene, S. & Brannon, E. M. (eds.) (2011). *Space, Time, and Number in the Brain: Searching for the Foundations of Mathematical Thought.* Oxford, UK: Elsevier Press.

Reviews focusing more specifically on numerical processing, both seminal and new, demonstrate the advances in knowledge made possible by a comparative framework.

Boysen, S. T., & Capaldi, E. J. (1993). *The Development of Numerical Competence: Animal and Human Models*. Hillsdale, NJ: Erlbaum.

Cantlon, J. F. (2012). Math, monkeys, and the developing brain. *Proc Natl Acad Sci USA*, **109**, 10725–10732.

Feigenson, L., Dehaene, S., & Spelke, E. S. (2004). Core systems of number. *Trends Cogn Sci*, **8**, 307–314.

Seminal work on counting within comparative and developmental framework has been completed by Gallistel and Gelman, and includes the following paper and book.

Gallistel, C. R., & Gelman, R. (1992). Preverbal and verbal counting and computation. *Cognition*, **44**, 43–74.

Gelman, R., & Gallistel, C. R. (1978). *The Child's Understanding of Number.* Cambridge, MA: Harvard University Press.

Some of the reseachers who have taught number symbols to animals have reviewed the work in papers and book chapters.

Matsuzawa, T. (2009). Symbolic representation of number in chimpanzees. *Curr Opin Neurobiol*, **19**, 92–98.

Matsuzawa, T., Asano, T., Kubota, K., & Murofushi, K. (1986). Acquisition and generalization of numerical labeling by a chimpanzee. In D. M. Taub and F. A. King (eds.). *Current Perspectives in Primate Social Dynamics* (pp. 416–430). New York, NY: Van Nostrand Reinhold.

Pepperberg, I. M. (2006). Grey parrot numerical competence: A review. *Anim Cognition*, **9**, 377–391.

7 Decision making

Background

Northwestern crows (*Corvus caurinus*) display an elaborate sequence of behaviors as they forage for shelled mollusks on the Canadian west coast. First, birds land on intertidal beaches that are littered with whelks (*Nucella lamellosa*). They walk across these sandy flats, 'testing' whelks, one at a time, by picking them up in their beak and then dropping them back on the ground. After several tests, birds select the largest whelk and fly with it above the nearby rocks. When they reach a certain height, birds drop the whelk and watch to see if it breaks. If it does, they retrieve the tasty inner morsels. If not, they pick it up again and repeat the process until they are successful in obtaining the food.

Observing this behavior, Zach (1979) wondered why particular whelks were selected, how crows determined when to drop them, and whether they would abandon a particular whelk that did not break after repeated drops. To answer these questions, he set up his own test site with a rope and pulley that allowed him to drop whelks of different sizes from different heights. Not surprisingly, the higher the drop, the more likely the break, with larger whelks breaking more easily than smaller or medium-sized ones. Bigger shellfish provide more calories, so it seemed intuitive that crows would select the largest whelks and then drop these from the highest possible height. The problem with this strategy is that carrying larger loads requires more energy and flying upwards becomes increasingly costly in terms of metabolic resources. Zach also observed that whelks dropped from greater heights were more likely to bounce or shatter, making it difficult to retrieve and consume them. Taking these factors into account, Zach determined that a crow's best strategy was to select the largest whelk and drop it from a height of approximately 5 meters, almost exactly what the crows were doing. He also concluded that repeated droppings did not increase the probability of a shell breaking so even if a particular whelk did not break after a certain number of drops, there was no advantage to abandoning it for a different one. This is particularly true as it would take more time and energy to go back to the beach and select another whelk, rather than retrieving the one that did not break. In line with this reasoning, crows continued to drop the same whelk until it broke or was lost.

Chapter plan

Foraging by northwestern crows appears to involve complex calculations of the costs and benefits associated with selecting and then dropping each whelk. There is no evidence that birds are 'aware' of these mental computations or that they begin each foraging bout with an understanding of how it will proceed (i.e. from searching to dropping to eating). The most likely explanation for this particular pattern of behavior is that cognitive processes guiding each step in the sequence were shaped by natural selection. The first section of this chapter develops this idea in detail, providing evidence that many species forage in a way that optimizes the return on their behavior. The same is true of other behaviors that involve a choice, such as habitat or mate selection. Even large groups of animals sometimes coordinate their behavior in a way that minimizes costs and maximizes benefits for the group. Taken together, these examples provide evidence that decision making (at least in these naturalistic examples) is guided by evolutionary pressures.

The chapter is not concerned with the particular choices that organisms make (e.g. which food, habitat, or mate to select), but with the processes that determine *how* these are made. This is examined in the second section of the chapter by reviewing experiments in which individuals select among various behavioral options. At least under certain conditions, humans and animals allocate their time and resources such that payoffs from these behavioral choices are maximized in the long term. This line of research developed into the field of behavioral economics in which decision making is described in terms of economic principles. These explain, at least partially, how humans and animals make choices between different types of reinforcers that are available simultaneously.

The chapter also covers mechanisms of choice behavior when outcomes are uncertain, particularly when risk is involved. Given these unknowns, many organisms rely on general purpose cognitive strategies to make decisions. Although these often lead to the 'best' choice (however that is defined), there are many examples of maladaptive decision making when animals or humans revert to these cognitive rules of thumb. On the other hand, when decision making is examined within an ecological framework using more naturalistic experiments, many of these seemingly poor choices can be reinterpreted as providing fitness advantages to both the individual and its species.

Contrary to popular belief and to most philosophical writing (at least from the past), emotions may facilitate, rather than disrupt, effective decision making. Evidence for this position is derived, primarily, from brain-damaged patients although the idea fits with evolutionary perspectives on cognition. The final section of the chapter describes the brain systems that mediate the interaction between emotions and decision making with a focus on understanding how connections between the cortex and striatum in mammals code for different aspects of choice behavior. Comparable mechanisms have yet to be identified in non-mammalian species although there is no doubt that these also involve physiological computations of fitness-related costs and benefits.

7.1 Evolution of decision making

The pattern of behavior displayed by northwestern crows in the chapter opening provides evidence for **decision making**: the cognitive process of selecting one course of action over a variety of options. **Choice** is the outcome of that process: a motor action or verbal response that indicates

Box 7.1 Free will

Humans have an unwavering belief that their actions are volitional, even if they acknowledge that some behaviors occur unconsciously. Indeed, faith in this principle is so strong that restrictions on freedom of choice in most cultures are considered to be a violation of human rights (as long as the subsequent behavior does not harm other individuals or society). Despite this belief, or perhaps because of it, the debate concerning free will, including whether it exists at all, is ongoing. Aristotle was probably the first philosopher to argue that decision making, at least in some instances, is guided by forces that originate within the individual (i.e. volition). In the thirteenth century, Thomas Aquinas formalized the idea into a doctrine of human action which stated that God is the first cause of any decision, but from there voluntary action may emerge (Healy, 2013). The debate is not limited to Western thinkers: Hindu, Buddhist, and Islamic philosophers have all queried both freedom of thought and freedom of action, particularly how these can be reconciled with unconscious acts. In contemporary writing, the philosopher Daniel Dennett argues that free will must exist (even if it cannot be identified) because some individuals behave in a way that is unpredictable (Dennett, 2003).

The main problem with these philosophical theories of free will is that they almost always invoke some metaphysical component, which makes it difficult to verify that free will exists. Even those who believe in the concept, such as William James, acknowledge that there is no evidence to support it. In the last 30–40 years, scientists have proposed methods to study free will using brain imaging technology. If anything, these data challenge the existence of free will. For example, Libet and colleagues (1983) demonstrated that brain activity related to behavioral choice occurs prior to conscious awareness of the choice. Although these early studies were criticized on methodological grounds, more sophisticated measures of brain activity confirmed the basic finding. Scientists were able to predict which of two buttons subjects would press up to 10 seconds before they were aware of making this choice (Soon *et al.*, 2008). Moreover, they could manipulate which hand an individual would use in a motor task by temporarily inactivating a portion of the frontal cortex (Brasil-Neto *et al.*, 1992). Subjects reported that *they* had selected the hand to move, suggesting that a conscious self does not initiate an action, but is 'informed' later of behavior that the rest of the brain and body are planning.

Many people are uncomfortable with these findings because they appear to reduce humans to mindless automata that respond predictably to environmental signals. But this is probably an exaggeration: even invertebrates behave in ways that do not depend, entirely, on external stimuli (Brembs, 2011). Moreover, when animals are trained in simple conditioning tasks, not all animals behave the same way and not every animal behaves the same on every trial. These individual differences in behavioral responses may be adaptive in that predictability is seldom advantageous, for predators or for prey. Animals must balance effectiveness and efficiency of responding with just enough variability to make their behavior appear random. Perhaps this flexibility in responding is a manifestation of an evolutionary strategy to introduce variability into behavioral choice; free will could reflect conscious awareness of this flexibility. Regardless of how it is defined, the debate surrounding free will is unlikely to decline soon as it has important implications for individual and legal responsibility. Most importantly, belief in free will gives people a sense of agency, which is an important determinant of life fulfillment.

selection of an option. The idea that decision making is an inherited trait, shaped by natural selection, *seems* to imply that organisms have no control over their own choices. This raises the question of whether humans (and animals) possess free will, a topic that has been debated for centuries (see Box 7.1).

7.1.1 Optimal foraging theory

The idea that decision making evolved to maximize fitness is often examined in terms of foraging behaviors: the amount of energy acquired during food consumption is offset by the energy expended to obtain it. Optimal foraging theory formalizes this idea by stating that organisms forage in a way that maximizes net energy gain per unit time (MacArthur & Pianka, 1966). This principle was demonstrated in a study examining seed gathering by eastern chipmunks (*Tamias striatus*) (Giraldeau & Kramer, 1982). Chipmunks are **central place foragers** meaning that they carry food back to a home base (e.g. a nest or burrow) where it is consumed or stored for later use. In the case of chipmunks, seeds are carried in their cheeks (which conveniently expand) and the further the seeds need to be transported, the more seeds the chipmunks carry on each journey (see Figure 7.1A). Humans exhibit a similar behavior when they drop by a corner store to pick up one or two items but reserve long drives to a supermarket for large shopping expeditions. At least for animals, transport becomes increasingly difficult with heavier loads. As with many animals, common starlings (*Sturnus vulgaris*) modify their foraging behavior to account for this factor (Kacelnik, 1984). First, they probe the grass with their beaks to uncover larvae of leatherjackets

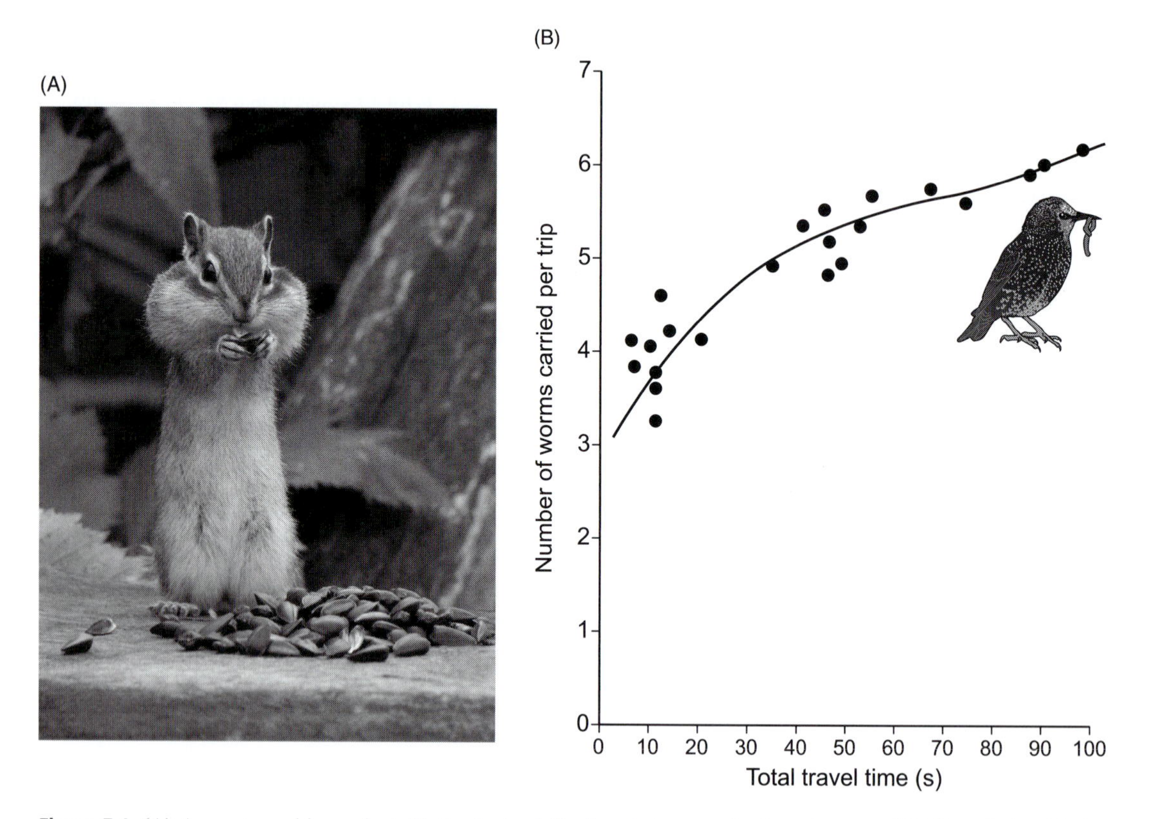

Figure 7.1 (A) An eastern chipmunk stuffs its cheeks full of seeds to carry back to the nest. (B) The relationship between load size and travel time in starlings. Birds were trained to retrieve worms from a feeder that was located at increasing distances from their nest. The number of worms carried on each trip increased with longer travel times between the two locations (Kacelnik, 1984).

(*Tipula palludosa*); then, they pick up these worm-like maggots to carry back to the nest. As their beak fills up, it becomes harder to find and pick up the next prey. As shown in Figure 7.1B, the birds' choice of how many leatherjackets to carry back to the nest took into account this 'diminishing returns' of larger loads. As predicted, these decisions maximized prey delivery to offspring per unit time.

Extensions of optimal foraging theory considered factors, other than travel time, that may impact net energy gain during foraging (Stephens & Krebs, 1986). One of the most important is **handling**, the time and energy required to extract a consumable food item from its source. This could include breaking a shell, sucking nectar from a flower, hulling seeds, or ripping open a carcass. Shore crabs (*Carcinus maenas*) feeding on blue mussels (*Mytilus edulis*) exhibit handling as they break the mussel outer shell to gain access to the inner flesh. The quantity of flesh extracted per handling time is maximized with intermediate-sized mussels, exactly those that are selected most frequently by crabs (Elner & Hughes, 1978). These data support the optimal foraging theory, but they do not fit the model perfectly. For example, when crabs were foraging in environments with an overabundance of large mussels, they consumed more intermediate-sized prey than expected, given that it took more time to find these than the readily available large mussels. This 'rate of encounter' was taken into account in a study of great tits (*Parus major*) feeding on mealworm beetles (*Tenebrio molitor*) of different sizes (Krebs *et al.*, 1977). Again, the general findings supported optimal foraging theory but choice behavior was not predicted *perfectly* by net energy gain per unit time. More specifically, when the relative abundance of small mealworms increased, birds did not select these more frequently. Krebs and colleagues proposed that some of these deviations from optimal foraging could reflect constraints on other cognitive processes. For example, animals may make errors when discriminating large from small prey or they may not remember the handling time of different types of prey. This would lead to less than optimal foraging, even if decision making was still intact.

Decision making during foraging may also be affected by factors not directly related to feeding, particularly predation. Many animals are in a vulnerable position when they leave the relative security of a home base to search for food. This explains why leafcutter ants (*Atta cephalotes*) forage primarily during the night when they are less likely to be attacked by parasitic flies (Orr, 1992). A similar effect was observed in common garden skinks (*Lampropholis guichenoti*) living in different environments (Downes, 2001). When these small reptiles were reared in enclosures that contained the scent of a predatory snake, they foraged at lower rates and grew more slowly than skinks reared in an environment without this scent. Foraging bouts by rats are also inhibited if predators are detected (Blanchard *et al.*, 1990). All of these findings emphasize that optimal foraging is compromised under conditions of high predation (see Section 7.3.2 for further discussion of this issue).

Although it is seldom tested directly, a fundamental assumption of optimal foraging theory is that net energy gain is associated with fitness. That is, in order for foraging efficiency to be favored during natural selection, it must be associated with higher reproductive success. This appears to be the case with captive zebra finches (*Taeniopygia guttata*) working for their daily ration of food (Lemon & Barth, 1992). In this study, the energy required to obtain the food was manipulated by increasing the amount of time required to search for the food. As expected, reproductive success over the birds' lifetime was positively correlated with net calorie gain per day and *not* with the total amount of food consumed. The relationship between optimal foraging and long-term fitness needs to be examined in other species, under both laboratory and natural environment conditions.

7.1.2 Marginal value theorem

A premise of optimal foraging theory is that food is distributed evenly throughout an environment. This is seldom the case. More commonly, consumable resources aggregate in **patches**, discrete areas within a habitat that contain food, separated by areas with little or no food. As animals forage in one patch, resources are depleted so the fitness benefits of staying in this one location decline over time. The **marginal value** of a patch is the point at which the net energy gain from foraging is lower than the average energy gain of the surrounding habitat (Charnov, 1976). Marginal value theorem predicts that animals will move to a new patch when foraging if the current patch declines below the marginal value. This idea fits an optimization model in that choice behavior (i.e. whether to move) is determined by the best payoff: staying in one patch versus moving to another. Figure 7.2 shows that the decision to move to a new patch is determined both by the amount of resources in the current patch and by the distance between patches (Cowie, 1977). In this experiment, great tits foraged for worms that were hidden in plastic cups filled with sand. The cups were set up in different locations (i.e. patches) in an aviary and were filled with different amounts of mealworms. In line with the marginal value theorem, birds spent more time in patches that had a greater number of worms and stayed longer when the next patch was further away.

The idea that animals choose to move to a new patch as resources decline suggests that they have some knowledge of the foraging payoff in different patches. Lima (1984) examined how this occurs in downy woodpeckers (*Picoides pubescens*) by setting up three separate habitats containing empty

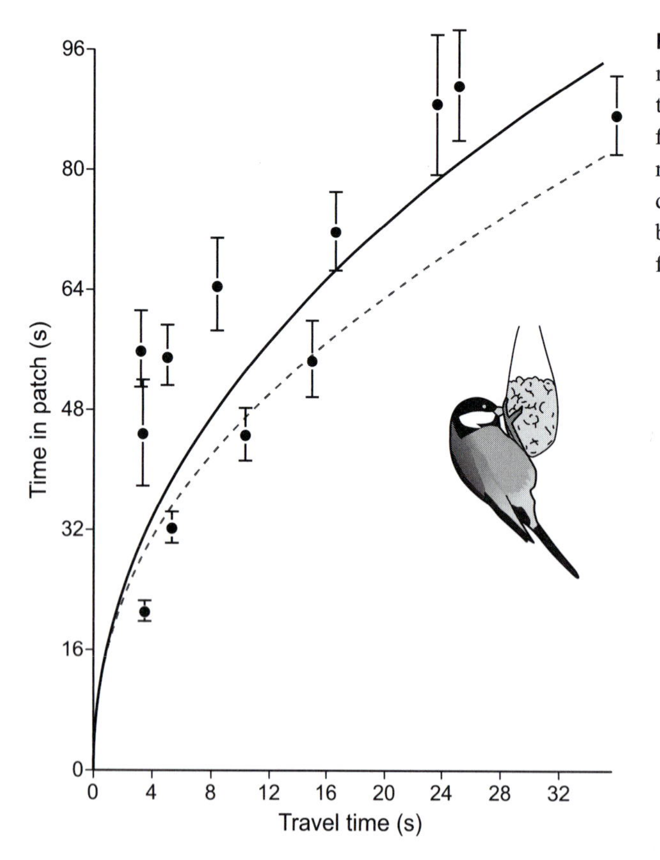

Figure 7.2 Decision making according to the marginal value theorem in great tits. Birds were trained to forage for worms placed in plastic cups filled with sand (patches) throughout an aviary. The number of worms in each patch varied, as did the distance between the patches. As the travel time between patches increased, birds spent more time foraging at individual patches (Cowie, 1977).

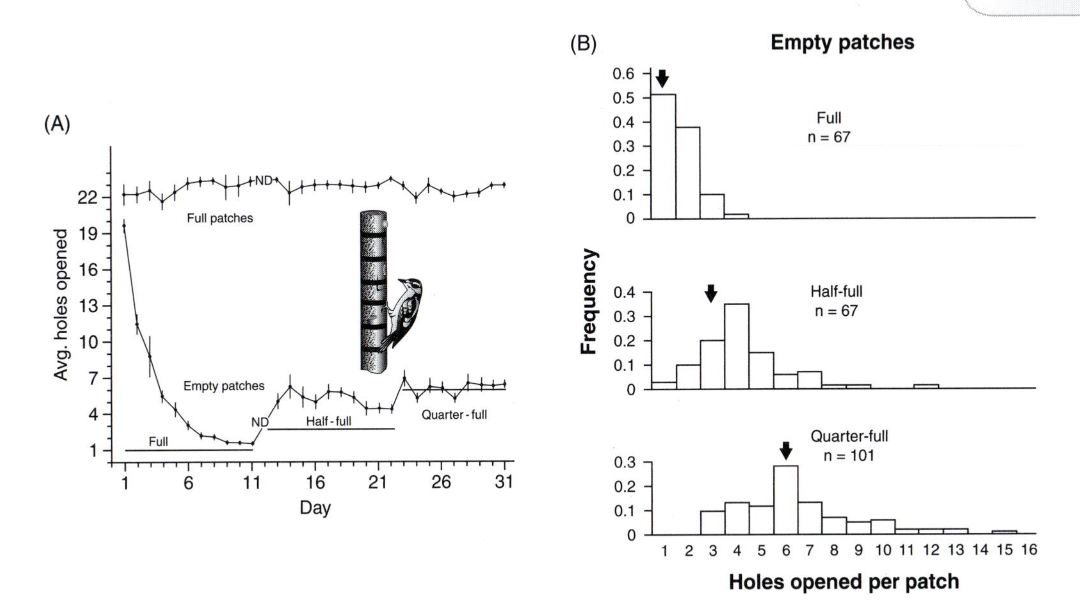

Figure 7.3 Foraging patterns of downy woodpeckers match predictions of the marginal value theorem. Birds had access to three habitats, each containing patches of poor and rich payoff (logs that were empty or filled with prey items). The number of logs filled with prey in the rich patches varied from full to half full to one quarter full. (A) The average number of holes that birds opened in full and empty patches for each day of the study. No data (ND) were collected on day 12. Horizontal lines indicate the time during which each environment was available and represent the predicted number of holes that birds should sample in the poor patches before giving up on a patch as empty. (B) The frequency distribution of the number of holes sampled before giving up on a patch as empty in each of the three habitats. Arrows indicate the predicted optimal number of holes a bird should sample, according to the marginal value theorem.

logs that were filled with a different number of prey items (birds accessed these by pecking through masking tape that covered 24 holes in each log). Each habitat had poor and rich patches: empty logs or those with food. Rich-patch logs in the first habitat had 24 prey items, those in the second had 12, and those in the third had 6. Woodpeckers had several days to sample each habitat and the question of interest was how quickly they would desert the poor patches (i.e. empty logs) when the payoff of foraging in rich patches varied. As shown in Figure 7.3, birds developed a different strategy in each habitat. In the first, they quickly abandoned a log with empty holes, appearing to reflect knowledge that these poor patches did not contain any prey. As the value of rich-resource logs decreased (i.e. from 24 to 12 to 6), birds sampled more holes in the empty logs before moving to the next patch. This suggested that the choice to move to a new patch was dictated by knowledge of patch payoffs within a habitat.

Woodpeckers in Lima's study were able to calculate habitat payoff by sampling resources in different patches. Sampling allows animals to update information on resource content in different patches, helping them to decide *where*, as well as when, to move. In other words, animals must divide their time between remaining in the most profitable patch and sampling nearby patches so that they can keep track of current levels of resources. In the lab, great tits balance sampling and foraging behaviors based on the difference between rich- and poor-quality patches; as this difference increases, birds spend less time sampling and more time foraging in the most profitable patch (Krebs *et al.*, 1978). This makes sense in that it is easy to determine which patch is the best if there is a large difference between them. Both great and blue tits (*Cyanistes caeruleus*) display the same pattern in the natural environment, foraging primarily in resource-rich patches but taking time out to sample nearby patches (Naef-Daenzer, 2000). Indeed, the foraging/sampling division of time is such that birds taking food back to their nest come close to the maximum possible rate of prey delivery in each patch.

Foraging patterns of many other animals fit the marginal value theorem, at least on a general level. When deviations occur, it is usually because animals stay in a patch after resources fall below the marginal value (Nonacs, 2001). This could reflect the fact that moving to a new patch introduces unknown risks, such as predation. Moreover, other behaviors, such as searching for mates, compete with foraging so it can be difficult to determine why animals do or do not move to a new location. Some researchers point out that these challenges (i.e. identifying all of the factors that impact foraging) minimize the utility of any optimality theory. Even so, marginal value theorem provides a good starting point for understanding decision making during foraging.

7.1.3 Ideal free distribution model

Marginal value theorem explains foraging behavior at the level of the individual but many species forage in groups. Under these conditions, the payoff in a resource-rich patch is lower if most of the animals congregate at this location. Rather than staying in the highest payoff patch, therefore, a better strategy is for animals to distribute themselves among patches. The **ideal free distribution model** takes this into account, proposing that the number of animals aggregating in a particular food patch is proportional to the amount of resources available at that patch (Fretwell & Lucas, 1970). Early support for this model came from an experiment with three-spined stickleback fish (*Gasterosteus aculeatus*) in which food was dropped into either end of the aquarium at different rates (Milinski, 1979). Sticklebacks distributed themselves accordingly: if twice as much food was dropped at one end, twice as many fish aggregated at this location (see Figure 7.4). Animals took time to sample resources at both 'patches' before they split into these groups. Individual animals continued to sample the other patch periodically, allowing them to update information about the rate of food delivery at both locations. Foraging bumblebees also conform to an ideal free distribution

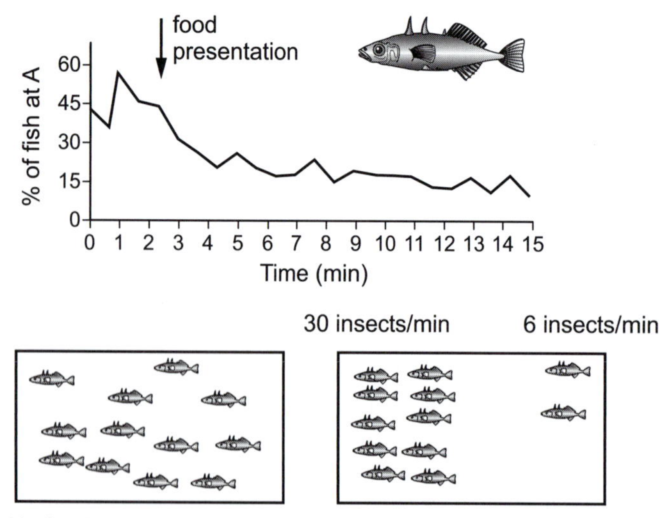

Figure 7.4 Ideal free distribution model in stickleback fish (Milniski, 1979). Food was dropped into each end of an aquarium (patches A and B) and the number of fish aggregating at each location was recorded. The arrow indicates the beginning of food delivery. When no food was available, fish were distributed randomly throughout the aquarium (bottom left). When food was delivered at a rate of 5:1 in patch B versus patch A, fewer fish aggregated at patch A (bottom right).

model in that they distribute themselves based on the density of flowers and the concentration of nectar in each plant (Dreisig, 1995). Ideal free distribution is difficult to assess in humans but one could imagine that a group of apple pickers would distribute themselves so that there are more individuals at the trees with the most apples. Moreover, people playing a competitive game in the lab selected colors that were associated with different payoffs by taking the choices of other players into account. The general pattern of decision making across the group fit the ideal free distribution model, even when the payoffs changed across trials (Sokolowski *et al.*, 1999).

Group foraging by many other species often fits the ideal free distribution model; exceptions occur, most commonly, when there are more individuals at the least profitable patch than the model predicts (Kennedy & Gray, 1993). Harper (1982) explained this discrepancy in an experiment in which researchers threw cubes of bread into opposite sides of a pond full of mallard ducks (*Anas platyrhynchos*). Ducks distributed themselves, *approximately*, according to the rate of food delivery at each site. Variability from this ideal value was due to a few dominant birds that monopolized most of the resources at the more profitable patch. This competitive advantage of higher ranked individuals also explains deviations from the ideal free distribution model in foraging long goldfish (*Carassius auratus*) (Sutherland *et al.*, 1988) and coho salmon (*Oncorhynchus kisutch*) (Grand, 1997). Under these conditions, dominant animals have greater access to food resources, further perpetuating their competitive advantage over other group members. Social foraging that does not fit an ideal free distribution model could also reflect differences in sensory and cognitive capacities. For example, cichlid fish (*Teleostei cichlidae*) that consume less food than conspecifics often spend more time sampling other patches, perhaps reflecting an inability to discriminate resource content at different sites (Godin & Keenleyside, 1984).

Finally, many researchers have noted that foraging groups may not distribute themselves according to an ideal free distribution because the model makes assumptions that are not always met in the natural environment. First, individuals are not always 'free' to move to a new location: dominant individuals, social restrictions (e.g. caring for young), or physical constraints may limit travel between food sources. Second, unlike Harper's experiment in which bread cubes were replenished continuously, resource content in a single patch is not stable. This has an obvious impact on an animal's decision to move to a new patch. Despite these limitations, the ideal free distribution model provides a general description of how individuals within a group distribute themselves during foraging. The model makes an important contribution to optimality theories by highlighting how social interactions impact choice behavior. The general idea that group decision making is an interactive process in which the optimal strategy for one individual takes into account the choices of other individuals is formalized in the Nash equilibrium (see Box 7.2).

7.1.4 Collective decision making

The impact of social interactions on choice behavior is particularly striking during collective decision making when large groups of conspecifics engage in coordinated activities such as foraging, nest building, or migration. Even though each group member only has access to a portion of the information needed to guide behavior, direct or indirect interactions between individuals lead to group cohesion (Jeanson *et al.*, 2012). These group behaviors, which seem to be independent of individual actions, include swarming bands of locusts, migrating flocks of birds, and marching colonies of ants (see Figure 7.5). Even crowds at large sporting events or concerts appear to behave as an organized unit. Collective decision making, also described as evolved coordinated behavior (Couzin, 2009), provides a fitness advantage to group members in that they benefit from the

Box 7.2 The Nash equilibrium

The Nash equilibrium is named in honor of the US mathematician John Forbes Nash (b. 1928), who won a Nobel Prize in economics for his work on the forces that govern group dynamics. Nash's life, particularly his struggle with mental illness, is profiled in the 2001 film, *A Beautiful Mind*, loosely based on the book by Sylvia Nasar (1998). Although many biographical details are fictitious, the film provides an accurate, although simplified, description of Nash's theories. In the film, Nash comes up with the idea of advantageous group choice when he and his graduate student friends at Princeton discuss how to approach a group of women in a bar. In contrast to the standard 'each man for himself' tactic, Nash proposes that they all have a better chance of success if they work together. It is unlikely that this incident was the origin of Nash's theory but it is true (as the film relates) that he was offered a position at MIT based on the strength of his theory describing group dynamics.

According to Nash, decision making of individuals in a group cannot be understood in isolation. Rather, the choices of each individual can be predicted *only* when the decisions of others within the group are taken into account. A premise of this idea is that all individuals know the strategy (or decision) of others. Group decision making has reached a Nash equilibrium if no single individual in the group can improve their payoff by unilaterally changing their strategy. The mathematical formulation of the Nash equilibrium provides a solution for these situations in which there is no advantage to changing strategies, given that other players remain consistent in their own actions.

The Nash equilibrium can be understood in terms of Harper's 'Cambridge pond' experiment with ducks. If ducks distribute themselves in a way that matches the rate of bread delivery, any subsequent movement between patches could only be disadvantageous to that individual. Specifically, if 30 ducks were in a pond with one researcher throwing in one bread ball every 10 seconds and another throwing one bread ball every 20 seconds, ducks should distribute themselves so that there are 20 at the first site and 10 at the second. The amazing thing about Harper's experiment is how quickly the ducks sorted themselves into an 'equilibrium.' Even after 90 seconds, when fewer than 10 pieces of bread had been tossed into one end of the pond or the other, ducks redistributed themselves according to the Nash equilibrium (or ideal free distribution model). During this time, many of the ducks had not even sampled one of the bread cubes! Moreover, even though individual animals moved unpredictably, the group maintained an equilibrium.

The Nash equilibrium has been used to analyze strategic decision making in competitive situations ranging from war (Schelling, 1980) to professional soccer (Chiappori *et al.*, 2002). It may also describe the outcome of cooperative efforts, such as those related to educational attainment (De Fraja *et al.*, 2010). Group dynamics of cooperative behaviors will be discussed at length in Chapter 11 under the broader category of game theory (see Section 11.1.2), also one of Nash's major theoretical contributions.

consequences of synchronized group activities, despite individual preferences or differences in the ability to access or process information.

Large-scale coordinated movements that emerge from collective decision making appear to be regular and predictable, suggesting that these cross-species behaviors are governed by general principles (Sumpter, 2006). One of the most important is that collective decisions are robust to individual perturbations in that a single defector will not alter the overall behavior of the group. For example, stickleback fish largely ignore movement of one neighbor but once a critical mass

Figure 7.5 Large-scale coordinated behavior displayed by milling fish.

('quorum') moves in a particular direction, all individuals in the group follow suit (Ward *et al.*, 2012). Moreover, increasing group size increases the accuracy of collective decisions (Sumpter, 2006) suggesting that this process may be more effective (and efficient) than individual choice. Similarly, the larger the group, the fewer individuals are needed to lead a behavioral choice (Couzin *et al.*, 2005). Leadership is not a prerequisite for coordinated group activity but when it occurs, leaders are usually 'informed' individuals who have access to pertinent and salient information. In the case of migration, these individuals have experience travelling along the navigation route, although there are many experienced individuals who do not act as leaders. At least with humbug damselfish (*Dascyllus aruanus*), leadership in swimming direction is not related to social dominance (Ward *et al.*, 2013) so it is not clear how leadership is determined in these collective decisions. In some cases, potential leaders (i.e. informed group members) reach a consensus decision that depends on the degree of disagreement between them. For example, when homing pigeons are released in pairs, they take a route partway between their trained routes when these are relatively close together, but select one route or the other if the distance between the two is great (Biro *et al.*, 2006). The movement of human crowds follows the same general pattern (Dyer *et al.*, 2009).

Recently, the processes underlying collective decision making have been examined within the framework of cognitive neuroscience (Couzin, 2009). According to this model, the 'behavior' of neural networks mimics those of ants working together to carry out essential tasks such as nest building or foraging. In both cases, individual units (neurons or ants) exhibit a low propensity to become spontaneously active, which is modified by encounters with local, discrete stimuli (electrical or chemical signals for a neuron and pheromone signals for an ant). This leads to low levels of excitement (intracellular changes or ant movement) and, once a threshold is reached, the ant or neuron changes from an inactive to an active state. This excitement spreads to adjacent units such that amplification of activity occurs rapidly. Both ants and neurons display an inactive period following activity (refractory period) during which there is a low probability of switching into an active state. In other words, both neural networks and ant colonies display synchronous patterns of activity even though the individual units are not intrinsically rhythmic. Neuronal networks are an efficient way for the brain to elevate activity close to threshold, providing discrete windows of high responsiveness to external stimuli (Buzsáki & Draguhn, 2004). In the same way, synchrony among ant colonies functions to efficiently allocate workers to required tasks when sensory cues are detected by individual insects (Bonabeau *et al.*, 2008). It is unlikely that this model will apply to all instances of collective decision making, but it is a thought-provoking attempt to explain how interactions among individuals scale to collective behavior with its own functional properties.

It is important to note that collective decision making is distinct from cooperation, when two or more individuals work together for a common goal (see Section 11.6 for details). Cooperative behaviors have an immediate gain for all actors, which is not necessarily true with collective decision making. Moreover, defection has no negative consequence in collective behaviors, but is clearly disadvantageous in cooperation because actors gain by working together. For these reasons, group behavior of social insects appears to fit the definition of collective decision making, rather than cooperation. Of course, this is not the final word on the topic: many researchers are convinced that these tiny organisms cooperate with conspecifics in a way that differs very little from other animals (including humans) (Cronin *et al.*, 2013).

7.1.5 Beyond foraging

Most of the discussion in this section relates to foraging but the same decision making processes can be applied to other naturalistic behaviors such as aggression (when and what to attack), habitat selection (when and where to move), or reproduction (whom to mate with and when). A simple rule of the lowest energy expenditure for the highest fitness benefit provides a general explanation of how decision making occurs in these situations (Pyke, 1984). These decisions are affected by cognitive processes, including memory, timing, and discrimination. For example, if animals move to a new foraging patch based on a marginal value, they must be able to calculate how resources in the current patch are changing, estimate the travel time to the next patch, and remember the quality of other patches in the habitat. Constraints such as travel costs, potential predation, and current metabolic state of the individual also affect decision making such that a particular choice may appear less than optimal at the moment, but results in reproductive or survival advantages in the long term (Nonacs, 2001). In sum, optimality theories all contain three components: a choice behavior, a currency that allows costs and benefits to be compared, and cognitive and environmental constraints that alter the relationship between these two.

All three of the optimality theories discussed in this section (optimal foraging, marginal value theorem, and ideal free distribution) assume that natural selection has shaped choice behaviors associated with foraging and other naturalistic behaviors. Although behavioral traits associated with these activities may be heritable, it is not a specific motor response (e.g. breaking open a shell to retrieve a mussel) that defines decision making but the rules by which animals make these choices. Moreover, it is not a particular choice that is inherited (e.g. select intermediate-sized mussels), but the cognitive process that leads to this decision (e.g. calculate foraging payoff given the relative abundance of different-sized mussels). In this way, animals can adapt their choices to changing environmental conditions so that they behave efficiently in a variety of situations.

7.2 Choice in the lab

As with research in the natural environment, there is a long tradition of laboratory work examining decision making in different animals. Most scientists working in this field are psychologists who examine how pigeons or rats respond in operant paradigms that involve a choice. While not ignoring functional explanations, these studies focus on proximate mechanisms of decision making by investigating factors (e.g. biological, environmental, or motivational) that influence the selection of one behavioral option over another.

7.2.1 The matching law

Choice experiments are set up in the lab using **concurrent schedules of reinforcement**, i.e. those in which two or more schedules are in effect simultaneous (see Section 4.2.2 for details on schedules of reinforcement). Using pigeons as an example, if pecking one key produces food under a fixed ratio (FR)5 schedule and pecking another key produces food under an FR2 schedule (two versus five responses produce a food pellet), pigeons respond (almost) exclusively on the FR2 key. The situation becomes more complicated when an all-or-none strategy does not produce the best payoff. This may occur if the two keys are reinforced under different interval schedules. FI-30s and FI-45s schedules would mean that pecking one key produces a reward after 30 seconds and pecking the other after 45 seconds. Because the intervals associated with each key continue to progress even when animals are not responding, pigeons receive the most rewards if they move back and forth between the two keys (see Figure 7.6). This is exactly what they do.

Concurrent schedules of reinforcement may be set up in a variety of ways using interval or ratio schedules, with either fixed or variable parameters. Most commonly, these are designed so that the

Figure 7.6 Design of a concurrent choice experiment in pigeons. Birds have the choice between two operant responses (left and right key peck) that deliver food rewards under different schedules of reinforcement.

maximal payoff occurs when animals divide their time between two or more behavioral options. The **matching law** explains how this division occurs by stating that the relative rate of responding on one alternative matches the relative rate of reinforcement for that alternative (Herrnstein, 1961). This can be written as a mathematical equation in which B=behavior, r=reinforcement rate, and $_A$ and $_B$ refer to two behavioral options (e.g. two different levers):

$$BA/(BA + BB) = rA/(rA + rB)$$

Choice behavior that fits the matching law has been observed in a variety of species under a number of different experimental conditions (Staddon & Cerutti, 2003).

The matching law also explains choice behavior when responding for each alternative requires a different amount of effort, when rewards vary in quality or quantity, and when there is a delay between the response and the reinforcer. In most cases, animals divide their responses in such a way that they maximize overall gains across sessions. When choice behavior deviates from the matching law, it usually does so in one of three predictable ways. Some animals exhibit a **bias** for one alternative or the other, regardless of the reinforcement payoff. For example, an individual rat may prefer to respond on the left, rather than the right, lever or a pigeon may prefer to peck red, rather than green, lights. If identified, this confound can sometimes be eliminated by restricting access to the biased alternative during training. Most commonly, animals exhibit **undermatching**, in which they appear less sensitive to reinforcement rates than the matching law predicts. In essence, subjects respond at higher rates on the lower payoff alternative than expected. **Overmatching** describes the opposite pattern: animals appear more sensitive to reinforcement schedules in that they respond at higher rates on the higher payoff alternative than expected.

It is not clear why animals have a tendency to undermatch, but it likely reflects the specifics of a lab environment. Unlike the natural world, food is constantly replenished in most operant choice experiments. In addition, the two response options are close together (levers or keys on the same wall of a chamber) so animals can readily switch back and forth between them. This contrasts with movement between patches in the natural environment that takes both time and energy, not to mention the possibility of predation. The costs associated with moving to a new patch can be simulated in the lab using a changeover delay: switching from one response option to the other delays the presentation of the next reinforcer. Under these conditions, undermatching is minimized and, in some cases, overmatching emerges. Overmatching resembles the scenario, described in Section 7.1.2, in which animals stay longer in a patch than predicted by the payoff provided by that patch (i.e. beyond the marginal value). Undermatching may also reflect sampling that occurs in the natural environment when animals 'test' reinforcement payoffs in different patches. This may explain why undermatching declines with increased training (Jozefowiez & Staddon, 2008): animals do not need to sample as frequently when they have more experience with the payoffs provided by each alternative.

The matching law has been applied to human behaviors as a way to explain how people direct their conversation to different individuals (Borrero *et al.*, 2007), how professional athletes choose one sports play over another (Reed *et al.*, 2006), how consumers make choices between food items (Foxall *et al.*, 2010), and how children divide their time among activities in a classroom (Billington & DiTommaso, 2003). Although these data are not as straightforward as those generated by concurrent choice experiments in the lab, they support the idea that humans, like many animals, possess the capacity to compute relative reinforcement payoffs and to adjust their behavior accordingly.

7.2.2 Behavioral economics

As research into the matching law progressed, many people recognized that the study of concurrent choice by lab animals parallels the study of economic choice by humans. This led to an emerging field of behavioral economics, in which operant behavior, specifically choice behavior, is analyzed using economic principles (Hursh, 1980). For example, economists explain human purchasing decisions by relating product demand to price: as the price of a commodity rises, demand goes down. Most people buy fewer electronic devices when prices increase, particularly if prices rise rapidly. Animals exhibit the same pattern in that operant responding decreases when response requirements for each reinforcer increase: rats reduce their rate of lever pressing when an FR schedule for food goes up (Collier *et al.*, 1992). As shown in Figure 7.7, however, this demand–price relationship is not consistent across reinforcers. Rats decrease their responding for a fat solution (a mixture of corn oil, water, and xanthan gum) much more rapidly than they do for standard food pellets that provide all of their required nutrients (Madden *et al.*, 2007). The same is true of humans who continue to buy essential food items when prices increase, although they may modify their choice of individual products. In contrast, purchase of non-essential items, such as alcohol, declines when the price is increased (Murphy *et al.*, 2009). In economic terms, demand for food is **inelastic** because changes in its price have relatively little effect on the amount purchased, whereas demand for alcohol is **elastic** in that it exhibits an inverse demand–price relationship.

Demand for food is inelastic in the lab when animals must access all of their food during operant conditioning sessions. In experiment terms, this is referred to as a closed economy. In open economies, food is readily available at other times, or at least available enough that rats can meet their nutritional requirements outside of the operant chamber. Under these conditions, responding for food drops off at high response requirements (i.e. demand is elastic). Open economies are uncommon for humans in that there are few, if any, items that are truly 'free,' although the relative price and availability of most items varies dramatically across different societies. Most economic choices, therefore, occur within closed economies with essential and non-essential items exhibiting inelastic and elastic demand respectively. This is the rationale behind increasing taxes on items with negative health and societal consequences such as cigarettes, high-fat food, and alcohol. These initiatives work for many people in that they reduce the purchase and use of these products. Sadly, for many people demand for drugs becomes inelastic: these items take on the properties of essential commodities and individuals go to extreme means to obtain them, regardless of the cost. In that

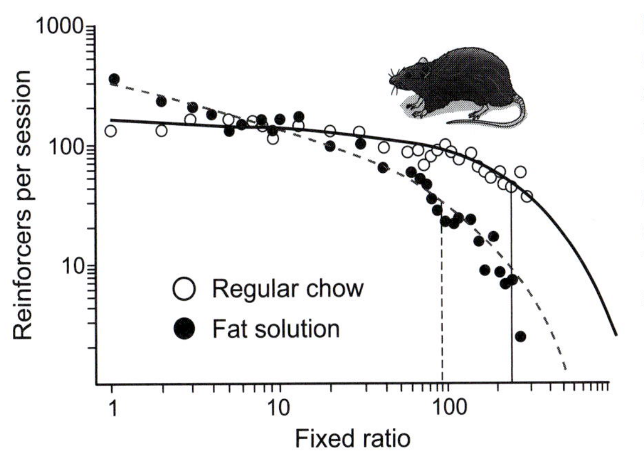

Figure 7.7 Demand–price curve for essential and non-essential items in rats. Animals were trained, under fixed ratio (FR) schedules of reinforcement, to lever press for regular chow or a fat solution. The number of responses required to obtain each reinforcer was varied across sessions, ranging from 1 to 200. At the lowest FR schedules, rats responded at higher rates for a fat solution but as the cost (i.e. responses) per reinforcer increased, rats responded at higher rates for food (Madden *et al.*, 2007).

sense, addiction may be defined by altered decision making regarding the relative costs and benefits of using these substances. Indeed, one of the most influential theories of drug addiction is that brain mechanisms of decision making are altered by chronic drug use, leading to a pathological focus on the drug (Jentsch & Taylor, 1999). There is increasing evidence that diets high in sugar and fat produce many of the same neurobiological effects, although of reduced magnitude (Volkow *et al.*, 2013).

7.2.3 Choice between alternative reinforcers

Choice in the natural environment almost always involves a decision between different types of reinforcement. Animals must distribute their behavior so that they have enough time to eat, drink, sleep, and reproduce. Depending on the species, they will also defend territory, search for mates, and care for young. In addition, many animals (like humans) engage in activities that are not directly related to survival, such as grooming or playing. Decision making, therefore, must support behavioral allocation between alternative reinforcers. This process can be examined in the lab by training animals to respond for different outcomes such as an experiment in which barbary doves (*Streptophelia risoria*) pecked different keys to obtain water or food (Sibly, 1975). Depriving birds of one commodity or the other prior to testing altered choice behavior; as expected, thirsty birds were more likely to respond for water and hungry birds for food. Interestingly, the doves tended to respond on one alternative for an extended period of time and then switch to the other, rather than alternating rapidly between the two. This probably reflects the cost of switching in the natural environment; even if this was not programmed into the experiment, many animals have inherited a tendency to adopt this behavioral strategy.

Sibly's experiment makes the obvious point that an animal's decision to select one behavioral option over another depends on its current motivational state. Behavioral choice is also determined by the extent to which a response can reduce a motivational deficit and the availability of the relevant commodity (e.g. does a response produce enough water to relieve thirst?) (Sibly & McFarland, 1976). Tinbergen (1951) expressed a similar idea when he described how motivational impulses, influenced by both internal states and external stimuli, determined which behavioral response would be dominant. For example, lengthening of the photoperiod induces hormonal changes which lead to nest building in many birds. This increased activity reduces time devoted to other behaviors, even feeding, but it cannot occur if nesting material is unavailable. So, the effects of hormones on behavioral choice (nest building or otherwise) will depend on environmental factors. Tinbergen's framework is useful for understanding the hierarchy of motivational needs, but it does little to predict what animals will do at any given moment. The main problem is that it is almost impossible to identify and quantify motivational states. Moreover, most activities are constrained in the lab so the choices that animals make under these controlled conditions may not reflect what they would do if other behavioral options were available. Nor is it easy to determine which reinforcer is the most powerful. Should one measure which is preferred in a choice trial? How hard animals will work for each over an entire session? Or how demand–price functions vary across reinforcers?

Despite these difficulties, there is little disagreement that behavioral choice is influenced by the availability and incentive value of alternative reinforcers. One telling example is that lab rats reared in enriched environments (e.g. large enclosures of group housed animals that contain balls, wheels, climbing apparatus, and space to play) administer less morphine than those reared in impoverished conditions (e.g. small spaces with little possibility for social interaction or other activities) (Alexander *et al.*, 1978). Findings such as this have led to the development of social policies that address maladaptive social behaviors by promoting alternative reinforcers, including social support systems and recreational activities. For example, unsafe sex practices among teenage girls may be reduced by increasing the opportunity to engage in other activities, such as sports or music (Bulow & Meller, 1998). Importantly, these alternatives must be immediately available as delayed reinforcers, particularly for adolescents, are frequently ineffective (Vuchinich, 1999).

7.3 Choice under uncertainty

The previous sections emphasized that many organisms make decisions based on optimal outcomes. Even if this strategy is advantageous, it can be difficult to implement because almost all choices involve an element of uncertainty. For example, the outcome of one behavioral option may be ambiguous or even unknown and a particular action may not produce the same effect every time it is performed. Decision making, therefore, cannot rely solely on what has happened most recently (or even most frequently) in the past.

7.3.1 Heuristics

In order to deal with choice in the face of uncertainty, humans and animals often rely on **heuristics**. These are strategies for solving problems that ignore or discount a portion of available information. Heuristics may be described as mental short cuts or cognitive rules of thumb. They allow organisms to make decisions when the problems are complex or information needed to make the decision is incomplete. Heuristics are often contrasted with **algorithms**, a specific procedure or set of instructions that leads to a correct solution. Mathematical formulas are algorithms. Decision making using algorithms is 'rational' in that the process proceeds through a logical set of steps to reach a conclusion. Compared to heuristics, algorithms are more likely to yield the 'correct' answer (e.g. an optimal outcome in foraging), but the cognitive process of progressing through an algorithm can be lengthy and mentally demanding. An algorithm to locate lost keys could be defined as a systemic retracing of all activities since the keys were misplaced (e.g. going through every behavior in reverse, if these could be remembered). A heuristic for the same problem might involve selective searches in places where keys are likely to be left (e.g. pant or jacket pockets).

Heuristics are particularly useful when the correct solution to a problem cannot be known. However, even when humans or animals *could* make a decision using an algorithm, they often rely on heuristics, saving both time and mental effort. In Section 7.1.2, great tits displayed a foraging heuristic when they applied the general rule 'spend more time sampling alternative patches as the differences between patch quality decrease.' Decision making heuristics such as this provide workable approximations of a solution by balancing behavioral and cognitive costs against accuracy. Despite their utility, cognitive heuristics often lead to biases and systematic errors. Tversky and Kahneman (1974) studied this phenomenon in humans, noting that heuristics are particularly prone to distortions, inaccuracies, and omissions. They argued that the consistency in these miscalculations sheds light on the adaptive value of decision making. This is exemplified by the sunk cost fallacy, one of the most common errors in human decision making (see Box 7.3).

Although he is a psychologist, Kahneman was awarded the Nobel Prize (2002) in economics for his work on decision making, testifying to both its importance and impact. (Tversky died in 1996 so could not share the award.) Behavioral ecologists took note of these ideas as evidence accumulated that animals, like humans, are biased in how they respond to ambiguous stimuli (Trimmer *et al.*,

Box 7.3 Sunk cost fallacy

A sunk cost is one that has already been incurred and cannot be recovered. From a purely economic perspective, money that is spent (or energy that is expended) should not factor into calculations of future payoffs. Many decisions defy this 'rational' assessment in that previous investments often influence future choices. This sunk cost fallacy is also known as the 'Concorde fallacy' in reference to the first commercial supersonic airliner built by French and British companies. From a purely rational perspective, the project was doomed to failure early on, but construction continued. Despite the large financial losses, not to mention the investment of time and effort, no one wanted to give up because they had already devoted so much to the project. The same phenomenon was observed in a lab experiment with humans using simulated decision making scenarios (Garland, 1990).

For many years, the sunk cost fallacy was considered to be a human-specific phenomenon (Arkes & Ayton, 1999). It was often discussed in reference to cognitive dissonance (i.e. the discomfort individuals feel when their ideas, beliefs, or emotions do not match reality) (Festinger, 1957). According to Festinger, people are motivated to reduce this internal tension so they adjust their beliefs or ignore relevant information under conditions of cognitive discord. Cognitive dissonance is difficult to measure in non-verbal subjects, but researchers have developed clever ways to demonstrate the same bias in animals. For example, European starlings (*Stumus vulgaris*) prefer stimuli associated with a high, compared to a low, work output (Kacelnik & Marsh, 2002); a human analog of this experiment produced the same result (Klein *et al.*, 2005). Similarly, pigeons persist in responding for one option even when an alternative with a higher payoff is available (Macaskill & Hackenberg, 2012). The behavior of these birds is consistent with the sunk cost fallacy in that knowledge of irrevocable, retrospective costs increases responding for the already time-invested choice.

There is still some debate as to whether these bird experiments provide evidence for a sunk cost fallacy in animals, but few doubt that the process affects decision making in humans. Even though it appears to defy optimality, the phenomenon has been explained in terms of cognitive evolution. If an individual abandons an activity in which they have already invested resources, they will suffer a large loss even if potential gains are not affected. Organisms that placed more urgency on avoiding losses than maximizing opportunities were more likely to pass on their genes. Over time, the prospect of losses became a more powerful motivator of behavior than did potential gains. In other words, humans do not put equal weight on potential losses and gains: they are set up to avoid losses, sometimes referred to as 'loss aversion.'

2011). Indeed, optimality models, such as the marginal value theorem, provide evidence for cognitive heuristics in that animals make decisions about what, when, and where to eat when information guiding these choices is limited. Reproductive behaviors also follow cognitive rules of thumb: peahens (*Pavo cristatus*) investigate three to four peacocks and then mate with the one that has the most spots (Petrie & Halliday, 1994). In other situations, animals must decide when to escape from a feeding site if a predator is approaching. Birds sometimes make this decision based on whether other birds in the flock are escaping, but flying away from a feeding site each time a conspecific does so would be inefficient. On the other hand, waiting to check for a predator may be fatal. A useful heuristic to solve this dilemma is to fly up whenever at least two other flock members do so (Lima, 1994). Modeling confirms this as an efficient strategy (Proctor *et al.*, 2001) and empirical studies support this general rule (Cresswell *et al.*, 2000).

In noting the ubiquity of these strategies, Girgerenzer and colleagues point out that cognitive heuristics must be adaptive (Gigerenzer & Brighton, 2009). They argue that Kahneman and Tversky focused too heavily on errors in decision making; in reality cognitive 'short cuts' often lead to the best choice of food, mates, or habitat. For example, a receptive female who uses the rule 'select the largest male of the first three encountered' may have better reproductive success than a female who compares this attribute across a much larger group. The second female may identify the highest fitness male but he could then be occupied with another, less choosy, female. Thus, rather than being a fall back strategy when information is lacking, it is possible that heuristics guide decision making in so many different situations because they provided an evolutionary advantage in the past (Hutchinson & Gigerenzer, 2005). For this reason, heuristics should not be considered as good or bad; nor should they be evaluated in terms of whether they lead to accurate or inaccurate choices, at least in the short term. It may be more appropriate to view cognitive heuristics as strategies likely to yield optimal outcomes in the long term.

7.3.2 Risk-taking

Uncertainty often arises in decision making when there is a risk associated with one or more of the available options. Although most people associate risk with danger, its broader definition is unpredictability. More precisely, choices for which the payoff from different outcomes is known, but the precise outcome is not, are defined as risky (Kacelnik & Bateson, 1997). Most organisms face risky decisions on a regular basis. People may choose to purchase lower-priced commodities knowing that these products may not last as long and many animals choose to forage in a patch with a low, but consistent, density of food versus one with a high, but unpredictable, density. These decisions can be modeled in the lab using concurrent choice experiments like the one for pigeons shown in Figure 7.8. Pecking one key yields a low, but reliable, payoff and pecking the other yields a high, but variable, payoff. There is no obvious advantage to responding on one option or the other

Figure 7.8 Design of a risk-taking experiment in pigeons. Birds have the choice between two operant responses (left or right key peck) that deliver food rewards at different rates. One key is associated with a low (2 pellets per key peck), but consistent (100% probability), payoff and the other with a high (8 pellets per key peck), but inconsistent (25% probability), payoff.

because the two yield the same number of reinforcers in the long run. Under these conditions, animals are more likely to be risk-averse in that they select the fixed, over the variable, payoff (Kacelnik & Bateson, 1996).

The experiments become more interesting when the schedules are set up so that the low payoff option is advantageous in the long run. As with gambling, selecting the risky option is less profitable, even if there are occasional opportunities for sudden, high wins. Although it defies optimality, pigeons exhibit a preference for the risky option (Zentall & Stagner, 2011). People display the same suboptimal decision making in a human version of the task, with the effect being more pronounced in individuals who gamble (Molet *et al.*, 2012). Rats also exhibit risk-prone behavior (Ito *et al.*, 2000), but few studies have been conducted with other species. Those that have show that individuals within a species display **risk sensitivity** in that they are not consistent in how they respond to risk at different times (Weber *et al.*, 2004). People may be more likely to select inexpensive products if they are feeling financial pressures and animals may be more likely to forage in a patch with potential dangers when they are food deprived. This fits with evidence that pigeons given a choice between alternatives with a low, but frequent, versus a high, but infrequent, payoff are more likely to select the latter when they are food deprived (Laude *et al.*, 2012). Paradoxically, the birds that need more food receive less. Stephens (1981) argues that organisms have a tendency to select the risky option when the safe option cannot meet their fitness needs (e.g. the low payoff option is below their metabolic requirements), but this explanation has not stood up to controlled experimental studies (Shettleworth, 2010). Another possibility is that high, but infrequent, outcomes are more memorable, so individuals assume that these occur at higher rates than they do. This cognitive bias is explained by an **availability heuristic** in which judgments about the frequency of an event are based on how easy it is to recall similar instances (Tversky & Kahneman, 1974). In other words, humans and animals select the risky option because they are over-calculating the rate at which it pays off. This process may explain some maladaptive decision making in humans as problem gamblers focus more on infrequent wins than on regular losses (Breen & Zuckerman, 1999).

As noted previously, unpredictability of a risky option may reflect potential danger. Predation risk during foraging is an obvious example: in choosing a feeding site animals must balance resource payoff with the likelihood that they will be detected and attacked in the vicinity. This would explain why gray squirrels (*Sciurus carolinensis*) eat small pieces of cookie left on a picnic table by an experimenter, but retreat to a nearby tree to consume larger pieces (Lima *et al.*, 1985). Black-capped chickadees (*Poecile atricapillus*) exhibit the same behavior, maximizing feeding efficiency but minimizing exposure time in locations where they may be vulnerable (Lima, 1985). The same behavioral tradeoff was observed in lab rats who were trained to retrieve food pellets of different sizes that were placed at various distances from a shelter (see Figure 7.9). Smaller pellets were eaten where they were found, whereas larger ones were carried back to the home base (Whishaw & Tomie, 1989). The decision to eat or carry the food was modified by a number of risk-related factors including travel distance, ambient lighting, the presence of a predator, and how hungry the animals were (Whishaw *et al.*, 1991). Interestingly, administering a drug that reduces anxiety in humans increased the likelihood that rats would eat the food, rather than carry it back to a safe home base (Dringenberg *et al.*, 1994).

Taken together, these lab and field experiments suggest that foraging decisions are controlled by a complex combination of environmental and motivational factors. Maximizing energy gain from feeding must be balanced against risks associated with unpredictable outcomes. The same situation characterizes other choices faced by animals (e.g. mate or habitat selection) as well as humans (e.g. accepting a job in a city that may or may not offer other lifestyle benefits). Given so many uncertainties, it is difficult to determine *which* choice would be adaptive in any given situation.

Figure 7.9 Design of a risk hording experiment in rats. Animals are trained to leave a home base and traverse a straight alleyway to find food. Food pellets of different sizes are placed at various distances from a home base. The decision to eat or retrieve the food depends on the size of the pellet and the distance it is located from the home base. Pellets found close to home base are carried back (top panel). Small pellets located further from the home base are eaten immediately (middle panel) whereas larger pellets at the same location are carried back to home base (bottom panel).

Nonetheless, the work is progressing as researchers continue to test models of decision making that are founded on evolutionary principles (Hammerstein & Stevens, 2012).

7.3.3 Impulsivity and self-control

In some choice situations, the payoff from one outcome is delayed, adding another type of uncertainty to the decision making process. Under these conditions, both humans and animals often exhibit a preference for an immediate payoff, even if this is smaller than a delayed payoff. For example, humans given the option (either hypothetical or real) of receiving $10 now or $100 in a year, often choose the $10. Pigeons and rats exhibit a similar preference if they are given the choice between different amounts of food delivered after short or long delays (Tobin & Logue, 1994). Even honeybees can distinguish between small immediate and large delayed rewards, frequently choosing the former (Cheng *et al.*, 2002) (not surprisingly, the delay intervals in insect experiments are much shorter than in bird or mammalian studies). This phenomenon, termed **temporal** or **delay discounting**, reflects the tendency for future rewards to decline in value. As expected, the longer the delay to the large reward, the more it is discounted and the larger the reward, the longer the delay before it is discounted. The point at which the immediate and delayed rewards are valued equally is called the **indifference point**. Temporal discounting can be used to measure **impulsivity** and **self-control**, defined as the preference for small, immediate and delayed large rewards, respectively.

In lab experiments, humans exhibit more self-control than pigeons, rats (Tobin & Logue, 1994), or other primates (Rosati *et al.*, 2007) although it is difficult to compare the level of reinforcement and the delay period that would be meaningful to different species. Nonetheless, when rates of temporal discounting are compared within a species, most animals exhibit very steep discounting functions (Luhmann, 2009). That is, even if they prefer a delayed to an immediate reward, this tendency rapidly declines as the delay is increased. In contrast, humans and other great apes maintain a preference for the delayed reward at longer intervals, possibly reflecting the cognitive capacity to consider future consequences (Rosati *et al.*, 2007). Even so, self-control eventually declines, suggesting that the propensity to view distant rewards as less valuable than immediate rewards is pervasive, even in humans.

From an evolutionary perspective, temporal discounting is a paradox because organisms would optimize their payoff by selecting the delayed reward. The phenomenon does not reflect an inability to time intervals as most animals can track temporal delays that are used in standard lab experiments (see Chapter 6 for more details). The most common explanation for temporal discounting is that delayed rewards decline in value because they are unpredictable; the selection of a 'small-now' over a 'large-later' reward is an adaptation to cope with this uncertainty. If an animal in the natural environment chooses to cache a food item rather than eat it, there is always the chance that the item will decay or be pilfered before it is recovered. Food that can be eaten now is worth more than food that may or may not be available at a later time. Humans could use the same uncertainty principle to justify choices driven by short-term gains over long-term goals: smoking rather than quitting or indulging in sweets rather than dieting.

Alex Kacelnik, who has conducted comparative studies of decision making for most of his career, questions the idea that discounting of delayed rewards reflects an evolutionary adaptation to risk. He points out that most self-control studies involve simultaneous and exclusive choice, a situation that rarely occurs in the natural environment. During foraging, animals generally encounter food items sequentially so cognitive processes of decision making (at least in terms of foraging) would have evolved to maximize payoffs under these conditions. In a series of experiments with birds (primarily starlings), Kacelnik and his colleagues demonstrate that the preference for immediate versus delayed rewards is predicted by an individual's preference for these items in risk-free situations (Vasconcelos *et al.*, 2013). They propose that animals assign a value to food based on the context in which it is experienced. Thus, the same food encountered in resource-rich and resource-sparse areas will be valued more highly in the latter. Similarly, food will be more valuable when an animal is hungry than when it is sated. These assigned values determine how quickly animals respond to each item in the future. In naturalistic experiments of bird foraging, Kacelnik and colleagues show that food preference in choice situations is predicted by the latency to respond to each item when it is presented on its own. Their interpretation of these data is that simultaneous encounters of food are too infrequent to act as a selection mechanism for decision making; when faced with this choice (e.g. in the lab), cognitive processes that evolved to deal with sequential encounters direct behavior.

Kacelnik does not deny that temporal discounting may occur, but suggests that it is unlikely to explain choice behavior in self-control experiments. He argues that an organism's prior experience is a critical factor in decision making, a fact that is often overlooked by evolutionary biologists searching for functional explanations of a behavior. According to Kacelnik, reinforcement learning provides a good general purpose decision strategy because it allows animals to adjust their behavior to current environmental conditions. Although most of his studies are with birds, Kacelnik believes that the principles governing choice behavior in these animals can be applied to other species, including humans.

Researcher profile: **Dr. Alex Kacelnik**

Figure 7.10 Dr. Alex Kacelnik.

When asked what motivated his interest in science, Alex Kacelnik is quick to respond 'bugs,' or 'bichos' in his native Spanish. He says that he showed the same natural curiosity for living things that all children display, but was limited to what he could find in urban Buenes Aires, an environment he describes as a concrete jungle. His interest was piqued enough to study zoology at university but he felt stifled, intellectually, by the military dictatorship and the strong religious influence that permeated Argentina in the 1970s. Ideas such as evolutionary theory were unofficially censored so he and his friends held private study groups to read Darwin and other important thinkers. According to Kacelnik, his move to Oxford University in 1974, when he was awarded a British Council Scholarship, was exhilarating. Both junior and senior scientists openly debated different theories with a passion and logic that suited Kacelnik perfectly. He acknowledges that he was attracted to evolutionary theory because it invokes no external authority in the designing of living things. The idea was liberating for someone from his cultural and academic background.

The work Kacelnik undertook for his doctoral thesis, decision making in birds, has remained a centerpiece of his research program for the last 25 years. These investigations have expanded to other species, including humans, with the goal of understanding how and why organisms make specific choices. After completing his PhD, Kacelnik conducted postdoctoral work at Groningen (Netherlands) and another spell at Oxford, and then Cambridge, UK, after which he returned to an academic position at Oxford where he founded the Behavioural Ecology Research Group in 1990. He and his students continue to study cognition and behavior in a variety of systems, including foraging choices in starlings, tool use in New Caledonian crows (*Corvus moneduloides*), and brood parasitism in South American cowbirds (genus *Molothrus*). A premise of their work is that theoretical models of cognition must be grounded in empirical findings and be formulated to take into account each species' ecology. Kacelnik's interests are primarily those of a biologist, but he pursues these with tools and concepts from diverse disciplines, notably behavioral ecology, experimental psychology, economics, and comparative cognition.

Kacelnik has received a number of international awards for his research including the 2004 Cogito Prize for interdisciplinary work and the 2011 lifetime contributions award from the Comparative Cognition Society. He is a Fellow of the Royal Society of London (UK), an organization which described him as an international leader in the field of decision making and cognitive processes in animals. In Kacelnik's words, the greatest challenge in behavioral research is to understand how physical events relate to mental experiences. Understanding this mind–brain connection would be daunting for any researcher but Kacelnik notes that scientific progress comes not only from finding the answers, but also from figuring out how to ask the right questions.

7.3.4 Ecological rationality

Kacelnik's ideas fit within a broader theoretical view, ecological rationality, which explains decision making as a cognitive heuristic applied to a particular environmental context (Todd & Gigerenzer, 2000). According to this position, information processing is limited in all species such that it is not possible (or at least not viable) to calculate the outcome of every behavioral option, particularly when there are so many unpredictable variables in the environment. Animals and humans, therefore,

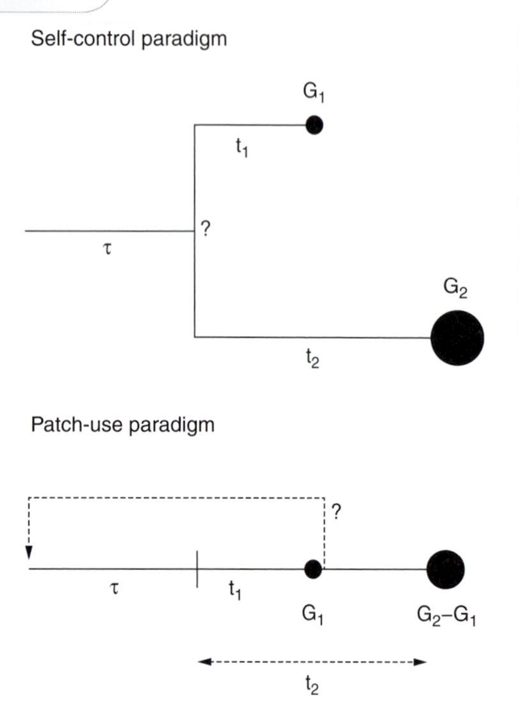

Figure 7.11 Schematic showing the design of self-control (top) and patch-choice (bottom) paradigms used to assess risk-based decision making in animals. In self-control experiments, subjects make a binary choice between a small, immediate and large, delayed reward. In the patch-choice paradigm, subjects encounter a small food item after a short delay (equivalent to the time delay and size of a small, immediate reward in the self-control experiment). The subject then chooses between waiting for the presentation of a large, delayed reward or continuing to respond for the small, immediate reward. G = goal; t = time

rely on heuristics to guide choices and these cognitive rules of thumb develop as organisms interact with their environment. The heuristic that is most adaptive in a given situation will depend on the species and the structure of its environment. Among other things, the sensory information that an organism receives and the ecological pressures it faces will determine the heuristic that should be applied to each decision. In other words, ecological rationality explains which strategy is better in a particular environmental context (Gigerenzer & Gaissmaier, 2011). These strategies are always described as better, not best, because the optimal strategy can never be known for sure.

Ecological rationality has been tested in the lab using a patch-choice paradigm that mimics foraging in the natural environment (Stephens & Anderson, 2001). In this task, shown in Figure 7.11, animals encounter food items sequentially. The first item is a small reward; subjects then choose between waiting for the next, larger reward or initiating a new trial that leads, once again, to the small reward. The time to the presentation of the small and then the large reward can be manipulated so that these intervals match those used in typical self-control experiments (i.e. when the choices are presented simultaneously). The delay times in the patch-choice paradigm model travel time to a new patch and the choice between waiting for the large reward or starting a new trial mimics the decision to stay or leave a particular patch. Stephens and colleagues, who developed the paradigm, argue that foraging animals choose between two strategies: they continue to search for food or they interrupt their searching to feed or attack prey. The decision to stop always occurs against the background of ongoing searching, which does not occur with the exclusive choices of self-control experiments. They also point out that options occurring within a brief temporal window are easier to discriminate so accuracy improves with short-term choices. Given these factors, Stephens (2008) concludes that impulsivity, as it is measured in typical self-control experiments, arose because selection of immediate rewards in the natural environment is beneficial in the long term.

Ecological rationality has also been applied to human decision making in choices ranging from the selection of a parking space to medical diagnoses (Todd & Gigerenzer, 2007). Although research in this area is sparse, the general idea is that cognitive heuristics that guide decision making are sensitive to patterns of information that occur in the natural environment. In other words, there is an adaptive fit between cognitive processes of information processing and how that information is structured in different contexts (e.g. looking for an empty spot in a parking lot and deciding whether to venture out from a hiding place would involve different processes). If this is true, it implies that decision making is an adaptive trait resulting from the fit between cognitive capacities and evolutionary pressures.

7.3.5 Individual differences

Ecological rationality helps to explain cross-species differences in decision making; not surprisingly, there are also large within-species differences in this cognitive process, particularly when choices involve some uncertainty. For example, many birds and mammals exhibit individual differences in their preference for risky versus non-risky options as well as in the selection of small immediate over large delayed rewards (Peters & Buchel, 2011). In humans, these behavioral measures are relatively stable across testing and correlate with self-reports of impulsivity (e.g. questionnaires asking how one would respond in different situations). At least in some individuals, these trait differences may contribute to pathological conditions such as drug addiction (see Box 7.4).

One of the greatest changes in individual responses to risk and uncertainty occurs across the lifespan. Young humans and animals display higher rates of impulsivity than their adult counterparts and often show high levels of risk-taking, particularly during adolescence (Peters & Buchel, 2011). In mammals, these developmental changes probably relate to maturation of the prefrontal cortex, a brain region associated with inhibitory control and executive functions such as long-term planning (Coutlee & Huettel, 2012). Given that impulsive and risky behaviors vary with both age and motivational state, many researchers are investigating the conditions under which these choices can be altered. This is particularly true of self-control in humans as the ability to delay gratification (i.e. select long-term over short-term rewards) is a strong predictor of academic performance (Mischel *et al.*, 1989) and social competence (Shoda *et al.*, 1990). Rewarding young children for appropriate behavior at a later time increases subsequent preference for delayed over immediate rewards, suggesting that experience with delayed gratification can improve self-control (Eisenberger & Adornetto, 1986). A common behavioral therapy technique to deal with impulsive behaviors is to institute precommitment strategies: decisions made in advance that are difficult or impossible to change at a later time. For example, binge eaters will be counseled to remove high-caloric snack foods from their house and compulsive shoppers to abandon credit cards and only carry enough cash to purchase necessary items. This technique is effective in pigeons (see Figure 7.12): these birds display higher levels of self-control if they can make a precommitment to selecting a large delayed over a small immediate reward (Ainslie, 1974).

In sum, all organisms face choices with unknown outcomes so it is not surprising that decision making is often discussed as an adaptation to cope with uncertainty (Stephens & Krebs, 1986). Cognitive processes that guide behavior under these conditions may lead to systematic biases that *appear* to result in suboptimal choices, at least in certain contexts. When decisions are examined from a broader ecological perspective, these choices may have adaptive value, both at the individual and the group level, particularly as these core processes emerge early in development and are shared broadly across species (Santos & Hughes, 2009).

Box 7.4 Impulsivity and addiction: what is the relationship?

Impulsivity is a primary feature of drug addiction in that affected individuals repeatedly choose the immediate gratification of drug intoxication over the long-term benefits of abstinence. Drug addicts are more likely to select small immediate over large delayed rewards in the lab when either drugs or money are used as reinforcers (Kirby & Petry, 2004), paralleling their choices in the 'real world.' One explanation for this relationship between impulsivity and addiction is that pre-existing impulsive traits increase the propensity to experiment with drugs and then to continue drug use once initiated (Wills *et al.*, 1994). Research with lab animals supports this idea in that high impulsive rats self-administer drugs at increased rates and progress more quickly to compulsive drug intake than do low impulsive rats (Jupp *et al.*, 2013). Moreover, a substantial portion of individual variability in impulsivity is attributable to genetic differences in both humans (Eisenberg *et al.*, 2007) and rats (Wilhelm & Mitchell, 2009), suggesting that this behavioral trait is inherited.

The idea that impulsivity is a predisposing factor for drug addiction is difficult to evaluate in humans because addicts are almost always tested after the disorder has developed. It could be that increased impulsivity in these individuals is a consequence, and not a cause, of substance abuse. According to this theory, chronic drug exposure alters brain regions associated with self-control, including the prefrontal cortex, such that impulsivity emerges with continued drug use (Jentsch & Taylor, 1999). Consistent with this idea, both heroin (Schippers *et al.*, 2012) and cocaine (Mendez *et al.*, 2010; Paine *et al.*, 2003) increase impulsivity in non-impulsive rats.

To sort through these contradictory theories, some research groups have mounted longitudinal and cross-sectional studies that assess a variety of personality variables and relate these to later drug use (and other behaviors). The studies are both time consuming and labor intensive so the findings may take years to summarize and interpret. Nonetheless, initial reports support the idea that impulsivity predates drug addiction, although it cannot explain the disorder entirely (MacKillop, 2013). In all likelihood, repeated drug use (particularly drug binges) promotes further drug use by increasing a pre-existing propensity to respond impulsively. If so, identification of trait impulsivity at an early age could provide a powerful tool for behavioral interventions to reduce subsequent drug use. Attempts to set up such a program with high-risk adolescents are showing initial success (Conrod *et al.*, 2013).

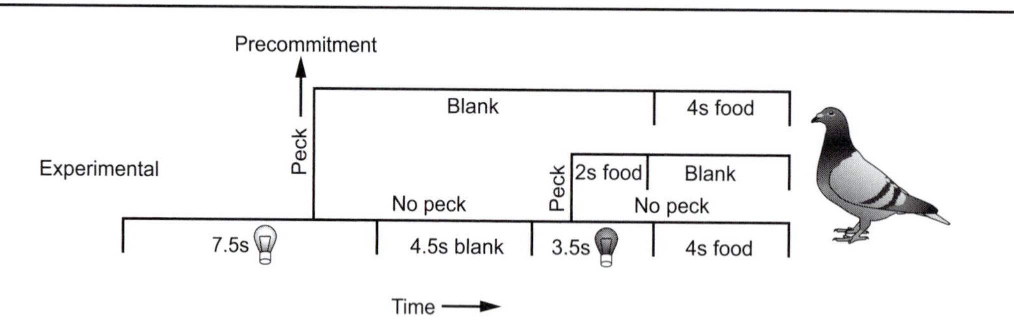

Figure 7.12 Schematic showing the design of a precommitment self-control experiment in pigeons (Ainslie, 1974). Each trial progresses across the horizontal plane, beginning with the presentation of a green light (7.5 seconds); if a bird pecks during this period, it receives a large, delayed reward (4 seconds access to food, following a 15-second blank period). If a bird does not respond during the green light, a blank screen is presented (4.5 seconds) followed by a red light (3 seconds). Responses during the red light lead to an immediate, small reward (2 seconds access to food). No response during the red light leads to the large, delayed reward. Under these conditions, pigeons often precommit to the large, delayed reward by responding during the green light. When this option is not available (i.e. the trials progress along the bottom line only), they are more likely to select the small, immediate reward when the red light is presented.

7.4 Emotional decision making

Optimal decision making is often equated with rational thinking, an antithesis to the irrational choices that are driven by emotions. Indeed, as far back as Plato, emotions were described as forces that needed to be tamed and guided by reason (Leahey & Harris, 2001). Over the centuries, the idea that emotions oppose or distort decision making was often queried by philosophers or psychologists, but it was not until the late twentieth century that the relationship between the two was examined scientifically. Damasio, a Portuguese-born neurologist working in Iowa, USA, was struck by a particular patient that he saw in his clinic. This individual scored at or above normal on most neuropsychological tests, including those purported to measure intelligence. Standard clinical assessments, therefore, were unable to reveal any cognitive deficit yet the patient displayed profound problems with making decisions related to everyday life. Damasio eventually identified other individuals with a similar profile. Some developed severe gambling problems, most had personal conflicts at work (if they were able to hold a job at all), and many were estranged from their family and loved ones. Often the patients recognized that their behavior was inappropriate but were unable to do anything about it. It was as if rational decision making was intact despite a consistent pattern of poor choices.

In searching for answers to this clinical paradox, Damasio uncovered the case of Phineas Gage, a railway construction foreman who suffered brain damage in 1848 when an iron rod shot through his head following an explosion. Remarkably, Gage survived the accident and remained conscious, even though the 6 kg rod went up through the side of his face, behind the back of his left eye, and out the top of his head (Damasio, 1994). Descriptions of Gage's behavior after he recovered are scant and many have been exaggerated but Damasio is convinced that Gage displayed many of the same behaviors as his own patients, most notably inappropriate decision making. He and his colleagues decided that they needed a clinical test to model this deficit so they developed the Iowa Gambling Task (IGT). The IGT is a 'game' in which individuals turn over cards, one at a time, that are set out in four piles. Each card displays a win or loss of points which are awarded to the player. Two decks are high risk (high wins but also high losses) whereas the other two are low risk (lower gains but also lower losses). The decks are set up so that the low-risk decks have a better payoff in the long run. Initially subjects select more cards from the high-risk decks but, over trials, control subjects (i.e. those with no neurological condition) shift their preference to the low-risk decks, even though they are unaware of the payoff contingencies. That is, they cannot identify which decks are better but say things like they have a 'gut feeling' about their choices. In contrast, those with decision making deficits continue to select cards from the high-risk decks, eventually losing all of their points.

Control subjects performing the IGT also exhibit increases in heart rate and galvanic skin responses (general measures of arousal) before they make a choice from one of the high-risk decks. Damasio's patients fail to show these anticipatory affective responses, which matches their blunted reactions to many life events (e.g. some appear completely detached from divorce or other emotional disruptions). Based on this evidence, Damasio (1994) developed the **somatic marker hypothesis**, the idea that decision making is informed by bodily reactions that are triggered by emotions. These emotional responses develop with experience, helping to guide appropriate choices in the future. According to Damasio and colleagues, disruption of somatic markers leads to poor decision making because emotional responses are unable to influence choice behavior. By scanning his patients, Damasio was able to relate their decision making deficit to damage in the ventromedial region of the prefrontal cortex (vmPFC) suggesting that somatic markers (i.e. the integration of emotion and decision making) emerge in this region. Reconstructions of Phineas Gage's brain,

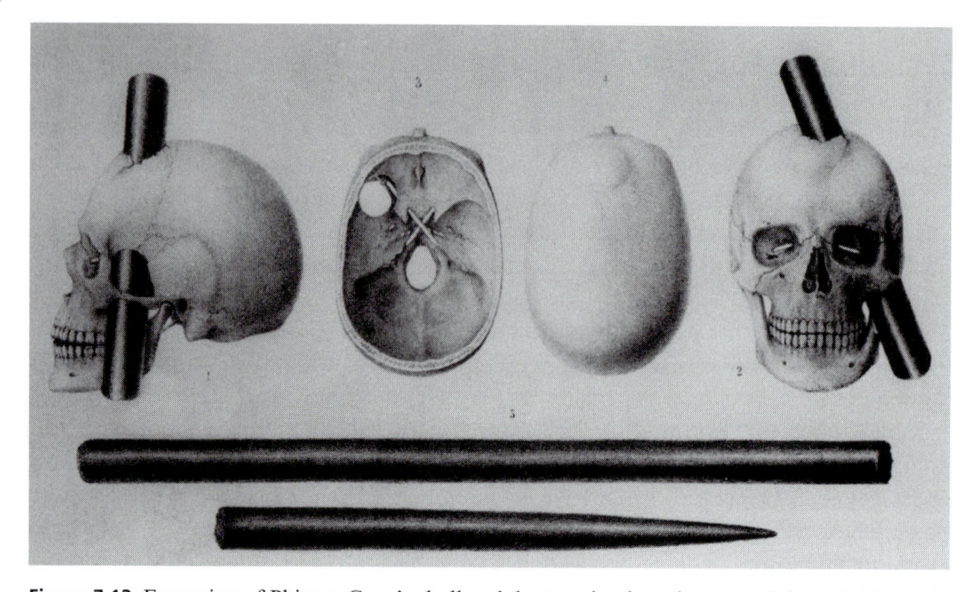

Figure 7.13 Engraving of Phineas Gage's skull and the tamping iron that passed through his head.

based on his exhumed skull, suggest that the same region of his brain was destroyed following his accident (Damasio *et al.*, 1994) (see Figure 7.13).

The somatic marker hypothesis challenges the notion that emotions disrupt decision making, but the idea fits well with evolutionary perspectives on cognition. For example, Cosmides and Tooby (2013) propose that emotions guide many cognitive processes including perception, attention, memory, and decision making. Fear is perhaps the best example. In a heightened fear state, one is more likely to attend to a sudden noise in the bush and energies can be directed toward responding to that stimulus. In the same way, emotions may influence decision making by drawing attention to relevant stimuli or making particular events more memorable (Brosch *et al.*, 2013). In addition, clinical anxiety disorders may be explained by enhanced attention to negative stimuli or threats, which then impacts decision making processes (Hartley & Phelps, 2012).

Most researchers now accept the idea that emotions may facilitate effective decision making, so the focus has shifted to understanding how these interact (Angie *et al.*, 2013). Many point out that neural systems mediating the two processes overlap and that regions such as the prefrontal cortex have strong connections to the amygdala, a brain structure implicated in emotional processing (LeDoux, 2007). This should not be surprising as emotions and cognition evolved together and effective communication between brain systems that control these two processes is likely to provide fitness advantages to a species. Investigations into emotional decision making advanced, recently, with the development of a rodent model of the IGT (Zeeb *et al.*, 2009); preliminary work with this paradigm supports the idea that emotional states guide behavioral choices.

7.5 Neuroeconomics

Research into emotional decision making, specifically how it is mediated at a neural level, fits within the broader field of **neuroeconomics**. Neuroeconomics, the study of how the brain interacts with the environment to enable decision making, is a relatively new endeavor, emerging just before the turn of the twenty-first century (Glimcher *et al.*, 2009). It combines theories and techniques from a variety of fields including economics, psychology, biology, and neuroscience. These complementary perspectives provide researchers with a novel framework for understanding decision making, at both a functional and mechanistic level.

7.5.1 Value-based decision making

Neuroeconomics is based on the premise that decision making is value-based. That is, the outcome of a choice has motivational significance to the organism making the choice. This is distinct from the type of classification decisions that will be discussed in Chapter 9, such as whether two sensory stimuli (e.g. two musical tones) are the same or different. Because it relates to the individual, an important concept in value-based decision making is **utility**: the subjective value of a behavioral outcome. The objective value of an outcome is relatively easy to measure, whereas the subjective value or utility will vary depending on the individual and the context. A monetary reward of $20 will not have the same value to a struggling student and to a wealthy professional. Similarly, even highly palatable food has a lower subjective value when an animal is sated compared to when it is hungry. The utility of a commodity, therefore, is an important determinant of choice behavior.

A series of studies examining how the brain computes utility (Shizgal, 1997) set the stage for scientific investigations in neuroeconomics (Glimcher *et al.*, 2009). In these experiments, rats had the choice between lever pressing for electrical brain stimulation or another reinforcer, such as sucrose or salt (when they were sodium deprived). Utility was measured as the preference for one outcome over another and by the rate at which animals would lever press for each reinforcer. The value of sugar and salt was varied by increasing the concentration of each in a solution that was injected directly into the rat's mouth or stomach. The value of brain stimulation was varied by increasing the frequency of the electrical trains. At low frequency stimulations, rats preferred the sugar or salt, a preference that was reversed when the frequency was increased. The point at which the preference shifted could be modified by food or salt deprivation, confirming that utility of each outcome is subjective. Most importantly, if two of the commodities were combined (e.g. rats received a sucrose reward and a train of electrical stimulation for a single lever press), the utility of this compound was equal to the combined utility of the two reinforcers. In other words, the subjective value of these very different outcomes summated. This suggests that the brain has a mechanism to convert reinforcement signals into a common currency of utility and that reward signals eventually converge on the same group of neurons. An alternative view is that the brain scales, but does not convert, utility so that specific information about different rewards (i.e. the sensory qualities of a food) is maintained in the brain (Grabenhorst & Rolls, 2012). In the latter model, utility is represented as the rate of neuronal firing (a common scale) in these different groups of neurons.

Regardless of whether utility is represented as a common or a scaled currency, the process likely involves the ventral striatum. In the common currency model (Shizgal, 1997), utility computation is centered on the medial forebrain bundle, a neuronal pathway running through the lateral hypothalamus that connects the ventral tegmental area to the ventral striatum (see Figure 7.14). This circuit mediates the rewarding effect of both natural and artificial reinforcers (Wise & Rompré, 1989) so it is not surprising that it has an important role in utility. The same region has been implicated in utility computation in humans (Breiter *et al.*, 2001). Participants in this experiment played a computer lottery game while they were in a brain scanner. The perceived desirability of a particular outcome was manipulated by changing the values of the two other possible outcomes. That is, the same number of points in the game would be viewed positively if it were the best outcome and negatively

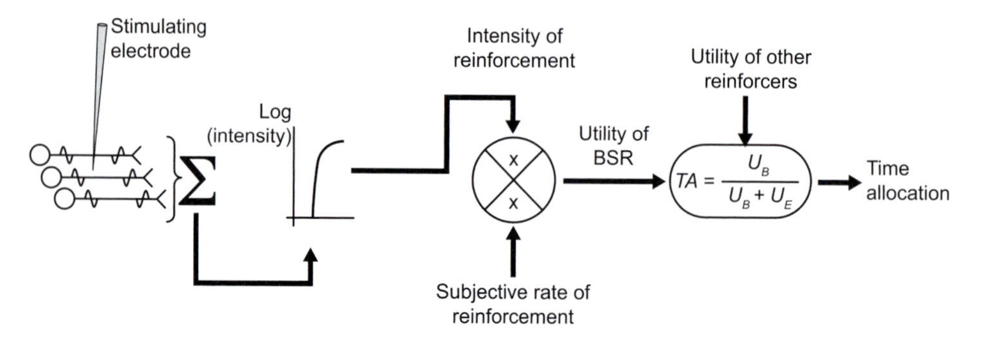

Figure 7.14 A model of the neural basis of utility estimation. The model is based on experiments in which rats respond for electrical stimulation, delivered through electrodes implanted in the medial forebrain bundle (MFB). Action potentials generated by the stimulation are accumulated by a hypothetical counter that converts the information into a signal representing the intensity of the reinforcer. The rate of reinforcement delivery is multiplied by the reinforcement intensity to yield a measure of utility. This value is compared to the utility of other reinforcers (e.g. sucrose); allocation of behavior is determined by the relative utility of available reinforcers according to the matching law. See Shizgal (1997) for details. BSR = brain stimulation reward.

if it were the worst. Brain activation in the ventral striatum matched these subjective values, fitting with animal studies showing that this brain structure has an important role in assigning subjective value to different reinforcers.

7.5.2 Cortical–striatal circuitry

The computation of utility is only one component of value-based decision making. This cognitive process also involves identification of possible choices, evaluation of both short- and long-term consequences of these choices, assessment of costs associated with each choice, and selection of an appropriate motor response. Most individuals (even humans) are unaware of these steps, but the brain is processing all of this information during decision making. Moreover, feedback from past choices (i.e. learning), as well as social and emotional factors, may interact with any one of these steps. Given this complexity, it is not surprising that decision making involves multiple brain regions. This has been studied, most thoroughly, in mammals, with a general consensus that decision making is mediated through interconnecting cortical–striatal circuits (see Figure 7.15).

The details are still being worked out, but the general idea of this schematic is that the orbitofrontal cortex (OFC) codes information about reward value, as well as the relationship between rewarding stimuli and sensory cues (Rolls, 2004). The OFC sends projections to the anterior cingulate cortex (ACC), which calculates the effort required to obtain different rewards (Hillman & Bilkey, 2012). This region, therefore, may be involved in a cost/benefit analysis of different outcomes (Peters & Buchel, 2011). Both the OFC and ACC send projections to the vmPFC, an area already implicated in emotional decision making (i.e. somatic marker hypothesis). According to a recent review of this work (Grabenhorst & Rolls, 2012), signals from the OFC and ACC are transformed into choices in the vmPFC. In truth, the precise function of these cortical regions is difficult to sort out, partly because neuroimaging studies in humans cannot always distinguish the anatomical boundaries between them and it is not clear if these brain regions are homologous, even in primates (Wallis, 2011). The issue is further complicated by the fact that the vmPFC, a relatively large brain structure, includes parts of both the OFC and ACC (the regions are separated based on cell morphology and anatomical projections). Despite these difficulties there is a

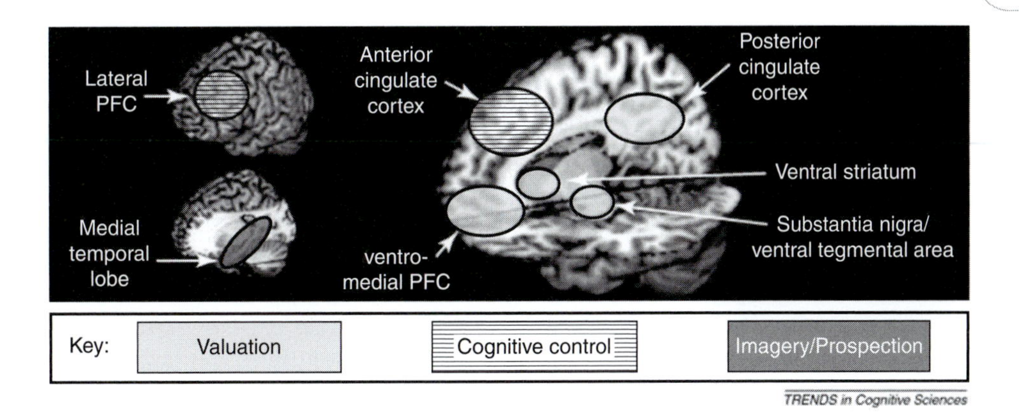

Figure 7.15 Cortical–striatal networks of decision making in the human brain. Shades of gray represent networks implicated in different processes: valuation, cognitive control, and imagery/prospection. The latter can be conceptualized as a mental representation of future events (e.g. reward outcomes).

general consensus that anatomically distinct regions of the prefrontal cortex (particularly medial and orbital regions) make dissociable contributions to value-based decision making, and that the process involves similar mechanisms in humans and animals (Floresco *et al.*, 2008).

The same cortical regions have been identified as playing an important role in naturalistic decision making. For example, when rhesus monkeys (*Macaca mulatta*) were trained to make hypothetical choices between foraging in the same patch or moving to a new one, the rate of firing in ACC neurons related to patch depletion and travel time (Hayden *et al.*, 2011). The authors suggest that the ACC was coding the relative value of choosing one patch or another, reflecting its role in cost/benefit analyses. Similarly, a human imaging study showed distinct patterns of activation in the vmPFC and ACC when individuals selected behavioral options related to foraging (Kolling *et al.*, 2012). ACC neurons signaled the cost of foraging as it related to the richness of the environment, whereas the vmPFC encoded the outcome value of a behavioral choice. Although these descriptions do not *exactly* fit those from the other experiments, the general pattern of functional specialization in the cortex is consistent across studies.

From the cortex, neural signals related to decision making are transmitted to subcortical regions, most notably the striatum (Balleine *et al.*, 2007). The ventral striatum computes utility (detailed above) whereas the dorsal striatum encodes action–outcome associations. Cortical connections to the dorsal striatum, therefore, control response selection based on the expected value of different options. The dorsal and ventral striatum were identified as playing an important role in operant conditioning (see Section 4.6.1), which makes sense in that an operant response (i.e. a behavioral choice) is the outcome of decision making. Finally, an important component of any decision making model is inhibition: signals related to self-control. This likely occurs through 'top down' projections from the prefrontal cortex to the striatum (Coutlee & Huettel, 2012), although it is not clear which subregions of the prefrontal cortex mediate this process. Regardless of the anatomical specificity, this connection is modified by experience in that learning to inhibit a response strengthens synaptic connections from the prefrontal cortex to the striatum (Hayton *et al.*, 2010). In other words, self-control can be acquired, or at least improved, with training and this behavioral change is reflected in neuronal modifications in cortical–striatal circuitry.

A premise of neuroeconomics is that environmental problems facing animals and humans shape the neural systems that control choice behavior. This helps to explain cross-species differences in decision making. Neuroeconomics also emphasizes that emotions impact decision making, an interaction that is manifested at the physiological (i.e. neural) level. At least in humans, this interaction varies across the lifespan such that emotions have a stronger impact on decision making in adolescence than at other ages (Blakemore & Robbins, 2012). These developmental changes likely reflect alterations in hormonal system function as well as increased personal autonomy and independence. In sum, research in neuroeconomics has advanced rapidly but there are still challenges in developing a theory that combines economic, psychological, social, and emotional factors (Blakemore & Robbins, 2012).

Chapter Summary

- Decision making, the cognitive process of selecting one course of action over others, is manifested as choice behavior, a motor action or verbal response that indicates option selection. The idea that decision making is an adaptive trait is supported by evidence that organisms forage in a way that maximizes net energy gain per unit time (optimal foraging theory), although factors such as predation alter this relationship. Marginal value theorem takes into account depleting resources within a patch, suggesting that animals will move to a new patch when foraging payoffs decline below the average payoff of the surrounding area. The ideal free distribution model describes how animals that forage in a group distribute themselves among patches with different amounts of food. Collective decision making occurs when large groups of conspecifics engage in coordinated behaviors, such as migration, that appear to have functional properties of their own. The general principle of these optimality theories of foraging (i.e. lowest energy expenditure for highest fitness benefit) can be applied to many other naturalistic behaviors.
- Mechanisms of decision making are frequently examined in the lab using concurrent schedules of reinforcement. According to the matching law animals maximize payoffs in these situations by matching their relative rate of responding to the relative payoff provided by that alternative. This line of research expanded into the field of behavioral economics which describes choice behavior in terms of economic principles. When alternative reinforcers are available at the same time (the most common situation in the natural environment), choice behavior is influenced by an organism's motivational state and by availability of alternative reinforcers.
- Almost all choices involve an element of uncertainty so humans and animals often rely on heuristics, or cognitive rules of thumb, to make decisions. These mental shortcuts save time and mental energy although they often lead to systematic biases in choice behavior. The ubiquity of this phenomenon suggests that cognitive heuristics may be adaptive even if they sometimes lead to less than optimal choices in the short term. Uncertainty in decision making frequently involves risk when the payoff from an outcome is known but its occurrence is not. It is still not clear why humans and animals are often risk-prone when the long-term payoff from a risky choice (high but infrequent wins) is lower than a low-risk alternative. Similarly, many individuals display impulsivity, preference for an immediate small over a delayed large reward, even if this produces lower long-term payoffs. This paradox may be explained by ecological rationality theory, the idea that cognitive heuristics for making decisions take into account the environmental context.
- Traditionally, emotions were viewed as opposing or disrupting decision making processes. Damasio's somatic marker hypothesis argues against this idea, stating that effective decision making relies on emotional signals; in their absence, individuals have no difficulty understanding the consequences of their choices but continue to make inappropriate decisions in everyday life. The deficits are associated with damage to the ventromedial region of the prefrontal cortex. The somatic marker hypothesis fits with evolutionary theories in that cognitive and emotional processes likely evolved together so effective interaction between the two would confer fitness advantage to an individual.
- Neuroeconomics examines how the brain interacts with the environment to enable effective decision making. It assumes that decision making is value-based, with the subjective value (utility) of an outcome dependent on the individual and the context. It is not clear whether utility is converted or scaled to a common brain currency so that the subjective value of different reinforcers can be compared. Either way, the process likely involves the ventral striatum, at least in mammals. Other

aspects of value-based decision making, such as selection of an appropriate motor response, are mediated through interconnecting cortical–striatal circuits that involve different subregions of the prefrontal cortex.

Questions

1. The psychologist and philosopher, Steven Pinker, argues that critics of reductionist views of psychological processes (such as decision making) often confuse explanation with exculpation. Discuss this statement in the context of free will.
2. Consider a predator animal in the natural environment that may encounter different prey. Each prey requires different time and energy to hunt and kill, but also yields different amounts of energy when consumed. What factors may alter the allocation of time devoted to hunting for each of these prey?
3. Large groups of humans, like animals, often exhibit coordinated behaviors that appear to take on a 'life of their own.' How do the cognitive processes of each individual relate to those of the group during this collective decision making? What do you think each individual is 'thinking' during these activities?
4. Using the matching law as a basis, write an equation describing how behavior would be allocated if two alternative rewards varied in quality. Identify two other variables that would impact this choice behavior and expand the equation to include these factors.
5. Explain how the principles of economic theory have contributed to an understanding of human and animal behavior.
6. Come up with a cognitive heuristic that is not described in the text and explain how this rule of thumb would be adaptive for a particular species.
7. Describe the difference between risky and impulsive behaviors. If necessary, use a graph to illustrate this relationship.
8. What is ecological rationality and how is it applied to lab studies of choice behavior?
9. What is the 'error' in *Descartes' Error* (Damasio, 1994)? If Descartes were alive today, how would Damasio try to convince him to change his ideas?
10. Describe the neural circuit of value-based decision making in mammals. Use a diagram to show the relationship between brain regions that mediate different steps in the decision making process.

FURTHER READING

The question of free will and determinism in decision making has fascinated philosophers, scientists, and ethicists for centuries. These positions are summarized and discussed in many contemporary texts and articles. Two of the most accessible and up to date are:

Dennett, D. C. (2003). *Freedom Evolves*. New York, NY: Viking Books.

Searle, J. (2007). *Freedom and Neurobiology: Reflections on Free Will, Language, and Political Power*. New York, NY: Columbia University Press.

Most cognitive science and cognitive psychology textbooks include at least one chapter on decision making. This information is expanded in a single text that provides detailed explanations of different theories and mechanisms of decision making from a cognitive perspective.

Hardman, D. (2009). *Judgment and Decision Making: Psychological Perspectives*. Chichester, UK: John Wiley & Sons Ltd.

The evolution of decision making is discussed in an edited volume that includes contributions from researchers across a broad range of disciplines (e.g. behavioral economics, cognitive psychology, biology, and neuroscience).

Hammerstein, P., & Stevens, J. R. (eds.) (2012). *Evolution and the Mechanisms of Decision Making*. Boston, MA: MIT Press.

Although more than a quarter century has passed since they were written, the original formulations of optimality theory provide an excellent overview of how ecology and evolution shape decision making processes.

Maynard Smith, J. (1978). Optimization theory in evolution. *Ann Rev Ecol Systematics*, **9**, 31–56.
Stephens, D. W., & Krebs, F. J. (1986). *Foraging Theory*. Princeton, NJ: Princeton University Press.

A contemporary overview of research examining choice behavior in the lab can be found in a special journal issue devoted to this topic. The articles in this volume focus on recent advances in the dynamics of choice with an increased emphasis on choice in unstable environments and how this is modeled in the lab.

Special edition. (2010). *J Exp Anal Behav*, **94**, Sept. No. 2.

The influence of economic theories on understanding human behavior, particularly decision making, can be found in:

Diamond, P., & Vartiainen, H. (eds.) (2007). *Behavioral Economics and its Applications*. Princeton, NJ: Princeton University Press.

Any study of emotional decision making should include Damasio's original description of the somatic marker hypothesis.

Damasio, A. (1994). *Descartes' Error: Emotion, Reason, and the Human Brain*. New York, NY: G. P. Putnam's Sons.

An introduction to neuroeconomics, including a summary of research linking emotional decision making to brain mechanisms, is presented in the two texts listed below. Neuroscience research has advanced dramatically since publication of the first, so some experimental findings are dated, but the book provides a general framework for understanding important issues in the field, most notably how emotions influence cognitive states. The second text, which is more up to date, is geared toward advanced undergraduate and graduate students.

Glimcher, P. W., Fehr, E., Camerer, C. F., & Poldrack, R. A. (eds.) (2009). *Neuroeconomics: Decision Making and the Brain*. London, UK: Academic Press.
Rolls, E. T. (2005). *Emotion Explained*. Oxford, UK: Oxford University Press.

Shiller's work has addressed how psychological factors influence decision making in the economic arena and the impact of group dynamics on financial markets. He explains why economies do not behave as predicted by 'rational' explanations. Economy is driven by 'psychological' forces – understanding the role of emotions in economic decisions; perhaps the reverse of economic principles affecting theories of decision making. Here knowledge of emotional decision making in humans influences how the 'behavior' of economies may be understood.

Akerlof, G., & Shiller, R. (2009). *Animal Spirits: How Human Psychology Drives the Economy, and Why it Matters for Global Capitalism*. Princeton, NJ: Princeton University Press.

8 Causality and tool use

Background

In the early 1960s, a former secretary who was working primarily alone in the African jungle made a scientific discovery that has continued to inspire research in the decades since. Jane Goodall observed that chimpanzees (*Pan troglodytes*) in Gombe National Park would modify sticks by removing leaves and then insert the sticks into termite mounds. As the termites grabbed onto the stick (presumably to 'attack' the intruding object), the chimpanzees would remove the stick and quickly slide the termites into their mouths. At the time, tool use was assumed by many to be the domain of humans, a behavior that was unique to our species. But, after this discovery, a flood of research documenting different types of tool use by wild chimpanzees was published.

As researchers continue to document wild tool-using behavior across a wide range of species (including primates, birds, and marine mammals), a question remains regarding the cognitive processes that underlie the behavior. For example, tool use could be the result of many processes, including: a predisposition to engage with certain objects using defined action patterns; predispositions coupled with operant conditioning; or a combination of predispositions, operant conditioning, and a system of learning by observing others. Enquiring even deeper, a second question remains as to whether a tool-using species is able to represent the causal relationships between the tool and the environment. Among other things, such an understanding might provide insight to solve additional problems that have an analogous causal structure.

Chapter plan

This chapter begins by considering causality from a historical and modern perspective. Many animals are sensitive to the consequences of their actions; rats, for example, can learn to press levers for food. Yet, can they reason about their actions, understanding that the actions cause certain events to occur (e.g. the appearance of food)? Then, the focus will shift to the causal nature of objects, including how they move and the properties that allow tools to help users achieve goals. Here, researchers address the topic of causal understanding by asking what, if anything, animals understand about the outcome of object interactions or why a tool can bring about a desirable outcome. Note that this chapter focuses on causality in the physical domain, but causality will be considered again in Chapter 10 in studies that examine whether animals recognize that mental states (e.g. beliefs, desires) can cause behavior.

8.1 Causality

It is difficult to imagine life without the ability to represent causality. Without causal reasoning, one might think of a light switch as simply something that covaries with the onset of light, not as a causal prerequisite for the onset. It would be difficult to realize, thus, that one could control the onset of the light by intentionally engaging the switch. Without causal representations, a billiard game would simply appear to be a series of sequential ball movements, not instances of motion caused by contact. Research and theory in psychology and philosophy has a rich history on the topic of causality, investigating everything from the question of its very existence to examining how causal knowledge is achieved. For the latter, comparative study has led to advances in our understanding of the extents and limits of causal representations in animals as well as our understanding of human causal cognition.

This section begins with early philosophical and experimental considerations of causality. Then, two frameworks detailing the nature of causal understanding in animals will be presented. In much of this work, comparisons are made between animals and humans, though the animal species examined have broad phylogenetic range. The work is based in the laboratory, a requirement because in many cases, experience and environment must be as controlled as possible. It will be seen in later sections, however, that questions of causal understanding also pertain to research in the wild in which naturally occurring tool use may be observed.

8.1.1 The illusion of causality

The philosopher Hume famously argued that causality is an illusion. We cannot observe causality, but we nevertheless hold causal beliefs about connections among things. Even in simple physical interactions such as the collision of billiard balls, he argued, causal inferences are unfounded: "Motion in the second billiard ball is a quite distinct event from motion in the first; nor is there any thing in the one to suggest the smallest hint of the other" (Hume, 1748/1977, p. 18, as cited in Newman *et al.*, 2008). That is, the events themselves do not present causality; the visual input is simply of coincident motion.

Others, like Kant and Michotte, proposed that causal relations do have a special form, or spatiotemporal pattern, that is detected. Contact and **launching events** like those of the billiard balls have a certain motion pattern that yields the perception of causality (e.g. Michotte, 1963). One does not simply perceive one ball stopping and another ball starting; one perceives causality because the brain automatically creates a 'causal impression.' This perception is an illusion, Michotte still admitted, but it is a perception of a causal interaction. Further, this perception is relatively unaffected by higher level beliefs (e.g. knowledge that the 'balls' are actually just computer-animated circles with no mass does not change the perception; Newman *et al.*, 2008).

Since Michotte's seminal work, many psychologists have examined the objective properties of actions that lead to the perception of causality as well as the development of causal perception in young human infants. Leslie and Keeble (1987), for example, showed 6-month-old infants video of a red brick that moved toward a green brick and stopped, at which point the green brick began to move (Figure 8.1). These movies were part of a looking-time habituation procedure, and after infants habituated (i.e. decreased their looking at the event to a predetermined criterion), test movies were shown in which the original movies were simply played in reverse. Importantly, two groups of infants were tested: one group saw habituation events in which the green brick moved immediately after being touched by the red brick (a launching event), and the other group saw the green brick

Direct launching

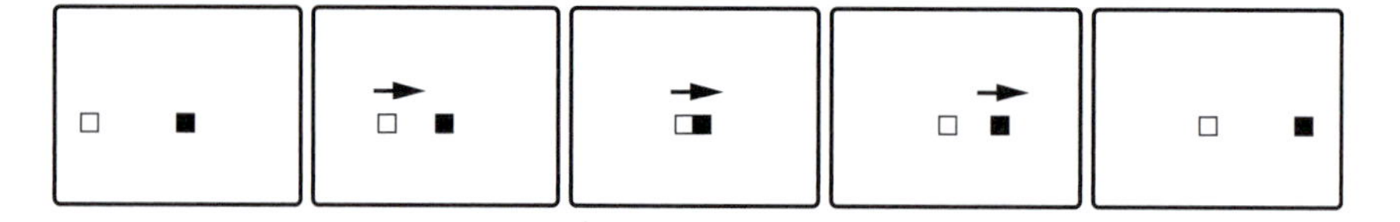

Delayed reaction

Figure 8.1 Launching and delayed reaction events as seen in Leslie and Keeble (1987). In launching events, the first brick made contact with the second, which then immediately started to move. In contrast, delayed reaction events included a pause before the second brick began to move.

move a short time after being touched (a delayed reaction event). Infants in the launching condition showed greater dishabituation (i.e. increased looking) to the reversed event in the test phase than infants in the delayed reaction condition, suggesting that the launching event was encoded as a causal interaction during habituation and the change in 'roles' during the test event warranted greater visual attention. Further, Leslie and Keeble proposed that causal perception may be an innate process of the visual system.

Although some researchers have remained skeptical about the innate nature of causal perception, much work has documented its early development during human infancy (e.g. Chaput & Cohen, 2001; Cohen & Oakes, 1993; Newman *et al.*, 2008). The perception of launching events has also been examined in animals, though little is known about the developmental processes. For adult pigeons (*Columbia livia*), Young *et al.* (2006) found mixed results. The pigeons showed some indication of discrimination between launching events and gap events in which the first object stopped prior to touching the second object (which nonetheless started to move). However, further analyses led the researchers to suggest that pigeons may attend to features of the events that human adults deem irrelevant (e.g. the duration of movement or the starting point of the launching object). In an intriguing follow-up experiment, Cabrera and colleagues found that pigeons visually tracked events in which an object appeared to approach and contact a food dispenser to a greater extent than they tracked movement away from the dispenser, a finding they humorously called the "lunching effect" (Cabrera *et al.*, 2009).

The causal perception of Michotte's launching event has been more thoroughly studied in humans, but studies like those above suggest that animals may share similar perceptions. It is possible, though, that the nature and familiarity of the stimuli (e.g. video versus real world) and the ecological validity (e.g. in relation to food, etc.) may influence experimental results. Physical causal events such as launching, however, are only one form of causality. Premack (1995), for example, contrasts the **natural causality** of launching with **arbitrary causality** in which the generative processes of the causal interaction are not as apparent (e.g. pecking a key 'causes' food to appear). Indeed, even Michotte (1963) suggested that the psychological mechanisms that underlie these two types of causal representation might differ. In the remainder of this section, the comparative research on 'arbitrary' causality will be presented, focusing on how such representations might be formed and which species might form them.

8.1.2 Association and causality

As noted in Chapter 4, operant conditioning is a change in behavior based on the consequences of that behavior. This may occur because organisms learn an association between stimuli in the environment and the response (e.g. a lever and a lever press), which is reinforced by the outcome (e.g. food). This stimulus–response (S-R) learning, also called habit learning, contrasts with learning to associate a response and an outcome (e.g. lever press and food). Response–outcome (R-O) associations suggest that an animal connects what they did with what happened (Adams & Dickinson, 1981b).

A question then arises as to whether it is accurate to characterize this form of association as a causal representation. It is clear that some aspects of conditioning parallel aspects of causality (Cabrera *et al.*, 2009). Dickinson and colleagues have proposed that humans and animals use associative mechanisms to learn about causal relations; causal learning is thus a byproduct of associative learning by this account (e.g. Dickinson & Balleine, 2000). The associative strength is considered functionally as causal strength. Under this framework, associationist theory can explain how information regarding causality is acquired (e.g. Wasserman & Castro, 2005).

However, it is not easy to explain all experimental findings with associative models. For example, Denniston *et al.* (2003) presented rats (*Rattus norvegicus*) with trials in which two stimuli (e.g. tones and lights) could occur with the availability of water, denoted here as AX+ (e.g. A = tone; X = light; + = water). Then, XY+ trials occurred in which the A stimulus was replaced with Y. Subsequently, some of the rats were presented with A in the absence of the water (A−), and these rats responded less strongly to later presentations of Y than rats who did not receive A− trials. Importantly, this change in response to Y occurred even though A and Y were never presented together. Some researchers have proposed that findings such as these support the idea that rats are engaging in higher order reasoning about causality (e.g. De Houwer *et al.*, 2005). That is, through the A− extinction trials, rats learned that A was not the cause and inferred that X, thus, must be the cause of water presentation. By extension, Y was also inferred to be non-causal. This inferential account is appealing because existing associative models cannot account for the change in behavior to Y. However, the authors of the original study have created an account of their findings that is based on an associative foundation, though it extends well beyond traditional accounts (Denniston *et al.*, 2003; see also Penn & Povinelli, 2007, for a summary). Their revised model does not assume the type of inference proposed by De Houwer and colleagues.

In sum, it is becoming apparent that some conditioning tasks may actually imply causal learning that cannot be explained through associations unless the associative model is modified from traditional forms. As this debate continues, alternative proposals for the creation of causal representations have been proposed. These accounts, described in the next section, generally propose that animals and humans *infer* the causal structure of events, and that causal learning strays from the tenets of associative theory.

8.1.3 Inferring causal structure

The study of human causal reasoning has been influenced greatly by the proposal that people understand causal structures in the world in a manner that can be modeled by **causal Bayes nets**. Causal Bayes net theory entails mathematical algorithms that go beyond the scope of this book,

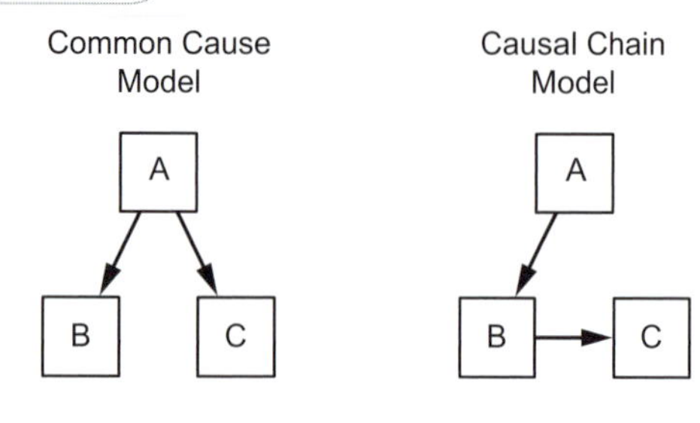

Figure 8.2 Two causal structures. In the common cause model, A can result in B and result in C, though B and C are not related casually. In the causal chain model, A causes B which causes C.

but much can be explained in basic terms. With causal Bayes nets, causal structures of the world are depicted with graphs such as those found in Figure 8.2 in which the letters denote events and the arrows denote causal relationships. In the 'common cause' model, for example, an event A can result in B and can also result in C, but B and C do not have a causal relationship. 'Causal chain' models, in contrast, represent instances in which event A causes B which in turn causes C.

There have been different approaches to the question of how an individual might infer one of these causal structures from input from the environment. One might experience events in the environment that covary, yet this information must be translated into a representation of the causal relationships; after all, as courses in statistics remind us, "correlation does not imply causation." According to some researchers, causal structure can be learned by attending to covariation and the conditional probabilities of events (e.g. Gopnik & Wellman, 2012). Other approaches additionally posit that pre-existing assumptions like "*causes precede effects*" help to constrain learning about causal structures (e.g. Waldmann, 1996) or that prior knowledge in a particular domain can affect causal inference (e.g. Tenenbaum & Griffiths, 2003). Over the past two decades, causal Bayes nets have greatly influenced the study of human causal cognition, including its early development (Box 8.1).

There is some evidence that rats can also understand causal relations in a manner consistent with causal Bayes net theories. In a study by Blaisdell and colleagues, rats experienced a flashing light that was sometimes followed by a tone and other times followed by the delivery of a sweet liquid (a common cause model). Then, one group of rats experienced the tone alone (the Observe Tone group), and one group experienced the tone each time they pressed a lever that had been introduced into the test chamber (the Intervene Tone group). The rats in the Observe Tone group often went to search for the sweet liquid after the tone occurred, but the rats in the Intervene Tone group searched much less often. The authors suggest that the rats had recognized the common cause structure, and those in the Observe Tone group reasoned that the light had likely occurred (though was missed) and so went to the food source. The rats in the Intervene Tone group, however, attributed the tone to their actions, and thus did not assume the light had occurred. In contrast, rats in another experiment were presented initially with a causal chain (i.e. a tone was experienced before the light, and the light was experienced before the sweet liquid). In the subsequent test, these rats did not act differently if the tone occurred alone or if their lever press was paired with the tone; they searched for the food with equal frequencies. Here, it seems that the rats recognized that the occurrence of the tone would lead to the sweet liquid, regardless of their intervention (Blaisdell *et al.*, 2006).

The rats in this study appeared to be creating predictions about the outcomes of their interventions (lever pressing) after learning about a causal structure purely through observation. This ability undermines most associationist accounts and presents a far more complex view of animal causal cognition than was previously available. Yet, there is reason to question whether

Box 8.1 Cognitive development and causality: detecting 'blickets'

Some researchers have proposed that even early in development, humans observe correlations and engage in further cognitive processes that lead to the creation of causal representations of events in the world. For example, if event Y causes event Z, then the occurrence of Y makes the occurrence of Z all the more likely. Observing and encoding this correlation is a cue to the actual causal structure in the world. Of course, this is only one cue, and correlation alone is not sufficient. For example, if Z also occurs in the presence of X and Y together, then all three are correlated, but it is unclear whether one or both caused Z. However, conditional information such as observing that Z does not co-occur in the presence of X alone would indicate that Z is only an effect of Y. Thus, a system that can attend to covariation and conditional probability can start to learn about causal structures in the world (e.g. Gopnik & Wellman, 2012; Sobel & Kirkham, 2007).

To examine whether young children engage in causal inference in this manner, researchers created a clever task involving a 'blicket detector,' a box that played music when a certain type of object (a 'blicket') was placed on top of it (e.g. Gopnik *et al.*, 2001). In a series of studies, 3- and 4-year-old children observed events in which objects, A and B, were placed on top of the box. For example, in one experiment, A was placed on the box alone, and music played, but there was no music when B was subsequently used. Then, both A and B were used together twice and music played each time. Children correctly inferred that only A was a 'blicket' (cause), even though B was involved in the music outcome more often than not.

However, these findings might be best explained by blocking; after learning that A is associated with the onset of music, the association with B is not learned (Sobel & Kirkham, 2007). In fact, some of the blicket detector studies can be explained by either traditional associative learning models or, in the case of complex blicket detector paradigms, more modern associative frameworks that were designed to account for laboratory animal studies that follow similar premises (e.g. Dickinson, 2001). That is, children may be solving these tasks based on the recognition of associative strength between certain objects and the box.

However, Sobel *et al.* (2004) propose that causal Bayes net theories more accurately explain the data. First, children appear to be making inferences after only a few trials, whereas even modern associative frameworks are based on many more opportunities for observation. Second, Sobel and colleagues found that children rely on **Bayesian inference**: they start by constraining the number of possible causal structures by considering prior probabilities and then update to determine the most likely structure based on new observations. Thus, children appear to integrate statistical knowledge into the process of creating causal structure. It remains to be determined whether humans are alone in this ability.

what rats are doing is similar to how humans approach such a task. Penn and Povinelli (2007) point out that humans engage in interventions in a deliberate and strategic manner, often to test their own hypotheses about causal structures, and no evidence exists at this time for this behavior in animals. Further, as will be seen in the next sections, it still unclear how the causal knowledge observed in studies such as that of Blaisdell and colleagues might be evidenced in other behaviors, such as during tool use.

Researcher profile: Dr. Ed Wasserman

Figure 8.3 Dr. Ed Wasserman.

As a child, Ed Wasserman surrounded himself with animals; his pets included turtles, frogs, fish, lizards, and a variety of birds. It seems rather unsurprising, then, that he ultimately became one of the most prominent researchers in the field of comparative cognition. Yet, as an undergraduate at the University of California, Los Angeles, he initially majored in physics. After taking a few courses in psychology, though, Wasserman became interested in the study of learning, both in humans and animals. The prospect of scientifically studying behavior, along with its challenges and controversies, was captivating. He switched his major to Psychology, and then moved to Indiana University to complete his PhD with Drs. Don Jensen, Jim Dinsmoor, and Eliot Hearst.

In 1972, Wasserman joined the faculty at the University of Iowa and has worked there ever since. During his career, he has studied cognitive evolution by examining the behavior of pigeons, rats, chickens, baboons, and humans. His contributions have been wide-ranging; indeed, his work has been cited in many different chapters of this book. In the early 1980s, Wasserman began a line of research most relevant to this chapter. His work examining causal reasoning across species has sometimes resulted in findings that do not fit most traditional associative learning theories. In response, he and a collaborator, Linda Van Hamme, developed a modified version of the Rescorla–Wagner model that provides an associative account of causal learning.

Wasserman's research framework is illustrative of the field of comparative cognition: naturalistic observation of behavior can be invaluable to generate hypotheses, yet laboratory experiments provide insight into the mechanisms underlying the behavior. His overarching goal is to characterize the basic components of cognition, which he thinks humans and other animals share. Wasserman proposes that even human behaviors that may be species-unique, like creating music or complex theories of physics, are the outcome of foundational cognitive processes that are widespread among animals.

He has encouraged the growth of the field as a co-founder of the Comparative Cognition Society, a scientific organization dedicated to the study of cognition across species, and many of his writings discuss the role of comparative cognition within psychology and ethology. Wasserman notes, "My enthusiasm has not waned and I look forward to further adventures in uncovering the basic laws of learning and cognition. This task will be all the more enjoyable because of the opportunity to work with so many talented students and colleagues." (Wasserman, 2007, *The Experimental Psychology Bulletin*).

8.2 Object physics

Many researchers, particularly in the field of cognitive development, have considered the topic of causality to include the understanding of basic **object physics**. As Baillargeon and colleagues state, causality can be construed at "a very general level in terms of the construction of conceptual descriptions that capture regularities in the displacements of objects and their interactions with other objects" (Baillargeon *et al.*, 1995, p. 79). This construal goes beyond the ideas put forward by Michotte in which one event is recognized to bring about another event through transmission of some type of force (e.g. a launching event). Instead, this broad meaning of causal reasoning considers questions such as: *Will small objects stop moving* *after making contact with larger, heavier objects? Can short screens hide tall objects? Can objects suspend in the air without support?*

Recognition of basic aspects of object physics allows for prediction of, for example, the end state of moving objects (a rolling ball will stop when it hits a solid wall) or the most efficient tool to achieve one's goal (a solid cane will pull in an object, but a flimsy one will not). Studies tapping into the understanding of object physics have a rich history in the study of infant cognitive development, and many of these paradigms have been modified for use with animals. This section will summarize some of the findings across two categories of object features: the movement of occluded objects and solidity/support.

8.2.1 Object permanence and tracking the displacement of objects

Behaviors such as foraging and hunting often rely on the ability to keep track of objects over space and time, and occlusion. Recognizing that objects exist even when they are out of view (or imperceptible via scent or sound) implies the concept of **object permanence**. Some of the work already presented in this book would suggest that some animal species represent the location of objects even after they undergo occlusion. For example, in order to show the ability to perform simple addition and subtraction in the tasks described in Chapter 6, monkeys would have to keep track of 1 to 4 objects that were hidden behind a screen (e.g. Flombaum *et al.*, 2005).

Other studies have examined object permanence with **object displacement tasks**. For example, Barth and Call (2006) tested chimpanzees, bonobos (*Pan paniscus*), gorillas (*Gorilla gorilla*), and orangutans (*Pongo pygmaeus*) with a task in which a piece of food was hidden under one of three opaque cups, and then the cup was moved to another location (i.e. a simplified "shell game"). To subsequently find the hidden food, the observer must possess object permanence and also recognize that the food moved with the cup, even though the only object observed to move was the cup. In general, the chimpanzees and bonobos performed better than gorillas and orangutans on these tasks, though on other related tasks involving the displacement of objects, all four great apes performed similarly (see also Beran & Minahan, 2000).

Some research has suggested that ape species may perform better on tasks like this than monkeys (e.g. de Blois *et al.*, 1998), but only a few monkey species have been tested and in some cases, the performance is actually similar to apes. Outside of the primate order, species such as African gray parrots (*Psittacus erithacus*) have also displayed object permanence and the ability to track hidden objects (Pepperberg *et al.*, 1997), but bottlenose dolphins (*Tursiops truncatus*) have only shown evidence for the former (Jaakkola *et al.*, 2010). A question remains, however, as to what cognitive processes underlie this ability. Do animals, for example, understand *why* the food moves along with the cup (e.g. due to containment within the cup and the contact with the cup's solid sidewall), or do they base decisions on other types of information, such as an association between the cup and food or an assumption that the two are somehow connected? These tasks alone cannot determine the basis for successfully tracking the food, but other research that more specifically examines animals' understanding of physical features such as solidity and support may shed some light on the understanding of the physical world.

8.2.2 Solidity and support

In 1992, Spelke and colleagues tested the development of the understanding of solidity in humans, specifically, how solidity constrains the path of a moving object. The task was conducted with infants and utilized a looking-time methodology in which longer looking to certain events is

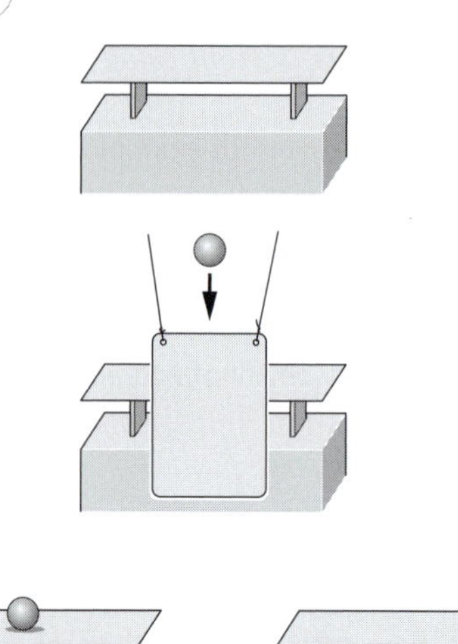

Figure 8.4 Testing events from Spelke *et al.* (1992). After observing the ball drop behind the screen, infants would see an outcome that was consistent or inconsistent with the property of solidity.

assumed to index an infant's recognition of inconsistency or unexpectedness. First, the infants were presented with a solid shelf that had been placed above a solid stage floor (Figure 8.4). An occluding screen was then placed in front of the display and an object was dropped from above the shelf. When the screen was removed, the infants either saw that the object was sitting on the shelf (an outcome consistent with the constraints of solidity) or saw the object on the stage floor under the shelf (an inconsistent outcome). Three- and 4-month-old infants looked longer at the inconsistent event, suggesting some level of understanding of how solidity affects the movement of objects. In contrast, no looking-time difference was found in a control condition in which the objects were not seen to drop, and thus the results of the experimental condition were likely not due to a simple visual preference for objects on the stage (Spelke *et al.*, 1992).

In an analogous task with rhesus macaques (*Macaca mulatta*), Santos and Hauser (2002) found that the monkeys looked longer when an object was revealed to be under a solid table, as if it had fallen through, than if it appeared to have stopped on the surface. Yet, a different picture emerges when the dependent measure is search behavior instead of looking duration. In the same study, rhesus macaques erroneously searched for the fallen object under the table. Chimpanzees, orangutans, gorillas, and bonobos made a similar error on this task when searching for the object, though performance improved over test trials (Cacchione *et al.*, 2009). Thus, it seems that both monkey and ape species have some understanding that solidity constrains object motion, but the knowledge is not evidenced in search behavior without experience. However, many researchers have posited that an additional feature of this task, the influence of gravity, may play a role in subjects' incorrect searches (see Box 8.2). This proposal is supported by additional experiments showing that performance improves in a horizontal version of the task (Cacchione *et al.*, 2009; Hauser *et al.*, 2001).

Another way to examine the understanding of support and solidity is to ask whether observers recognize that objects should fall if not adequately supported. For example, Baillargeon and colleagues found that during infancy, humans incrementally develop an understanding of support such that at 3 months, any form of contact with a platform is assumed to keep an object from falling,

Box 8.2 The 'gravity bias'

Often, the verbal, pointing, and reaching errors that participants make on experimental tasks shed light on their underlying cognitive representations. One particularly informative error was observed in a study of children's understanding of the interplay between solidity and gravity. In Hood (1995), 2-year-old children observed a ball as it was dropped down an S-curved opaque tube and slid into an opaque cup. Of interest was the particular cup, of a possible three, that children would search to obtain the ball. Children frequently erred by searching directly beneath the point at which the ball was dropped, evidencing a "gravity bias." The tendency to search in a straight-line path from the release point of the ball was not observed when the apparatus was placed in a horizontal position, suggesting that the bias is specific to vertical motion (Hood *et al.*, 2000).

Studies using a similar procedure with cotton-top tamarins (*Saguinus oedipus*), common marmosets (*Callithrix jacchus*), and dogs have documented the same pattern of errors (Cacchione & Burkart, 2012; Hood *et al.*, 1999; Osthaus *et al.*, 2003), though results with chimpanzees are mixed (e.g. Cacchione & Call, 2010; Tomonaga *et al.*, 2007). It is possible, then, that instead of basing search behavior on knowledge of how gravity and solidity interact in the bent tube, the human and animal participants were relying on a rule they had formed through experience that objects tend to fall straight down.

What remains in question is whether the species tested in this task actually have knowledge of the physical constraints involved yet cannot override a prepotent, rule-based expectation, or if an understanding of the interplay between solidity and gravity in this event is beyond their ability. One clue comes from a study with young children that replicated the gravity-biased search, yet also found that children looked predictively to the correct location when the ball was dropped (Lee & Kuhlmeier, 2013). Children may have understood the object physics of the event, yet searched based on a gravity bias because the cognitive demands of holding a representation of the occluded ball, producing an answer, and inhibiting the bias were overwhelming. In contrast, predictive gaze required only two of the three demands – representation of the ball's location and inhibition of the gravity bias. Alternatively, children may have only achieved a weak abstraction of object motion regularities that underlie movement in the tube, and this limited understanding perhaps could guide eye gaze but not overt search behavior.

Either interpretation is possible for the animal species tested as well, as at least one species, the common marmoset, has also shown looking behavior (but not reaching behavior) consistent with a representation of the object's movement in the tube (Cacchione & Burkart, 2012). Additionally, it is possible that some of the species tested do not have knowledge beyond a gravity 'rule.' A more complex understanding would presumably require experience with analogous physical events in their environment (e.g. objects sliding down an inclined plane, etc.), and for some species, these events may simply not occur with sufficient regularity.

but at 12.5 months there is a more mature understanding of the importance of the amount and distribution of contact (see Baillargeon, 1995, for review). An analogous study was conducted with juvenile and adult chimpanzees, and results suggested that for this species, it is the amount of contact between the platform and the object (regardless of where the contact is distributed) that determines whether the object should fall (Cacchione & Krist, 2004). In one of the experiments in this study, for example, the chimpanzees did not increase looking to an event in which the side of the object contacted the side of the platform yet the bottom was not supported. The authors note that

Köhler (1927) found that some chimpanzees would press a box against a wall and then try to climb onto it to reach food, a seemingly analogous error. It is thus possible that chimpanzees differ from humans by basing judgments of support on simple rules regarding amount of contact rather than a conceptual understanding. However, the authors admit that further studies are needed to rule out other factors, such as assumptions the chimpanzees may have made that the object was stuck to the side of the platform (which it actually was in order to create the effect).

Research on the understanding of object physics in animals has primarily been conducted with primates, owing perhaps to the motivation to compare cognitive processes with humans. Other species, such as dogs (*Canis familiaris*), dolphins, and some bird genera, however, have also been tested. In sum, there is evidence that many animals are sensitive to regularities in the way objects move in relation to environmental features. As noted in the introduction to this section, this is a broad consideration of causality, though attention to the constraints of object physics can lead to predictions regarding the outcome of events. In the next section, the focus will switch to tool use, a behavior for which the physical features of objects, and the outcomes of those features, play an important role in the efficacy of a tool to cause completion of a goal.

8.3 Tool use defined

Tool use is one of the central areas of interest within the field of comparative cognition. It has provided a window into the cognitive processes of social learning (Chapter 13), communication (Chapter 12), and relevant to this chapter, causal reasoning. Beyond comparative cognition, insights from animal tool use have influenced other disciplines such as human neuroscience, human–computer interaction, and robotics (St Amant & Horton, 2008). Researchers realized early on that the study of tool behaviors would require clear definitions of what did and did not characterize true tool use. Often, the pursuit of defining features of a category of behavior helps enhance both communication among scientists and the understanding of the issues at hand. What is it, for example, that might make us include a hammer as a tool, but not the desk it allowed us to build? This section will discuss the leading definitions of tool use and consider the species that have been observed to use tools in the wild. Consideration of the neural regions underlying tool use is presented in Box 8.3. With this foundation in place, the later sections will consider how tool use develops and what animals understand about tools.

8.3.1 What is a tool?

Motivated by her observations of chimpanzee behavior in Gombe, Goodall famously defined tool use as "the use of an external object as a functional extension of mouth or beak, hand or claw, in the attainment of an immediate goal" (van Lawick-Goodall, 1970, p. 195). Most modern definitions of tool use do not contradict this early definition, but they have greatly elaborated it. Alcock (1972), for example, focuses on the act of manipulating one object to improve one's ability to affect another object. Perhaps one of the more nuanced definitions, and the most seminal, came from Beck (1980):

> Thus tool use is the external employment of an unattached environmental object to alter more efficiently the form, position, or condition of another object, another organism, or the user itself when the user holds or carries the tool during or just prior to use and is responsible for the proper and effective orientation of the tool. (p. 10)

Box 8.3 Tool use in the brain

In clinical studies, injuries to the human brain appear to support the proposal that at least two factors of tool use are dissociable: the skill of manipulating a tool and the conceptual knowledge regarding a tool (e.g. its intended function, etc., see Box 8.4). The brain areas involved in each of these aspects of tool use are different such that damage to certain cortical regions of the left hemisphere can lead to problems in using tools, but damage to other areas within the left hemisphere can result in conceptual errors (e.g. using a toothbrush to shave). Often, however, clinical studies of injured patients are limited; for example, the extent of injury can vary and encompass large areas of the cortex. Yet, there has been great interest in uncovering the brain areas implicated in tool use, stemming from diverse motivations such as treating injured patients, examining the use of prostheses, and understanding the basic cognitive processes of artifact use.

Studies using functional magnetic resonance imaging (fMRI) have converged on cortical areas that are part of a system that underlies human tool use (summarized in a meta-analysis by Lewis, 2006). These studies often ask participants to pantomime using a certain tool (to limit head movement within the scanner), which appears to involve activity in the left superior parietal lobule (SPL) among other areas. Importantly, the hemispheric localization is not simply due to participants using their right hands to enact the tool use; even left-handed participants showed more activity in their left hemisphere. The SPL is thought to be involved in representing body schema, or the ability to track the location of one's trunk and limbs, while the body is in motion (e.g. Wolpert *et al.*, 1998). These findings are consistent with the idea that tool use can lead to integration of a tool into the mental representation of one's body, a suggestion that has been made for humans as well as a species of macaque monkey (*Macaca fuscata*; Iriki *et al.*, 1996)

A separate area, the inferior parietal lobule (IPL), appears to play a role in the creation of action schema, or preparing and planning motor actions. Some researchers have proposed that activation within a specific area of the IPL (the anterior supramarginal gyrus, or aSMG) is unique to humans; though rhesus macaques and humans appear to activate many similar neural circuits while observing actions with tools, humans alone showed enhanced activation of the aSMG (Peeters *et al.*, 2009). Even after some of the macaques were trained to use the tools, this area was not implicated when observing tool use.

The findings from Peeters and colleagues are intriguing, but comparative research on the neuroscience of tool use is only just beginning. Imaging awake animals during cognitive tasks is still a relatively new technique, and improved paradigms are likely to develop. With advances, it may also be possible to better understand how cortical processing occurs over the duration of a tool use action.

Under this definition, certain behaviors are clearly not tool use (e.g. a rat pressing a lever to obtain food or an elephant spraying water on its back using its trunk).

However, this definition leads to some difficult categorizations (St Amant & Horton, 2008). By Beck's definition, if a gull drops a mussel onto a stone surface, it is not tool use, yet dropping stones onto mussels would be. Behaviors like nest building (e.g. adding a twig to a growing structure) are not traditionally considered tool use, but the 'form' of another object has been 'altered'. Further, since the publication of Beck's volume, new behaviors have been documented that are widely

Box 8.4 The functional understanding of tools

In 1945, a problem-solving experiment was reported in which adult humans were slower to devise novel uses for known objects due to pre-existing knowledge of the objects' conventional functions (Duncker, 1945). For example, if provided with candles, a box, tacks, and a book of matches, participants had to invent a way of attaching a candle to the wall (solution: create a platform for the candle by emptying the box and attaching it to the wall with tacks). If participants had been initially provided with the box full of tacks, they were slower to reach the solution than if they had seen the box empty. That is, the adults were 'fixed' on the function of the box when they were reminded of its conventional role as a container and subsequently had difficulty thinking of it as a possible platform. In an analogous task in which a tower had to be built using objects like boxes and blocks, 6- and 7-year-old children showed **functional fixedness** if the original design function of the box was primed prior to the problem, but 5-year-olds did not (i.e. younger children outperformed older children; Defeyter & German, 2003; German & Defeyter, 2000). Functional fixedness is not just observed in technologically sophisticated cultures; cultures like the Shuar of Equador that have fewer artifacts also are slower to use objects in a manner different from their typical functions (German & Barrett, 2005).

Thus, humans focus on an object's **function**, or what it was made for. Though functional fixedness for some objects might appear around 6 years of age, even younger children do show attention to function. After being shown a successful action with a previously unknown tool, 2-year-olds will use that same tool a few days later, even though other tools that are physically capable of completing the task are available. But, they also use it for other tasks; it is not seen as exclusive for one action (Casler & Kelemen, 2007). Exclusive functionality starts to be seen after 24 months of age (Casler & Kelemen, 2005) and may then lead to functional fixedness.

Hernik and Csibra (2009), among others, have proposed that animals do not form functional representations of tools as existing for the achievement of specific goals even though some species may show an understanding of the relationship between a tool's structure and an outcome of an action. It is possible that humans are the only species that represent tools in terms of function, even when the tool is not being used. That is, animals may not think about what a tool is for, particularly if the problem is not currently presented. One recent finding may be an exception; in a laboratory setting, bonobos and orangutans were found to save a tool for later use (Mulcahy & Call, 2006b). Yet, the 'problem', an apparatus that dispensed food, was still visible while they waited. Two subjects, however, saved tools for 14 hours without visual access to the apparatus, suggesting that further research is warranted to determine whether a functional understanding of tools is unique to humans.

accepted as tool use but do not fit the definition. For example, bottlenose dolphins have been observed to place sea sponges on their rostrums, presumably for protection, before searching for food on the ocean floor (Krützen *et al.*, 2005; Figure 8.5), but in doing so, they are not truly altering other objects in their environment. Some researchers have responded by updating the working definition of tool use. St Amant and Horton (2008; see also Shumaker *et al.*, 2011), for example, propose that a tool can be said to mediate sensory input (as the sponge does for dolphins), mediate communication (as a stick might do when a chimpanzee waves it in the air while displaying to another chimpanzee), or change another object (as a stone might do when a chimpanzee uses it to crack a nut).

Figure 8.5 Bottlenose dolphin carrying a marine sponge.

8.3.2 Ecology and tool use

Very often, animals use tools to obtain food. It is more likely that a species that engages in **extractive foraging** will be found to engage in tool use than a species that does not have to physically extract a food source from the environment (e.g. a shell or crevice; Shettleworth, 2010). Ecological conditions, thus, can favor tool use in some species over others. Shettleworth also suggests that for some species, a lack of competition from other species that have anatomical adaptations for extracting food (e.g. the stout bill of a woodpecker) may create a niche that can be exploited with a specialized behavior (e.g. the use of probing sticks for insects by some finches).

Capuchin monkeys of South America provide another clue to the role of ecology in tool use behavior. Even though Brazilian folklore and reports from the sixteenth century spoke of capuchin monkeys using stones to crack open nuts, it was not until recently that field researchers were able to document the behavior (Ottoni & Izar, 2008). For some time, researchers had thought that this type of tool use would be unlikely for capuchins; most members of the genus *Cebus* are primarily arboreal, a niche that does not readily support stone tool use. However, some capuchin species live in savannah-like environments, spending much time on the ground. One species of capuchin monkey (*Cebus libidinosus*), for example, has been observed frequently using stones for nut-cracking by placing a nut on a solid surface, lifting a stone high, and then dropping it forcefully onto the nut (e.g. Waga *et al.*, 2006; Figure 8.6). The behavior of this species of capuchin monkey thus suggests that ecological variables such as terrestrial habitat can influence the evolution of tool use; life on the ground allows for encounter with stones and fallen nuts, as well as potentially a scarcity of other types of foods. Some researchers have additionally posited that some forms of tool use may depend on social factors such as tolerance of the presence of others to allow for social transmission from generation to generation, a topic that will be discussed in greater detail in Section 8.4.

Figure 8.6 Capuchin monkey using a stone to crack open a nut.

8.4 Development of tool use

Examining the development of a behavior can lead to insight about the cognitive processes that underlie the behavior. For example, some species will modify objects in their environment to produce tools (**tool manufacture**), but the development of this behavior can take many forms. It is possible that an individual comes to make effective tools via trial-and-error learning; some actions may lead to a tool that yields a food reward and thus will be repeated. Alternatively, some aspects of the behavior may be learned by observing others, implying a type of social learning (see Chapter 13). It is additionally possible that through the learning process, a representation of the causal structure is formed.

To explore the development of tool manufacture, this section will first detail research that has focused on the creation of probing tools by New Caledonian crows (*Corvus moneduloides*). Then, a second 'case study' of development will be presented, that of nut-cracking in tufted capuchin monkeys. While nut-cracking does not involve the manufacture of tools, it is a complex behavior that takes years to master. Observations of this behavior have led some researchers to propose that learning to use tools from older members of a group may require certain other behaviors and dispositions that optimize social learning.

8.4.1 Tool manufacture in New Caledonian crows

New Caledonian crows are endemic to the islands of the New Caledonia archipelago. They have been observed in the wild to use a variety of tools including sticks and pieces torn from leaves to forage in holes for insects (e.g. Hunt, 1996; Hunt *et al.*, 2001). Two observations in particular have suggested that young birds learn to create and use tools at least in part through observing others: the shape of tools varies across New Caledonia, and juveniles remain near parents (Bluff *et al.*, 2007).

Documentation of behavior in the wild has been combined with laboratory experiments to start to elucidate how young birds become effective tool users. In one study, young birds were raised in the laboratory, with one pair able to observe human caretakers using sticks to probe for food (though not tool manufacture) and another pair that had equivalent interaction with humans but no exposure to tools. Both pairs began to use tools at the same time, but the two birds that saw actions with tools spent more time in tool-related activities (Kenward *et al.*, 2005, 2006).

Thus, at least with stick tool use, New Caledonian crows do not require input from tool use observations before beginning to use tools themselves. Interestingly, the authors also noted that effective food retrieval was preceded by **precursor behaviors** that contained elements of mature tool use behavior. For example, the young birds would hold a stick in their beaks and move their heads as if probing a hole, even though no hole was present and they were simply touching the stick to a hard surface. It is possible that these behaviors play a role in the learning of tool use, though it should be noted that no food reward for the behavior was occurring. The existence of precursor behaviors has suggested to some researchers that the development of tool use in New Caledonian crows may start from an inherited trait or behavioral template, a proposal further supported by examining the manufacture of tools by tearing pieces off leaves.

In the wild, New Caledonian crows tear off sections of leaves from trees of the genus *Pandanus* and fashion probing tools (e.g. Hunt & Gray, 2003). Kenward *et al.* (2006) found that in the laboratory, birds that had not observed tool manufacture from leaves still engaged in tearing behavior, but the resulting tools were crude relative to those created by mature birds in the wild. It is possible, then, that for this species, tool manufacture and use rest on a foundation of biological predispositions to engage in activities with the tool materials, yet elaboration occurs through subsequent trial-and-error learning, social learning from others, or some combination of both.

A remaining question concerns what aspects of New Caledonian crow tool use, if any, are represented in a causal manner. Does mature tool manufacture result from the development of a causal representation of the relationship between the tool and the probing action? Much research is still needed to answer this question, though the novel tool manufacture behavior of a New Caledonian crow named Betty engendered interest. During a laboratory study, one member of a pair of birds flew to another part of the enclosure, taking with him the hook-shaped wire that was being used to pull a container holding food out of a vertical tube. The remaining bird, Betty, was then observed to take one of the straight wires that remained and try, unsuccessfully, to lift out the food container. At this point, she bent the wire to make a hook-like shape and proceeded to obtain the food (Weir *et al.*, 2002; Weir & Kacelnik, 2006). Although this would appear to be good evidence for causal understanding, the fact that on subsequent trials Betty tried to lift the food container with straight sticks before changing her behavior to bend the material questions that interpretation.

8.4.2 Social tolerance and capuchin nut-cracking

As noted in Section 8.3, wild capuchin monkeys have been observed to use stones to crack open nuts. The nut is placed on a hard substrate (the 'anvil'), and a 'hammer' stone is lifted high and then pushed onto the nut. de Resenda *et al.* (2008) report that proper, effective nut-cracking is not achieved until around 3 years of age, yet earlier in development, juveniles attend to nut-cracking by members of their group. These observations have led researchers to propose that certain behaviors might help to create an optimal situation for social learning of tool use (e.g. Ottoni & Izar, 2008). First, young capuchins are drawn to the nut-cracking behavior of others likely because of the small pieces of food that become available to scrounge. The association of the tool-use activity of others and the presence of food would likely encourage continued attention to the location.

Second, the nut-cracking individual must tolerate the presence of a scrounging juvenile. Young capuchins attend to nut-cracking not just by their mothers, but also by adult males and other juveniles. A certain degree of **social tolerance** is necessary, then, to allow for young members of the group to engage in social learning of tool use, a proposal that has been generalized to other tool-using species (e.g. van Schaik *et al.*, 1999). That said, some researchers have also argued that an opportunity for innovation occurs when at least some of the individuals of a group are foraging alone or are less tolerated by others (e.g. Boinski *et al.*, 2003; Kummer & Goodall, 1985). Thus, a tool-using species like the tufted capuchins might represent a balance between social toleration for scrounging/proximity of naïve individuals coupled with factors that create opportunities for innovation.

8.5 Causality and tool use

The observation that many animal species frequently use tools has motivated researchers to examine whether animals infer the causal relationships among a tool's structure, the aspect of the environment acted upon, and the outcome of the action. An alternative that has most commonly been contrasted with this possibility is that a tool-using individual might engage in these behaviors as an outcome of some form of associative conditioning. As discussed earlier, some researchers have proposed that organisms use associative mechanisms to learn about causal relations, so in some ways, the dichotomy is false. Yet, when considering complex physical problems with tools, the question is whether the tool-user goes beyond learning that a tool, manipulated in a certain manner, causes the availability of a reward to realizing *why* it causes the food outcome. This latter understanding would, for example, allow for detection of analogies between tasks and thus quick determination of solutions (e.g. understanding why a hammer can insert a nail may help you also to realize that it can break a window).

Two main types of research approaches have been used to examine this understanding. The first, 'trap-tube tasks', examine the understanding of the potential outcomes of using a tool given particular environmental features, which in turn explains something about how the tool-user construes the causal relations of the situation. The second approach simply asks subjects to pick one tool among many to solve a particular task. When the tool options are systematically controlled to either possess or not possess effective physical features, researchers can examine whether these features are relevant to the subject. Both research approaches are discussed in the next two sections.

8.5.1 Trap-tube tasks

The **trap-tube task** has been used extensively in the study of physical causal reasoning. Food is placed in a horizontal, transparent tube that has a short, vertical tube (a 'trap') attached to it (Figure 8.7). Subjects must use a stick to push or pull the food, from one side or the other of the tube, in a manner that avoids the hole that leads to the trap. The dependent measure is typically how long it takes until the trap is avoided consistently across trials. (Follow-up tasks are then used to determine what underlying learning and decision processes led to the correct use of the stick tool.) The hole and the base of the trap are two causally relevant features; the former is relevant because objects move horizontally over continuous surfaces, and the latter contains the food because of its

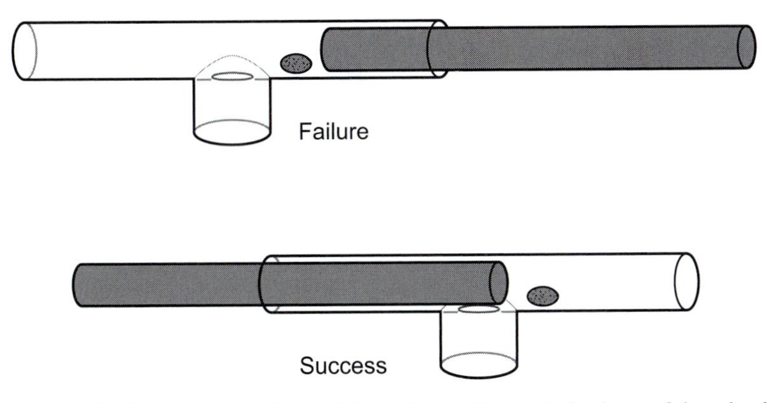

Figure 8.7 The trap-tube task. A stick can be used to push food out of the tube, but to avoid the trap, it is critical to insert the stick in the correct end of the tube.

solidity (Taylor *et al.*, 2009). In versions of the task in which multiple sticks were provided, the length and width of the stick are also relevant causal features.

In some of the first experiments using this task, capuchin monkeys and chimpanzees learned to successfully avoid the trap; they pushed the food away from the trap and out of the tube. However, in follow-up experiments in which the trap-tube was inverted (and thus the trap was no longer relevant), they continued to insert the stick as if to avoid the trap, suggesting that they were not using causal reasoning to solve the task (Limongelli *et al.*, 1995; Povinelli, 2000; Visalberghi & Limongelli, 1994). Instead, they may have learned through operant conditioning that certain behaviors yielded food (e.g. pushing the food away from the hole/trap).

Later studies using the trap-tube task have pointed to some of its limitations. First, it is possible that the action itself (i.e. pushing food away from you) may pose additional inhibitory challenges above and beyond causal reasoning. Studies with chimpanzees, orangutans, and woodpecker finches (*Cactospiza pallida*) suggest that this might be the case; if the tool allowed the food to be pulled out of the tube, the trap was not avoided in the follow-up inverted tube trials (Mulcahy & Call, 2006a; Tebbich & Bshary, 2004). Second, 'correct' performance when the tube is inverted is simply random behavior, which can be interpreted in many ways, including fatigue or disinterest. There is also no reason to alter behavior when the trap is no longer functional; there is no cost to inserting the stick in either end of the tube. In fact, even adult humans tend to avoid the inverted trap when tested in an analogous manner (Silva *et al.*, 2005).

To address some of these concerns, other studies have approached the trap-tube paradigm in a different manner. The method involves examining the ability to transfer performance from one task to another task that has an analogous causal structure but non-causal stimuli are changed. If transfer between tasks occurs, subjects perform correctly very soon after being presented with the new task, and the hypothesis that animals are engaging in causal reasoning is supported because rule-based behavior or reliance on stimulus generalization is less likely (e.g. '**triangulation**'; Heyes, 1993). That is, transfer implies some abstraction of the similar causal structure of the two tasks. Using this paradigm, however, has yielded mixed results. Rooks (*Corvus frugilegus*), a bird species that does not use tools in the wild, did not quickly solve a new task after reaching reliable performance on the trap-tube task (except for one subject out of the seven tested; Seed *et al.*, 2006). In another study, none of the four ape species showed quick transfer between tasks (Martin-Ordas *et al.*, 2008).

New Caledonian crows showed slightly better transfer, with three out of six subjects quickly solving a new task (Taylor *et al.*, 2009). In that study, crows were presented with a two-trap task in

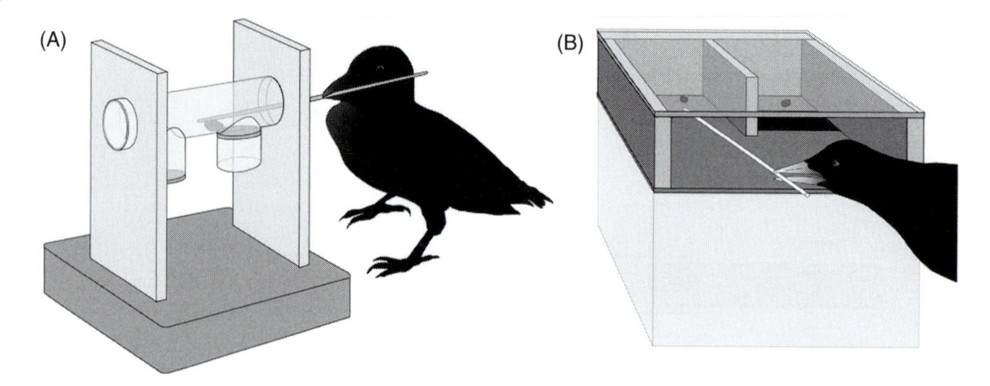

Figure 8.8 Drawing of the apparatus used with New Caledonian crows. (A) The two-trap task. (B) The trap-table task.

which one of the 'traps' was non-functional because it was covered with a solid surface (Figure 8.8). After demonstrating consistent successful performance on the tube task, the crows were presented with a 'trap-table' task in which they had to choose one of two tools with which to pull in a piece of food. Pulling one tool would result in the food falling into a trap while pulling the other tool brought the food toward the crow over a continuous surface. Thus, the visual features of the trap-table task differed from the trap-tube task, but the tasks were analogous in terms of the causal relations. The fact that three of the birds showed transfer by quickly solving the trap-table task is intriguing, yet the differences in performance across the sample of birds suggest that more research is needed.

8.5.2 Choosing the right tool

Another approach to addressing the question of what, if any, causal understanding underlies animal tool use requires subjects to pick one tool among many to solve a particular task. In experiments like this, the tool options are controlled such that they possess or do not possess physical features that allow the attainment of goals. Some have proposed that these types of tasks are less demanding than the trap-tube tasks and more comparable to the actual discriminations that tool-using species must make in the wild (e.g. Fujita *et al.*, 2003). For example, without the trap and with various sticks to choose from, the tube task can examine whether animals recognize the relevant tool features that enable access to the food (e.g. length, width, rigidity). Visalberghi, Fragaszy, and Savage-Rumbaugh (1995) provided a baited tube along with a bundle of sticks and a stick with an inappropriate 'H'-shape to capuchin monkeys and three ape species (bonobos, chimpanzees, and orangutans). Each of the species ultimately took one of the sticks from the bundle to solve the task, but only the capuchin monkeys also attempted to use ineffective tools such as trying to push the entire bundle into the tube. In a similar task, New Caledonian crows picked appropriately long sticks to obtain food from a tube and would modify sticks by removing horizontal branches to create a tool if necessary (Bluff *et al.*, 2007).

Other studies have examined what tools animals will choose after initial training with one type of tool. After learning to use a certain cane-shaped tool to pull in an out-of-reach piece of food, tamarin monkeys, which have not been observed to use tools in the wild, reliably chose other cane-shaped tools, even if the canes differed in texture or color from the training tool (Figure 8.9; Hauser, 1997). These results suggest that the tamarins recognized which feature was relevant to the task (shape) and

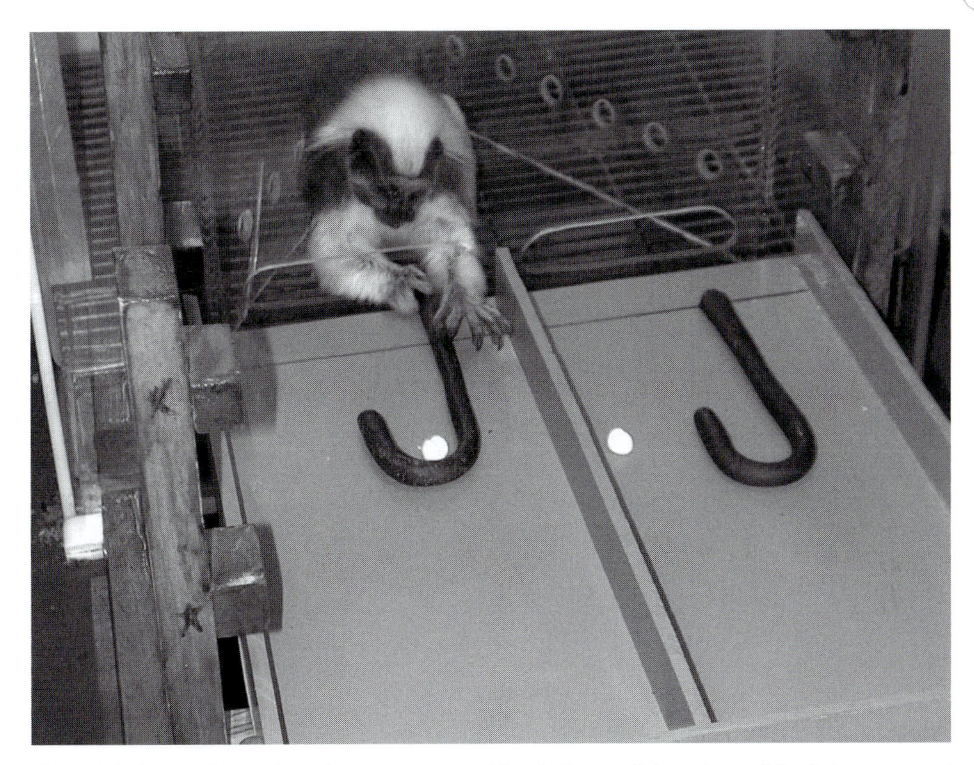

Figure 8.9 A tamarin monkey choosing a cane with which to pull in a piece of food. Across experiments, tamarin monkeys reliably chose other cane-shaped tools, even if the canes differed in texture or color from the training tool.

disregarded changes to other, non-causal features like color and texture. Since this study, other primate species have been tested and similar results obtained (e.g. Santos *et al.*, 2006).

8.5.3 Conclusions

The research presented in this chapter has addressed, but not solved, the question of causal, physical understanding in animals. Many researchers have proposed that the either/or framework (i.e. either associative learning or human-like inference) is not an effective way to attempt to categorize the causal cognition of other species (e.g. Chappell, 2006; Penn & Povinelli, 2007). After all, some causal understanding may result from initial associative learning, and even when it cannot be attributed to conditioning, it may be premature to assume it is like human inference. As humans, we deliberately intervene to diagnose causal dependencies; we experiment. Of course, cognitive scientists are still working out the best way to model human causal understanding, and even causal Bayes nets appear to have their limitations (e.g. Tenenbaum *et al.*, 2007).

The research field now accepts that animals understand that certain features of objects are important in tool-using, and when subjects show the ability to transfer knowledge from one tool-using situation to another, it would appear that some analogy has been made between the causal structure of each situation. However, there is still little documentation of this transfer ability, and further research is needed. In contrast, humans generalize freely from one situation to another and improvise; a textbook can be used as a doorstop or a rock as a hammer (though when a tool has conventional status in human culture, improvisation is slowed down by knowledge of its intended function). Regardless of potential differences between humans and animals, one important conclusion from the present research is that many animals have evolved complex behaviors and understanding of the physical world that are far beyond our initial assumptions. Even if these abilities are determined to be the result of relatively 'simple' cognitive processes, it would be cause for updating our understanding of the basic properties of cognition.

Chapter Summary

- The philosopher Hume famously argued that causality is an illusion. We cannot observe causality, but we nevertheless hold causal beliefs about connections among things.
- Some aspects of conditioning parallel aspects of causality, and it has been proposed that humans and animals use associative mechanisms to learn about causal relations. By this account, causal learning can be thought of as a byproduct of associative learning.
- However, it is not easy to explain all experimental findings with associative models. There is some evidence that rats can also understand causal relations in a manner consistent with causal Bayes net theories. This ability undermines most associationist accounts and presents a far more complex view of animal causal cognition than was previously available.
- Many researchers, particularly in the field of cognitive development, have considered the topic of causality to include the understanding of basic object physics. Recognition of basic aspects of object physics allows for prediction of the outcome of object interactions. Studies tapping into the understanding of object physics have a rich history in the study of infant cognitive development, and many of these paradigms have been modified for use with animals.
- Tool use is one of the central areas of interest within the field of comparative cognition. Researchers realized early on that the study of tool behaviors would require clear definitions of what did and did not characterize true tool use.
- Ecological conditions can favor tool use in some species over others. It is more likely, for example, that a species that engages in extractive foraging will be found to engage in tool use than a species that does not have to physically extract a food source from the environment.
- Some species will modify objects in their environment in order to manufacture tools, but the development of this behavior can take many forms. The existence of precursor behaviors has suggested to some researchers that the development of tool use in some species may start from an inherited trait or behavioral template, yet elaboration occurs through subsequent trial-and-error learning, social learning from others, or some combination of both.
- A certain degree of social tolerance may be necessary to allow for young members of a species to engage in social learning of tool use. That said, some researchers have also argued that an opportunity for innovation occurs when at least some of the individuals of a group are foraging alone or are less tolerated by others.
- Animals appear to recognize that certain features of objects are important in tool-using, and when subjects show the ability to transfer knowledge from one tool-using situation to another, it would appear that some analogy has been made between the causal structure of each situation. However, there is still little documentation of this transfer ability, and further research is needed.

Questions

1. Why did the philosopher, David Hume, propose that causality is an illusion?
2. What differentiates natural causality and arbitrary causality?

3. Draw a causal model in graph form of the following fictitious scenario derived from Waldmann and Hagmayer (2005): Researchers hypothesize that insect bites cause the production of the substance Pixin. Pixin causes the substance Xanthan, which causes the rise of the levels of both Sonin and Gastran. Pixin is also assumed to increase the level of Histamine, which generates Gastran.
4. Describe the procedure of an object displacement task and define the cognitive process that it has been designed to measure.
5. To examine the understanding of solidity, you first show rhesus macaques a table with a bucket on top and directly below. You then hide the table and buckets with an opaque screen and drop a ball over the top bucket. Upon removal of the screen, where, on average, will the monkeys likely search for the ball?
6. Why might dolphin sponge-wearing behavior not have fit Beck's original definition of tool use?
7. What ecological conditions might favor the evolution of tool use?
8. Why have some researchers proposed that aspects of social behavior are key factors in a species' tool use?
9. Design a task that has an analogous causal structure to the trap-tube task but has different non-causal features.
10. What reasons might you propose for why animals may not exhibit functional fixedness?

FURTHER READING

Key readings in causal cognition include:

Penn, D. C., & Povinelli, D. J. (2007). Causal cognition in human and non-human animals: A comparative, critical review. *Ann Rev Psychol*, **58**, 97–118.

Sperber, D., Premack, D., & Premack, A. J. (eds.) (1995). *Causal Cognition: A Multidisciplinary Debate*. Oxford, UK: Oxford University Press.

A range of papers using associative and cognitive frameworks on predictiveness and causality can be found in a special issue of the journal *Learning and Behavior*:

Allan, L. G. (2005). Learning of contingent relationships [Special issue]. *Learning & Behavior*, **33**(2).

Other models of causal learning that were not included in this chapter include the ΔP model and the Power PC model:

Allan, L. G. (1980). A note on measurement of contingency between two binary variables in judgment tasks. *Bull Psychonomic Soc*, **15**, 147–149.

Cheng, P. W. (1997). From covariation to causation: A causal power theory. *Psychol Rev*, **104**, 367–405.

Tool use has been the focus of many literature reviews, books, and chapters. Some key readings include:

Beck, B. B. (1980). *Animal Tool Behavior: The Use and Manufacture of Tools by Animals*. New York, NY: Garland STPM Press. *See also*: Shumaker, R. W., Walkup, C. R., & Beck, B. B. (2011). *Animal Tool Behavior* (Revised and updated Edn.). Baltimore, MD: Johns Hopkins University Press.

Povinelli, D. J. (2000). *Folk Physics for Apes: The Chimpanzee's Theory of How the World Works*. Oxford: Oxford University Press.

Sanz, C., Boesch, C., & Call, J. (eds.) (2013). *Tool Use: Cognitive Requirements and Ecological Determinants*. Cambridge: Cambridge University Press.

9 Categorization and concept formation

Background

Children sort blocks into different piles, birds cache food in different locations, and ants attack ants from different nests. All of these examples show how organisms attempt to organize stimuli and events in their environment. The tendency of humans to systematically arrange their natural world can be traced back to the ancient Greeks, evidenced by Aristotle's biological classification systems. This basic schematic was revised and redrawn over the centuries as scientists and amateurs grouped both plants and animals according to different sets of rules. This often resulted in large catalogue-like descriptions that were completely unworkable, particularly as an increasing number of specimens were brought back to Europe in the seventeenth and eighteenth centuries from Asia, Africa, and the Americas. The classification process was revolutionized by Carl von Linné, a Swedish naturalist who is better known by the Latinized form of his name, Carolus Linnaeus. In 1735, Linnaeus published his first edition of *Systema Naturae* which categorized the hundreds of plants and animals that he and his students had identified. The tenth and final version of this work (1758) described more than 4000 animals and 7000 plant species.

Linnaeus' taxonomy was neither complete nor flawless, but his contribution to modern science was enormous. Although others had used the binomial system of species names, Linnaeus was the first to do so consistently. This allowed scientists to communicate about the species they were studying, and facilitated the development of many disciplines, most notably zoology and botany. Another of Linnaeus' innovations was to categorize plants and animals based on similarities, rather than differences, creating a hierarchical arrangement of species that is a template for contemporary schematics. For example, even though he was writing one hundred years before Darwin (and believed that species were fixed entities created by God), Linnaeus produced the first draft of an evolutionary tree pointing to genetic relatedness. As historians have noted, "one of the most impressive things about Linnaeus was his willingness to change his mind when new information reached him" (Nicholls, 2007, p. 256). This led to a constant revising of his classification system such that some animals (including his own pet raccoon) were moved repeatedly from one genus or species to another.

Linnaeus' taxonomy underwent major modification with the rise of evolutionary theory and the emphasis on grouping organisms based on common descent. The entire process was revolutionized in the twentieth century when it became possible to sequence the DNA of living (and sometimes extinct) species. Using this evidence, chimps replaced gorillas as the closest living relative of humans. Despite these advances, conflicts arose when similarities in observable traits did not match genetic or evolutionary data. Nonetheless, the validity of Linnaeus' original premise remains: groups of animals at both the top and bottom of the taxonomic hierarchy share similar traits.

Chapter plan

Besides the obvious connection to comparative research, Linnaeus' classification system is profiled in the opening of this chapter because his system of taxonomy reflects the cognitive processes of categorization and concept formation. In its simplest form, categorization groups similar things together, such as different animals that look the same. The chapter opens with a brief description of important terms and experimental paradigms in categorization research, and then highlights the evolutionary significance of this cognitive process. The bulk of the chapter is devoted to describing four types of categorization: perceptual, functional, relational, and social. Grouping items together using one of these processes allows newly encountered stimuli to be placed within an existing framework (an important outcome of Linnaeus' work). As with biological taxonomy, categorization is an ongoing process: if new items do not fit into one category or another, individuals may redefine how the categories are (just as scientists continually modify the criteria that define a particular species). This chapter attempts to explain how this cognitive process operates, highlighting both similarities and differences across species.

One of the greatest appeals of Linnaeus' classification system is that other individuals (both scientists and non-scientists) can immediately grasp information about an organism, even one that is completely unfamiliar to them. Knowing that a particular animal is a mammal allows one to infer that it has hair, mammary glands, and three (rather than one) inner ear bones. This ability to extrapolate unobserved, or even unobservable, characteristics of novel stimuli is discussed under the topic of concept formation. Three prominent theories of concept formation (elemental, exemplar, and prototype) are then used to explain different aspects of perceptual, relational, functional, and social categorization. The chapter concludes with a description of category-specific semantic deficits, a human condition produced by brain damage or atrophy of particular regions of the cerebral cortex. The relationship between cognitive impairments and affected neural regions in this disorder provides insight into brain mechanisms that mediate both categorization and concept formation.

9.1 Fundamentals

Researchers from many different disciplines examine how and why organisms systematically arrange the world around them. As with any scientific endeavor, the work is facilitated by a common terminology and agreement on how to measure these cognitive processes. Although debates are ongoing, research in this area has helped scientists to formulate both proximate and ultimate explanations for categorization and concept formation.

9.1.1 Definitions

Categorization is the cognitive process of classifying items or events into groups based on one or more common features. The converse, **discrimination**, is distinguishing items or events based on one or more distinct features. A critical aspect of categorization is that items within a category must be distinguishable, even though they share common features. In other words, a category is not a collection of identical elements, but a group of separate units that are held together by common characteristics. For example, stimuli that fall into the category 'circle' may come in an infinite number of colors and sizes, whereas a collection of circles that are exact replicas of each other would not constitute a category.

The ability to categorize is intimately linked to **concepts**, the abstract set of rules that define membership in a category, or the mental criteria which allow stimuli, events, or even ideas to be grouped together. Categories, and the concepts that define them, often change with experience. This experience may involve direct interaction with the items that make up a category and/or increased knowledge of the criteria that define group membership. Consider how children learn to divide animals into pets, farm animals, and wildlife. They may draw on their own experience in different situations (e.g. visiting a farm or zoo), use information acquired from picture books or TV, or rely on adult explanations of 'where' these animals live. At a later age, most children learn the list of criteria that define the subgroup, mammals. The subsequent ability to categorize animals as mammals or non-mammals depends on knowledge of mammalian characteristics (e.g. mammary glands and hair on body), acquired through specific instruction. These very different means of learning new information allow children to refine their understanding of animals and how they may be subdivided into separate groups. This cognitive process of establishing and updating the abstract set of rules for category membership is referred to as **concept formation**.

Importantly, categorization may occur in the absence of concept formation. This may sound counterintuitive, however the ability to group similar items together is not the same thing as understanding the criteria, both necessary and sufficient, that define that group. Forming a concept of a particular category allows an individual to infer properties about items in that category which may or may not be apparent on initial exposure. For example, on any given day, most people encounter a number of new objects that resemble large steel boxes with wheels. With only brief visual exposure to the exterior, these items are categorized as cars. Immediately, an assumption is made that each object has a steering device, an internal mechanism that requires an energy source, and that it is used to travel from one place to another. It is not necessary to see the interior or to view the object moving to make these attributions. In some cases, the categorization may be incorrect – the characteristics that define a category may be vague or require further refining. Perhaps the object is a minivan and not a car; the concept of 'cars' would have to be modified accordingly (e.g. does not include larger multi-seat vehicles with hatchback openings). An important distinction between categories and concepts, therefore, is that the latter involves mental representation of group attributes that do not depend on immediate interactions with the environment.

9.1.2 Measuring categorization

Assessing categorization in humans can be as simple as asking them to sort playing cards according to some rule (e.g. suit or face value) or to press the space bar on a computer when presented with a specific class of stimuli (e.g. low versus high tones). The same principle is applied to animal experiments in which subjects display categorization by responding in the same way to items within a category. Numerous examples of this experimental design were presented in previous chapters, such as blue jays (*Cyanocitta cristata*) pecking an image on a computer screen when they saw a moth (Chapter 2). Figure 9.1 shows an auditory version of this setup in which a zebra finch (*Taeniopygia guttata*) is waiting for the next stimulus presentation. The bird will indicate how it categorizes recorded conspecific song notes by flying (or not) to a perch located at the opposite end of the enclosure. These go/no-go experiments allow animals to respond 'yes' or 'no' to the question 'does this stimulus belong in category X'? The task may be modified by training animals to make different responses for each category: fly to the left perch if the stimulus belongs in category X and to the right one if it belongs in category Y.

Laboratory experiments such as these help researchers to understand how animals are categorizing stimuli. Frequently, a novel **exemplar** (a distinct item within a category) is presented in a small subset of trials; this manipulation tests whether animals are classifying stimuli based on common features or simply memorizing each item in a category. In Figure 9.1, zebra finches correctly classified notes that they had never heard, suggesting that they were able to identify similarities between novel and familiar exemplars. Carefully controlled experiments may also distinguish cognitive mechanisms based on item classification from those based on broad discrimination of a category boundary. For example, Polynesian field crickets (*Teleogryllus oceanicus*) respond differently to auditory stimuli that are above or below a frequency of 16 kilohertz (Wyttenbach *et al.*,

(A)

Figure 9.1 Experimental design (A) and results (B) from an auditory discrimination task in zebra finches. (A) Birds were trained to wait on a perch until a zebra finch call note was presented through a speaker. Notes were classified as A, B, C, or D according to acoustical properties. Each bird was trained to fly to a perch at the other end of the enclosure when it heard notes from one category (but not the others). (B) Over trials, birds correctly classified notes in each category, responding to only those notes within the assigned category.

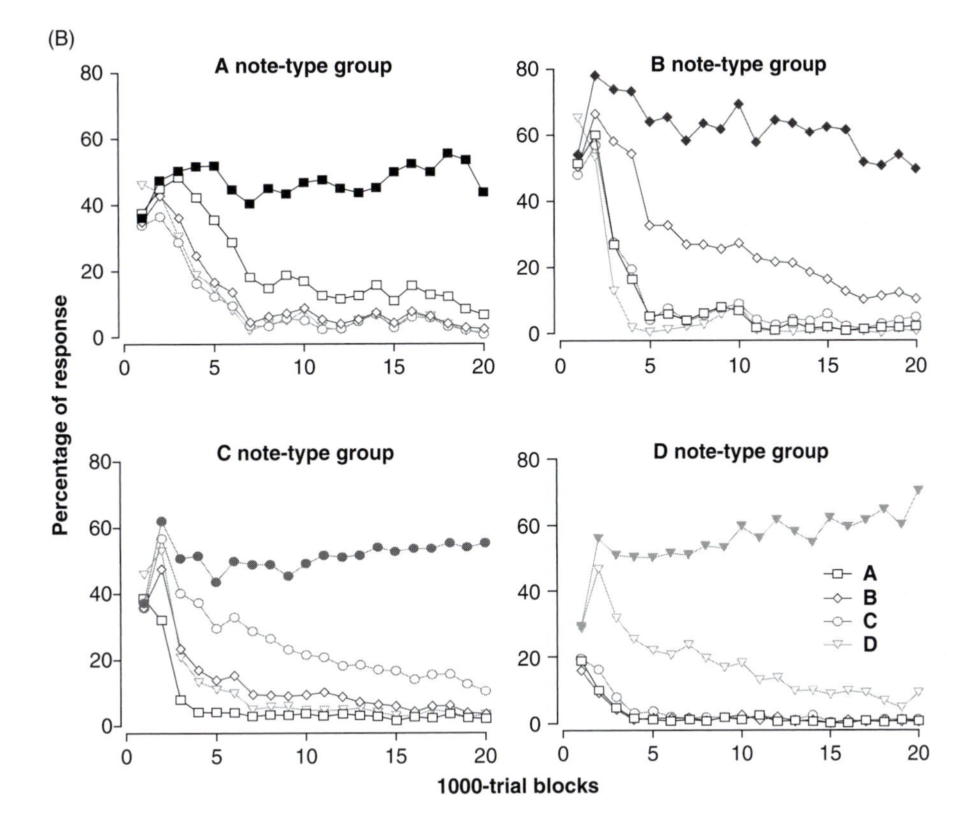

Figure 9.1 *(cont.)*

1996), but the experiment did not test whether these animals could differentiate tones within each group. Thus, it is not clear whether crickets were classifying sounds based on similarities (i.e. categorization), or dividing stimuli into two large and amorphous groups using a single threshold value. More tightly controlled experiments revealed categorization abilities in other insects: honeybees, ants, and wasps categorize visual stimuli (including novel exemplars) using specific features such as color, pattern, or motion contrast (Avarguès-Weber *et al.*, 2011).

Another experimental method for assessing categorization is to test whether items within a class can be substituted for each other. The assumption behind the test is as follows: if items within a category elicit the same response, altering the response to one exemplar should shift responding to all others within the category. Children who are bitten by a dog are likely to exhibit fear responses to other (perhaps all) dogs, not just the one that bit them. This shows that they have a mental representation of the category dog that includes distinct items within the group. Many animals 'pass' both the novel exemplar and stimulus substitution tests of categorization, but whether this provides evidence for concept formation is much more difficult to establish.

9.1.3 Evolutionary significance

Despite debates on concept formation, most researchers acknowledge that almost every species tested displays some level of categorization (Zentall *et al.*, 2008). Even young infants possess this cognitive ability (Eimas, 1994). The fact that categorization is pervasive in the animal kingdom suggests that it has some adaptive value. The most plausible is that mental resources are reduced if organisms can group stimuli or events together that require the same response. As for the cricket, a simple rubric that divides auditory stimuli into two classes makes it much easier to decide whether to approach or retreat from a sound. A further advantage of categorization is that updating cognitive representations of one item will be transferred to other items in the same category. This allows organisms to use experience acquired in one situation to modify behavior in related ones. Because two stimuli or events will never be identical (if nothing else, the time at which they occur is different), categorization provides a framework for interpreting new experiences. Animals that live in cities often find food scraps in the garbage, allowing them to classify garbage and refuse bins as good food sources. Some theorists go so far as to state that memory, without categorization, is useless (Jackendoff, 1987). Not everyone agrees with this extreme statement, but there is little doubt that an ability to remember specific details, but an inability to extract generalities from these details, is disadvantageous (see Box 9.1).

Box 9.1 A downside to details

Some people have an amazing capacity to remember almost every detail of past events and are often surprised to learn that other people do not share this ability. One of the best known is Solomon Shereshevsky (1886–1958), a Russian journalist who was studied for over 30 years by the neurologist Luria. Shereshevsky ended up in Luria's office quite by accident: his boss was annoyed that he never took notes during meetings and confronted Shereshevsky by suggesting that he was not fulfilling his work obligations. Shereshevsky confided that he was baffled by his colleagues' need to keep written records and revealed that he could recount every detail of his editor's conversations from previous meetings. Fortunately, his boss recognized these cognitive abilities as unique and exceptional, so he referred Shereshevsky to Professor Luria.

Luria documented Shereshevky's aptitude for remembering details, including the fact that he could recite (both backwards and forwards) a string of numbers, close to 100 digits long, more than 15 years after he originally heard it! It goes without saying that he could memorize long lists of words, entire texts in foreign languages, and complex mathematical equations. Amazingly, none of this required any effort on his part. Shereshevsky was different from the hyperthymesic individuals, described in Box 3.6, in that his memory abilities were not limited to episodic events. But like Jill Price, the cost of maintaining all of these details in cognitive store was enormous. Try as he might, Shereshevsky could not forget information and he complained that his mind often felt like a cluttered junk box. Perhaps because of this, he had difficulty using retrieved memories in other cognitive processes. For example, he claimed to have problems recognizing people because their faces were constantly changing. He had vivid mental images of facial details, but most people do not look exactly the same at every encounter, particularly if considerable time has passed between meetings. Shereshevsky also had problems extrapolating from the specific to the general – understanding that dogs could all belong to the same species despite their overt differences. In a similar way, he could understand the literal, but not the abstract, meaning of a text.

Luria (1968) provides a detailed and very sympathetic account of Shereshevsky's cognitive abilities, as well as the emotional and personal challenges he faced. Similar profiles are presented in clinical reports by the contemporary author and neurologist Oliver Sacks, and in the film *Rain Main*, which is loosely based on the life of Laurence Kim Peek (1951–2009). The striking point in all of these portrayals is that an extraordinary memory for detail, with a limited ability to generalize beyond these, is frequently associated with poor social and cognitive adjustment.

Despite the obvious advantages of categorization, this cognitive process comes at a cost. For one thing, the speed and ease with which new information is categorized may lead to incorrect classification or a failure to update the criteria which define a category. In other words, individuals who make quick category decisions based on minimal information may overlook unique and defining features of items within a category. This is reminiscent of perceptual rules of thumb, discussed in Chapter 2 (Section 2.1.4): a cognitive strategy that allows organisms to make cost-effective decisions sometimes results in inappropriate response to, or judgments about, individual stimuli. The same is true of categorization. Stereotypes are a prime example. People often make snap judgments about others based on their sex, ethnicity, or religious displays. The outcome can range from personal insult to widespread bloodshed. As deplorable or maladaptive as these responses may seem, they are offset by the advantages (including a reduced need for computational processing) that come with categorization.

Finally, the evolutionary advantages of categorizing stimuli are magnified when cognitive processing extends to concept formation. A mental representation of a concept, which includes group attributes, provides information about stimuli or events that are not immediately apparent. This allows individuals to predict non-evident from evident properties and to make judgments about events in their environment. As the preceding discussion indicates, categorization facilitates decision making as well as many other cognitive processes that are discussed throughout this book. Some researchers argue that the ability to extend abstract rules about group membership to new situations is a fundamental component *only* of human cognition. As expected, not everyone agrees and the debate is far from settled.

9.2 Perceptual categorization

One of the simplest ways to categorize stimuli is to group them together based on shared physical features. This **perceptual categorization** uses sensory input in one modality, such as vision or olfaction, to identify similarities between different stimuli and to group them together accordingly. Note that perceptual categorization is not defined by the sensory information that organisms can detect (species differences in sensory detection are discussed in Chapter 2, Section 2.1.1). Perceptual discrimination, the reverse of perceptual categorization, uses sensory information to distinguish between items. Because perceptual categorization and discrimination are manifested as similar and different responses to sensory stimuli, the topics are sometimes discussed as stimulus control of behavior.

9.2.1 Stimulus generalization

Experimental investigations of perceptual categorization often rely on stimulus generalization tests. In the simplest of these, subjects are trained to respond to one sensory stimulus (e.g. a light of a particular wavelength) and then tested with a range of stimuli that vary across this sensory dimension. Unlike categorization tests in which subjects make a single response to indicate group membership (i.e. they are rewarded for pecking an image that portrays a moth), generalization tests measure the *strength* of responding to different stimuli. The result is a stimulus generalization gradient (see Figure 9.2). As expected, the training stimulus elicits the strongest response with a gradual decline in responding as the similarity between training and test stimuli is reduced (Guttman & Kalish, 1956). Stimulus generalization gradients, such as those shown in Figure 9.2A, have been generated for many different species using a variety of sensory stimuli. It appears, therefore, that animals have a natural tendency to respond in a similar way to stimuli that share perceptual features.

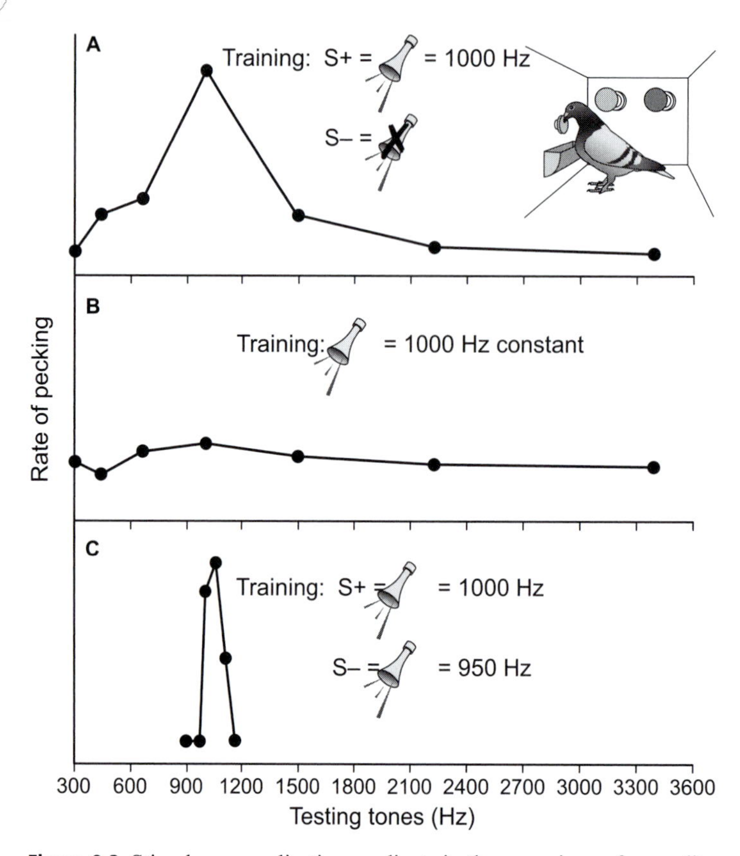

Figure 9.2 Stimulus generalization gradients in three versions of an auditory discrimination test in pigeons. (A) Birds were trained to peck a key for food reward in the presence of a 1000-Hz tone. If the tone was turned off, responding was not reinforced. (B) Pigeons were trained to peck a key for a food reward; a 1000-Hz tone was turned on throughout the entire session. (C) Birds were trained to peck a key for food reward in the presence of a 1000-Hz tone and to withhold responding in the presence of a 950-Hz tone.

A key component of stimulus generalization tests is that they allow researchers to determine how variations in one sensory dimension alter responding. This, in turn, provides clues as to how subjects categorize stimuli with common features. A completely flat stimulus generalization gradient (see Figure 9.2B) indicates that subjects cannot distinguish the sensory properties of different test stimuli OR that they have not learned to differentiate training and test stimuli. The latter possibility can be tested by introducing discrimination training: responses to one stimulus are reinforced whereas responses to another are not (see Figure 9.2C). For example, training pigeons to peck a key when they hear a 1000 Hz tone but not when they hear a 950 Hz tone produces a very steep stimulus generalization gradient (Jenkins & Harrison, 1962).

In essence, discrimination training is providing feedback on how to categorize stimuli according to sensory properties. Over repeated trials in which stimuli from different categories are presented, individuals acquire the ability to detect subtle differences in sensory features. This process forms the basis of many forms of expertise in humans. For example, wine connoisseurs often spend years comparing the sight, smell, and taste of wines from different regions, different grapes, and different vintages. This experience allows them to classify wines according to characteristics that vary along each sensory dimension. The more similar two wines are, the more difficult it is to tell them apart, and the more expertise is required to do so. The same principle can be applied to expert birders who identify species using subtle differences in visual or auditory traits, such as beak shape or call note. As with discrimination training in lab animals, these individuals are using a two-stage process of categorization. First, they focus on one stimulus modality (e.g. the smell of the wine or the sound of the bird call); second, they identify variations within this modality. Perceptual categorization, therefore, illustrates the complementarity of cognitive processing: attention, memory, and sensory discrimination must all work together to accomplish the task of correctly classifying novel stimuli.

9.2.2 Categorization or feature recognition?

The ability of many animals to categorize stimuli based on a single sensory feature, such as the pitch of a tone, often extends to more complex stimuli. Close to 50 years ago, Herrnstein and Loveland (1964) provided evidence that pigeons can discriminate photos that show a human from those that do not (Herrnstein & Loveland, 1964), an effect that was replicated with dogs (Range *et al.*, 2008), monkeys (D'Amato & Van Sant, 1988), and orangutans (Vonk & MacDonald, 2004). Pigeon categorization studies were later extended to photos of water and trees (Herrnstein *et al.*, 1976), with subjects correctly classifying novel exemplars of all three categories. An experiment using **pseudocategories** (random collections of stimuli that have no obvious cohesive feature) indicated that pigeons had not simply memorized which stimuli would be reinforced and which would not (Bhatt *et al.*, 1988). As shown in Figure 9.3, pigeons learned to categorize photos of chairs, cars, flowers, and humans much more quickly than if the items from each category were mixed together to form arbitrary groups (i.e. pseudocategories). As expected, memorizing individual items within a category is much more difficult than learning to group stimuli together based on common features.

It is tempting to conclude that pigeons are solving the categorization task in the same way that most humans would: by identifying the object in each photo. However, even if the two species place items in the same categories, it does not imply that they are using the same criteria to do so. Indeed, by manipulating computer images of the photos, Huber and colleagues revealed that pigeons were relying on visual features, such as texture or shape, to correctly categorize images containing humans (Huber & Aust, 2006). Similarly, monkeys focus on the presence or absence of red patches on facial images to classify photos of humans versus non-humans (D'Amato & van Sant, 1988) and orangutans use the color and placement of eyes to distinguish subclasses of primates (Marsh & MacDonald, 2008). Experiments such as these led many researchers to conclude that perceptual categorization in pigeons (and probably many other animals) results from stimulus generalization of specific sensory elements (Shettleworth, 2010). In other words, the data *cannot* be taken as evidence that animals are grouping photographs into categories, such as one defined as 'human.'

9.2.3 Natural categories

If animals rely on feature recognition to group stimuli together, this cognitive mechanism may be the basis for identifying different classes of prey, predators, and mates. From an evolutionary perspective, therefore, it makes sense to examine which aspects of sensory stimuli animals are predisposed to attend to, and how they use this information to judge group membership (Troje *et al.*, 1999). The same principle emerged as a guiding force in human research, inspired by the pioneering work of cognitive psychologist Eleanor Rosch. Rosch pointed out that many lab experiments of categorization bore little resemblance to how subjects classify stimuli or events in the real world. Based on her cross-cultural work (a different type of comparative approach), Rosch hypothesized that categorization is a fluid cognitive process that depends on interactions with the environment.

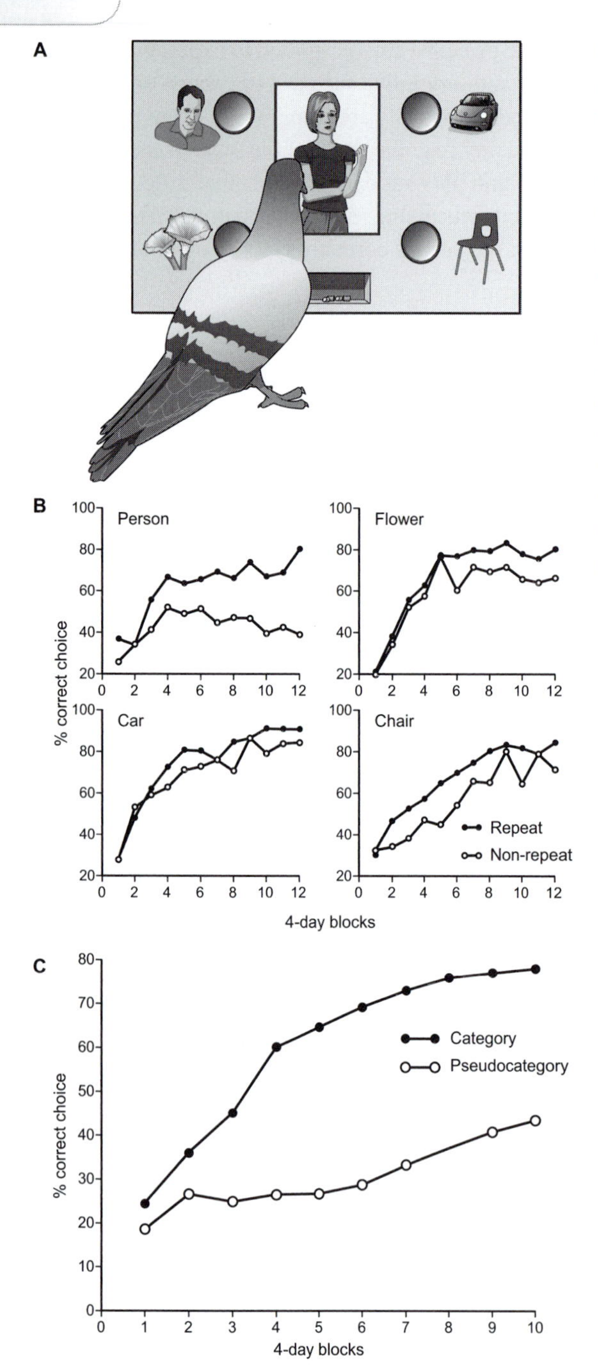

Figure 9.3 Experimental design (A) and results (B) of a perceptual categorization experiment in pigeons (Wasserman *et al.*, 1988). (A) On each trial, a visual stimulus was presented in the center of a screen. Birds were required to peck one of four keys, located in outer corners of the screen, depending on the category of stimulus presented (people, cars, flowers, or chairs). (B) Over training, birds responded at higher than chance levels (i.e. 25%) for each stimulus category. Stimuli that were presented previously (i.e. repeated) were categorized at higher rates than non-repeated stimuli. (C) Pseudocategories were formed by placing the same number of stimuli, randomly selected from each category, into different groups. In other words, each category was made up of a random collection of people, cars, flowers, and chairs. Compared to natural categories, pigeons were impaired at correctly classifying items within pseudocategories.

Individuals may have a mental representation of the criteria that define a category, but these criteria will be updated, constantly, by their experience with real-world events. This led Rosch and her colleagues to advocate a focus on natural, rather than experimenter-defined, categories as a means to understand how organisms make sense of their world (Mervis & Rosch, 1981).

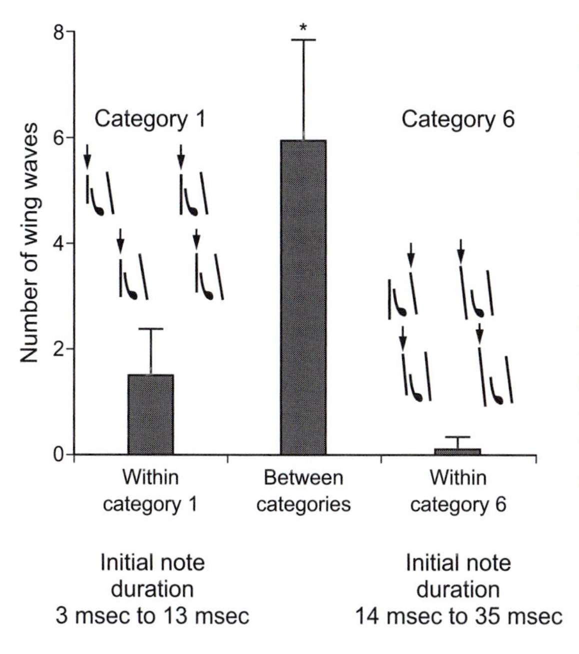

Figure 9.4 Natural categories of birdsong in swamp sparrows (Nelson & Marler, 1989). Song phrases were classified into categories based on the length of the initial note in the phrase (category 1: 3–13 msec; category 6: 14–35 msec). Territorial responses, measured as number of wing waves, were recorded across repeated presentations of phrases. When the phrases were all within one category or the other, response declined (i.e. habituated). In contrast, responses remained elevated when song phrases from categories 1 and 6 were presented across trials suggesting that birds did not treat these as a single category. Importantly, song phrases were selected for the between category presentation such that the differences between the length of the initial notes was less than the difference in the initial note length of song phrases presented in either category 1 or category 6.

One of the best examples of natural category research in animals is the analysis of songbird communication by the neuroethologist Peter Marler (see Chapter 12 for extended discussion of this research). Birdsong, as the name implies, is a complex sound pattern that varies across time (see Figure 9.4). Each song can be described by features such as the frequency range of notes in the song, the number of notes produced, as well as the duration of notes and phrases. Although researchers knew that songs conveyed information about territoriality and reproduction, they did not understand how song features allowed birds to classify this information. For example, which aspect of the song signals whether the singer is an aggressor or a potential mate? Marler and his colleagues addressed the question by testing how alterations in specific stimulus features (e.g. range or number of notes) alter naturalistic behavior. In a sample experiment, they played recordings of the 'average' conspecific song from the area (i.e. they selected the median of each of the identified features) and then modified one feature at a time until a response, such as aggression, was elicited.

The difference between this approach and the categorization experiments described above is that Marler was asking which stimulus variants produce a *meaningful* difference to the animal, rather than testing whether birds are capable of distinguishing two or more variants. In much the same way that crickets divide notes into high and low groups, swamp sparrows (*Melospiza georgiana*) used the frequency of initial notes to classify a song. In other words, as shown in Figure 9.4, birds responded differently to songs beginning with notes that fell above or below a threshold. Most importantly, two notes that were the same distance apart, but did not cross this threshold value, elicited the same response (Nelson & Marler, 1989). This suggested that birds were classifying notes based on a criterion for group membership, and not simply responding to differences along a stimulus continuum. A similar process occurs with speech perception in that humans make categorical judgments about phonemes (the smallest unit of sound in a language) based on threshold values. For example, if recorded consonants are synthetically altered, humans perceive a sudden shift from phonemes such as 'b' to 'g'. Small changes that cross a consonant boundary are labeled as different phonemes, whereas large changes within one category are not (Liberman *et al*., 1957).

Marler's work also revealed that the features songbirds use to categorize other songs depend on the auditory environment. Within a given habitat, birds are likely to hear songs and calls from their own species, as well as those of other species. They must be able to classify the conspecific vocalizations without getting these confused with signals from other birds. Field sparrows (*Spizella pusilla*) living in the Northeastern United States rely on alterations in note frequency, as opposed to other features such as note duration, to categorize conspecific songs (Nelson & Marler, 1990). This makes sense in terms of the local soundscape as note frequency is the thing that distinguishes field sparrow song from other birdsong in the area. Songbirds in other regions may use a different criterion to classify songs, depending on the features of songs emitted by surrounding populations. An important point about this work is that the birds' classification system is based on experience with their natural environment. The cognitive process of grouping meaningful stimuli together is inherited, whereas the specific categories that animals respond to are learned. In humans, perceptual categorization is also a combination of inherited and acquired abilities, although the distinction between the two can be difficult to decipher (see Box 9.2).

The classification of auditory stimuli by songbirds is a prime example of how a continuous variable is partitioned into units that are meaningful to that species. As with all perceptual categorization, this process reduces the demand on cognitive resources and increases the speed and accuracy of responses to environmental stimuli. These evolutionary advantages (particularly in terms of natural categories) help to explain the prevalence of perceptual categorization across species. Although many questions remain, the ability to group stimuli together based on perceptual features appears to be a cognitive trait that is shared by humans and a variety of animals (Freedman & Miller, 2008).

Box 9.2 Perfecting absolute pitch

Absolute pitch (AP), often referred to as perfect pitch, is the ability to identify or produce the pitch of a musical note without an external reference. Most people, even those with no musical training, can distinguish very high from very low notes and many can divide the range of audible frequencies into six or eight different categories (Zatorre, 2003). For those with AP, the skill is much more refined: they can specify *exactly* which note they heard, usually with very little effort. Many can name the pitch of sounds in the environment, such as a fog horn or a bird call. In order to compute these values, individuals with AP must have a mental representation of the Western musical notation system which they use as a reference to classify notes. In that sense, AP is a highly developed form of pitch categorization.

The incidence of AP in the general population is relatively low, roughly 1 in 10 000. It is much higher in musicians, particularly those that began their training at a young age (Baharloo *et al.*, 1998). This correlation could be taken as evidence that AP develops with experience, particularly during a sensitive period, *or* that those with a finely tuned ear are more likely to display an aptitude for music, and to excel as musicians. The 'pre-existing trait' argument is supported by a high concordance rate of AP among siblings and a higher incidence of AP in Asian populations, even when sociocultural factors are accounted for. The current consensus on AP is that it is due to a complex (and as yet unspecified) interaction between genetic and environmental factors (Zatorre, 2003).

AP is studied, almost exclusively, in humans but the ability to identify pitches in isolation may have evolved in other species, particularly if it provided some evolutionary advantage in the past. Songbirds

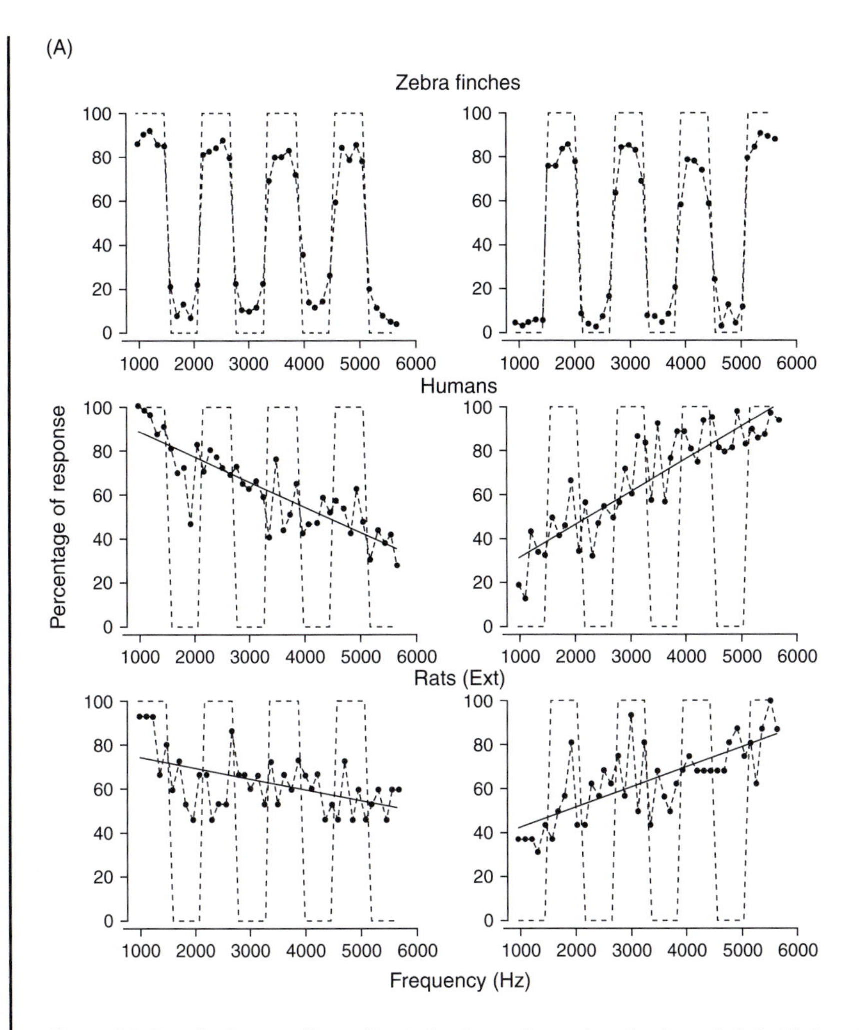

Figure 9.5 Results from auditory discrimination tasks testing absolute pitch in birds and mammals. (A) Mean percentage of responses for zebra finches, humans, and rats classifying tones, equally spaced on a frequency continuum from 1000 to 6000 Hz of different frequencies. The dashed line shows perfect performance with subjects responding to reinforced stimuli (S+) and not responding to non-reinforced stimuli (S–). Rats were tested during extinction (Ext). Reprinted with permission from Weisman *et al.* (2004). (B) The same test conducted with humans who do (right) and do not (left) possess absolute pitch.

are a likely candidate as they have a highly developed auditory communication system and songs are sung within a relatively narrow species-typical range (Nelson & Marler, 1990). Based on this evidence, Weisman and colleagues (2004) developed the hypothesis that pitch identification would be more accurate in songbirds than in mammals. Rather than training zebra finches, rats, and humans to identify specific notes (as in a typical AP test), they developed a task in which subjects were required to classify notes that were equally spaced across the pitch frequency continuum (Figure 9.5). The interesting thing about this paradigm is that it removes any reference to Western notation because the pitch frequencies did not always match a scale note. As predicted, birds outperformed mammals but a later study revealed one notable exception... humans with AP performed as well, in some cases better, than birds

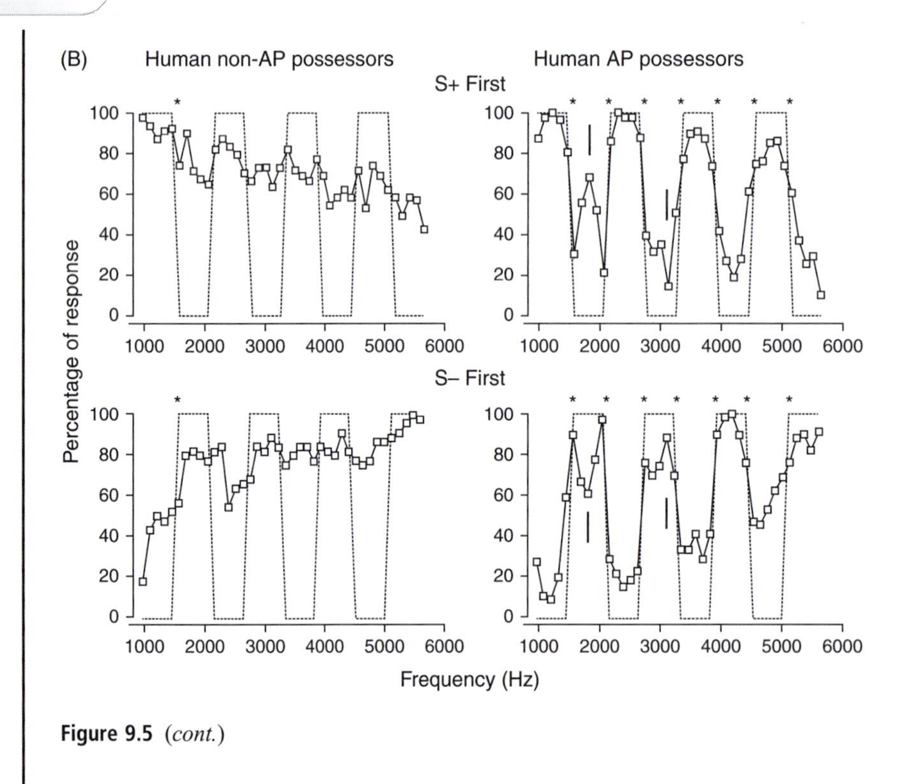

Figure 9.5 *(cont.)*

(Weisman *et al.*, 2010). The authors describe the findings as "puzzling and intriguing," noting that individuals with AP are "an exception to a phylogenetic rule." Why this extraordinary ability emerges in some individuals is still a mystery.

9.3 Functional categorization

Perceptual categorization is a straightforward means of classifying sensory stimuli, but many categories are defined by more abstract criteria. For example, **functional categorization** groups items or events together according to a shared meaning or association. Forks, knives, and spoons belong in one category whereas hammers, wrenches, and screwdrivers belong in another. Although functional categorization does not depend on perceptual properties, items within a functional category may share perceptual features. For example, a 'dangerous' category could include large moving objects but not small stationary ones. On the other hand, the rules for group inclusion in functional categories do not necessarily reflect physical similarities between exemplars. Items of cutlery are similar in size, shape, and color, but if the category were expanded to 'meals' it could include such diverse items as tables, cooking odors, and dinner bells. Because functional categorization often involves a common associate (in this case food), it is sometimes called associative categorization.

Functional categorization is more cognitively demanding than perceptual categorization so it is not surprising that the latter precedes the former in human development (Rakison & Poulin-Dubois,

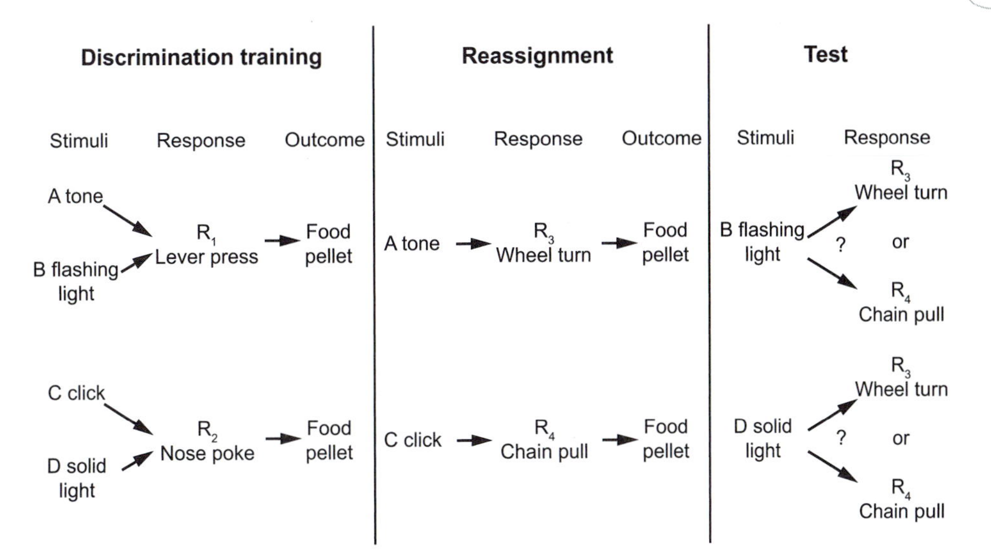

Figure 9.6 Experimental design for equivalence testing used to assess functional categorization in non-verbal subjects. During discrimination training, subjects associate two different stimuli (A and B or C and D) with one response or another (R_1 or R_2). During the reassignment phase, one of the original stimuli from each group (i.e. A and C) is associated with a new response (R_3 or R_4). During testing, the remaining two stimuli are presented (B and D) and subjects have the option of performing R_3 or R_4.

2001). Determining when this ability emerges across phylogeny is more difficult because many functional categories (as defined by human experimenters) will not have the same meaning to other species. Despite the challenges, there is strong evidence that monkeys and other non-human primates have functional categories for food (Fabre-Thorpe *et al.*, 1998; Inoue *et al.*, 2008) and animals (Roberts & Mazmanian, 1988), presumably due to the biological significance of each. Other animals may exhibit a consistent response to groups of stimuli such as 'predators' or 'prey,' but the response is often elicited by a single sensory feature (e.g. a red belly on stickleback fish that elicits aggression) so it does not reflect true functional categorization (i.e. grouping distinct items together based on meaning or association).

Given these difficulties, researchers have developed alternative ways to examine functional categorization in animals. One approach is to test functional equivalence: the understanding that different forms of an item represent the same thing. In other words, a picture of a dog, a description of a dog, and a real dog belong in the same category (Sidman, 2000). Pigeons (Watanabe, 1993) and non-human primates such as monkeys (Zuberbühler *et al.*,1999), baboons (Bovet & Vauclair, 1998), and chimpanzees (*Pan troglodytes*) (Savage-Rumbaugh *et al.*, 1980) all exhibit picture–object equivalence for food versus non-food categories. Along with chimpanzees, California sea lions (*Zalophus californianus*) (Kastak *et al.*, 2001) and common bottlenose dolphins (*Tursiops truncatus*) (Marino, 2002) are capable of extending this functional equivalence to arbitrary symbols, which is the basis for written and spoken language in humans.

Functional categorization may also be examined using a stimulus equivalence test (see Figure 9.6) in which stimuli within a group are associated with a common outcome, either the same response or the same reinforcer. For example, rats may be trained to press a lever whenever an exemplar from groups A or B are presented and to nose poke whenever items from groups C or D are presented. In a subsequent reassignment phase, they learn a new response for items in groups A and C. The critical test occurs following reassignment: do subjects now make the new response to items in groups B and D? If so, they have demonstrated stimulus equivalence in that changing the meaning of one category member alters the significance of all members. Experiments using this basic set up have revealed stimulus equivalence in humans and some animals (Urcuioli, 2006), with most of these tests using pigeons as subjects. In any case, the acquisition of stimulus equivalence does not depend on language competence (Carr *et al.*, 2000), supporting the idea that this ability need not be limited to humans.

In many ways, functional categories have more ecological relevance than perceptual categories because the groups are defined by meaningful relationships. It seems reasonable, therefore, that this cognitive ability would evolve in different animals although the capacity may vary across species. The majority of research on functional categorization uses humans, other primates, or pigeons (most commonly the latter); comparative studies that expand this list would help to trace the evolution of this cognitive process. As an example, functional categorization in humans and some other primates extends to arbitrary symbols, but the paucity of data on this topic makes it difficult to establish whether other species possess the same ability.

9.4 Relational categorization

Both perceptual and functional categorization are based on some property of the stimulus to be classified. In contrast, **relational categorization** groups items together that exhibit the same connection between them. The properties of relational categories, therefore, are relative rather than absolute: determining whether a stimulus is bigger, louder, or brighter depends on what it is being compared to. Close to one hundred years ago, Révész (1924) demonstrated that chicks understand these relationships in that they can learn to peck the smaller or larger visual stimulus, regardless of its shape. Other animals tested since that time (including some insects) have all 'passed' this relational test (Reznikova, 2007).

A more stringent measure of relational categorization is to test whether subjects can group stimuli based on a same/different association. The question is not whether individuals can detect similarities or differences between stimuli, but whether they can identify a relationship between stimuli (i.e. same or different) and then use this rule to classify new exemplars. For example, subjects may be presented with a red circle during a sample phase and then see an identical red circle and blue circle during the testing phase. If the categorization rule is 'same,' the correct response is to select the red circle over the blue one. Animals ranging from insects to birds to primates quickly learn this discrimination, but the critical test occurs during the transfer phase in which new stimuli are presented that vary on a different dimension. Subjects may see a triangle (of any color) during the sample phase and then a triangle plus a square (again, of any color) during the testing phase. If they have learned to categorize stimuli based on relational, rather than absolute, properties, subjects will select the triangle over the square. That is, they know to select the stimulus that shares a critical feature with the sample (in this case shape), regardless of how the two differ on other features (such as size or color). Because it relies on understanding rules for combining stimuli, relational categorization is often called 'rule-based' learning.

Monkeys (Wallis *et al.*, 2001), chimps (Thompson *et al.*, 1997), and some corvid species (Wilson *et al.*, 1985) are able to match stimuli based on similarities or differences and, most importantly, to transfer these rules to new sets of items. Originally, it appeared that pigeons had difficulty learning relational categories (Edwards *et al.*, 1983), but later work contradicted this finding (Cook *et al.*, 1997). That is, birds were able to transfer same/different judgments from one visual element to

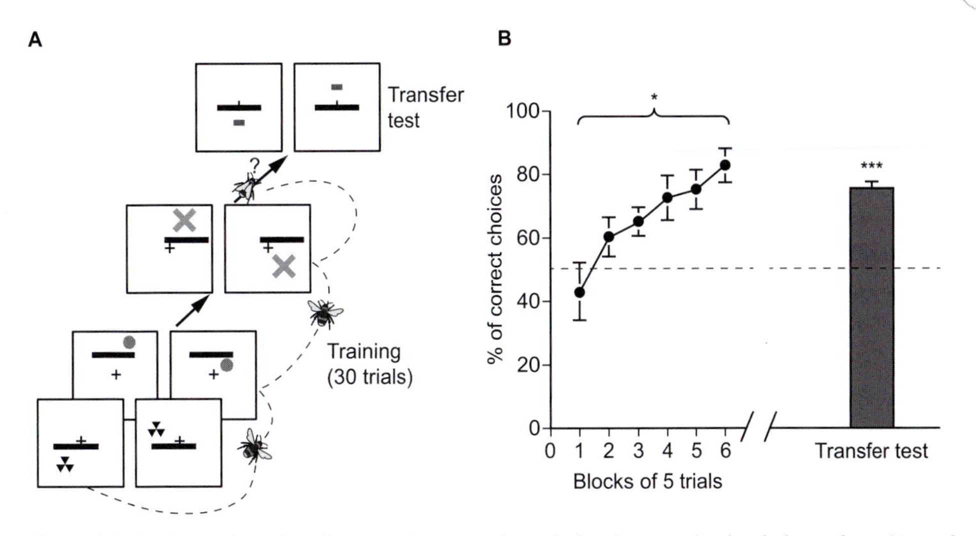

Figure 9.7 Design and results of an experiment testing relational categorization in honeybees (Avargèus-Weber *et al.*, 2011). (A) Bees were trained to fly along a path from one visual stimulus to another. Bees were rewarded if they flew to the same stimulus category at each choice point: target above the bar or target below the bar. (B) Performance improved across trials indicating that bees had learned the relational rule. When presented with a novel stimulus pair (transfer test), bees made more flights to the stimulus that matched their training category (i.e. 'target above the bar' or 'target below the bar'). The dotted line indicates chance performance.

another (e.g. texture to feature). Indeed, as long as the experiment is set up appropriately (see Figure 9.7), honeybees are able to transfer a rule for selecting same or different stimuli from one sensory modality to another (Giurfa *et al.*, 2001).

Not all researchers agree that these findings provide evidence for relational categorization across species. Many point out that humans, including young children, transfer knowledge of category relationships to new situations almost spontaneously. In contrast, most animals undergo hundreds or even thousands of training trials before they can perform correctly in transfer tests (typically non-human primates require less training than rats or pigeons). Similarly, humans and the other great apes can form same/different categories with exposure to only two stimuli, whereas baboons must be exposed to three or four different pairs, and pigeons to at least 16 (Penn *et al.*, 2008). Thus, even though many animals can make same/different judgments during a transfer test, this may not reflect the formation of relational categories. One suggestion is that animals are relying on previously learned associations and generalization of perceptual properties to perform these tasks (Urcelay & Miller, 2010).

Grouping stimuli together based on higher order relationships, or relationships among relationships, also appears to be limited to great apes (Thompson & Oden, 2000). These 'nested relationships' are tested by presenting subjects with a pair of stimuli that are, or are not, matched on one dimension (e.g. if color is the dimension to be matched, a red circle and a red triangle would be the same, whereas a blue circle and a red circle would be different). During the test phase, subjects must categorize a new pair of stimuli by indicating whether the relationship between these is the same or different as the relationship between the pair presented in the sample phase. This ability to construct higher order relationships, and then apply the information to novel situations, is a form of analogical reasoning often ascribed only to humans (Penn *et al.*, 2008).

(A)

(B)

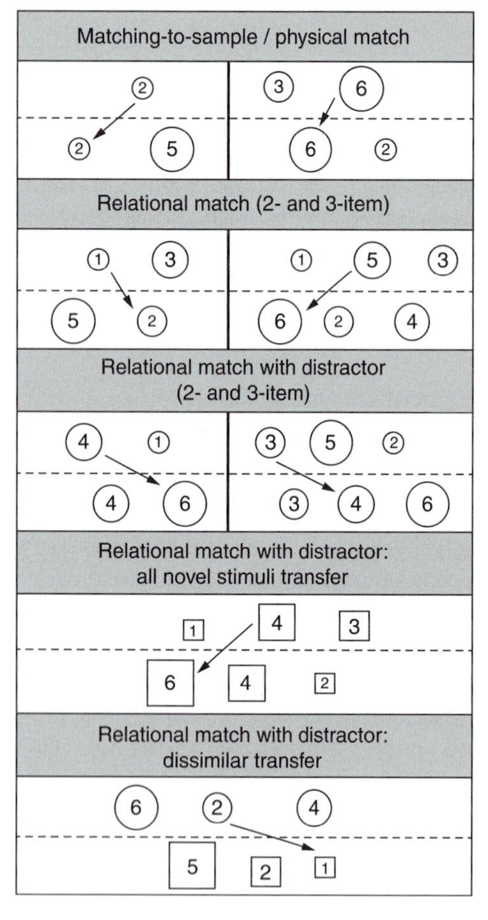

Figure 9.8 Schematic presentation of an experiment testing analogical reasoning in chimpanzees. (A) An experimenter hides food under one cup on the upper level (e.g. medium-sized). To receive the reward, the chimpanzee must point toward a cup on the lower level that bears the same relationship to the cups on this level. (B) Different task conditions in the analogical reasoning task. Numbers represent the absolute size of each cup. The top row in each phase represents the experimenter's set of cups. The bottom row represents the subject's choice set. Arrows connecting cups indicate correct matches. Distractor trials included a cup in the subject's level that matched the absolute size of the cup where the food was hidden in the experimenter's level. The dissimilar transfer trial used stimuli with different shapes in the experimenter and subject levels.

There is little, if any, evidence that animals form relational categories of this nature (Shettleworth, 2010), with the exception of chimpanzees (Flemming & Kennedy, 2011). The experiment demonstrating this effect used an analogical reasoning paradigm adapted from developmental psychology (see Figure 9.8). Subjects watched an experimenter place food under one cup in a set, and then searched for a hidden food reward in another set of cups. If chimps were using an analogical strategy, they should look under the cup that bears the same relationship to the cups in their set as does the cup with the hidden food. For example, regardless of absolute sizes, if food was hidden under the experimenter's smallest cup, chimps should look under the smallest cup in their own set. Rather than using information about relative size, however, two of the three chimps in the experiment relied on relative position to search for the food. The experimenters concluded that, although they had not performed as expected, chimps had developed their own analogical reasoning strategy. The same ability emerges in humans between 3 and 4 years of age (Gentner & Rattermann, 1991), but whether it exists in monkeys is still debated (Flemming *et al.*, 2007; Kennedy & Fragaszy, 2008). Thus, it is not clear whether the cognitive process of forming higher order relational categories is restricted to the family of great apes, or the broader order of primates. Nor do we know whether corvids, who display many of the same cognitive traits as great apes, possess this ability.

9.5 Social categorization

Many behavioral ecologists remain skeptical of results obtained from lab-based studies of categorization. It is not that they doubt animals can form perceptual, functional, or even relational categories; rather, they question the significance of findings from experiments in which human researchers define how stimuli are to be classified. In other words, the question is not whether animals can categorize, but whether they do so in the same way as humans. One of the most telling examples, described above, is the study in which chimps used information about relative position, as opposed to relative size, to look for a food reward (Flemming & Kennedy, 2011). For chimps foraging in the natural environment, location is a more relevant attribute than size so these animals were probably displaying an inherited tendency to use one type of cue over another. Once again, these findings emphasize the importance of using ecologically relevant information in comparative cognition research. For most animals, ecologically relevant information includes cues that facilitate **social categorization**: the process of classifying organisms into groups such as conspecifics versus heterospecifics, familiar versus unfamiliar individuals, and kin versus non-kin.

9.5.1 Conspecific recognition

The first level of social categorization, **conspecific recognition**, is the identification of one member of a species by another member of the same species (also called intra-species recognition). As with all categorization, this recognition is manifested as differential responses to members of the two groups. Conspecific recognition is critical for reproduction as individuals who attempt to mate with another species (heterospecific) will expend time and energy with no evolutionary gain. It is also an important factor in territorial defense because heterospecifics who inhabit the same area are less likely to compete for the same resources than are conspecifics.

Different species use various combinations of sensory cues to recognize conspecifics: humans rely, primarily, on vision and focus on facial features. Even newborns prefer to look at human faces over any other visual stimulus (Valenza *et al.*, 1996), although they do not spend more time looking at human versus other primate bodies until they are 3 to 4 months of age (Heron-Delaney *et al.*, 2011). According to Darwin (1872), focusing on facial features is adaptive because facial expressions convey emotions, making these an important means of intra-species communication. Newborn rhesus monkeys (*Macaca mulatta*) show the same preference for conspecific faces, which is shifted to humans if they see humans, but not monkeys, early in life (Sugita, 2008). Facial recognition in primates, therefore, involves two mechanisms. The first is innate, selected for attention to biologically relevant information (i.e. conspecific faces). The second is more flexible, involving

Figure 9.9 Male banded demoiselle.

learning during a sensitive postpartum period (i.e. *which* face to attend to). It is unlikely that the latter is completely open-ended as monkeys may not develop a preference for other animal faces (e.g. elephants), even if they do not see any primates during the first few months of life.

This combination of inherited and acquired processes mediates conspecific recognition in many other species, including some insects. For example, the ability of female banded demoiselle (*Calopteryx splendens*), a type of damselfly (see Figure 9.9), to recognize males of their species depends on environmental experience (Svensson *et al.*, 2010). Different groups of these insects live in areas that do, or do not, overlap with the habitat of a phenotypically similar species. Females from the overlapping habitat can discriminate between males of the two species (a distinguishing feature is the amount of wing melanization), whereas females from the isolated habitat cannot. Thus, the probability of encountering another species with similar physical traits modifies the criterion for social categorization, in this case conspecific recognition. The authors speculate that a learned component of conspecific recognition is important early in speciation. In other words, certain ecological conditions support robust and reliable conspecific recognition.

This flexibility in conspecific recognition may help to explain the surprising number of incidents in which animals respond to both territorial and mating signals of heterospecifics (Ord *et al.*, 2011). From an evolutionary perspective, these are costly mistakes as they divert resources away from activities that promote reproduction and survival. In reviewing situations in which this occurs, Ord and colleagues noted that animals are more likely to misidentify heterospecifics and conspecifics if the density of heterospecifics is low in their natural environment. Under these conditions, there is no pressure for a refined conspecific recognition system to evolve, either in terms of signals or detection mechanisms. Frequently, however, environments are altered (sometimes far too rapidly) or animals move to other territories. The cost/benefit ratio of responding to heterospecifics may change in these situations, so species that can adapt to new ecological surroundings will have an evolutionary advantage. Thus, like most cognitive processes, conspecific recognition involves both inherited and experience-dependent mechanisms.

9.5.2 Kin recognition

Conspecific recognition is only the first step of social categorization. Most organisms also possess the ability to recognize individuals within their species, particularly family members. This process of **kin recognition** categorizes individuals based on genetic relatedness. An obvious evolutionary advantage is avoidance of inbreeding, but kin recognition also plays a role in many social behaviors: individuals can increase the probability that copies of their genes will be passed through successive generations if they promote the reproduction and survival of closely related individuals (Hamilton, 1964). For example, non-human primates preferentially groom family members over unrelated conspecifics (Gouzoules, 1984), a social behavior that helps to promote hygiene and maintain body function. As noted later in the book, kin recognition also facilitates cognitive processes such as social competence (Chapter 10), cooperation (Chapter 11), and communication (Chapter 12).

Parent–offspring recognition

One of the most important kin recognition mechanisms, filial imprinting, was described in Chapter 1. As with conspecific recognition, this process of parental recognition has a high degree of flexibility, evidenced most dramatically by the trail of goslings that followed Lorenz. Filial imprinting occurs in other species, even if they do not follow a parent with such determination. In humans, the process begins in the womb when an unborn baby hears voices (primarily its mother) which it later recognizes (Kisilevsky *et al.*, 2003). This recognition is reciprocated in that parents also display an extraordinary ability to identify their offspring. For example, after a night of foraging, mother bats can immediately locate their own baby in a cave crowded full of young bats (Kolb, 1977) (see Figure 9.10). Given the evolutionary advantages of rapid and effective parent–offspring recognition, it is not surprising that this ability is particularly pronounced in animals that live in tight social groups.

Figure 9.10 Bats inside a cave, Ankarana National Park, Madagascar.

Even in social animals, however, kin recognition is not perfect, as imprinting on the wrong 'parent' (e.g. Lorenz) demonstrates. Common cuckoos (*Cuculus canorus*) take advantage of this by laying their eggs in the nest of other birds, who then raise these non-related chicks. Hatched cuckoos make things worse by pushing the eggs or other young birds out of the nest, thereby securing the resources of their foster parents (Davies & Brooke, 1988). From the parent's perspective, the possibility of rejecting an egg or chick that carries their genes must be weighed against the resources required to raise an adopted bird. If nesting birds encounter few other bird species in their habitat and resources in the environment are plentiful, parental rejection is low. As the probability that a nest will contain non-related offspring increases (e.g. there are more cuckoos in the vicinity), parents are more likely to reject foreign looking eggs (Davies *et al.*, 1996). Thus, the categorization mistake in which parents misidentify an offspring can be explained, once again, by a cost/benefit ratio.

Despite notable exceptions like the parasitic cuckoo, parent–offspring recognition is often accomplished rapidly and (seemingly) effortlessly. Both humans and animals use sensory cues to identify their relatives, with the effectiveness of different cues reflecting sensory system evolution in each species (see Chapter 2, Section 2.1). For example, mother bats use both auditory and olfactory information to find their babies, whereas humans rely primarily on vision. Within these sensory systems, however, the specific cues that will facilitate kin recognition depend on experience. Chicks inherit the tendency to follow the first moving object that they see, but the imprinted object itself may vary. The same process appears to occur in humans: babies have evolved to focus on faces, which allows them to recognize individuals that they see frequently (usually a relative). The idea that learning is an important component of kin recognition is supported by cross-fostering experiments in which newborn animals are raised by a non-related mother along with her own offspring: the fostered and biological offspring treat each other as kin when they are adults (Mateo & Holmes, 2004). It is almost impossible to know whether animals are using the rule 'this individual is related to me' or 'I spent a lot of time with this individual in the past,' but the latter is probably an overgeneralization of the former. That is, in most cases, babies raised together *are* genetically related so evolution would favor a cognitive process that categorizes (and then treats) these individuals as kin.

Recognition of unfamiliar kin

Kin recognition may be influenced by experience, but it is not entirely dependent on this process as both vertebrates and invertebrates can recognize relatives that they have never encountered, and many can discriminate closely from distantly related kin (Sherman *et al.*, 1997). Kin are often unfamiliar to each other if they are siblings from different breeding seasons because one is weaned and independent before the next is born. Even if these first degree relatives meet, they may have no more interaction with each other than with other conspecifics in their social group. In order to be recognized, therefore, one individual must be emitting some kind of signal that the other relative can detect. Most evidence points to chemical signaling in mammals (Sherman *et al.*, 1997) and insects (Howard & Blomquist, 2005), originating from either genetic or environmental factors. Genetic transmission of chemical signals has been traced to a DNA locus that is highly polymorphic, so individuals who are close kin have similar olfactory profiles (Rinkevich *et al.*, 1995). Environmental cues that are used to identify kin may come from odors associated with early life experience, such as the natal nest.

Some avian species also display a remarkable ability to recognize unfamiliar kin but, for many years, the process mediating this ability was unknown. Biologists speculated that evolutionary

Figure 9.11 European storm petrels.

pressure against inbreeding would favor a mechanism of kin recognition in birds that exhibit **natal philopatry** (i.e. they nest in the vicinity of their birthplace), particularly if they form long-term monogamous relationships. Initial attempts to reveal this trait, however, produced mixed results (Komdeur & Hatchwell, 1999). Originally, researchers assumed that kin recognition in birds would depend on vision and audition as these sensory systems are highly developed in avian species, and conspecific as well as parent–offspring recognition is mediated by visual and auditory cues. Probably for these reasons, the role of olfactory signaling in kin recognition by birds was overlooked.

This oversight was rectified by Bonadonna and colleagues, who have spent many years studying the behavior of small sea birds, known as petrels, in their natural environment (see Figure 9.11). They provided the first evidence that blue petrels (*Halobaena caerulea*) in the Kerguelan Archipelago of the southern Indian Ocean can use olfactory cues to recognize their mate (Bonadonna & Nevitt, 2004). Petrel olfactory cues are secreted through glands onto their feathers. Researchers collect samples of the scents by rubbing cotton swabs on birds' feathers. Analysis of these secretions by gas chromatography revealed that individual birds emit a chemical signal that is consistent across years (Mardon *et al.*, 2010). The researchers then set out to determine whether kin related information is conveyed in these individual olfactory profiles. To do so, they set up an experiment with mice using a habituation procedure like the one described in Chapter 3 (Box 3.5). First, mice were presented with one bird odor until they habituated to the smell (i.e. they stopped exploring a small hole that was the source of the odor). In the discrimination phase, two new odors were placed in different holes: one from a parent or offspring of the first odor donor and one from an unrelated conspecific. Mice spent more time exploring odors from non-kin than from kin, suggesting that they perceived a high degree of similarity between kin odors (Célérier *et al.*, 2011). In other words, petrels possess a distinguishable family odor that may be used to indicate genetic relatedness. Kin

Researcher profile: **Dr. Francesco Bonadonna**

Figure 9.12
Dr. Francesco Bonadonna.

Francesco Bonadonna's advice to young scientists is to "take time to observe the world around you." His own observations of the natural environment have taken him from his native Italy to various parts of the globe including other European countries, Malaysia, Iceland, and the Antarctic. During these travels, Bonadonna built on his early training in animal biology and ethology (Pisa University, Italy) to investigate both the function and evolution of animal behavior. This work has included an examination of cognitive processes, such as navigation and communication, in a wide range of animals such as pigeons, green turtles (*Chelonia mydas*), thick-billed murres (*Uria lomvia*), and fur seals.

In 2001, Bonadonna moved to Montpellier, France where he is currently team leader and senior researcher at the Centre d'Écologie Fonctionnelle Évolutive (CEFE), part of the Centre Nationale de la Recherche Scientifique (CNRS). He continues to investigate how evolution shapes behavioral traits, with a focus on understanding olfactory signals in birds. An important aspect of Bonadonna's work is the development of testable hypotheses. He observes behavior in the natural environment, develops explanations for why the behavior occurs, and then designs an experiment to test this idea. A prime example of this synergy (i.e. behavioral observation and controlled experimental testing) is Bonadonna's work on kin recognition in sea petrels. By documenting mate choice and nesting site in the natural environment, he and his team developed the idea that these birds can recognize unfamiliar kin using olfactory cues. They tested this hypothesis using a Y-maze that was adapted from standard laboratory equipment for rodents. They also brought samples of the individual bird scents from the Kerguelan islands to France where they were used as stimuli in a mouse discrimination task. This approach of combining behavioral ecology and experimental psychology techniques allows Bonadonna to examine both proximate and ultimate explanations of kin recognition.

recognition by odor was confirmed in birds of a related species, Humboldt penguins (*Spheniscus humboldti*), that had lived their entire life in captivity (Coffin *et al.*, 2011).

Phenotypic matching

An animal's ability to identify chemical (or other) signals from a genetically related individual is only the first step in kin recognition. They must then *use* this information in some process of self-referent mapping. One possibility is that organisms observe a conspecific's **phenotype** (the combination of an organism's observable characteristics, including behavioral traits and chemical secretions) and compare it to their own, a process known as **phenotypic matching** (Holmes & Sherman, 1982). Phenotypic matching is based on the premise that a closely related **genotype** (an organism's unique genetic code) will be reflected by similarities in phenotype. It could explain how blue petrels identify unfamiliar kin, an effect that has also been observed in several rodent (Mateo, 2003) and fish (Brown *et al.*, 1993; Gerlach *et al.*, 2008; Quinn & Busack, 1985) species. Many scientists speculate that phenotypic matching plays an important role in kin recognition by primates, but it is difficult to disentangle this process from familiarity and early learning because offspring of these species spend an extended period of time with their mother and other genetic relatives (Widdig, 2007).

Mateo and Johnson (2000) used a cross-fostering experiment to compare the contribution of phenotypic matching versus learned cues in kin recognition by golden hamsters (*Mesocricetus auratus*). Because they were reared by foster mothers in the presence of foster siblings, test animals had extensive exposure to the odors of their non-genetic family. In contrast, the only genetically related odor that they experienced was their own. Despite this difference, animals classified the odor of an unfamiliar genetic relative as less novel than the odor of an unfamiliar relative of their foster family (the authors make a cogent argument against the idea that *in utero* or post-birth odors would have contributed to this effect). This suggests that the golden hamster's own odors are more important in kin recognition than are the odors of early rearing associates (i.e. foster siblings and mother). The results, supporting phenotypic matching, may explain how some animals are able to discriminate between unfamiliar kin based on degree of genetic relatedness (Sherman *et al.*, 1997).

Phenotypic matching should facilitate nepotism, the favoring of family members over non-related individuals, but the two do not necessarily go together. Mateo (2002) revealed this apparent paradox by studying two species of ground squirrels: Belding's ground squirrels (*Spermophilus beldingi*) that exhibit kin favoritism in the natural environment and golden-mantled ground squirrels (*Spermophilus lateralis*) that do not (with the exception of mothers and dependent offspring). Surprisingly, in an experimental lab situation, both species were able to discriminate kin from non-kin odors and showed greater familiarity of odors from more closely related individuals (e.g. mother over grandmother over half aunt). This clever combination of behavioral observation in the natural environment and controlled experiments in the lab suggests that kin recognition and kin favoritism may have evolved independently.

Inherited versus environmental cues

Kin recognition studies, such as those cited above, suggest that animals can categorize conspecifics based on genetic relatedness as well as shared environmental experience. The relative importance of each set of cues likely depends on the ecological context of a particular species. For example, social insect colonies contain individuals from a diversity of genetic backgrounds because queens have multiple matings and many colonies have multiple queens. Under these conditions, evolution would favor a nestmate, not a kin, recognition system in that nepotism may be detrimental to overall colony functioning (Martin *et al.*, 2012). In line with this idea, early environmental cues play a critical role in social categorization by these animals: golden paper wasps (*Polistes fuscatus*) that are removed from their cone when they emerge from a pupa, or are exposed only to nestmates but not their nest, lose the ability to distinguish kin from non-kin (Shellmann & Gamboa, 1982). Indeed, there is no evidence that social insects can recognize unfamiliar kin (Martin *et al.*, 2012) suggesting that this mechanism did not evolve in these species.

Finally, kin recognition in this section focuses entirely on the ability of individuals to identify their own genetic relatives. Some animals have a broader understanding of kin relationships in that they are able to identify familial connections between two or more other individuals. For example, monkeys can categorize photos, including novel exemplars, of individuals from their social group based on whether the photo shows a mother–offspring pair (Dasser, 1987). These findings provide evidence for relational categorization at the social level, a topic that is discussed in great detail in the next chapter (see Section 10.5).

9.5.3 Individual recognition

In addition to kin recognition, social categorization includes the ability to identify non-related (or at least very distantly related) conspecifics. This process of **individual recognition** does not depend on familiarity in that humans and many animals can recognize a conspecific that they have encountered

Box 9.3 Prosopagnosia

Like many medical terms, prosopagnosia comes from the Greek: literally translated, it means 'not knowing faces.' The contemporary diagnostic criterion for prosopagnosia is a deficit in face recognition with no apparent loss in the ability to recognize objects. In other words, prosopagnosia is not caused by an impairment in sensory system function. Prosopagnosics have no problem telling the differences between cats and dogs or frogs and fish; nor do they show impaired performance on perceptual or relational categorization tasks. The deficit is linked, specifically and uniquely, to human faces. Many prosopagnosics cannot even identify their own face in the mirror. (Theoretically animals could exhibit a similar deficit in the ability to recognize individual conspecifics, but this has never been reported or even examined in the scientific literature.) Some prosopagnosics learn to use other cues, such as hair color or voice, to recognize individuals but this process is never as automatic as 'normal' facial recognition, and often less reliable.

Prosopagnosia may be congenital or acquired, the latter produced by brain damage, such as stroke or head trauma, to parts of the occipital and temporal lobes. Patients with this condition can no longer recognize other people, even close relatives and friends, which leads to awkward and inappropriate social interactions. The situation is far more devastating than trying to identify the person who strikes up a conversation at a party; prosopagnosics have no sense of whether an individual is familiar to them or not. The neurologist and popular science writer Oliver Sacks presents a comical but touching portrayal of one of these individuals in The Man Who Mistook His Wife for a Hat. (In truth this patient is not a pure prosopagnosic because he exhibits other perceptual deficits, but the point that his life is turned upside down by the impairment is well made.)

Congenital prosopagnosia is much more difficult to detect because individuals born with this condition have no comparison: they grow up without the ability to recognize faces and may not realize that other people 'see' things differently. They often develop compensatory mechanisms for individual recognition (i.e. relying on voice) but many continue to have difficulties with social interactions and often become reclusive. There is no evidence (to date) for neurological alterations in congenital prosopagnosia (Behrmann & Avidan, 2005) but even if these individuals are identified at a young age, they cannot learn to recognize other people by their facial features. The deficit, therefore, may be inherited, a hypothesis that is supported by evidence that congenital prosopagnosia runs in families (Duchaine *et al.*, 2007) and that variations in face recognition abilities of normal controls may be explained by genetic differences (Wilmer *et al.*, 2010). Importantly, facial recognition is *not* correlated with scores on other visual and verbal recognition tests, confirming that it is a highly specific cognitive trait. Most likely, it evolved in humans to facilitate the identification of other individuals as aggressive, cooperative, or trustworthy, information that could be used advantageously in future social interactions.

only once. Humans excel at this type of categorization; indeed, those who are unable to recognize other individuals exhibit dramatic deficits in social functioning (see Box 9.3). As with other types of categorization, individual recognition is accomplished in different ways, with humans relying, primarily, on visual cues in the face. Giant pandas (*Ailuropoda melanoleuca*) can also discriminate facial features of conspecifics (Dungl *et al.*, 2008) although it is not clear if these provide the most important information for individual recognition in this species. Other animals, such as white-throated sparrows (*Zonotrichia albicollis*) (Falls & Brooks, 1975) and bullfrogs (*Rana catesbeiana*) (Bee & Gerhardt, 2002), use auditory cues to identify familiar conspecifics.

Individual recognition extends to insects (at least some species of social insects) in that hairy panther ants (*Pachycondyla villosa*) recognize each other using olfactory cues (D'Ettorre & Heinze, 2005) and paper wasps identify conspecifics by the yellow/black patterns on the abdomen and face (Tibbets, 2002).

Even if animals can recognize cues associated with a particular conspecific, careful experimentation is necessary to show that they can group different cues together that belong to the same individual. The question of interest is whether animals can form a category defined by individual specific traits. At least for some species, the answer appears to be 'yes.' When presented with photos of the face or body of familiar conspecifics, long-tailed macaques (*Macaca fascicularis*) correctly put the two together (Dasser, 1987) indicating that they categorize these features as part of the same individual. Rhesus monkeys perform a similar categorization with photos of other monkeys and recordings of their vocalizations (Matyjasiak, 2004). Finally, great tits (*Parus major*) trained to discriminate songs of conspecifics show generalized responding to new songs of the same individual suggesting that they are categorizing the singer, not the song (Weary & Krebs, 1992). Few other animals have been tested in these types of experiments, partly because it is difficult to establish which cues different species rely on to recognize other individuals.

Recognition of nonrelated individuals differs from both conspecific and kin recognition in that it depends *entirely* on experience. The tendency for humans to focus on facial features may be inherited, but the specific face that individuals come to recognize is learned. Newborns show a preference to look at their mother's face (Bushnell *et al.*, 2011), but the ability to recognize other individuals does not develop for several months (some researchers argue that parental recognition is also due to learning that begins *in utero*). Individual recognition occurs when the unique traits of a conspecific are paired with prior experience of that individual. In animals, these experiences often involve fitness-related interactions such as territorial defense, mating, or social dominance. For example, American lobsters (*Homarus americanus*) can recognize individuals based on previous aggressive encounters (Karavanich & Atema, 1998); dominance interactions also play a role in individual recognition by different species of insects, fish, and mammals (Tibbets & Dale, 2007). In a different type of social interaction, hooded warblers (*Setophaga citrine*) recognize their neighbors by associating individual songs with the song location (Godard, 1991). If recordings of familiar songs are played from a different location, resident birds respond aggressively, suggesting that they are specifically targeting conspecifics who do not observe territorial boundaries. Recognition of territorial neighbors is also observed in frogs, fish, and invertebrates (Tibbets & Dale, 2007).

Individual recognition may also occur through associations with positive fitness-related encounters such as mating, food sharing, or co-parenting. Indeed, the existence of long-term relationships in many species (see Figure 9.13) implies that these animals are able to recognize individual unique cues associated with reproductive activities. Studies of individual recognition based on positive social interactions are more difficult to conduct than those based on aggressive interactions because it is almost impossible to control for familiarity effects in the natural environment (e.g. animals tend to spend more time with their mate than with other conspecifics). Nonetheless, mate recognition is observed in polygamous species with no shared parenting (Amos *et al.*, 1995; Caldwell, 1992) suggesting that, at least in some species, familiarity alone cannot explain individual recognition of reproductive partners.

Some people argue that individual recognition in humans involves a different process because most people can identify other individuals, even when their social interactions are not fitness related. There are two responses to this criticism. First, as detailed in Chapter 1, adaptations may serve a

Figure 9.13 Heart of pink flamingos.

purpose for which they did not originally evolve (see Section 1.5.2). The ability to recognize conspecifics provides an evolutionary advantage to species that live in social groups characterized by hierarchies and bonding (Troje *et al.*, 1999); humans inherit this cognitive trait which is then applied to contemporary interactions (e.g. this is the person who tried to steal my food as opposed to this is the person whom I met at the cinema last week). Second, fitness-related interaction is a broad category that includes a number of subtle social cues. Even with very little information to go on, humans quickly (and seemingly accurately) categorize other people as kind, cruel, friendly, or aggressive (Funder & Sneed, 1993). This suggests that evolution has favored a cognitive process that links personality traits to identifying characteristics, information that is used to guide future interactions (Klein *et al.*, 2002). Many people can relate to this scenario, having experienced an immediate feeling of unease when they meet someone who looks or smells like another person that they dislike. On the other hand, it can be difficult to recognize someone after repeated, but innocuous, interactions.

In sum, social categorization is a cognitive process that allows organisms to identify and then classify other individuals. It provides an organization framework to the social environment that allows humans and many animals to recognize conspecifics, kin, and unrelated individuals. These species are born with a mechanism that draws attention to evolutionary relevant stimuli, including conspecifics, and a predisposition to link an individual's traits with prior experience of that individual (Heron-Delaney *et al.*, 2011). The ability to categorize individuals based on previous interactions is particularly important in species with stable social structures. Indeed, as group size increases, identity traits of individuals become more pronounced but *only* in animals that live and interact with the same group of individuals (Pollard & Blumstein, 2011). Presumably those that are part of large, but fluid, social groups have less need to categorize conspecifics based on previous interactions or identified behavioral tendencies.

9.6 Concept formation

Categorization occurs with concept formation but the former is not proof of the latter. Humans or animals may classify novel exemplars correctly, even if they have not formed a mental representation of the rules that define group membership. In addition, concept formation implies an understanding of group properties beyond those in the presenting stimulus (see Section 9.1.1); categorization may occur in the absence of this knowledge. Because organisms are constantly provided with feedback about what does and does not belong in a particular group, concept formation is an ongoing cognitive process in which the criteria for categorizing new information are continually refined and updated.

9.6.1 Theories of concept formation

Most contemporary theories of concept formation can be traced to the mid-twentieth century work of Jerome Bruner, a leading figure in early cognitive psychology research. According to Bruner and colleagues, humans have a mental representation of attributes that define a concept. Categorization is accomplished by comparing novel exemplars to these stored representations, noting both similarities and differences in concept relevant features (Bruner *et al.*, 1956). The basic premise of this theory was extended to other species as research in animal cognition expanded. The theories were refined over the next 50 to 60 years and can now be grouped into three broad categories: elemental, exemplar, and prototype theories. Each of these was formulated to explain data from perceptual categorization studies although at least some aspects of the theories can be applied to functional, relational, or social categorization.

Elemental theory

As the name suggests, an elemental theory of concept formation hypothesizes that the categorization rules are based on specific features of group items. These are not always simplistic rubrics that depend on one element; a concept may be defined by a large number of elemental features, some necessary and some sufficient for group membership. Indeed, a challenge for elemental theories is to explain the relative importance of different features in categorization decisions. Radiologists face this task on a regular basis. Should they focus on the size, shape, or intensity of a dark spot on an x-ray to decide whether they are viewing normal tissue variation or the beginning of a tumor? This example makes it obvious that perceptual categorization is often explained by an elemental theory of concept formation. The same theory can apply to functional categories when elements are defined in a different way. For example, food has the features of being both biological material and consumable, although there is considerable variability in the second criterion. As Figure 9.14 illustrates, not everyone agrees on which biological material can and should be eaten. Different individuals, therefore, may develop their own concept of 'food' based on the material's functional elements.

Relational categorization can also be explained by an elemental theory of concept formation because both are defined by specific rules for group membership. In this case, the relationship between items makes up the elements or features of a category. For example, bigger or smaller things go together in a relational concept of size. Social categorization is not typically discussed in terms of elemental theories of concept formation, although conspecific, kin, and individual recognition frequently involve identification of perceptual elements (e.g. a particular scent, song note, or facial feature). This raises the issue, once again, of how animals conceptualize social

Figure 9.14 Insects sold as snacks at a market in Thailand.

categories. Do they respond differently to kin because they identify features that link an individual to themselves, or because they have a cognitive representation of familial connections? This question is unlikely to be answered in the near future, at least not for all animals.

Exemplar theory

On a general level, exemplar theory is a more specific form of Bruner's original hypothesis, proposing that concept formation involves storing representations of distinct items within a group and then noting similarities and differences between these stored exemplars and newly encountered stimuli. Exemplar theory also assumes generalization between group properties such that a new exemplar does not have to be an *exact* match to a stored exemplar in order to be classified in the same group. The criteria for generalization (i.e. how similar the stimuli are within the same group) depend on an individual's previous experience with that category and with related categories. For example, the ability to categorize bird calls and to use these to discriminate bird species becomes more refined with experience and with feedback on correct and incorrect classifications.

An exemplar theory of concept formation was appealing, initially, because it is simple. The idea that individuals store representations of perceptual stimuli fits with how most people think they 'see' the world. The theory fell out of favor when it became apparent that perception and memory are both active cognitive processes (Chapters 2 and 3 emphasize this point). That is, forming an exemplar collection of category items may not make sense if information that is stored and then retrieved is a modified version of stimuli or events from the past. Another drawback to exemplar theory is that it places high demands on cognitive resources by assuming that all exemplars have a separate mental representation. Moreover, each of these exemplar representations must be available as a comparison to novel stimuli during categorization. For these reasons, the current view of exemplar theory has shifted to one in which it explains the initial stages of concept formation. When humans or animals have encountered relatively few items within a category, they may keep mental representations of these novel exemplars. With experience, information from the accumulating examples can be extracted to develop elemental rules (i.e. necessary and sufficient features) for categorization. In that way, exemplar theory is a precursor to an elemental theory of concept formation.

Prototype theory

The same is true for prototype theory in that a **prototype**, the typical or most representative member of a category, develops from exemplars. A prototype theory of concept formation proposes that categorization is the cognitive process of comparing novel exemplars against a mental representation of the category prototype. The closer a new exemplar is to the prototype, the more likely it is to be categorized in the same group. With feedback (i.e. knowledge that the exemplar does or does not fit into the category), the prototype is modified accordingly. A prototype is not just an average of the exemplars in a category; rather, it reflects features of the category that best represent the criteria for category membership. Thus, the prototype is sometimes viewed as the 'model' or 'ideal' member of a category. Birds are often used to illustrate this point: when asked to name a bird, few people come up with penguin although almost everyone knows that penguins are birds. The reason for this is that the most common and distinguishing feature of birds is that they fly. Thus, the prototypical bird is one that flies, which penguins do not.

Prototype theory was developed in cognitive psychology to explain evidence, such as the bird example, that humans rank exemplars within a category in terms of how well they represent the ideal criteria for group membership (Rosch, 1973). Prototype effects are also evident when humans classify stimuli that represent the central tendency of a group more quickly than those on the extremes. As an example, most people can answer the question 'Is a dog an animal?' much more quickly than they can answer the question 'Is a fruit fly an animal?'

Prototype theory can be tested in animals using a categorization paradigm in which the category prototype is presented as a novel exemplar during testing (see Figure 9.15). The prototype is

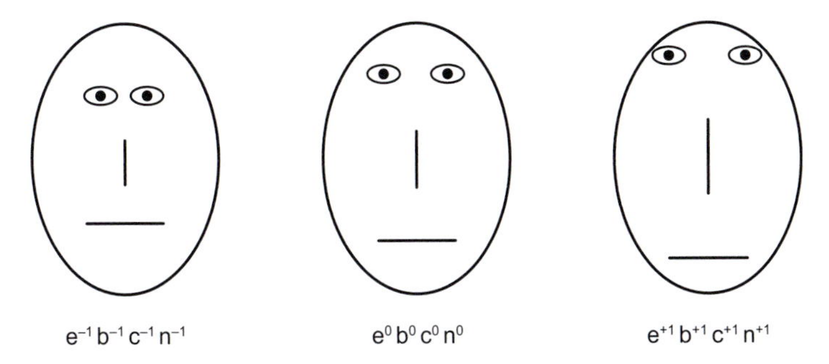

Figure 9.15 Stimuli used in a prototype categorization experiment with pigeons (Huber & Lenz, 1993). Pigeons were required to classify drawings of faces that varied along four dimensions: the distance between the eyes (e), the height of the forehead above the brow (b), the length of the nose (n), and the position of the mouth in relationship to the chin (c). The middle configuration represents a value of 0 for each dimension with the faces on the left and right side representing −1 and +1 variations of these dimensions, respectively. Varying each dimension independently yields 81 different face configurations. Taking the sum of the four dimensions produces 31 faces with negative values and 31 faces with positive values. Pigeons were trained to categorize faces into these positive and negative categories; when presented with novel face configurations, they correctly categorized these using the positive/negative rule. This suggests that they had formed a prototype of each category in that no single feature could be used to determine membership in either category.

constructed as a central tendency of items in the category such that it shares the most features with other members of the category and the fewest features with members of the opposite (i.e. to be discriminated) category. Prototypes can be produced with visual stimuli by morphing the individual exemplars into one image. In experiments with this setup, pigeons (Huber & Lenz, 1993), rhesus monkeys (Jitsumori, 1994), and baboons (Depy *et al.*, 1997) classify novel prototypes at better rates than exemplars that they were exposed to during training. In other words, the animals exhibit more rapid and accurate classification of a prototype stimulus that they have never seen, compared to less typical exemplars that were previously reinforced.

Despite this evidence, some authors disagree that animals can form cognitive prototypes. In the case of pigeons, categorization could be explained by an exemplar theory if subjects focus on unique features of stimuli within a category, as opposed to features that can be generalized across a category (Jitsumori *et al.*, 2011). (Pigeons are particularly good at memorizing visual exemplars which is one reason that they are frequently used in these experiments.) A more stringent test of prototype theory suggests that both monkeys and humans categorize complex visual stimuli according to a prototype model (Smith *et al.*, 2008), but no other species have been tested using this procedure. Nor is it known whether monkeys and other non-human primates can form prototypes for functional or relational categories.

In contrast, prototype theory is frequently discussed in terms of social categorization, particularly kin recognition. Phenotypic matching, described in Section 9.5.2, indicates that many animals have a stored template of genetic relatedness; this may be modified as they interact with familiar kin and with unrelated conspecifics. The resulting prototype would be a ranked list of distinguishing features that could be used to discriminate kin from non-kin: traits that are represented frequently in the familial group are likely to show up in other kin and self-identified traits are stronger predictors of genetic relatedness than are traits expressed by distantly related kin. Based on a prototype model, therefore, kin recognition may emerge through a process of self-referent mapping in which an individual stores information about their own phenotype, as well as those of known relatives (e.g. parents or littermates), and then compares these to the phenotype of newly encountered conspecifics. Importantly, the kin prototype that is used for phenotypic matching is a combination of inherited and acquired information.

A prototype theory of concept formation assumes a higher level of cognitive processing than either elemental or exemplar theories because a prototype emerges when organisms extract commonalities across exemplars and rank the importance of criteria that determine group membership. This process does not involves consciousness: few individuals can verbalize why they say sparrow or robin, rather than penguin, when asked to name a bird, and prototype learning of visual categories is not disrupted in amnesic patients (Knowlton & Squire, 1993). Even so, the level of cognitive abstraction necessary for prototype formation may be restricted to primates; at least there is no conclusive evidence to suggest otherwise.

9.6.2 A synthesis

Although they are presented as separate topics, elemental, exemplar, and prototype theories are not mutually exclusive: it is unlikely that one theory or the other explains all instances of concept formation. If items in a category share one or more distinct features, concept formation may be explained by an elemental theory. When categories are poorly defined, subjects may rely on exemplars to classify new items. If category members include complex stimuli that vary on multiple dimensions, prototype formation may be the most parsimonious solution to categorization. In addition, there may be an interaction between different mechanisms of concept formation, at least in humans: participants use rule-based strategies (e.g. elemental) to categorize perceptual stimuli, but may augment these with exemplars if their performance on certain stimuli is weak (Rouder & Ratcliff, 2006).

A commonality in all three theories is that concept formation is modified by experience. Concepts are continually updated by new information, provided in the form of feedback on correct and incorrect classifications. This creates wide individual differences in concepts (as in the food example shown in Figure 9.14), regardless of the theory that explains how they are formed. In addition, elemental, exemplar, and prototype theories all assume that concept formation involves mental representation of group attributes that may not be apparent when stimuli are being classified.

This last point highlights a contentious issue in comparative cognition research. There is little doubt that most animals can categorize stimuli and events in their natural environment (at least to some degree), but how this relates to concept formation is still debated. Some authors argue for a cognitive discontinuity across evolution, noting that there is no solid evidence that animals share the human ability to make connections between abstract mental representations (Penn *et al.*, 2008), a necessary prerequisite for concept formation. Other researchers approach the issue in a different way: rather than asking 'can animals form concepts?', they compare *how* different species classify the same type of information. This methodology revealed that rhesus monkeys, unlike humans, use a stimulus generalization strategy to categorize perceptual stimuli (Smith *et al.*, 2004). Further studies of this nature are needed to uncover similarities and differences in the cognitive systems that support concept formation across species, information that could inform broader questions of cognitive evolution.

9.7 Neural mechanisms

Given the debate surrounding categorization and concept formation in animals, it is not surprising that the neural mechanisms of these abilities have been studied, most thoroughly, in humans. A large part of this work involves the study of brain-damaged patients, with neuroimaging and electrophysiological research complementing the findings.

9.7.1 Category-specific semantic deficits

In the early 1980s, Warrington and her colleagues published clinical reports of **category-specific semantic deficit**: the inability to access information about one semantic category while leaving others intact (McCarthy & Warrington, 1994). Patients in these studies had difficulty identifying and categorizing living, but not non-living, items. For example, asked to describe a specific animal (e.g. a giraffe), they were at a loss to come up with defining characteristics for this species. Likewise, provided with a photo or description of a particular animal, they could not generate the species' name. The results were the same regardless of whether the information was presented in written or verbal form, suggesting that the deficit was not sensory related but based on conceptual knowledge of the category. The same patients had few problems identifying manufactured objects such as tools or furniture, confirming that they did not suffer a general cognitive impairment. Within a decade, the opposite pattern was described in a different group of patients: intact conceptual knowledge of biological categories with an absence of conceptual knowledge of manmade categories. Since that time, more than one hundred different cases of category-specific semantic deficits have been reported. The most common impairment is an inability to categorize living things, with some patients displaying more specific deficits, such as an inability to identify animals or fruits and vegetables.

The most common cause of category-specific semantic deficit is stroke or head trauma. Brain damage in these situations is seldom restricted to one structure, but relating specific impairments to patterns of neural changes across patients revealed that the ability to identify living and non-living entities is mediated in distinct brain regions (Damasio *et al.*, 1996). Damage to the anterior regions of the left inferior temporal cortex disrupts conceptual knowledge of biological categories, whereas damage to posterior and lateral regions produces deficits in recognizing and identifying inanimate objects. The same brain regions are activated in normal controls when they view photos of animals versus tools, supporting the idea that neural organization reflects these categorical distinctions. Further research extended this work by identifying, more precisely, the brain regions associated with conceptual knowledge of biological and non-biological categories (Mahon & Caramazza, 2011).

The obvious interpretation of these studies is that cognitive representations of living and non-living items are separated in the human brain. This fits with evidence that one of the first conceptual distinctions babies exhibit is between animate and inanimate objects (Rakison & Poulin-Dubois, 2001). This domain-specific hypothesis was challenged by Warrington and colleagues who pointed out that living and non-living items are also distinguished by how they are identified. Living items are recognized, almost exclusively, by perceptual features and manmade items by their functional properties (e.g. tools or furniture). Thus, rather than living versus non-living, the brain may organize conceptual knowledge along a perceptual/functional distinction (McCarthy & Warrington, 1994). The debate continued with further, more detailed, investigations of category-specific semantic deficits: based on this evidence Caramazza and Shelton (1998) proposed that the brain differentiates categories that have evolutionary significance (e.g. food and animals) from those that do not (e.g. tools, furniture). The controversy is ongoing as researchers continue to examine the mechanisms that mediate categorization of different classes of items (e.g. perceptual and/or functional).

9.7.2 Facial recognition

In contrast, there is general agreement on the mechanisms underlying social categorization, at least in terms of conspecific recognition by humans. Acquired prosopagnosia (see Box 9.3) provided some of the first evidence that specific neural systems process facial information in that deficits emerge with damage to circumscribed brain regions. Further work with these patients along with neuroimaging in the intact brain and electrophysiological recordings in non-human primates produced an updated facial recognition circuit (see Figure 9.16). This cortical network links brain structures in the temporal and frontal lobes with other regions involved in visual processing or emotion (Atkinson & Adolphs, 2011). Importantly, different structures within the circuit mediate different aspects of facial recognition. For example, the fusiform face area contributes to lower level processing such as discriminating faces from objects, whereas the occipital face area is involved in higher level processing that helps to identify unique features of the individual. These features become associated with emotional states through neural connections that arise from subcortical structures, such as the amygdala. In this way, facial recognition in humans (and probably other primates) involves a dynamic interaction between several brain regions.

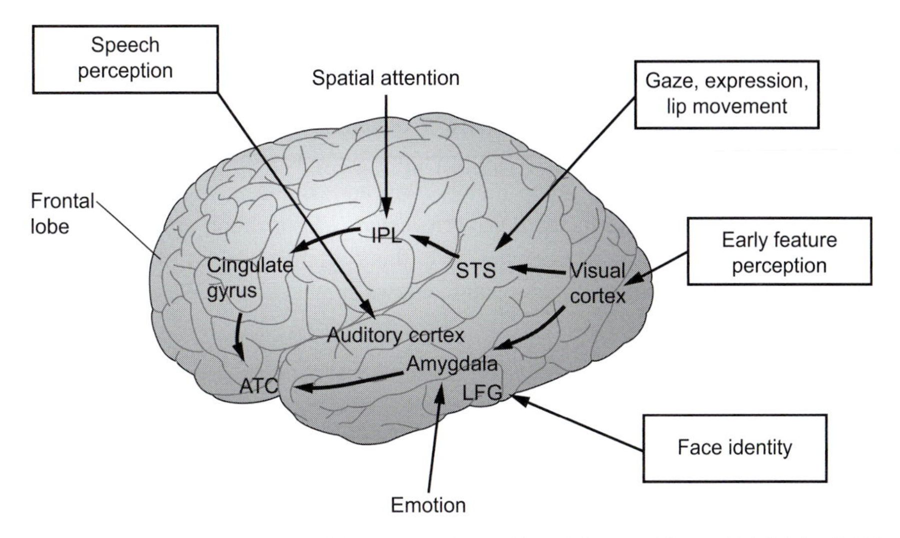

Figure 9.16 Neural substrates of face processing. Adapted from Atkinson & Adolphs (2011). See text for details. ATC = anterior temporal cortex, IPL = inferior parietal lobe, LFG = lateral fusiform gyrus, STS = superior temporal sulcus.

9.7.3 Concept formation

As with facial recognition, the neural mechanisms of concept formation involve an interaction between multiple brain sites. The most sophisticated view of this circuit was developed from electrophysiology studies in which rhesus monkeys learned to categorize visual stimuli (Freedman & Miller, 2008). In these experiments, specific features of a stimulus activated neurons in one cortical region or another (e.g. shape was encoded in the inferior temporal cortex and motion direction in the medial temporal cortex), regardless of how the stimuli were categorized. In contrast, prefrontal cortex (PFC) neurons fired when stimuli within a category were presented, regardless of the specific visual features. This suggests that neurons encoding basic sensory features of a stimulus are separated from those that encode its abstract meaning (i.e. group classification). The PFC receives input from a number of cortical regions (including inferior and medial temporal cortices) that process signals from different sensory modalities so it is a likely candidate for using this lower level information to develop categorization rules. The authors acknowledge that their model is not the final word on the topic: concept formation (even for perceptual categories) undoubtedly involves many other brain regions, but these studies provide an initial framework for further investigation of this process.

An alternate, although not necessarily contradictory, view is that concepts are represented by patterns of activation across several neural sites (Tyler & Moss, 2001). That is, abstract information about category membership may be coded as differences in how neuronal firing patterns are linked across time. This idea reflects the fact that the rules defining category membership can be independent of specific category attributes, and that there are different ways of organizing conceptual information. In other words, individuals may possess knowledge of both functional and sensory properties of the same object, and they may use different criteria for classifying the same item into different categories. These cognitive principles of conceptual knowledge must be represented in the brain (at least for those species that exhibit concept formation). Currently, it is almost impossible to examine whether concept formation reflects dynamic patterns of neural integration, although this may change with technological advances in the neurosciences. Ultimately, this work has implications for understanding how conceptual knowledge emerges, both developmentally and ontologically, as well as how it is structured at both the functional and neural levels.

Chapter Summary

- Categorization and concept formation are closely linked, although the former may occur in the absence of the latter. The cognitive process of grouping items together based on similarities as well as the mental criteria for defining group membership are examined, most effectively, using controlled laboratory studies. These methods have revealed categorization abilities across the animal kingdom, pointing to an evolutionary advantage to this cognitive process. Although it may lead to poor judgments or misattributions, categorization allows organisms to make cost-effective decisions based on minimal information.
- Humans and most animals exhibit some form of perceptual categorization, the grouping of stimuli based on similarities in sensory features. Even if different species group sensory stimuli into the same categories, they may not be using the same criteria to do so. Natural category research helped to reveal how different species are adapted to attend to specific features of sensory stimuli. The tendency to use this information in perceptual categorization is inherited, whereas the specific categories that an individual forms may depend on environmental experience.
- Functional categorization groups items together based on a shared meaning or association. It is studied, most commonly, using functional or stimulus equivalence tests. Functional categorization, particularly as it relates to natural categories such as food, has been observed in a variety of species, but the evolution of this ability is still debated.
- Relational categorization, grouping items based on connections, depends on the relative, not absolute, properties of stimuli to be categorized. An understanding of the rules that connect items or events is a form of analogical reasoning. Many animals exhibit the ability to form relational categories, but relatively few (at least of those tested) can transfer this information to higher order relationships.
- Social categorization, the classification of other organisms into groups, includes conspecific recognition, kin recognition, and recognition of non-related individuals of the same species. Conspecific and kin recognition involve both inherited and environmental factors; the relative importance of each depends on the evolutionary history and ecological constraints of a particular species. Recognition of unfamiliar kin may occur through a process of phenotypic matching in which individuals compare their own behavioral and physiological traits with those of another individual. Individual recognition develops through experience, and may be based on an evolutionary adaptation to link individual unique traits with fitness-related interactions.
- Concept formation, an understanding of the mental rules that define group membership, is explained most commonly by elemental, exemplar, or prototype theories. These are not mutually exclusive in that exemplar theory may be a precursor to either elemental or prototype theories. Moreover, humans or animals may use a different strategy to classify new information, depending on their prior experience with group membership. Despite extensive research on animal categorization abilities, there is currently no resolution as to which species share the human ability to form concepts.
- Category specific semantic deficits and brain imaging studies reveal that conceptual knowledge is organized in the human brain, although the cognitive separation of this information (e.g. living/non-living or perceptual/functional) is still debated. In contrast, the neural circuit mediating social categorization, specifically conspecific recognition in humans, is described in a network that connects fusiform and occipital face areas in the temporal and frontal lobes with brain regions

involved in visual processing or emotion. Concept formation also involves a dynamic interaction between multiple brain regions, although the precise mechanisms that mediate this process are yet to be established.

Questions

1. Explain how experiments that use novel exemplar or stimulus substitution manipulations provide information on the cognitive processes mediating categorization.
2. Generalization gradients are fine tuned by discrimination training in that the slope of the curve becomes much steeper on either side of the highest point. Surprisingly, the peak of the curve may correspond to a stimulus that was *not* presented during training. In other words, the generalization gradient may be shifted so that the greatest responding is to a stimulus that is presented for the first time in test sessions. This 'peak shift' is always away from the stimulus that was not reinforced during training. For example, if pigeons are reinforced for responding to a tone of 1000 Hz and not reinforced for responding to a tone of 900 Hz, the peak shift would be close to 1100 Hz. Provide both proximate and ultimate explanations for this phenomenon.
3. Explain how perceptual categorization may enhance or disrupt functional categorization. Could this enhancement and disruption happen at the same time?
4. Based on evidence presented in previous chapters (particularly Chapters 6, 7, and 8), defend the position that corvids can form higher order relational categories. Design an experiment to test this hypothesis.
5. Filial imprinting occurs in many animals, including humans, but it is displayed most dramatically in precocial species, such as geese. Why would this behavioral trait be more pronounced in some species than in others?
6. Olfactory signals that are emitted by an individual and used by conspecifics in social recognition do not develop until weaning. Provide both proximate and ultimate explanations for the delayed emergence of this individual chemical signal.
7. How does prototype theory apply to perceptual, functional, and relational categorization?
8. Use the three primary theories of concept formation to explain kin recognition.
9. Design an experiment to test whether category specific semantic deficits reflect a dissociation between perceptual and functional categories or between items that have evolutionary significance and those that do not. How would this paradigm be adapted to a brain imaging experiment with normal controls?

FURTHER READING

A thorough and updated summary of experimental work examining categorization and concept formation in lab animals is presented in:

Zentall, T. R., Wasserman, E. A., Lazareva, O. F., Thompson, R. R. K., & Rattermann, M. J. (2008). Concept learning in animals. *Comp Cogn Behav Rev*, **3**, 13–45.

For convincing arguments that insects possess many of the same cognitive abilities, see:

Avarguès-Weber, A., Deisig, N., & Giurfa, M. (2011). Visual cognition in social insects. *Ann Rev Entomol*, **56**, 423–443.

The mechanisms underlying categorization by humans (specifically visual object recognition) are compared to those of computer vision, with the idea that similar problems and solutions have arisen in these two fields.

Dickinson, S. J., Leonardis, A., Schiele, B., & Tarr, M. J. (eds.) (2009). *Object Categorization: Computer and Human Vision Perspectives*. New York, NY: Cambridge University Press.

Naturalistic studies of categorization and concept formation are built on the legacy of Peter Marler's groundbreaking work on birdsong. His influence is apparent in an edited volume that explores both the proximate and ultimate explanations of this behavior. In addition to straightforward explanations of research findings, the book includes beautiful illustrations of birds in their natural environment and birdsong recordings that 'bring the science to life.'

Marler, P. S., & Slabbekoorn, H. (eds.) (2004). *Nature's Music: The Science of Birdsong*. San Diego, CA: Elsevier Academic Press.

This edited volume uses a multidisciplinary approach (e.g. psychology, biology, and sociology) to examine how and why individual animals respond differently to kin and non-kin.

Hepper, P. G. (ed.) (2005). *Kin Recognition*. Cambridge, UK: Cambridge University Press.

The following text presents a coherent overview of concept formation from the perspective of human cognitive psychology. The book is clearly written so that many seemingly confusing issues appear straightforward. Importantly, the author acknowledges that there is no 'answer' to the debate of whether exemplar or prototype theories provide a better explanation for category learning.

Murphy, G. L. (2002). *The Big Book of Concepts*. Boston, MA: MIT Press.

A more challenging analysis of the same topic, with specific reference to the study of linguistics and how metaphors work, can be found in:

Lakoff, G. (1987). *Women, Fire and Dangerous Things: What Categories Reveal About the Mind*. Chicago, IL: University of Chicago Press.

In a series of entertaining and enlightening books, Oliver Sacks explains how case histories of neurological patients have advanced cognitive neuroscience research. One of the most informative (at least in terms of categorization and concept formation) is Sacks' description of his own case in which eye cancer left him with an altered form of prosopagnosia. This can be found in the longer work:

Sacks, O. (2010). *The Mind's Eye*. New York, NY: Alfred A. Knopf.

10 Social competence

Background

Primate researcher Hans Kummer (as reported in Byrne, 1995) once observed a female baboon spend 20 minutes slowly moving over a short, 2-meter distance while remaining in a seated position. She inched over to a location behind a rock, and when she finally arrived there, she began to groom a young, subordinate male. This action is one that would not have been permitted by the dominant male who was resting nearby. Kummer observed that from the vantage point of the dominant male, because of the rock, only the back of the female could be seen, and thus the illicit behavior went unnoticed. Why did the female move to a location behind the rock? It is possible that over time, she learned that engaging in behaviors while near large solid objects like rocks did not result in threatening behavior from the dominant male. Alternatively, it is possible that she recognized that the dominant male would not be able to see her actions behind the rock, and thus he would be unaware of them, allowing her to groom in safety. These options represent two different cognitive processes, yet both result in the female's effective navigation of her social environment.

These processes have been the focus of research attention for the last three decades. For example, in 1978, Premack and Woodruff posed the question "Does the chimpanzee have a theory of mind?," by which they meant, like humans, "Does the chimpanzee interpret the behavior of others in terms of mental states such as goals, intentions, desires, or even beliefs?" (Premack & Woodruff, 1978). This can be a very difficult question to answer. How can one conclude that the behavior experimenters are observing is the result of an animal attending to patterns of behavior and creating associations that allow for prediction of future action (e.g. grooming a subordinate behind a large object does not lead to violence), *or* that animals are going beyond the surface characteristics of an action and positing an underlying mental state (e.g. grooming behavior cannot be seen if we are behind a large object)? It is a challenge to design experimental methods that allow for this distinction; strong conclusions are difficult to make and disagreements among researchers frequently arise.

Inherent to the examples above is another challenge of living within a group: keeping track of the relative social rank of others and yourself. The female baboon's grooming behavior would likely have looked much different if she was intending to groom the dominant male. She must have recognized something about the difference between the two males, and how she did it is a focus of active research. Did she need to see direct interactions between the dominant and subordinate males to learn their relationship, or were their separate interactions with other males informative enough for the female to create a representation of her group's hierarchy?

Chapter plan

Most animal species live in social groups and social living results in particular cognitive challenges that must be overcome in order to interact effectively. This chapter examines these social cognitive processes. The majority of this chapter will focus on the manner in which animals interpret the behavior of others. The research is ongoing, and debates among researchers will not likely abate anytime soon. The chapter begins by examining the detection of social (animate) entities, and then considers the understanding of others' intentions, perception, and knowledge. Then, there will be consideration of another fundamental aspect of effective social interaction: understanding of the social group structure in terms of relations and hierarchies. There are other topics that one might consider integral to group living, such as social learning, cooperation, and communication. Given the depth and extent of experimentation and theorizing in these areas, they will be discussed in separate chapters.

10.1 Detection of animacy

The detection of animacy would seem a basic, foundational ability for the ability to interact in a social environment. Recognizing conspecifics, predators, and prey is critical for adaptive behavior. For example, 'dark spot detectors' in the frog retina are cells that respond to stimuli that have the characteristics of small flying insects, prey for the frog. Upon their excitation, and with information from other areas regarding the speed, distance, and direction of the movement, the frog jumps after the insect.

This section will focus on two other, more complex systems that are important for the differentiation of animate, biological entities from inanimate objects in the environment: biological motion and face perception. There are likely other important characteristics that humans and animals use to categorize the animate world from the inanimate world (e.g. self-propulsion, presence of morphological features such as hair, contingent reactivity, etc.). We focus here on these two because of their history of comparative theory and research and because of the connections that have been made to social cognition in general.

10.1.1 Biological motion

The movement of animals, particularly vertebrates, is in most cases non-rigid, yet it is constrained by the configuration of the body and gravity itself. **Biological motion** offers an abundance of information to the adult human observer; individuals can extract information regarding species classification, gender, attractiveness, and emotion from how something moves. Intriguingly, humans do so even when the motion is shown in simple **point-light displays** created by depicting the movements of individual joints with single lights, effectively producing an animation of movement. These displays convey biological motion in the absence of key morphological features like faces, skin, or hair (e.g. Johansson, 1973, 1976; Troje, 2002). The perception is robust and so fast that fewer than 200 ms of motion is sufficient to identify a point-light human walker.

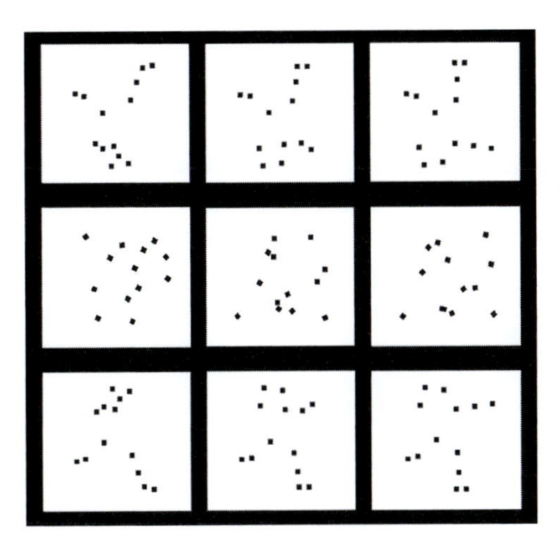

Figure 10.1 Examples of point-light displays. The top row depicts frames from a movie of a walking, point-light hen (biological motion display). The middle and bottom rows depict random (non-biological) and inverted point-light displays, respectively (Simion *et al.*, 2008).

The efficiency of this ability has suggested to researchers that the detection of biological motion is a fundamental process that is part of an early developing and evolutionarily endowed mechanism shared across species. As such, there has been an interest in examining biological motion detection from a comparative and human developmental perspective. For example, early research demonstrated that by 4 months of age, human infants distinguish upright from inverted human point-light walkers (e.g. Bertenthal *et al.*, 1984) and show a preference to attend to the former (Fox & McDaniel, 1982). Even infants as young as 2 days old differentiate between biological and random motion point-light displays and look longer at biological motion, even when the displays depict non-human animate motion (i.e. that of a chick; Simion *et al.*, 2008; Figure 10.1).

Humans are not the only animals to show this precocious ability to detect biological motion. Newly hatched chicks prefer to attend to point-light animations of a walking adult hen as compared to a depiction of random motion (Vallortigara *et al.*, 2005). One might be tempted to assume that the attention to biological motion was based on an innate template for hen motion instead of biological motion in general. However, this was dismissed by a second experiment in which the chicks were shown a point-light display of a cat and a display of non-biological motion; the chicks showed an attentional preference for the cat.

This ability to recognize and preferentially attend to the motion of biological entities has been hypothesized to underlie developing social cognition. At least in the case of humans, early differentiation of animate entities and attention toward them may provide necessary opportunities for learning about others' actions, goals, and later, beliefs. That is, by closely attending to those entities around it that are alive, an organism is provided with important experiences which can guide learning. Indeed, similar hypotheses have been presented in regard to face perception.

10.1.2 Faces

Northern mockingbirds (*Mimus polyglottos*) populate trees around the University of Florida campus in Gainesville, Florida, and females will defend their nests from intruders by swooping out of the nest while engaging in alarm calls and attacks (Figure 10.2). Levey and colleagues at the university were interested in whether mockingbirds would learn to distinguish humans who approached their nests posing a threat. Over 4 consecutive days, a human 'intruder' approached a nest and stood for 30 seconds. For 15 of these seconds, he or she placed a hand on the rim of the nest. A total of ten

Figure 10.2 A northern mockingbird. It is possible that these mockingbirds use the recognition of individual features, possibly within the face, to identify intruders.

intruders engaged in this activity, each at a different nest. The female mockingbirds flushed from their nests toward the approaching intruder increasingly faster as the days progressed. Then, on the fifth day, a new person approached the nest, acting just as the original intruder had. Now, mockingbirds reacted exactly the same way they did on the first day with the original intruder, as if this second person was novel to them. Together, these results suggest that over the 4 days (and with only 30 seconds of exposure each day), the birds very quickly came to distinguish and identify the particular human who approached their nest and learned that he or she was an intruder, a finding made all the more surprising considering just how many other students on this busy campus walked in the vicinity of the nest on their normal treks to class (Levey *et al.*, 2009).

It is likely that the mockingbirds used the recognition of individual features, possibly within the face, to identify and distinguish intruders from other humans. This conclusion is supported by two subsequent studies, one in which wild American crows (*Corvus brachyrhynchos*) directed vocal 'scolding' toward facial masks that had been worn by intruders even when they were worn by new people (Marzluff *et al.*, 2010) and one in which wild magpies (*Pica pica*) learned to recognize intruders even when hair style, hair color, skin color, height, and walking style were controlled as tightly as possible (Lee *et al.*, 2011). Faces are good predictors of the presence of an animate social entity (predator, prey, or conspecific), provide identifying information for individuals, and provide informative cues regarding underlying emotional states. This section expands on the earlier discussion of social categorization in Chapter 9, highlighting some of the growing body of research and theory on the perception of faces and emotion-reading in animals. This is an area of study in which many researchers are interested in comparisons with human abilities, and thus the work with animals has often focused on non-human primates.

Perceiving and discriminating faces

Human newborns look longer at and orient more toward images of faces (both real and schematic versions) than non-face patterns (e.g. Johnson *et al.*, 1991). This suggests that a specialized cognitive system is in place and working very early in life. One candidate system (**CONSPEC**) has been proposed by Morton and Johnson (1991; and updated in Johnson, 2005). This visual system in the brain works through innately specified parameters that orient the newborn to face-like patterns. The purpose of the system is to maximize the amount of input regarding faces that can be obtained during the first months of life, after which a second system, **CONLEARN**, becomes active and allows for further learning about the characteristics of conspecifics. (This may sound a little like the proposals made for the early orientation toward displays of biological motion. In fact, researchers have used the CONSPEC/CONLEARN models as analogies to explain the early attention to, and learning about, biological motion. A similar system, they argue, may be in place for motion as for faces.)

There has been a simultaneous interest in early face perception in non-human primates. If the systems proposed for humans are advantageous adaptations for social living species, they might also be in other social primate species. For example, similar to humans, is there a system that orients individuals to faces from very early in development? Further, how does face perception develop over time? Sugita (2008) tested a group of rhesus macaque monkeys (*Macaca mulatta*) that had never seen a human or monkey face, or even a picture of a schematic of one (this was achieved by raising monkeys apart from each other and having the human caregivers wear masks while interacting; Figure 10.3). In a preferential looking procedure, these monkeys looked longer at black and white faces (both human and other macaques) over non-face objects that were also novel (e.g. cars, houses, clocks). Thus, it is likely that macaques possess a system, possibly innate, that orients them to the characteristics of faces.

Interestingly, monkeys in a control group who had received normal exposure to faces (which consisted mostly of the faces of monkeys in their captive group) also showed a preference for faces, but only for the monkey faces. They preferred the faces of conspecifics over the non-face objects and over the faces of humans. Perhaps, then, exposure to faces affects face processing by 'tuning' it

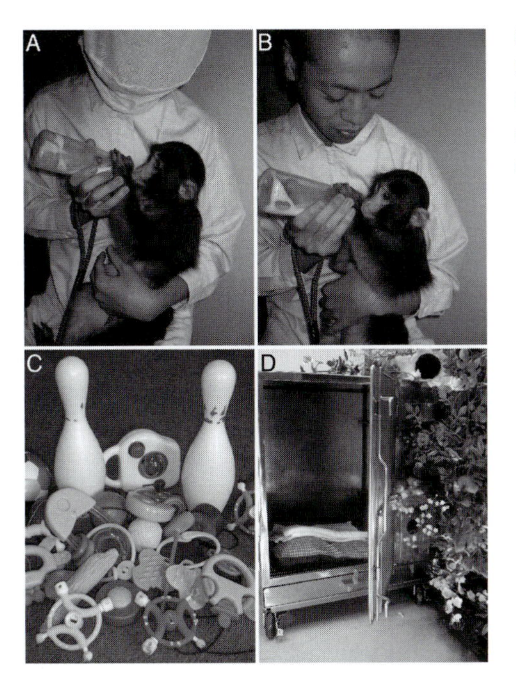

Figure 10.3 Stimuli and procedure in Sugita (2008). Rhesus macaques were raised without exposure to human faces (A) though had exposure to toys (C) and decorations (D). Later, they were exposed to human caregivers who did not wear a mask (B).

to the types of faces that are most prevalent in the environment. A further experiment followed up this finding. The monkeys who had been deprived of face stimuli were given 1 month of exposure to either human or monkey faces. After this period, they still showed a preference to attend to face stimuli, but only to the types of faces to which they had been exposed. For example, if they had been exposed to human faces, they looked longer at human faces than objects, but they looked at monkey faces for the same amount of time as non-face objects.

These findings are consistent with **perceptual narrowing**, a process by which experiences with certain stimuli result in a system that is tuned to those features, allowing for increased recognition and distinctions within those stimuli as compared to other stimuli. Human infants show a similar process. Pascalis and colleagues, for example, found that 6- and 9-month-old infants recognized and distinguished among human faces, but only the younger group could also do so with a set of monkey faces (Pascalis *et al.*, 2002). This suggests that the face system becomes tuned via experience to human faces between 6 and 9 months of age; humans effectively lose the ability to easily distinguish among non-human faces without adequate exposure, presumably to gain expertise with those of their own species. There is suggestion that perceptual narrowing for faces exists in species beyond humans and macaques, including chimpanzees (*Pan troglodytes*) and sheep (*Ovis aries*). Indeed, these abilities may have played an integral role in the development of social structures that are based on relationships between individuals in which attention to individuals is important.

The findings from perceptual narrowing studies are also interesting in light of the inter-species facial discrimination that appears to be occurring for mockingbirds, crows, and magpies in the research presented earlier. If mockingbird face processing, for example, was tuned to conspecific faces, how would they make such a discrimination among humans? This is an area of future research, but there are a few possibilities: (1) perhaps face perceptual tuning does not occur in mockingbirds; (2) these mockingbirds live in an urban environment in which humans are prevalent, and perhaps their systems are tuned for humans as well; (3) the faces of the intruders were learned over the exposure trials just as adult humans can learn to distinguish, say, monkey faces. In any case, the propensity of animals to attend to faces and to recognize and distinguish among individuals would seem to be fundamental for social species (see Box 10.1 for a consideration of the ability to recognize oneself).

Emotion reading from faces

Another ability that can effectively aid social interactions is the ability to gather information beyond identity from faces. In this section, the ability to read emotion from faces, a topic that has been studied in depth with chimpanzees by Parr and her colleagues, will be considered. Several facial expressions in chimpanzees, like those that occur during play (a relaxed open-mouth face), submission (a bared-teeth display), screaming (an open mouth with teeth showing), and pant–hoot vocalizations (an open mouth with extended lips), are readily noticeable by human observers. Images of these expressions have been used in a match-to-sample procedure (MTS, similar to the MTS procedures described in Chapter 3). Here, a sample image is matched to one of two other images on a single particular stimulus dimension (e.g. type of facial expression). For example, an image of chimpanzee A with a scream face could be matched to chimpanzee B's scream face, but not chimpanzee C's play face (Figure 10.4; Parr, 2003).

Chimpanzees could match photographs of different chimpanzees who were emitting the same expression, but then the task was modified slightly such that chimpanzees had to match images in terms of likely associated contexts. Chimpanzees watched videotaped events depicting either

Figure 10.4 Facial expression match-to-sample task based on Parr (2003). An image of a scream face could be matched to another scream face, but not a play face.

positive scenes (e.g. favorite toys, foods) or negative scenes (e.g. veterinarians with tranquilizer darts) and were presented with two images of facial expressions (e.g. for positive, a play face, and for negative, a scream face). Chimpanzees made appropriate matches, suggesting that they were linking the likely underlying emotions of the scenes with actual depictions of facial expressions. The mechanism underlying the correct matching still requires additional study, but it could be based on a form of emotional contagion (see Chapter 11) in which the situations engendered certain emotions in the participants which were then matched to the facial expression. It is also possible that the chimpanzees' choices were based on a learned association between these types of events and the accompanying facial expressions of conspecifics or even an understanding of the meaning of the facial expression that allows consideration of possible causes (Parr, 2003).

Box 10.1 Mirror self-recognition

In the 1970s, Gallup conducted a series of studies with the aim of determining whether chimpanzees have a sense of self, which he proposed was fundamental to also understanding the mental states of others. To test this, chimpanzees were observed for 10 days after a mirror was placed in their enclosure. Upon first seeing the mirror, chimpanzees often responded to their own image as if it were another chimpanzee – they would threaten it, greet it, and engage in other social responses. Over the course of a few days, though, they began to decrease their social responses and increase self-directed responses such as grooming parts of their body that they could not otherwise see. These observations suggested that chimpanzees recognized that the image in the mirror corresponded with their own bodies (Figure 10.5). To further test this interpretation, Gallup designed the **Mark Test** (Gallup, 1970). While the chimpanzees were anesthetized, they were marked on one eyebrow and one ear with an odorless, colorful dye. After recovering from the anesthesia, the chimpanzees were observed for 30 minutes during which they were without access to a mirror. When the mirror was placed back in the enclosure, chimpanzees observed themselves and touched and rubbed the marks, something they had not done before the mirror was available. Chimpanzees appeared to recognize themselves in the mirror and through this, learned about the presence of the marks on their bodies. Interestingly, unlike the chimpanzees, rhesus monkeys did not pass the Mark Test.

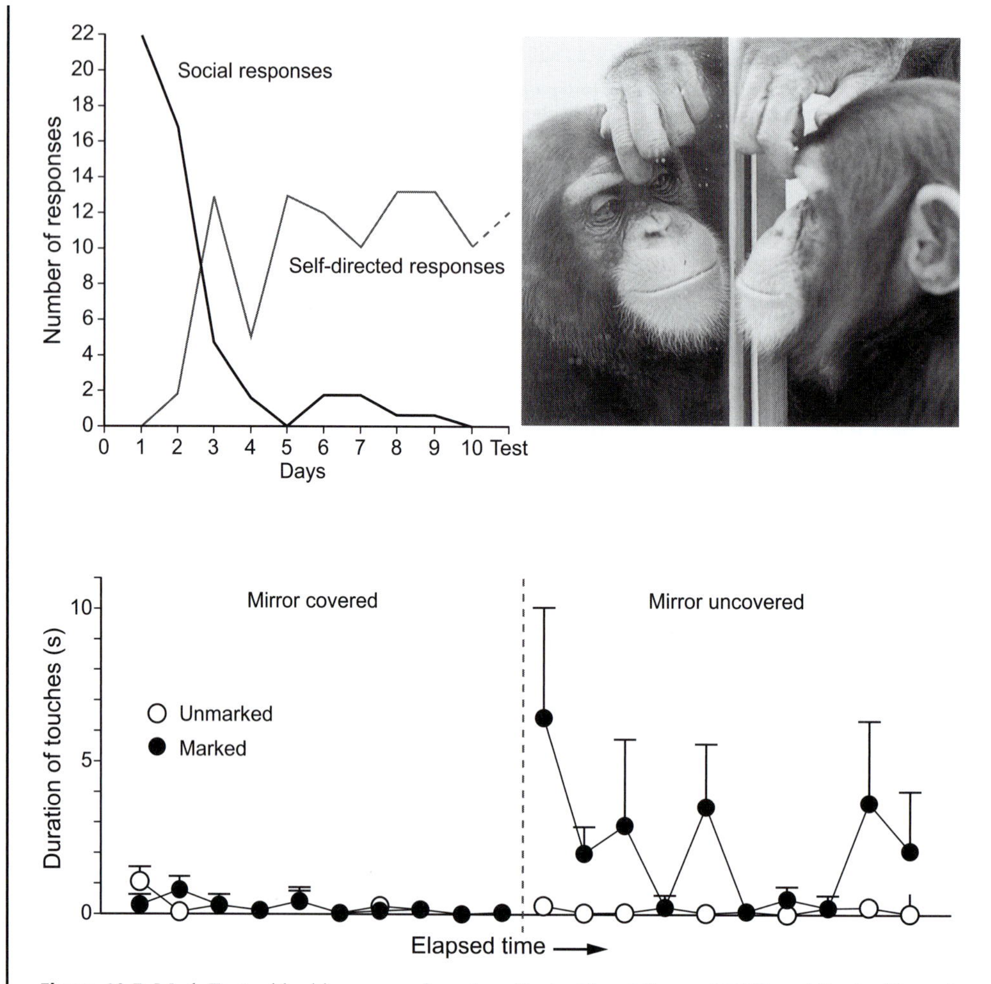

Figure 10.5 Mark Test with chimpanzees based on Povinelli and Pruess (1995) and Povinelli *et al.* (1997). In the top row, social and self-directed responses to initial mirror exposure are graphed. In the bottom row, responses to marked and unmarked areas of the face during the Mark Test are graphed.

Gallup interpreted the data as indicating that chimpanzees have a 'self concept,' which caught the attention of quite a few critics. First, many criticized the use of the term 'self concept.' Some researchers were hesitant to consider it as akin to human's self concept, complete with a sense of uniqueness, dispositions, and mortality (e.g. Heyes, 1994). Others thought it more appropriate to conclude that the chimps were simply responding to an interesting image that had movement patterns that correlated with their own. This interpretation, in turn, seemed a bit flawed as it still required chimpanzees to be aware of their own body movements. At the very least, many researchers agreed that chimpanzees show self-perception (matching between kinesthesis and vision) instead of self-conception, meaning that they recognize their own bodies in the mirror, but might not have a concept of self that is similar to our own (e.g. Heyes, 1994).

In recent years, many different species have been tested and new methodologies have been used which have suggested that the number of species that show mirror self-recognition may be greater than previously assumed (e.g. Mitchell, 2012) and that, in some cases, the Mark Test is not an appropriate test of self-recognition (Broesch *et al.*, 2011). Particularly relevant to this chapter, though, is the fact that Gallup originally proposed that passing the Mark Test is also indicative of having a conception of others, particularly of others' mental states. The link here is a bit tenuous, however. As will be seen in this chapter, researchers are still actively debating whether chimpanzees or any other species interpret the behavior of others in terms of mental states.

10.2 Thinking about thinking

In 1978, Premack and Woodruff tested whether an adult female chimpanzee, Sarah, had a **theory of mind**. That is, did Sarah recognize that those around her had minds and mental states, like intentions and beliefs, which guided their behavior? Sarah was presented with videotaped scenarios of a human actor who was trying to solve a variety of problems. In one event, for example, an actor was struggling to get food that was placed just out of reach. The tape was paused before the actor solved the problem, and Sarah was provided with two photographs to choose from, one of which depicted the solution to the problem (e.g. the person picking up a long stick to rake in the food) and one that did not (e.g. picking up a short stick). In 21 out of 24 trials with different scenarios such as this, Sarah chose the photograph depicting the correct solution.

This is truly a fascinating finding, but one obvious interpretation is that Sarah's behavior might not have reflected an understanding of the actor's mental states (his intentions and desires) but what she knew about the laboratory environment. It is possible that, in each case, she chose the picture that fit with a familiar sequence of events that she had witnessed before. For example, Sarah may have observed long tools being used to reach food in the past and associated the two. Still, it would be wrong to dismiss the study altogether, especially considering the research it inspired during the decades that followed and up to present day. Not only is social cognitive research on theory of mind with animal species being completed, but also human developmental psychologists have made great strides in charting the ontogeny of this ability, and researchers studying developmental disabilities such as autism have come to recognize a core deficit in mental representation in the disorder.

At this point, there is no straightforward 'yes' or 'no' answer to whether chimpanzees, or any species other than humans, have a theory of mind, but the research has suggested that there are likely many ways in which animals can interpret the behavior of others (Call & Tomasello, 2008). A recurring debate in this and the next two sections, for example, will be whether animals respond in experimental tasks based on past experience and learned behavior rules or whether they are also able to represent certain mental states.

10.2.1 Mental attribution

When adult humans observe someone's behavior, a representation is created that includes such elements as the physical actions, the setting, and the internal **mental states** (e.g. feelings, desires, emotions, beliefs, and intentions). These mental states are the assumed causes of the behavior: *he reached for the ice cream sundae on the counter because he wanted to eat it*, or *she looked in the* *cupboard because she believed that the coffee mug was in there.* Often, mental state attribution will allow us to make predictions about future actions: *he will likely eat ice cream again when given the chance*, or *she will look for the coffee mug elsewhere if she does not find it in the cupboard.* In this way, mental state attribution is very powerful as a means of understanding the actions of others and interacting effectively with them; it allows people to act kindly (offer him a sundae after our next dinner) or even deceptively (secretly move the coffee cup to a new location after she puts it away in the cupboard). Indeed, difficulties in mental state attribution can have dramatic effects, as observed in individuals with autism (see Box 10.2).

Box 10.2 Autism

Imagine a silent, animated movie starring the shapes seen in the figure below (Figure 10.6). There is a rectangle with a slit that can open and close, a big triangle, a small triangle, and a small circle. The objects are animated in such a way as to depict a story. For example, provided with a 50-second clip, a typically developing teenager might describe what he saw as the following:

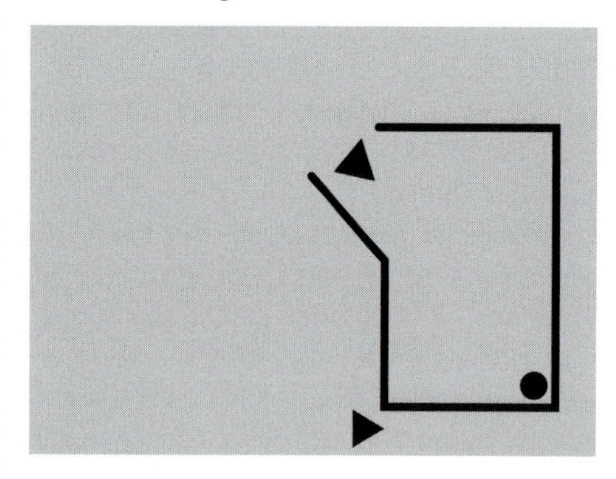

Figure 10.6 An example based on a still frame from the movie by Heider and Simmel (1944).

"What happened was that the larger triangle – which was like a bigger kid or a bully – and he had isolated himself from everything else until two new kids came along and the little one was a bit more shy, scared, and the smaller triangle more like stood up for himself and protected the little one. The big triangle got jealous of them, came out, and started to pick on the smaller triangle. The little triangle got upset and said like 'What's up?', 'Why are you doing this?'. . ." (Klin, 2000).

In contrast, an adolescent with **autism** might describe the scene in this manner:

"The big triangle went into the rectangle. There were a small triangle and a circle. The big triangle went out. The shapes bounce off each other. The small circle went inside the rectangle. The big triangle was in the box with the circle. The small circle and the triangle went around each other a few times. They were kind of oscillating around each other, maybe because of a magnetic field. . ." (Klin, 2000).

The movies were originally designed by Heider and Simmel in 1944, and have since been used by autism researcher Klin to examine spontaneous mental attribution in individuals with autism. Autism is a 'spectrum' disorder, and individuals may have a wide range of characteristics from low to high IQ, and little language to full language abilities (and a very small subset have savant-like capabilities in a specific domain). One fairly common characteristic, however, is poor social interactions (e.g. Baron-Cohen, 1985). Many researchers subscribe to the theory that individuals with autism have limitations in their theory of mind abilities, and one can see this play out in the descriptions of the movie scene above as well as difficulties in false belief tests described later in this chapter.

The extent of the limitations and the affected cognitive mechanisms are still being determined. Indeed, this research must progress alongside human developmental psychology research as it attempts to chart typical theory of mind development. Further, comparative research can aid in characterizing human social cognitive mechanisms via comparison with other species. For example, by considering differences between species in terms of social cognitive processes, we become more aware of possible cognitive systems that are alternatives to mental state attribution (e.g. Apperly & Butterfill, 2009).

Importantly, mental states are internal to an individual. So, to attribute a mental state means that, for example, one might offer a person a sundae not solely because that person was observed to reach for one last week (observable behavior), but because one thinks that person likes and wants ice cream (invisible, posited mental state). Many developmental psychologists assume humans readily reason about the mental states of others by at least 4 years of age (e.g. Wellman & Liu, 2004). Humans are a verbal species and can thus report our attributions directly and 'pass' a wide variety of experimental procedures designed to test mental attribution. It is much harder to examine whether any non-human animal species, or even human infants, engage in mental state reasoning. Verbal answers to test questions are, of course, not available. Further, as with Premack and Woodruff's study with Sarah the chimpanzee, it is difficult to determine whether the subjects are actively attributing mental states or coming up with the correct answer by some other means.

10.2.2 Alternatives to mental attribution

As noted above, Sarah the chimpanzee could have performed well on the experimental tasks by means of associative learning. She may have linked certain situations and elements of her environment and responded accordingly (e.g. a picture of a person holding a hose is associated with a picture of a hose attached to a spigot more than a hose not attached, because the former events have been common in her environment). In this sense, Sarah may not have been considering the behavior of others at all. Other cognitive processes that also do not involve the attribution of invisible, internal mental states can lead to correct performance as well. Animals may, for example, only represent and reason about behaviors yet perform on a task in a manner that would otherwise suggest that they also engaged in mental state reasoning (e.g. Penn & Povinelli, 2013).

Most experimental procedures involve tests designed with the idea that in a given situation, if a subject responds in a certain way to the behavior of others, it must be via mental state attribution. A problem for researchers, however, is that perhaps the subject is just responding on the basis of the behavior it was observing. To take an example from Povinelli and Vonk (2004) that will be expanded upon in Section 10.4, imagine a chimpanzee observing two experimenters, one wearing a blindfold and one not blindfolded. The chimpanzee must decide whom to approach with a food begging gesture. If the chimpanzee approaches and begs from the experimenter without the blindfold (which is actually not what was found in the real study, but this is just a thought experiment), a possible interpretation is that the chimpanzee gestured to this experimenter because he or she knows that only people who can see them can respond appropriately to begging. Here, the mental state of seeing is being attributed. However, it is also possible that the chimpanzee gestured to this experimenter because he or she knows that only people with visible eyes respond to begging. This latter possibility is an example of reasoning based solely on representations of behavioral contingencies and regularities without reference to mental states.

In the next two sections, there will be discussion of how animals, and in some cases, human infants and young children, reason about the behavior of others. Of primary interest is whether they reason directly from behavior or whether they consider both behavior and the mental states that drive it. Three mental states will be discussed: intention, perception, and knowing. There is good reason for considering these mental states separately within a more general consideration of theory of mind. Developmental psychologists often posit that understanding certain mental states (e.g. seeing) emerges earlier in development than reasoning about other mental states (e.g. beliefs; Wellman & Liu, 2004). Additionally, some comparative researchers have proposed that during the course of primate evolution, mental state reasoning may have evolved in discrete steps, thus making it possible to find a shared understanding of, say, seeing, across species, but differences in reasoning about other mental states (Povinelli & Eddy, 1996).

10.3 Understanding intentions

For humans, the ability to understand others as intentional agents is critical for many aspects of our social lives. This section will consider both the developmental research completed with infants and young children, as well as analogous work with animals, primarily primate species, on understanding intentions. By 'intentions,' many researchers are referring to mental states like 'wanting' or 'desiring' that lead agents to complete certain actions (e.g. Call & Tomasello, 1998). In turn, these actions are often referred to as 'goal-directed' actions. Of course, sophisticated mental attribution can include additional mental states (e.g. she wants the toy and *believes* it is in the box), but here the focus will be on intentions.

10.3.1 Infants' detection of goal-directed action

Human infants recognize the 'boundaries' in an action stream that are related to the initiation and completion of intentions; 10-month-olds look longer at events in which action is paused in the middle of a goal action as opposed to events in which the pause occurs after a goal has been completed (Baldwin *et al.*, 2001). Infants also seem to have interpretations about the goals themselves. After witnessing a person repeatedly reach and grasp one of two toys, 6-month-old infants expect the person to continue to reach for that particular goal toy, even if its location has been moved (Woodward, 1998).

Infants will also attribute goals to non-human agents. Using computer-animated objects (e.g. balls) with features such as non-rigid transformations and self-propulsion, Gergely, Csibra, and colleagues (1997, 1999) found that 12-month-olds develop expectations about the trajectories of objects based on their apparent goal-directedness. In that study, looking-time measures suggested that infants assumed that a ball that had approached another ball in a rational, efficient manner (e.g. jumped over an obstacle blocking its path) would subsequently approach the goal ball in a straight-line trajectory if the obstacle was no longer there, as opposed to repeating its jumping action.

Many developmental researchers have interpreted these results and others to mean that by at least 6 months of age (and perhaps earlier), infants view certain actions as goal-directed. But what is meant by 'goal-directed' can vary; some researchers claim that infants are only reasoning about behavior, and some claim they are reasoning about mental states such as the desires that cause behavior. In the task by Woodward described above, for example, infants may be reasoning that the person "grasped the toy" (behavior) or they might be reasoning that the person "grasped the toy because she wanted it" (behavior and mental state). Another, slightly humorous way of thinking about this distinction is to consider the joke, "Why did the chicken cross the road?" The standard answer is based only on behavior: "To get to the other side." The mentalistic reasoning answer, however, is different: "It wanted to get to the other side." Both types of reasoning lead an observer to make predictions based on the actions observed – the person might grab for the same object in a new situation or the chicken my cross other roads it comes upon – but only one is considered mental attribution (Csibra & Gergely, 1998).

Both types of reasoning tend to be called 'goal attribution' in infant literature. To be sure, infants do seem to be encoding the end state of actions as important, and they are not just considering actions in terms of physical movement alone. That is, infants at least form the behavioral representation that, for example, "the ball went to the other ball" as opposed to simply "the ball is moving toward the right." The question that developmental researchers often face is when and how the infant comes to represent both behavior and underlying mental states. Research with animals is currently facing the same conundrum (see also Box 10.3 for further considerations).

10.3.2 Interpretation of goal-directed action by animals

Some of the experimental tasks completed with human infants have been slightly modified to be used with non-human primates. Uller, for example, has tested infant chimpanzees using the procedure that Gergely, Csibra, and colleagues developed. Like human infants, looking-time measures suggested that the young chimpanzees interpreted the actions of an animated ball as directed toward a goal, and the chimpanzees assumed that the ball would continue to approach the goal in a rational, efficient manner in a new context (Uller, 2004). Additionally, Wood and colleagues tested three primate species in a task based on the Woodward (1998) study. Two monkey species, tamarins (*Saguinus oedipus*) and rhesus macaques, as well as chimpanzees, observed a human actor reach for and grasp one of two potential food containers (Figure 10.7). When subsequently given access to the containers, the majority of observers, regardless of species, chose to inspect the container that had been grasped (Wood *et al.*, 2007; Wood & Hauser, 2011). Interestingly, not just any type of action on a container would lead to it being chosen; if the actor accidently made contact with a container, subsequent choices were at random.

Both of these tasks can be solved by reasoning about behavior without reference to mental states (e.g. "the ball moved toward the other ball" and "the person grasped that container"), but some have pointed out that tasks including unintentional actions like the 'accidental' condition in Wood *et al.* (2007) may allow for conclusions based on mental state attribution (Call & Tomasello, 2008). The logic of this argument is that when an agent's action leads to successful goal completion, the desire/intention matches the outcome of the behavior, and thus, it is virtually impossible to distinguish when a participant is reading behavior or reading behavior and positing a mental state. In accidents and failed actions, however, the intention and the behavioral outcome do not match.

Using this logic, and tests incorporating accidental actions, Call and Tomasello have suggested that chimpanzees do understand the goals of others with reference to mental states. In one study, chimpanzees responded differently to situations in which an agent did not give them a food treat because he was unwilling to do so or because he was unable and failed to do so. Note that in this case, the outcome is the same: the chimpanzee receives no food. However, the underlying intentions

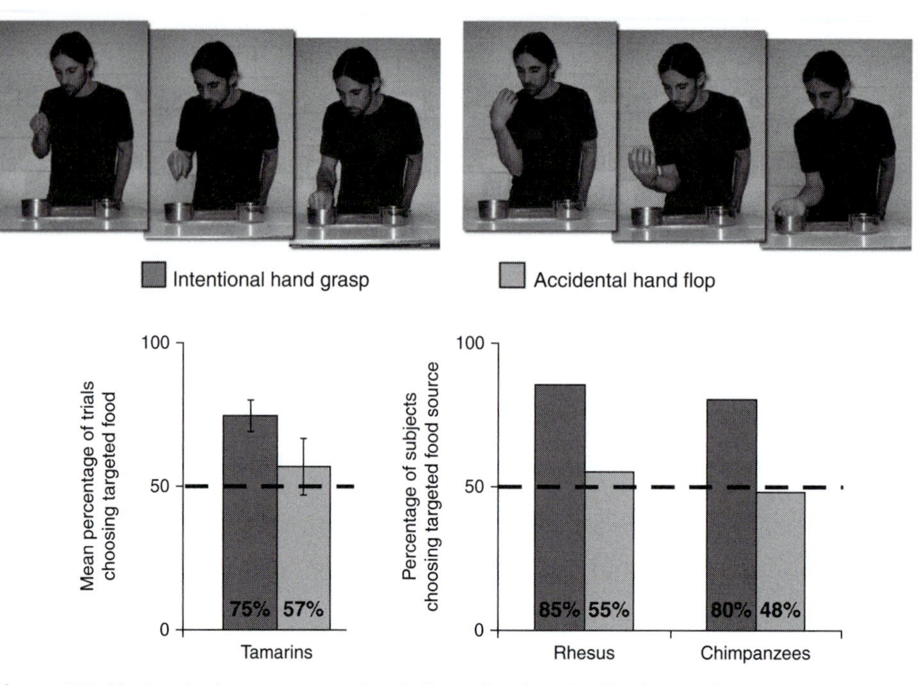

Figure 10.7 Testing the interpretation of goal-directed action. In Wood *et al.* (2007), three primate species observed a human actor reach for and grasp one of two potential food containers. With access to the containers, the majority of observers chose to inspect the container that had been grasped. When the actor's behavior was portrayed as accidental, subsequent choices were directed equally to each container.

of the agent differ, and appropriately, chimpanzees left the testing arena earlier and became more agitated when the actor was unwilling to give food (Call *et al.*, 2004). Other studies have also suggested chimpanzees' understanding of intentions, showing that chimpanzees react differently to accidental versus intentional actions (Call & Tomasello, 1998), that they help people who are unable to reach an object (Warneken & Tomasello, 2006), and that they will copy intentional actions on novel objects more than accidental actions (Tomasello & Carpenter, 2005). The authors of these studies propose that the novel nature of the actions and the objects that the actions work upon, as well as the variety of tasks that are used, suggest that something akin to the spontaneous attribution of intentions is present (Call & Tomasello, 2008).

Box 10.3 Mirror neurons

Great excitement followed the discovery of **mirror neurons** in the rhesus macaque (Gallese & Goldman, 1998). Neurons in the premotor cortex fire when monkeys reach for and grasp objects, and intriguingly, some of these neurons also fire when the monkey observes another monkey doing the same action. Human brains seem to have analogous regions (Figure 10.8).

Some researchers have theorized that mirror neurons are the means by which we attribute goals, intentions, and beliefs to others (e.g. Gallese *et al.*, 2004). In one experiment, researchers took advantage of the fact that monkey mirror neurons respond to actions that are directed toward objects (i.e. a grasp) but not to a mimed version of the same action (i.e. a grasp with no object present). Monkeys were shown events in which an object was or was not placed behind an occluding screen, followed by a hand reaching behind the screen. Thus, the actions were the same, but in one case, they were directed toward an unseen object. Mirror neurons fired when the occluded object was grasped, but not when the object was not present, suggesting that the neurons respond to more than just the external, physical features of an action. Instead, they only seem to respond to actions that are directed toward goals.

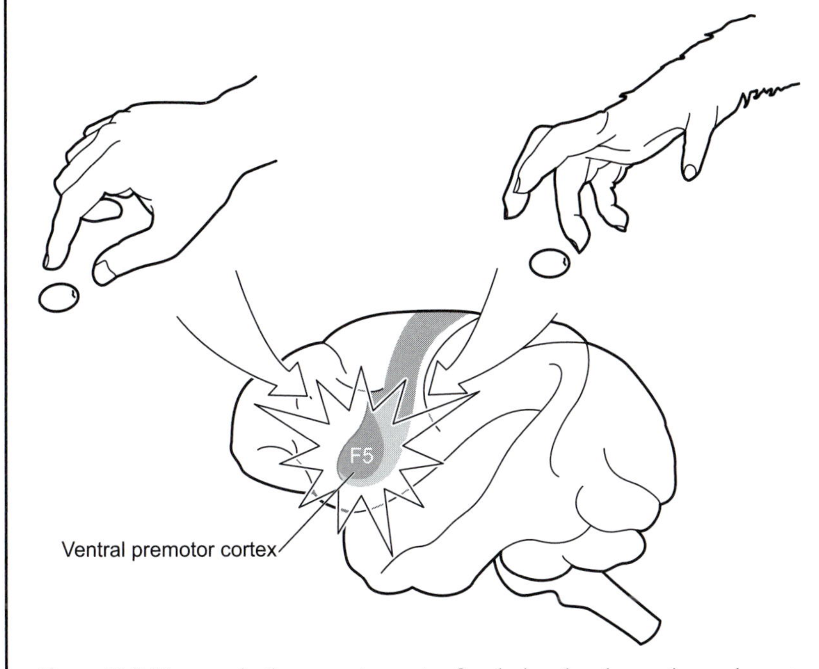

Figure 10.8 Neurons in the premotor cortex fire during the observation and enactment of certain actions.

A question still remains, however. Do these mirror neurons really represent the internal states (e.g. *wanting* the object) of the actor? Further research is required, and for now, many researchers question whether these neurons are responsible for anything more than interpreting the physical movement of an action, and perhaps making predictions as to probable end states of the action. In any case, there will no doubt be many future studies of the mirror neuron system, elucidating its role in social cognitive processes.

10.4 Understanding seeing and knowing

In their original paper described above, Premack and Woodruff proposed several methods for testing whether an animal's theory of mind extends to the appreciation that seeing leads to knowing. Using terminology from Call and Tomasello (2008), this research question can be rephrased as: Can animals understand what someone else *sees* (what he registers from the environment), or even what he *knows* (what information he has previously registered)? In this section, you will see that in the decades after Premack and Woodruff's paper, the field of comparative cognition has come closer to answering this question, but it still remains elusive.

10.4.1 Gaze following

One of the best ways to figure out what a person is seeing, and thus, becoming aware of, is to follow their eye gaze. For this reason, gaze following behavior in animals, particularly chimpanzees, has been the focus of much research. Across a range of experimental tasks, chimpanzees have demonstrated that they can follow a person's line of sight and subsequently look toward what the other person is looking at. Call and Tomasello (2008) review experiments showing that chimpanzees can follow gaze to locations behind them, 'check back' with a person if the agent does not seem to be looking at anything relevant, and even move to see the side of an opaque barrier that the person is looking at. In some cases, chimpanzees will also use a person's gaze to cue the location of hidden food (Barth *et al.*, 2005).

The range of gaze following that chimpanzees engage in indicates that the behavior is not caused by orienting reflexes that simply respond to the direction of eyes and faces. Still, it is almost universally accepted that gaze following behaviors do not definitely demonstrate that a chimpanzee understands what an agent sees. That is, gaze following experiments do not provide evidence that chimpanzees actively represent that a person is seeing something and thus having a mental experience (e.g. Penn & Povinelli, 2013). What they do suggest is that at the very least, chimpanzees are inferring the target of a person's looking behavior in a sophisticated manner that takes into account physical constraints such as opacity. The experimental attempts to examine the attribution of seeing and knowing thus have had to move beyond the documentation of gaze following and instead rely on different types of measures, which are described next (see also Box 10.4 for a simplified task used with dogs).

10.4.2 Early experimental attempts

In the early 1990s, Povinelli and colleagues conducted a series of studies with chimpanzees and rhesus monkeys that followed a general paradigm in which subjects could use the information provided by actors to find hidden food. In the first stage of the experiment, there were four food containers. At the beginning of a trial, the containers were hidden behind a screen, and the subject observed an experimenter hide food in front of another experimenter (the Knower). During this process, a third experimenter was out of the room (the Guesser). The Guesser then entered the room, and the screen was removed. The Guesser and Knower pointed toward different containers, but the Knower pointed to the site containing the hidden food, while the Guesser randomly chose from the remaining three (Figure 10.9). The chimpanzee or monkey could then choose a container and would receive the food reward if they chose the baited site.

If subjects understand the knowledge states of others, they should choose the container indicated by the Knower. Three of the four chimpanzees tested consistently chose the container advised by the Knower. The rhesus monkeys, however, were never able to discriminate between the Guesser and Knower. In the case of the chimpanzees, however, the task can simply be solved by learning a behavioral rule that choosing the person who remains in the room while the container is baited yields a reward. After all, multiple trials were completed (approximately 300 for the chimpanzees), allowing time to learn such a rule. In fact, the tendency to choose the Knower increased over all the trials, as if they were learning a rule, not showing an understanding

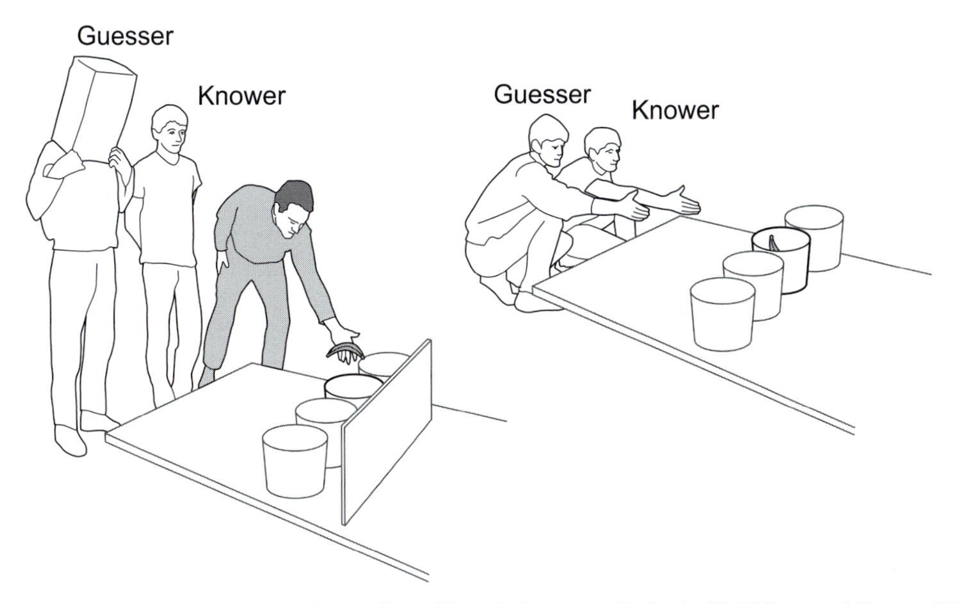

Figure 10.9 Examining the understanding of knowledge states. In Povinelli, Nelson, and Boysen (1990), during the hiding of the food, the Guesser was either outside of the room or present but unable to see (i.e., bag over the head). Subsequently, only the Knower pointed to the correct food location.

of knowledge. This stood in stark comparison with the performance of 4-year-old human children who choose correctly from the first trial.

Subsequent tests also called into question the chimpanzee's ability to understand perception and knowledge. Other experimental designs in which chimpanzees could beg for food from one of two experimenters, one who could see the chimpanzee and the begging gesture and one who could not due to a blindfold or turned-away posture, also failed to reveal sensitivity to the perception of others. Here, chimpanzees begged from both experimenters equally (Povinelli & Eddy, 1996).

Findings such as these led to the general conclusion that neither chimpanzees nor other non-human primates are able to understand that others have the mental states of seeing or knowing. They may be capable of rather sophisticated behavioral interpretations, but in the studies just described, these are the result of learned behavior rules. Of course, not all researchers agreed with this conclusion. As described in the next section, some argued that the experimental procedures were simply not appropriate for the species being tested (e.g. Hare *et al.*, 2000).

Box 10.4 Domestication

Domestic dogs perform better than chimpanzees on tasks in which they must choose between two odorless containers, one of which hides food, using only the gaze and point of a human experimenter as a cue (e.g. Hare *et al.*, 2002). Though the potential behavioral outcomes of domestication have been a focus of study for a long time, the finding that dogs respond better to human communicative gestures than even the chimpanzee has spawned much interest in the connection between domestication and social cognitive processes.

Figure 10.10 A specialized breed of the domestic dog: the Labrador retriever.

It is estimated that sometime between 14 000 and 135 000 years ago, humans and dogs began sharing the same ecological niche (e.g. Druzhkova *et al.*, 2013; Sablin & Khlopachev, 2002). It was likely a mutually beneficial arrangement; humans provided waste areas that contained food remnants and dogs were allowed to stay, perhaps even guarding the area. Over time, humans began to domesticate dogs, effectively creating a new species (*Canis familiaris*), as well as specialized breeds (Figure 10.10).

Researchers are still examining why dogs appear to be so well-tuned to human communicative gestures, but recent data are starting to limit the hypotheses. One possibility, for example, is that dogs learn to be responsive to these cues through conditioning. Dogs are raised around humans and no doubt have much exposure to human gestures; learning behavioral contingencies leading to reward (e.g. food, praise) would seem possible. Two findings, however, call this into question. Wolves (*Canis lupus*), even if socialized with humans from a young age, do not perform as well as domestic dogs. Second, dog puppies typically do well at following human points, even as young as 9 weeks of age and regardless of whether they were raised with a human family or in a kennel with littermates (Hare *et al.*, 2002; though see Wynne *et al.*, 2008). Thus, exposure to learning opportunities with humans cannot fully explain this unique ability of domestic dogs. In fact, the failure of wolves to follow human gestures as readily as dogs also discounts another hypothesis. One might have assumed that since wolves live in packs that cooperatively hunt, they also may be unusually flexible in the use of social information from others, and dogs may have inherited this trait. But again, wolves do not perform as well as dogs on the experimental tasks, making this explanation unlikely.

It is instead possible that the domestication process itself led to dogs' social-communicative skills. During domestication, there was selection for systems that mediated aggression and fear of humans, and the social skills we see today might be a byproduct of this process (Hare & Tomasello, 2005). Indeed, further research with domestic foxes (*Vulpes vulpes*; Hare *et al.*, 2005) and goats (*Capra hircus*) (Kaminski *et al.*, 2005) provides converging results; they too follow the social cues of humans.

10.4.3 Recent research using alternative procedures

Tasks like those described above, in which subjects must use the cooperative gestures of humans to find the location of hidden food, might not be appropriate for chimpanzees, who are more likely to compete for food (Hare *et al.*, 2000). Competing for food, not sharing information about food, is a more apt characteristic of primate societies (e.g. Wrangham, 1980). Secondly, in the earlier tasks, chimpanzees were being tested on the ability to attribute knowledge to humans, not to conspecifics, and the latter is more likely to occur in their natural environment.

To address these potential experimental limitations, Hare and colleagues developed a novel testing procedure in which chimpanzees would have to compete with other chimpanzees for food (Hare *et al.*, 2000; Figure 10.11). A subordinate and a dominant chimpanzee were in rooms that were on opposite sides of a third room. Each room contained a door that could be opened wide enough to allow the chimpanzee to see into the third room, but not leave. An experimenter placed two pieces of food at various locations while the doors were closed. In one condition (Dominant–Door), one piece of food was visible to both chimpanzees, but the other was only visible to the dominant chimpanzee. In another condition, both pieces of food were visible to both chimpanzees (Door–Door), and in a third condition, one piece was visible to both, while the other one was only visible to the subordinate chimpanzee (Subordinate–Door). As subordinate chimpanzees do not take food from dominant chimpanzees, the measure of interest was whether the subordinate would go to a piece of food in any of the conditions. That is, in the condition in which the subordinate could see the food, but the dominant could not, would the subordinate enter and retrieve it?

Subordinate chimpanzees did obtain more food in the Subordinate–Door condition during which the dominant did not see one of the pieces of food; they reliably took the food that was visible only to them. Several control conditions were tested as well, including a test of whether the subordinate chimpanzee was actually considering what the dominant one could see, or whether the subordinate was simply monitoring the dominant's behavior and going to the location that he did not go to. In this case, the procedure was similar to the Subordinate–Door condition except that the dominant chimpanzee's door did not open all the way, and the subordinate was released before the dominant. Again, the subordinate went for the food that the dominant could not see.

A subsequent study using a similar procedure also examined whether chimpanzees can know what others know (i.e. what they have seen in the past, instead of what they are presently seeing; Hare *et al.*, 2001), and the results were positive, suggesting to the researchers that chimpanzees would consider the knowledge states of other chimpanzees when competing for food. Such a claim could be made stronger, however, if analogous findings were obtained in situations with completely different surface features and interactions. Thus, Hare and colleagues conducted an additional series of experiments. Here, chimpanzees concealed how they approached food so as to influence what a competitor could or could not see and hear. For example, a chimpanzee was in competition for food

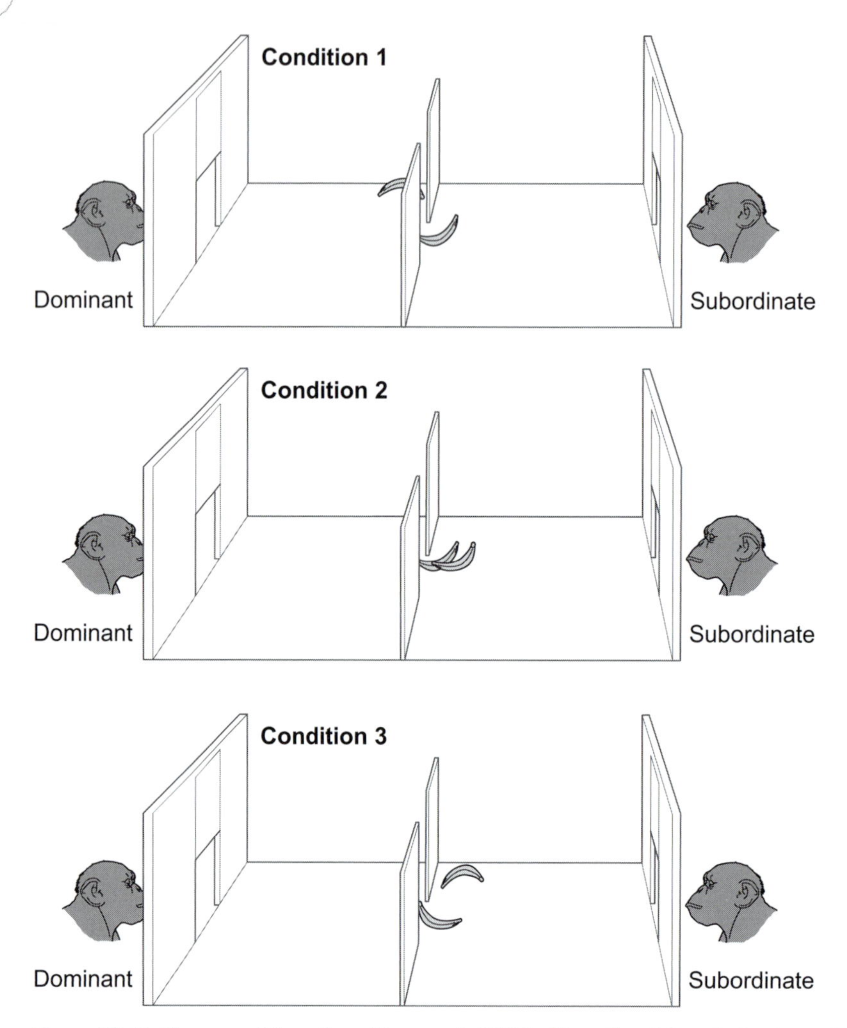

Figure 10.11 Three conditions from Hare *et al.* (2000). Here, Conditions 1–3 refer to Dominant–Door, Door–Door, and Subordinate–Door, respectively, as described in the text.

with a human experimenter who was inside a windowed booth. Chimpanzees chose to only approach food that was closer to a side of the booth that was blocked with an opaque screen, effectively hiding their approach from the experimenter.

Thus far, this section has focused on chimpanzees; however, other experiments have led some researchers to suggest that rhesus macaques may also recognize seeing and knowing in others in competitive tasks (Flombaum & Santos, 2005). Additionally, ravens (*Corvus corax*) and scrub-jays (*Aphelocoma californica*) may also take others' perception into account when protecting their food caches (e.g. Bugnyar & Heinrich, 2005; Dally *et al.*, 2006b). Western scrub-jays are socially living, food-storing birds that are faced with the challenges of remembering where their caches are and protecting their caches from thieves. A primary strategy that corvids use to protect their caches is 're-caching,' that is, moving their food to another location when the original cache location was observed by a competitor. Work by Clayton and colleagues (e.g. Dally *et al.*, 2006a; see Chapter 3) has argued that constituent episodic encoding (i.e. encoding what, where, when, and perhaps who) allows scrub-jays to engage in strategic re-caching (Dally *et al.*, 2006a; Emery & Clayton, 2001).

In these studies, birds cached food in distinctive trays while observers in an adjoining cage looked on. When subsequently given the opportunity to recover food in private, jays re-cached food items more often if they had been previously observed by a dominant group member than if the observer had been a partner or a subordinate (Dally *et al.*, 2006a).

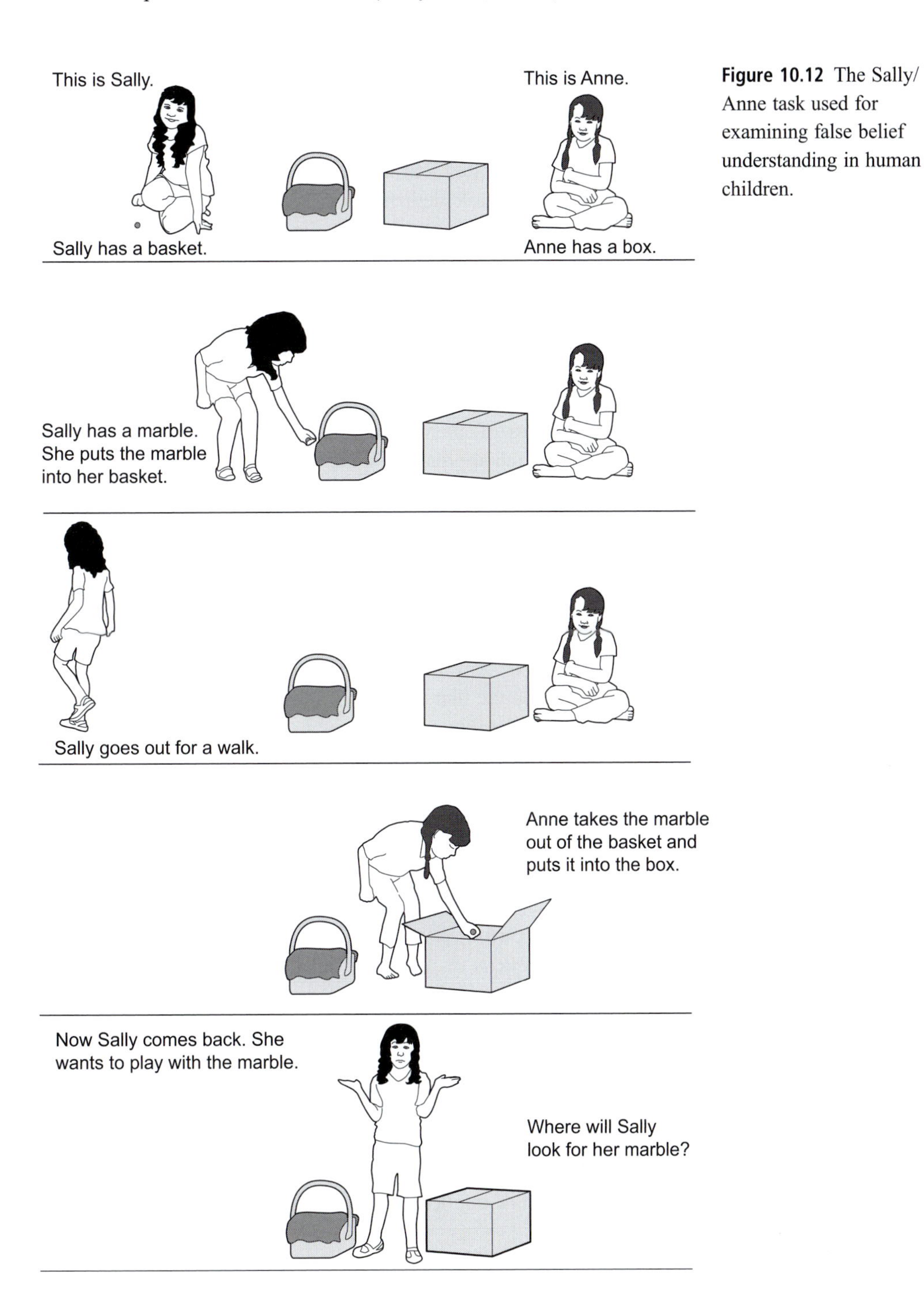

Figure 10.12 The Sally/Anne task used for examining false belief understanding in human children.

Understanding false beliefs?

Humans show an additional aspect of knowledge attribution that appears unique to our species (see also Box 10.5). We can predict what another person will do based on what he or she knows, even when we know something else to be true. That is, we can represent others' **false beliefs**. Verbal tests of false belief understanding such as the popular "changed location" or "Sally/Ann" task are typically not passed until 4 years of age (Figure 10.12; Wellman & Liu, 2004). Here, children are asked to report where a character will look for her toy in a situation in which the character is ignorant of the fact that her toy has been moved to a different location. The child is aware of the correct, present location of the toy, but must reason about the character's (incorrect) knowledge of the situation. Recent studies have suggested that infants and toddlers may also have the ability to consider false beliefs in others when task demands are altered, though this remains a point of exciting debate (Apperly & Butterfill, 2009; Onishi & Baillargeon, 2005; Surian *et al.*, 2008).

Box 10.5 Imaging the human social brain

The incorporation of brain activity imaging techniques such as **fMRI** and **EEG/ERP** into the study of human social cognition is still in its infancy, but is already having an impact on our understanding of the neural bases of intention and emotion reading, as well as belief and knowledge attribution. The research approaches some important questions: Do humans have evolutionarily designed, specialized mechanisms for reasoning about other minds or do human social cognitive abilities share domain general systems with other cognitive abilities such as numerical competence or physical causality? Also, given the research reported in this chapter, there may be social cognitive abilities that humans share with apes and monkeys, and those that are not – are there distinct regions implicated in the human-specific aspects of social cognition such as false belief attribution?

Research thus far has provided answers of 'yes' to both of these questions, though the neuroscientists involved would be the first to admit that the work is only just beginning. As seen in Figure 10.13, detecting animate agents appears to be associated with the extrastriate body area (EBA); it responds selectively to human bodies, body parts, and biological motion, as compared to other familiar objects in the world. Interestingly, stories about a human body do not cause a response in this area (Saxe & Kanwisher, 2003).

The superior temporal sulcus region (STS) responds to goal-directed actions; activity is increased when the observer witnesses a mismatch between an action and its target. For example, the STS responds when a character gazes in the opposite direction to an interesting stimulus, or when a character picks up an object that she previously frowned toward (Pelphrey *et al.*, 2003; Vander Wyk *et al.*, 2009). It has also been suggested that this effect is absent in participants with autism. A region adjacent to the STS, the temporal parietal junction (TPJ), appears to be involved in the attribution of beliefs. Response is high, for example, when participants read stories about a character's beliefs, but not when they read about the character's appearance, or even the character's internal states such as hunger (e.g. Saxe & Powell, 2006).

There is ongoing debate regarding the role of the medial prefrontal cortex (mPFC). Some argue that this region is involved in emotional perspective-taking and shared goals, but not belief attribution, while other work implicates the mPFC in false belief attribution and its development (Liu *et al.*, 2009).

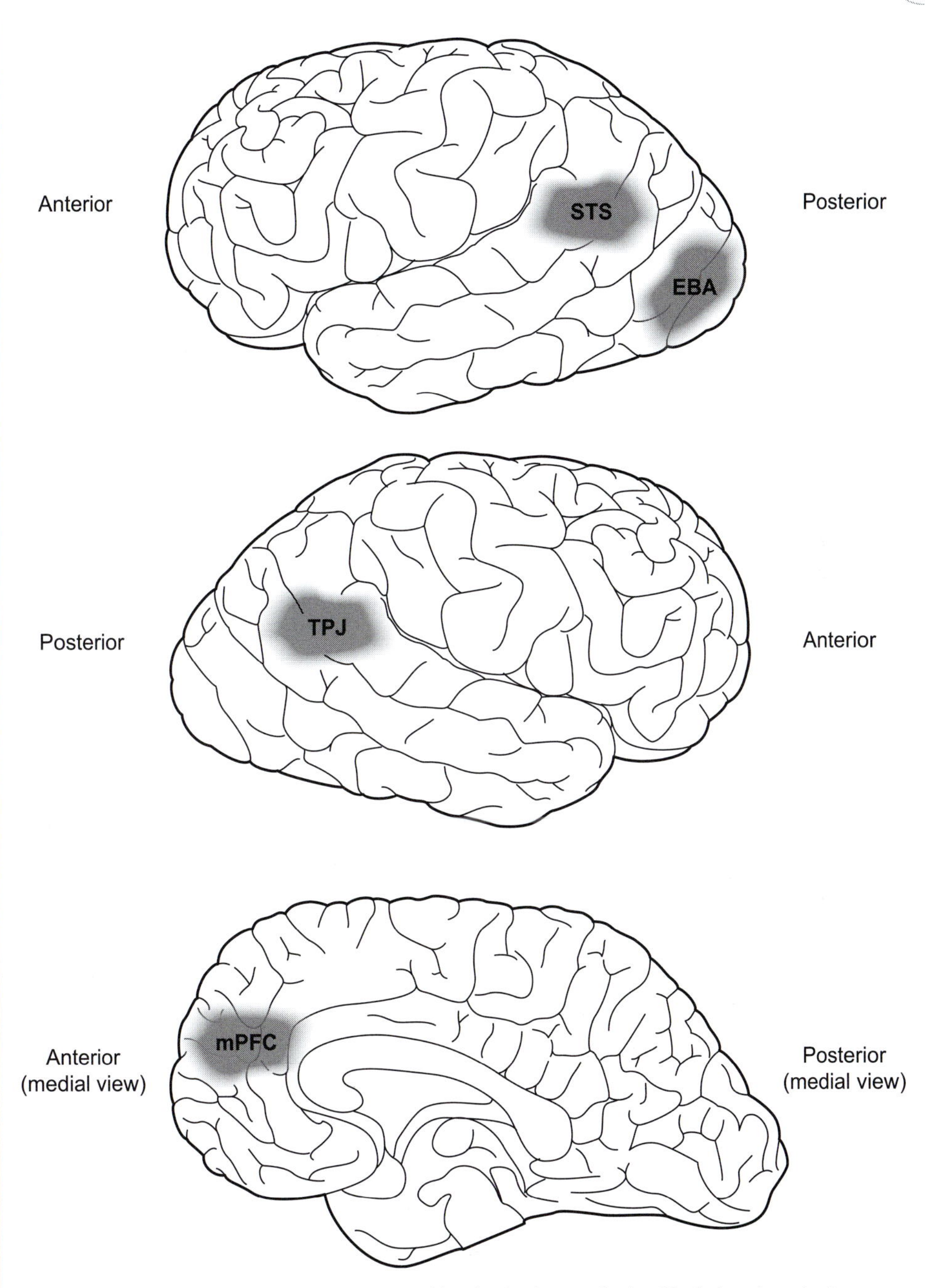

Figure 10.13 Areas implicated in social cognition in the human brain. Shaded regions in the top brain depict the extrastriate body area (EBA) and the superior temporal sulcus region (STS, posterior). The shaded region in the middle brain depicts the location of the temporal parietal junction (TPJ). The shaded region in medial view shown in the bottom brain depicts the medial prefrontal cortex (mPFC).

Even those researchers who suggest that chimpanzees can recognize what others see and know conclude that there is no evidence to date that chimpanzees recognize that others may hold false beliefs. At least three experiments have directly compared human children and chimpanzees and it is only the former who pass (e.g. Call & Tomasello, 2008). One further example comes from a particular condition in the study by Hare and colleagues described above. In this condition, the dominant chimpanzee always observed the location of the food, but then the food was moved to another location. On some trials, the dominant saw the food being moved, but in others he or she did not ('Misinformed' condition). Thus, the dominant chimpanzee held a false belief as to the current location of the food. If subordinate chimpanzees understood that the dominant chimpanzees had a false belief, they should have tried to get the food, assuming that the dominant would search at the wrong location. However, performance did not differ from the trials in which the dominant chimpanzees observed the moving of the food, thus providing no indication of false belief understanding.

10.4.4 The ongoing debate

It has been proposed that given the range and variety of situations that chimpanzees have been tested in, it would be *less* parsimonious to claim that chimpanzees might have behavioral rules for each novel situation than to claim that, like humans, they understand what others see and know. By this argument, any one experiment can be discussed in terms of behavioral rules, but it becomes harder to accept that chimpanzees create them within minimal trials in novel situations as the number of tasks increases (Call & Tomasello, 2008). This claim, however, is part of a larger debate regarding the social cognitive processes of animals, and to some extent, of humans.

Povinelli and his colleagues often present a counter argument to claims of mental state reasoning about seeing and knowing in animals, proposing instead that no current study definitively demonstrates it. Interestingly, the two sides agree that low-level behavioral rules are likely not the best explanation for the 'successes' seen in experimental tasks. There is, however, another option that does not commit one to concluding either that, for example, chimpanzees are solely operant learners or that they are actively representing mental states (e.g. Penn & Povinelli, 2013). Penn and Povinelli argue that chimpanzees are likely not creating a long list of unrelated, statistically learned facts or rules about behavior and slogging through them to solve experimental tasks. Instead, they may have a system that allows for past experience of others' behavior to be organized by biases or heuristics that would allow them to recognize causally relevant aspects of current behavior and reason about likely future behavior. For example, when passing some of the tasks created by Hare and colleagues, chimpanzees may be demonstrating an understanding of the causal relation between a dominant chimpanzee's line of sight, the type of object (food versus a rock, which chimpanzees do not compete over), and the actions the dominant is likely to engage in. This is not mental attribution, but it is still a very powerful cognitive system. In fact, Penn and Povinelli further propose that humans share this system with chimpanzees and often use it when navigating the social world, yet humans also have an ability to 'reinterpret' the behavior information into mental state terms when required, which in turn allows humans to engage in unique social cognitive reasoning.

Of course, only time and subsequent study will settle this debate and determine the correct interpretation. The procedures are beginning to broaden (as are the types of species that are being tested), and the exciting challenge is to create tasks that cannot be solved by a system that reasons about behavior alone. Through this process, we may also come to better understand our own, human social cognitive mechanisms.

Researcher profile: **Dr. Michael Tomasello**

Michael Tomasello holds an enviable position for a comparative cognition researcher. In his role as the co-director of the Max Planck Institute for Evolutionary Anthropology in Leipzig, Germany, Tomasello leads the Developmental and Comparative Psychology Department. Here, researchers examine the cognitive abilities of the four great ape species housed at an affiliated primate research center, dogs, and human children. His research topics are far-ranging, including cooperation, theory of mind, social learning, communication, and specifically with children, the acquisition of language.

Tomasello was trained primarily in the United States, receiving a bachelor's degree in Psychology from Duke University and a PhD in Experimental Psychology from the University of Georgia. Prior to moving to Germany, Tomasello spent 18 years at Emory University in Georgia and was affiliated with the nearby Yerkes Primate Center. His vast body of research is testament to the power of the comparative approach, in this case, one that intersects phylogeny, ontogeny, and cultural approaches. In his theoretical papers and books based on his and others' empirical studies, Tomasello often considers the critical differences between humans and animals (most often between human children and non-human primates, specifically apes). He and his co-authors have stated the following about human cognition:

> It is one form of primate cognition, but it seems totally unique as people go around talking and writing and playing symphonies and doing math and building buildings and engaging in rituals and paying bills, and surfing the web and creating governments and on and on. Also unique in the animal kingdom, human cognition is highly variable across populations, as some cultures have complex foraging and navigational techniques whereas others have very few of these, and some do algebra and calculus whereas others have very little need for complex mathematics. And so the biological adaptation we are looking for is one that is rooted in primate cognition but then provides humans with the cognitive tools and motivations to create artifacts and practices collectively with members of their social group – that then structure their and their offspring's cognitive interactions with the world. We are thus looking for a small difference that, by creating the possibility of culture and cultural evolution, made a big difference in human cognition. (Tomasello *et al.*, 2005)

What is this small difference, then? Tomasello proposes that humans have an adapted trait for engaging in collaborative activities and sharing intentions with others. As children, shared intentionality develops in humans as the ability to understand intentionality (that, he argues, humans share with apes) intertwines with human-unique motivations to share mental states, emotions, and activities with other people, ultimately leading to the development of human-unique symbolic communication, artifact use, and a culture that supports and encourages both.

10.5 Social knowledge

The sophisticated understanding of others' behavior that may be present in animals would be advantageous to an individual's ability to predict the behavior of others in order to interact effectively with other members of the group. There is another aspect of social cognition, however, that is likely a key element to the maintenance of smooth social interactions as it too provides information from which one can predict behavior. This aspect is often referred to as **social knowledge**, the recognition of individuals and their relationships to one another.

Many animal species are thought to live in socially complex societies, though what this "complexity" refers to may differ. In the case of social insects, for example, thousands of individuals may live within the same colony, interacting with each other differently based on sex, caste, or reproductive state. Many monkey species live in considerably smaller groups that can range from 10 to 30 individuals, yet their interactions appear to be sensitive to dominance, kinship, sex, and even previous interactions, suggesting a complex system of social interactions. Baboons (*Papio anubis*) and vervet monkeys (*Cercopithecus aethiops*), for example, will engage in **redirected aggression**; after being threatened by one individual, a monkey may redirect his own threatening behaviors to an individual related to the one who originally threatened him, suggesting some sort of knowledge of the relationships between other individuals in the social group (e.g. Cheney & Seyfarth, 1989; Smuts, 1985; see Engh *et al.*, 2005, for examples from non-primates).

Yet, social complexity does not *necessarily* imply the underlying workings of complex cognitive mechanisms. Using the example of baboons: "Does a baboon that apparently knows the matrilineal kin relations of others in her group have a 'social concept,' as some have argued (e.g. Dasser, 1988), or has the baboon simply learned to link individual A1 with individual A2 through a relatively simple process like associative learning, as others believe (e.g. Thompson, 1995)? At present, the preferred explanation depends more upon the scientist's mind than upon any objective understanding of the baboon's" (Seyfarth & Cheney, 2003b, p. 208). This section will consider proposals regarding the organization and structure of social knowledge, primarily in vervet monkeys, as this species has been followed closely by psychologist Robert Seyfarth and biologist Dorothy Cheney in the wild. In particular, there has been focus on kin relations and dominance rank, both of which will be described below.

10.5.1 Kin relations

In vervet society (Figure 10.14), males emigrate to other groups when they reach maturity while females remain in their natal group. Subsequently, the hierarchy follows the stable matrilineal families with offspring ranking right below their mothers. There is ample evidence that individuals recognize members of their own families, given their propensity to groom and form alliances more often with kin. However, to navigate effectively through the social group, one should also know about the relationships among members outside of one's own family, that is, **third-party relationships**. (It would seem important, for example, to avoid threatening an individual who is part of a more dominant matriline.)

One piece of evidence for this type of social knowledge comes from the example of redirected aggression presented above, but redirected aggression can be even more complicated. Cheney and Seyfarth (e.g. 1986, 1989) found that a female (A1) is more likely to aggress toward another monkey (B1) if a relative (A2) was recently involved in an aggressive encounter with her opponent's relative (B2). This suggests that A1 recognizes a similarity between her relationship with A2 and the relationship between B1 and B2.

10.5.2 Dominance rank

As mentioned, the vervet hierarchy follows matrilineal lines, and to human observers, it follows a linear, transitive order: if 1 is dominant to 2, and 2 is dominant to 3, then 1 is dominant to 3. Monkeys could potentially recognize the ranking differences among the other members in their group or simply keep track of those who are dominant or subordinate to them. Observations from wild vervet monkeys suggest that this species does, in fact, recognize other individuals' relative

Figure 10.14 Vervet monkeys. In this species, the hierarchy follows matrilineal families, with offspring ranking right below their mothers.

ranks. Cheney and Seyfarth took advantage of the fact that females often compete for grooming partners to guide their observations. If two females are grooming and a third one arrives, the arriving monkey will supplant one of them and resume grooming (Seyfarth, 1980). Interestingly, if the female who approaches is dominant to the other two, it is the less dominant of the two groomers who leaves. For this to occur with any regularity, it would appear that the more dominant of the groomers would have to recognize her rank relationship with both the approaching female and the present grooming partner, as well as the relationship between the approaching female and the grooming partner (Cheney & Seyfarth, 1990a). That is, for her to remain seated, she would have to recognize her 'middle' position between the approaching female and the grooming partner, and that the partner is ranked lower than both herself and the approaching female.

Underlying cognitive mechanisms

There are at least two, if not more, potential cognitive mechanisms that might underlie the ability to recognize third-party relationships and hierarchies. One suggestion is that animals form **equivalence classes**, that is, sets of stimuli that have been grouped together via experience. Importantly (and different from stimulus generalization) the grouping is not based on physical similarity but prior association. Thus, it is possible that vervets form an equivalence class consisting of members of a matriline based on their prior association. A vervet may then threaten the relative of an opponent because of the class membership. Additionally, this allows for an understanding of the transitivity in the hierarchy because if A1 and A2 are equivalent and B1 and B2 are equivalent, then A1>B1 would imply that A2>B2 (Schusterman & Kastak, 1998; Seyfarth & Cheney, 2003b; Figure 10.15).

However, there are researchers who suggest that this type of associative process cannot adequately account for all forms of social knowledge. For one, within a matriline, aggression still

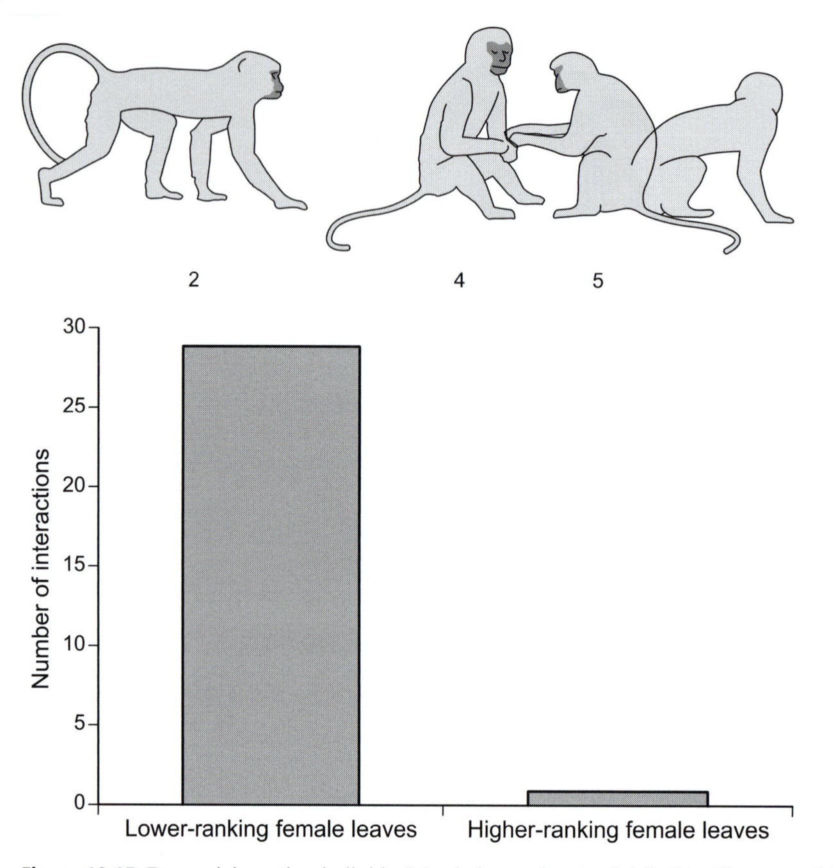

Figure 10.15 Recognizing other individuals' relative ranks. As detailed in Cheney and Seyfarth (1990b), if a second ranked individual approaches a grooming pair consisting of a fourth ranked individual and a fifth ranked individual, the fifth ranked individual leaves.

may occur at relatively high rates and grooming may occur in different amounts between different members. Additionally, friendships with non-kin exist and can fluctuate even on a daily basis. Thus, the formation of equivalence classes through associative relationships may be a difficult task (Seyfarth & Cheney, 2003b). That said, there is very likely a role played by associative learning in social knowledge; however, there may be other processes as well. One strong candidate is a memory system that allows for chunking, as discussed in Chapter 3. Individual members can be linked through associations, and this knowledge can be organized into hierarchical groups (chunks). Monkeys who must remember numerous dyadic and even triadic relationships would be greatly aided by a system that decreased memory load and organized information in a hierarchical manner.

Chapter Summary

- Social living results in particular cognitive challenges that must be overcome in order to interact effectively.
- The detection of animacy is a basic foundational ability for the ability to interact in a social environment. Recognizing conspecifics, predators, and prey is critical for adaptive behavior. Mechanisms appear to be in place that allow animals to detect living entities, perhaps even from birth, in turn allowing rapid learning about important individuals in the environment, including conspecifics, and their behavior.
- Perceptual narrowing is a process by which experiences with certain stimuli result in a system that is tuned to those features, allowing for increased recognition and distinctions within those stimuli as compared to other stimuli. Exposure to faces, for example, may affect face processing by 'tuning' it to the types of faces that are most prevalent in the environment.
- Adult humans observe someone's behavior and can create a representation that includes such elements as the physical actions, the setting, and the internal mental states (e.g. feelings, desires, emotions, beliefs, and intentions). These mental states are the assumed causes of the behavior.
- It is very difficult to examine whether any non-human animal species, or even human infants, engage in mental state reasoning. Verbal answers to test questions are, of course, not available. Further, it is difficult to determine whether the subjects are actively attributing mental states or coming up with the correct answer by some other means.
- Most developmental researchers will state that by at least 6 months of age, human infants view certain actions as goal-directed. But what is meant by 'goal-directed' can vary; some researchers claim that infants are only reasoning about behavior, and some claim they are reasoning about mental states such as the desires that cause behavior.
- Some of the experimental tasks completed with human infants have been slightly modified to be used with non-human primates. Many of these tasks, however, can be solved by reasoning about behavior without reference to mental states. Tasks that include unintentional, 'accidental' actions may allow for conclusions based on mental state attribution.
- Using experimental paradigms that place chimpanzees in a competitive context and specifically examine their mental attributions toward conspecifics, some researchers have suggested that chimpanzees reason about certain mental states like seeing and even knowing, yet perhaps not false beliefs.
- Alternatively, some propose that chimpanzees have a cognitive system that allows for past experience of others' behavior to be organized by biases or heuristics that would allow them to recognize causally relevant aspects of current behavior and reason about likely future behavior.
- Social knowledge refers to the recognition of individuals and their relationships to one another. Relationships that may be of particular social importance to a social species are kinship and dominance.
- The cognitive mechanisms that allow for representing third-party relationships are still being examined, but may involve aspects of associative learning and memory systems that allow for chunking.

Questions

1. What is the empirical evidence that biological motion detection is shared across species?
2. Why might the perceptual narrowing seen in face discrimination be adaptive?
3. How has the understanding of facial emotions been tested with chimpanzees?
4. Imagine that a young infant observes a person reach out and grasp an object, and then in a subsequent test, the infant seems to expect that person to grasp that object again. How might this test outcome come about via a behavior-reading cognitive system that does not attribute mental states?
5. What experimental designs have been used to test animals' understanding of others' intentions?
6. What are two reasons that have been proposed for why it may *not* be appropriate to test chimpanzee mental state reasoning with tasks in which they must use the cooperative gestures of humans to find the location of hidden food?
7. Characterize the arguments within the ongoing debate regarding whether animals like chimpanzees understand mental states like seeing and knowing.
8. What might the study of animal theory of mind ultimately tell us about human theory of mind?
9. What is the evidence that some monkey species understand the kin relationships between other members of their groups?
10. What are the cognitive challenges faced by animals in figuring out a group's dominance hierarchy? Why is it important to figure this out in the first place?

FURTHER READING

For discussion of specialized systems for face and biological motion detection, see:

Morton, J., & Johnson, M. H. (1991). CONSPEC and CONLERN: A two-process theory of infant face recognition. *Psychol Rev*, **98**, 164–181.

Troje, N. F., & Westhoff, C. (2006). Inversion effect in biological motion perception: Evidence for a "life detector"? *Curr Biol*, **16**, 821–824.

The debate over mental state attribution in chimpanzees is played out in the following papers:

Call, J., & Tomasello, M. (2008). Does the chimpanzee have a theory of mind? 30 years later. *Trends Cogn Sci*, **12**, 187–192.

Povinelli, D. J., & Vonk, J. (2003). Chimpanzee minds: Suspiciously human? *Trends Cogn Sci*, **7**, 157–160.

Tomasello, M., Call, J., & Hare, B. (2003). Chimpanzees versus humans: It's not that simple. *Trends Cogn Sci*, **7**, 239–240.

Tomasello, M., Call, J., & Hare, B. (2003). Chimpanzees understand psychological states – the question is which ones and to what extent. *Trends Cogn Scis*, **7**, 153–156.

Detailed consideration of social knowledge can be found in the following book:

Cheney, D. L., & Seyfarth, R. M. (2007). *Baboon Metaphysics: The Evolution of a Social Mind*. Chicago, IL: The University of Chicago Press.

For theories on the uniqueness of human social cognition, see:

Lurz, R. W. (2011). *Mindreading Animals: The Debate Over What Animals Know about Other Minds*. Cambridge, MA: The MIT Press.

Penn, D. C., Holyoak, K. J., & Povinelli, D. J. (2007). Darwin's mistake: Explaining the discontinuity between human and non-human minds. *Behav Brain Sci*, **30**, 109–130.

Tomasello, M., Carpenter, M., Call, J., Behne, T., & Moll, H. (2005). Understanding and sharing intentions: The origins of cultural cognition. *Behav Brain Sci*, **28**, 675–735.

11 Prosocial behavior

Background

The hungry female left the nest at dusk, and after hours of searching for food, finally found a good location and ate as much as she could. When she returned to the nest, another member of the group, unrelated to her but one of the many who spent time there, approached. Unsuccessful in her own search for food, this starving female begged, and the successful female quickly obliged, sharing some of her recent collection. This behavior is quite generous. Every calorie counts; there is no guarantee that the sharing female will find food again tomorrow night or the next, and she will not live for long without a meal.

It might seem surprising that this rather heartening example of prosocial behavior should come from a species that strikes fear in many of us, owing in part to a connection with Bram Stoker's *Dracula*. The vampire bat (*Desmodus rotundus*), however, will readily share, through regurgitation, a recent blood meal taken from sleeping livestock or other animals (the bites do not kill the victim, but do leave a wound). There are over 1000 species of bat, but only the vampire bats (consisting of three species) feed on blood, and their sharing behavior has been studied by Wilkinson (1984, 1985) as a model of prosocial behavior.

Why would any species demonstrate such a behavior? How could a behavior that clearly comes at a survival and reproductive cost be selected and transmitted across generations? Even Darwin was initially puzzled by prosocial behavior and expressed his concerns in *On the Origin of Species*. Social insects like bees, for example, in which some members work only for the benefit of the hive and do not reproduce, were considered "one special difficulty, which first appeared to me insuperable, and actually fatal to my theory" (Darwin, 1859, p. 236). Still, prosocial behavior can be seen across many different animal species, with one species in particular, humans, demonstrating a wide range of behavior aimed to benefit others, often from a very early age.

Chapter plan

Biologists, psychologists, mathematicians, and economists have all examined the manner by which prosocial behavior can evolve. This chapter will consider many of these theories and then discuss the various prosocial behaviors observed in animals and humans, including helping, sharing, comforting, and cooperation. There would likely be very few who would disagree that humans have a unique and ubiquitous suite of prosocial behaviors, despite our species' ability to engage in violence and immoral acts. Thus, many of the comparisons that will be considered in this chapter will be between the prosocial behaviors (and the underlying cognitive mechanisms) of humans and those of other animals.

11.1 Evolution of prosocial behavior

If behaviors that lead to increased survival and reproductive success are behaviors that will evolve, then a question arises as to how a behavior that comes at a cost to the actor might ever take hold. How might self-interest be trumped by self-sacrifice given the basic manner in which natural selection works? This section will focus on the proposals of many theorists, across disciplines, to explain how prosocial behavior could not only appear but actually flourish. The ideas stem at least from the mid-1800s and Darwin's writings, but even today some of the top science journals continue to publish new studies and accompanying theories. In subsequent sections, examples of animal behaviors that fit some of these models will be presented.

11.1.1 Kin selection

In the 1960s, Hamilton, a biologist, considered the fact that a relative's offspring are part of an individual's **reproductive fitness**, or contribution to the gene pool. Kin carry a subset of an individual's genes, and thus, Hamilton argued, prosocial behavior that increases the probability of the survival and reproduction of kin can further spread an individual's genes. Prosocial behavior toward relatives can thus evolve if reproductive success of the kin is enhanced enough to outweigh the costs to an individual's own reproductive success (Hamilton, 1964).

Hamilton (and also Price) expressed this mathematically through a formula that considered **inclusive fitness**, which is simply a measure of an individual's fitness that takes into account his or her own offspring (direct fitness) and the offspring of relatives (indirect fitness). Of course, the amount that kin contribute to an individual's fitness depends on just how closely related they are. An individual shares 50% of their genetic material with their own child, but less (25%) with a nephew or niece, and thus Hamilton's formula includes a measure of the degree of relatedness. In sum, 'Hamilton's Rule' states that a prosocial behavior can evolve via natural selection if the benefit of the action in relation to the degree of shared genes is greater than the costs to an individual's own reproductive success.

There are examples of animal behavior that fit the model such as the case of sun-tailed monkeys (*Cercopithecus solatus*) that show preference for kin (e.g. more affiliation and less aggression) and a bias that decreases with decreasing relatedness (Charpentier *et al.*, 2008). Examples from everyday human experience suggest that we selectively help close kin (e.g. will tend to proscribe resources according to relatedness), and laboratory experiments concur. In one such study, participants were told that more money would be given to a relative the longer the participant held a painful "skiing" pose. Participants stayed in the position for longer durations if the recipients were siblings and less if they were nieces/nephews or cousins (Madsen *et al.*, 2007).

As research progresses, it may actually be that some of the basic tenets of kin selection theory are altered. Some researchers like Nowak, Tarnita, and Wilson have argued that the predictions of Hamilton's rule almost never hold (Nowak *et al.*, 2010), in turn starting a lively debate in the literature. For now, the debate is ongoing, but it is important to keep in mind that Hamilton's model is only addressing one of Tinbergen's questions, namely, how prosocial behavior toward kin could evolve. It does not, however, address the more proximate questions of how the behavior could develop or what sensory and cognitive processes are required. It is unlikely that animals that engage in prosocial behavior toward kin are consciously calculating relatedness before making decisions. Yet, as Chapter 9 details, many animal species have means of identifying kin, which is a necessary prerequisite for kin selection.

11.1.2 Prosocial behavior toward non-kin

The evolution of prosocial behavior cannot be completely explained by kinship since much of the observed behavior, both in humans and animals, is directed toward non-kin. In 1971, Trivers' seminal paper, "The Evolution of Reciprocal Altruism," presented a solution to the puzzle of prosocial behavior. His proposal was that genes for these behaviors could be selected for if individuals do not indiscriminately act to benefit others but instead direct prosocial behavior to those who have been prosocial themselves.

Trivers' idea was presented using **game theory**, a mathematical study of simulated interactions between two or more individuals. In these simulations, the payoff to an individual is dependent on the actions of the other individual. Trivers' proposal was that the evolution of prosocial behavior could be modeled by a particular game called '**the prisoner's dilemma**.' (The game is usually explained in terms of human behavior, but later in this chapter, animal behavior will be considered.) Imagine a situation in which two criminal suspects are arrested, and the police wish to get a confession from at least one of them as a means of gathering enough information for conviction. The police offer each man, separately, the following deal: "If you testify against the other man (defect), and the other man keeps quiet (acts to aid you), then you go free and the other man gets a 5-year prison sentence. If you both remain quiet, you'll each get only 1 year in prison. If you both 'rat out' each other, then you'll each get a 3-year sentence."

Figure 11.1 shows the 'payoffs' for each possibility. Because each suspect individually receives the greatest payoff if he defects, then each should testify against the other. The dilemma is if they both defect, they actually receive more prison time than if they both aid each other by remaining quiet. Thus, here is a situation in which an animal or human will individually receive more payoff to defect, but two individuals will each do better when they both aid each other versus both defect. In sum, this game can model the evolution of prosocial behavior. One can also consider an iterated version of the prisoner's dilemma in which individuals encounter each other repeatedly, as they often do in social groups. Now, the options are not just to aid or defect; other strategies can be devised, like "aid only if he aids me first, otherwise defect." In fact, Axelrod and Hamilton mathematically demonstrated that in iterated prisoner dilemma games, the most successful strategy is to aid the other person in the first encounter, and then simply copy the moves that the partner makes (e.g. defect if they defect) in subsequent interactions. This strategy is commonly referred to as "tit-for-tat" and demonstrates how the rewards for back and forth benefitting can come to outweigh defection.

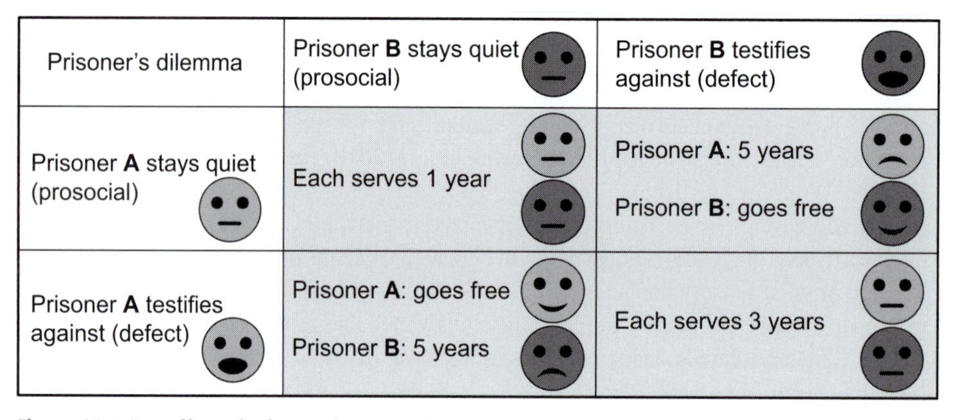

Prisoner's dilemma	Prisoner **B** stays quiet (prosocial)	Prisoner **B** testifies against (defect)
Prisoner **A** stays quiet (prosocial)	Each serves 1 year	Prisoner **A**: 5 years Prisoner **B**: goes free
Prisoner **A** testifies against (defect)	Prisoner **A**: goes free Prisoner **B**: 5 years	Each serves 3 years

Figure 11.1 Payoff matrix for a Prisoner's Dilemma game. The matrix presents a situation in which a player will individually receive more payoff to defect, but two individuals will each do better when they both aid each other versus both defect.

Punishment

Since the proposal of the tit-for-tat strategy, researchers have elaborated on the manner in which individuals might aim to 'control' individuals with whom they interact. One particularly interesting behavior is the fact that humans also engage in punitive behavior; we punish cheaters and freeloaders, and not simply by defecting at the next opportunity. For example, in another game, '**the ultimatum game**,' one player can give any part of an endowment of cash to a second player. This potential recipient, however, can choose whether to accept or reject the proposer's offer. If the recipient accepts, then the money is distributed accordingly. If the offer is rejected, neither player gets any money. In truth, the recipient should accept any offer (except $0) because rejecting would lead to no monetary gain at all. Yet, when this game is played by people in many different countries all over the world, recipients typically reject offers of less than 20% of the proposer's endowment, meaning that neither proposer nor recipient gains any money (e.g. Henrich *et al.*, 2006). (In Section 11.4, applications of this game with chimpanzees will be discussed.) Additionally, in versions in which a third person is able to spend some of his or her own endowment to punish a proposer for making a low offer to a recipient, the third person often pays to have the proposer's endowment reduced (also see Box 11.1). People seem to be willing to incur costs to punish others, even if they were not harmed themselves (Fehr & Fischbacher, 2004; though see Pedersen *et al.*, 2013).

Reputation and partner choice

The previous game theory scenarios are all cases in which partners are provided, not chosen. Yet, social interactions often are the result of active partner choice, and in these cases the emphasis might not be on reacting to a cheating partner by cheating or punishing (partner control) but by simply starting a new relationship with someone else (partner choice) (e.g. Baumard *et al.*, 2013). Here, cheating becomes costly because of the lost opportunity to benefit from interactions. A general preference for trustworthy partners thus selects for prosocial behavior within a species. Relatedly, Nowak and Sigmund (1998) have proposed that for humans, having the reputation of being a prosocial individual has its benefits. They proposed that individuals who engage in prosocial behavior develop a reputation for doing so, and will potentially be selected by others as interaction partners.

Box 11.1 Reciprocity and punishment in the human brain

In recent years, the research field of neuroeconomics has been growing, representing the collaborative, interdisciplinary efforts of biologists, psychologists, and economists. The general approach is to examine the neurological underpinnings of decisions such as those made in reciprocal altruism games. Most work has been completed using brain imaging via functional magnetic resonance imaging (fMRI).

In one such study, participants played an iterated prisoner's dilemma game with an unknown partner in an adjacent room. All the while, the participant's brain activation was measured via fMRI. When mutual prosociality occurred (i.e. neither player defected), strong activation of brain areas associated with reward (ventromedial/oribitofrontal cortex, anterior cingulate cortex, caudate nucleus, and the nucleus accumbens) was reported. In fact, the activation of these areas during mutual prosociality rivaled activation during situations in which the participant defected and the partner shared, which would lead to greater monetary reward for the participant (Rilling *et al.*, 2002). The researchers thus suggest that the activation of these brain areas reflects the rewarding effect of experiencing a mutually prosocial interaction.

Additionally, areas including the caudate nucleus are activated during the punishment of 'cheaters,' or those who violate trust in an economic game. In the 'trust game,' player A must trust player B to send money back after an exchange. If B fails to send the money back, A often punishes B, even if this action is costly to A. In fact, players often express that they enjoy punishing in this way, and the activation of reward areas of the brain during these actions concurs. Enhanced activation of these areas did not occur, however, when player A was told that player B did not have control over whether the money was sent back (de Quervain *et al.*, 2004). In sum, neuroeconomic research is ongoing, and there is still much to learn, but even these early findings point to possible cognitive and emotional mechanisms that underlie prosocial behavior.

11.2 Types of prosocial behavior

Thus far in this chapter, the term 'prosocial behavior' has referred to instances of bats sharing a blood meal and humans sharing a monetary endowment. These two behaviors do have a similarity (individuals with a material resource give some of that resource to individuals who are lacking), and both are actions that benefit another (prosocial). Other types of behavior besides sharing, however, fit under the heading of prosocial behavior, such as helping, comforting, and cooperation. In the sections that follow, we examine instances of prosocial behavior in animals and humans, and each of these types of prosocial behavior will be considered separately.

Traditionally, textbooks on animal behavior do not separate these types of prosocial behavior, but the present text will do so for two reasons. First, research on animal prosocial behavior has grown in the last 10 years, and innovative methodologies have captured different types of prosocial responses to others' needs. Second, research in developmental psychology that considers distinct varieties of other-oriented action like helping, comforting, and sharing has demonstrated unique ages of onset and distinct, uncorrelated developmental trajectories, suggesting that different prosocial behaviors may require different cognitive processes (e.g. Dunfield *et al.*, 2011; Warneken & Tomasello, 2009). As of now, the evolutionary models presented in Section 11.1 are likely still valid across all types of prosocial behavior, and in particular, sharing. However, as the body of research grows, existing theory may be elaborated.

One additional consideration is the use of the term 'cooperation.' In many texts, this term is used in much the same way as the present text uses 'prosocial behavior.' That is, 'cooperation' is often used as a broad category that encompasses many types of behaviors that serve to benefit others. However, here the use of the term is reserved for instances in which two or more individuals act together to achieve a joint goal, such as multiple lions engaging in a hunt or even amoebae joining together to move through substrates that an individual amoeba cannot traverse alone (also see Box 11.2). By limiting the use of the term, the door is opened to theories from developmental psychology that consider the unique nature of human cooperation.

11.3 Helping

In the field of developmental psychology, 'helping' typically refers to behaviors that respond to an instrumental need, such as retrieving an out-of-reach object for someone, opening the door for someone whose hands are full, or even providing needed information (e.g. Tomasello, 2009). When the species under consideration is human, researchers often speak of the recognition of others' needs as the recognition that someone has a particular intention that is currently unachievable (e.g. wanting to obtain an out-of-reach object). As detailed in Chapter 10, however, it is not always clear whether other species recognize intentions and needs in this manner. Thus, this section will consider behaviors that *function* to assist others with certain instrumental needs. In most cases, authors may remain non-committal as to whether helpers are responding solely to stimuli in the environment and the behavioral states of others (e.g. an outreached hand toward a far away object), or recognizing behavior states and also recognizing the intentions underlying those behavioral states (a goal to obtain the object).

Many different species have been observed to engage in helping behavior, and the contexts in which it is seen vary widely. Two naturally occurring contexts will be considered here, mating/offspring care and providing information, followed by consideration of experimental examinations of helping in the laboratory setting.

11.3.1 Helping for mating and offspring care

Helping behaviors that serve to increase the reproductive success of others clearly have important benefits to the recipient. A particularly interesting example comes from a perhaps unexpected source: the slime mold (*Dictyostelium discoideum*). These single-cell amoebae often merge to form a 'fruiting body,' which in turn produces spores. The amoebae that form the stalks of the fruiting bodies serve to help position other amoebae to produce spores while losing their own chance at reproduction. This system is also prone to cheaters, since some cells may avoid becoming part of the stalk and instead gain access to the fruiting body directly (e.g. Santorelli *et al.*, 2008). Amoebae, however, may be able to keep cheating in check. In soil in the wild, a cell-adhesion gene (csaA) is necessary to get into the initial fruiting body formation. Thus, those with the csaA gene are actually helping others with this gene, and cheaters (those without the gene) are left out of the formation. The cell-adhesion gene is an example of a **green-beard gene**, or a gene that codes for mechanisms that ultimately allow organisms to 'recognize' a fellow prosocial individual. The term 'green-beard gene' comes from the anthropomorphic analogy that, at least for humans, one individual would easily notice a fellow green-bearded individual (Dawkins, 1976; Pennisi, 2009); though in general, 'noticing' or 'recognizing' a prosocial individual need not be the result of cognitive processes.

Assisting in reproductive success is also seen in larger-brained species. In mammals, for example, behaviors have been observed in the wild that aid another individual's opportunity to mate with females. Often, this is done through the creation of alliances of males who help each other gain access to receptive females. Research by Packer (e.g. 1977) with olive baboons (*Papio anubis*) illustrates this point. Males in contest with an opponent will sometimes enlist the help of another male by alternating looking between the opponent and the solicited male. If the opponent was consorting with a female, the coalition often results in the female leaving the opponent and approaching the soliciting male, resulting in a reproductive advantage. The enlisted male thus helped the other male to gain access to a female, but what, if any, benefits exist for helping? Packer reported that a system of reciprocity seems to be in place in which males who were aided by another male will be more likely to help the enlisted male in subsequent interactions.

A similar form of helping exists in bottlenose dolphins (*Tursiops truncatus*). Dolphins are notoriously difficult to study in the wild, but work by Connor, Krutzen, and others has documented instances of males helping males to keep reproductive females nearby. Alliances of two to three males will help each other to keep a female from leaving their group. These male alliances appear to remain stable over years, offering opportunity for reciprocal altruism. It is presently unclear what cognitive processes might underlie the reciprocity among alliance-forming baboons and dolphins and whether individuals remember specific personal interactions with those who helped them over long time periods. A simpler interpretation might be that when helped, an individual assigns a positive value to the helping individual, and that subsequent behavioral decisions (approach, affiliate, or defend) are made on the basis of that positive association (Schino *et al.*, 2007; Shettleworth, 2009).

Helping behavior is also seen in relation to the outcomes of mating: taking care of offspring. This behavior, often termed "**cooperative breeding**," refers to situations in which individuals – who are not parents themselves – are actively engaged in raising others' offspring. It is a behavior seen in about 9% of bird species (Cockburn, 2006), including the western bluebird (*Sialia mexicana*), which has been the focus of much research (Figure 11.2). In this species, a small proportion of breeding pairs are helped by a related male (mostly sons and brothers). Help by feeding the nestlings and defending the nest increases the breeding success of the pair. Yet, many studies have suggested that the inclusive fitness benefits of helping do not actually make up for not breeding independently, raising the question of how this behavior might have evolved. To address this question, Charmantier and colleagues (2007) recently used a genetic analysis of 1593 breeding individuals and found that

Figure 11.2 The western bluebird, a cooperative breeding species. In some instances, breeding pairs are helped at the nest by a related male.

Box 11.2 Inter-specific mutualism

Figure 11.3 An example of a cleaner/client relationship. Here, striped cleaner wrasse clean a potato cod (*Epinephelus tukula*).

The term '**mutualism**' has traditionally been used to describe situations in which unrelated individuals simultaneously benefit from an interaction. To distinguish mutualism from reciprocity, consider the difference between these two phrases: "you scratch my back and I'll scratch your back" (reciprocity) and "you scratch my back while I scratch your back" (mutualism). Note also that different from the definition of cooperation that is used in this chapter, here the behaviors of the two individuals are not necessarily directed toward a joint goal that is held in common.

One of the best and most studied examples of mutualism is a case of *inter-specific* mutualism in which two different species are each benefitting from an interaction: cleaner fish and their clients (Figure 11.3). Cleaner fish eat the ectoparasites found on the surfaces of other fish (clients), who, in turn, benefit from the parasite removal. The clients are of another species and are often larger and potentially predatory, yet, these clients let the cleaner fish approach and feed off of their surface, even allowing the cleaners to dart in and out of their mouths. The benefits of the interaction are both simultaneous and mutual: a meal for the cleaner and a healthy cleaning for the client. One can see how the helping behavior of the cleaner (removing dangerous parasites that the client cannot remove otherwise) and the tolerant behavior of the client could evolve – both behaviors have adaptive value.

The cleaner–client mutualistic interactions are also characterized by partner control and partner choice. In one well-studied pair, cleaner wrasse (*Labroides dimidiatus*) and Australian reef fish, cleaners sometimes cheat by eating the client's mucus, which is preferred over the parasites. The clients, however, find this aversive and may react in two ways. Sometimes clients punish by going on the attack, chasing and driving the cleaners away (partner control). Clients may also engage in behavior that exhibits partner choice, such as swimming away and finding other cleaners. In fact, partner choice is also evidenced by observations of clients preferentially approaching cleaners who were previously observed cleaning other fish without conflict. These cleaners evidently develop something like a prosocial reputation by being associated with peaceful interactions.

helping behavior was a heritable trait. In turn, this means that in this species, individuals who once helped others – and now have broods of their own – have a good chance of being the recipients of help by their relatives (e.g. a son from a previous brood helping with a new brood). At a proximate level, it is possible that the observation of helping behavior by relatives plays a role by leading to social learning of the behaviors (see Chapter 13); for example, a member of a brood who was helped may help a new brood in the future.

11.3.2 Providing information

The transmission of information to an individual who requires it can be thought of as a helpful act. Even young children label individuals who provide them with information to help solve a puzzle as 'helpful,' and they selectively help these individuals in subsequent interactions over individuals who have withheld information (Dunfield & Kuhlmeier, 2013). Chapter 12 will detail instances, across many species, in which a member of a group forgoes eating and instead watches for predators, emitting an alarm call that functions to warn others of danger if necessary. This act of informing seems much more costly than simply providing the answer to a child's puzzle; in order to stand guard, one is missing out on food intake, and emitting the alarm might direct the predator's attention to oneself. Traditionally, it was suggested that the behavior evolved through kin selection or reciprocal altruism (another individual may switch positions and take over the watch). This is likely the case for some species like dwarf mongooses (*Helogale parvula*) in which members regularly risk predation when on guard duty, but other mechanisms may be in place for other species.

In meerkats (*Suricata suricatta*), for example, the potential costs and benefits of providing information have been examined by researchers such as Clutton-Brock and his colleagues (e.g. Clutton-Brock *et al.*, 1999). Some members of the meerkat group serve as 'sentinels,' watching for danger while other group members eat. At least in this species, though, the dangers of acting as a sentinel actually may be less than the danger for those distracted by foraging. Additionally, there is no evidence of reciprocal altruism as there is no pattern of rotation for sentinels. Instead, what seems to occur is that satiated members of the group guard when they are done eating; it is the optimal activity since they are actually less vulnerable when on guard. A remaining question is how to interpret the fact that sentinels who observe a predator will produce a call, which potentially is costly to themselves due to predator detection. The researchers suggest that this may be the one aspect of the sentinel behavior that is driven by kin selection or reciprocity.

11.3.3 Helping in the laboratory

In the previous examples, naturally occurring helping behavior was discussed. Among a suite of behaviors any one species might typically enact in the wild, one or two may seem to stand out to observers as instances of helping. Yet, when one thinks of human behavior, one can think of all types of helping acts, and can even imagine helping occurring in a completely novel situation as long as the potential helper has the knowledge and physical ability to enact a solution. One of the possible reasons for this is that many acts of helping require complex social cognition and a good understanding of the physical world. At least for humans, the assumption is that the observer must recognize the (failed) intention of another individual and devise an effective solution.

Warneken and Tomasello (2006) have compared the helping behavior of captive chimpanzees (*Pan troglodytes*) and 18-month-old children in novel laboratory situations with particular focus on the ability to recognize another individual's instrumental need. Importantly, the study included experimental trials in which an actor was in need of help and control trials that were superficially similar, but no need was present. For example, subjects would see an actor reach out for a far away object in an experimental trial, yet in a control trial, the actor simply looked at the out-of-reach object with a neutral expression. Both chimpanzees and children helped by retrieving the object for the actor in the experimental conditions, but refrained from picking up the object when no need was present (Figure 11.4). This is not behavior that has been observed in the wild, yet suggests that at least captive chimpanzees will respond to others when there is instrumental need, even without immediate benefit to themselves. There are some limitations, though. Chimpanzees did not show as many helping actions as the toddlers (e.g. opening doors), which may indicate that in those situations, the actor's need and/or the solution was not immediately apparent.

Follow-up laboratory studies have shown that chimpanzees will help other chimpanzees by unlocking a door to allow them access to a room with food (the helper did not see or receive the food; Warneken *et al.*, 2007). Additionally, chimpanzees will provide needed tools to other

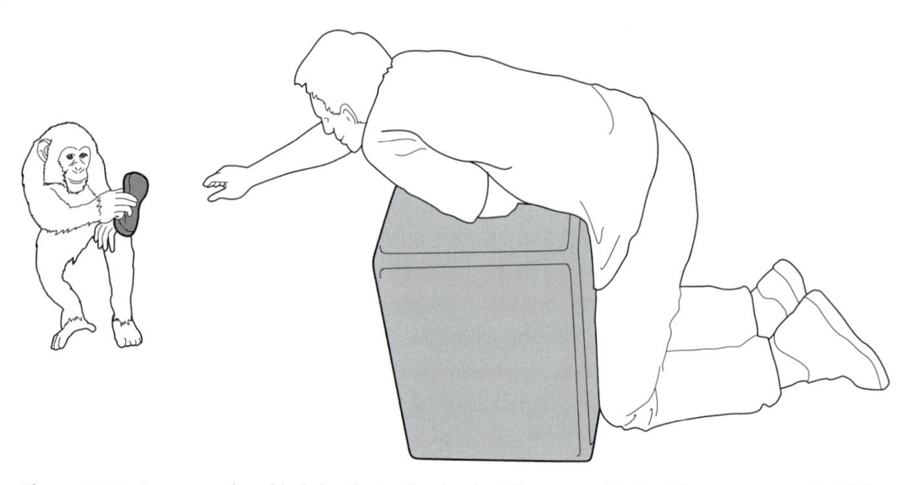

Figure 11.4 An example of helping behavior in the laboratory. Both chimpanzees and children will help by retrieving an object for an actor, but will refrain from picking up an object when no need is present (e.g. Warneken & Tomasello, 2006).

chimpanzees. In this study, chimpanzees in adjacent enclosures were presented with food puzzles that required a specific type of tool. However, in some cases, each subject was provided with the other subject's required tool. The chimpanzees were more likely to swap tools when they were given the wrong tools than in conditions in which they were provided with the correct tools (Yamamoto *et al.*, 2009).

For any of these studies, though, a question remains as to how the underlying cognitive processes of the helper might be characterized. Consider the distinctions made in Chapter 10: Are the chimpanzees recognizing the intentions that cause the actor's behavior (e.g. he *wants* the out-of-reach toy), or are they making decisions based solely on representations of behavioral contingencies and regularities without reference to mental states? Unfortunately, as detailed in Chapter 10, there is currently no resolution to this debate. Whichever manner in which they represent the behavior, though, what is particularly intriguing is that the helping chimpanzees seem to be subsequently motivated to act on the behalf of others.

Some researchers have proposed that certain animals may have a sense of empathy that motivates helping behavior. For example, in a carefully controlled study, Bartal and colleagues (2011) found that rats will open a restrainer for a trapped rat that is emitting alarm calls and trying to escape. In control conditions, empty restrainers or restrainers holding a toy rat were opened at significantly reduced rates. The authors propose that rats freed their trapped cagemates as a means of ending the cagemate's distress or even their own distress. If this is true, it would seem that the rats recognized the distress calls and also came to the solution that opening the restrainer would end the distress. Yet, how would they come upon that particular solution? Over trials, helping rats may have learned the contingency between the door opening and the cessation of distress (which could occur with or without attention to the actions of the trapped rat itself). Of course, considering helping behavior as a response to overt distress is specific to this type of experimental paradigm. In the tasks above in which chimpanzees retrieved an out-of-reach object for an actor, distress cues were not present, yet the subjects readily helped. It would appear that more work on the cues that encourage helping behavior across species is warranted. Further, the concept of empathy in animals has itself garnered a lot of attention, and this topic will be more fully discussed later in this chapter.

11.4 Sharing

When vampire bats give part of their blood meal to a hungry conspecific or when human adults give a portion of a monetary endowment to someone else, the prosocial behavior consists of giving some of one's own material resources to benefit another individual who is in need. Sharing is, in fact, one of the more commonly studied prosocial behaviors in humans (particularly with economic games), but there are also some instances of this behavior in animals. In this section, naturally occurring sharing behavior in animal species, specifically the sharing of food, will be considered, followed by laboratory experiments using economic games with humans and other animals.

11.4.1 Food sharing

It may seem surprising, but great apes do not show the greatest similarity with humans in terms of the propensity for sharing, particularly food sharing. Instead, species that engage in cooperative breeding are more likely to engage in food sharing than those who breed more independently. In cooperative breeders (including some species of birds, callitrichid monkeys (Figure 11.5), and social carnivores such as meerkats), not only do members of the group who have no offspring help other members defend offspring as mentioned above, but they also share food with these offspring (Jaeggi *et al.*, 2010). Similarly, humans are cooperative breeders; family members such as older siblings, aunts and uncles, as well as grandparents often engage in the care of infants, and of course, often share food. In contrast, the great apes raise offspring in a much more independent manner than humans, and they seldom engage in food sharing, even among kin. This pattern of observations has

Figure 11.5 Pygmy marmosets (*Cebuella pygmaea*). In this small New World monkey, some individuals help in the care of others' offspring and engage in food sharing.

led some to propose that prosocial behavior such as food sharing may be explained by convergent evolution among species with larger kin networks.

What cognitive mechanisms might underlie food sharing? The act of sharing should be directed toward those in need of a resource, and thus a sensitivity to need would be necessary. In some cases, this sensitivity might be built in reflexively as a response to a signal (e.g. an open mouth at the nest). In other cases, 'reading' a need might require more cognitively complex abilities, such as instances in which the potential recipient is showing no outward signals yet previous history and/or assumptions about his beliefs and desires lead to the representation of need. These distinctions have led some researchers like Jaeggi and colleagues (2010) to propose different forms of sharing: proactive (offering food in the absence of any form of begging), reactive (sharing in response to signals of need), and passive (simply allowing or tolerating others to take one's own food). At least among primates, only humans and callitrichid species have thus far been observed to engage in proactive food sharing. For the latter, parents and other individuals in the group give food calls and then offer food to infants (e.g. Feistner & Chamove, 1986). In future research, the cognitive processes underlying sharing behavior may be elaborated by our growing knowledge of the underlying biological mechanisms (see, for example, Box 11.5).

11.4.2 Dictator games

As discussed in Section 11.1, sharing of material resources by humans is often studied in the laboratory with games such as the ultimatum game, and a variant, the 'dictator game.' The **dictator game** is quite similar to the ultimatum game except that the recipient of the endowment does not have the opportunity to reject offers as she or he does in the ultimatum game. Behavior during the two games is similar as well; typically proposers allocate 20–30% of their cash to the other player. Games such as these are useful for examining sharing behavior in animals because subtle manipulations in the task allow for insight into the underlying motivation and cognitive processes. Relatedly, researchers may have greater control over and knowledge of past interactions among the animals and the level of genetic relationship.

Most of the research in this area has been completed with primates, particularly chimpanzees, with the goal of understanding the evolutionary forces that shaped human sharing behavior. Silk and colleagues (2005), for example, created a game scenario for pairs of chimpanzees that had a similar structure to the dictator game. A "proposer" chimpanzee had access to a machine that, dependent on the experimental condition, could be made to deliver food to the proposer and the recipient, or only one of them. In a condition in which proposers were presented with the choice of one handle that forced the machine to provide food to both the proposer and the recipient, and another handle for which food would only be delivered to the proposer, a generous response would be to choose the first lever. Yet, there was no evidence that chimpanzees would systematically choose the handle that provided food to the other chimpanzee, and further, there was no evidence that the handle choice differed between conditions in which a recipient was actually present or if the adjoining room was empty. Similar findings have since been obtained repeatedly in experiments with slightly altered designs (Jensen *et al.*, 2006; Vonk *et al.*, 2008; Yamamoto & Tanaka, 2010).

These findings generated much discussion, with questions centered on why sharing behavior was not more prevalent, especially since some evidence of passive sharing behavior has been documented for wild chimpanzees (Muller & Mitani, 2005). It is possible that chimpanzees did not fully understand the experimental apparatus. However, some studies included conditions designed specifically to determine whether chimpanzees understood how the apparatus worked, and while it seemed they did understand, sharing behavior was still not seen (e.g. Yamamoto & Tanaka, 2010). Confusing matters all the more is that even cooperatively breeding primates do not systematically choose options that provide others with food in this experimental paradigm (marmosets: Burkart *et al.*, 2007; tamarins: Cronin *et al.*, 2009). Indeed, across five primate species, studies using this task never clearly show instances of the provision of food to another individual, even when doing so is at no cost to the donor (Silk & House, 2011). Although these studies are difficult to interpret, researchers have continued their work. In fact, recent studies hint that perhaps one aspect of the experimental design might play an important role; when food is not present and chimpanzees are in a position to choose a token indicating a prosocial choice versus a token associated with a selfish choice, they systematically choose the prosocial option (Horner *et al.*, 2011). In the next section, other task designs will be presented that were created with the aim of examining factors that might influence sharing behavior.

11.4.3 Punishment and fairness

While playing the ultimatum game, human participants across cultures will regularly reject offers of less than 20% of the proposer's endowment, meaning that neither proposer nor recipient gains any money. That is, players often punish other players for not sharing to a 'meaningful' extent, even if by doing so, they receive nothing at all. Whereas the dictator game and the variant just described in the previous section examine proposals to share, the ultimatum game considers the behavior of the recipient and how it may serve to encourage sharing by punishing those who are not generous. Testing animals with such a game may, then, provide further information about sharing behavior and the factors that encourage it.

In 2007, Jensen, Call, and Tomasello created a version of the ultimatum game for chimpanzees (Figure 11.6). One animal played the role of the proposer and chose between two ropes: pulling one rope would provide eight pieces of food to the proposer and two to the second chimpanzee, and another rope would provide, on some trials, five food pieces and five food pieces, or some other distribution of the total ten pieces. Importantly, the rope the proposer chimpanzee chose would bring the food closer, but not all the way, to both chimpanzees. The second chimpanzee played the role of the responder and could then complete the pull by pulling on a rod, bringing the food piles within reach. Given this apparatus, the responder could choose not to pull at all, effectively rejecting the proposer's offer. Strikingly different from human players of the ultimatum game, the chimpanzee responders typically pulled their rod for any non-zero offer (Jensen *et al.*, 2007). That is, they did not reject, even though offers were quite often strongly in favor of the proposer (e.g. 8 versus 2 food pieces).

It may be, then, that 'spite,' or punishment for not sharing equally between self and other, is not a common aspect of chimpanzee social life. Another way to probe an animal's sense of fairness, however, might be to not focus on punishment behavior toward those who do not share, but on other behavioral responses such as **inequity aversion**. To be averse to inequity involves a negative reaction to situations in which there is 'unequal pay for equal work'. (Note that this is actually different than simply being averse to *inequality*, which would simply imply a negative reaction to unequal distributions, regardless of any sense of whether each individual is deserving of their share.) In 2003, Brosnan and de Waal examined inequity aversion with tufted capuchin monkeys (*Cebus apella*) who had been trained to trade tokens for pieces of food. Each capuchin would readily trade a token for a piece of cucumber, but when some capuchins saw that other group members received a better food reward for a trade (e.g. grapes), the observing capuchins would refuse to trade their token, even for the once desired cucumber.

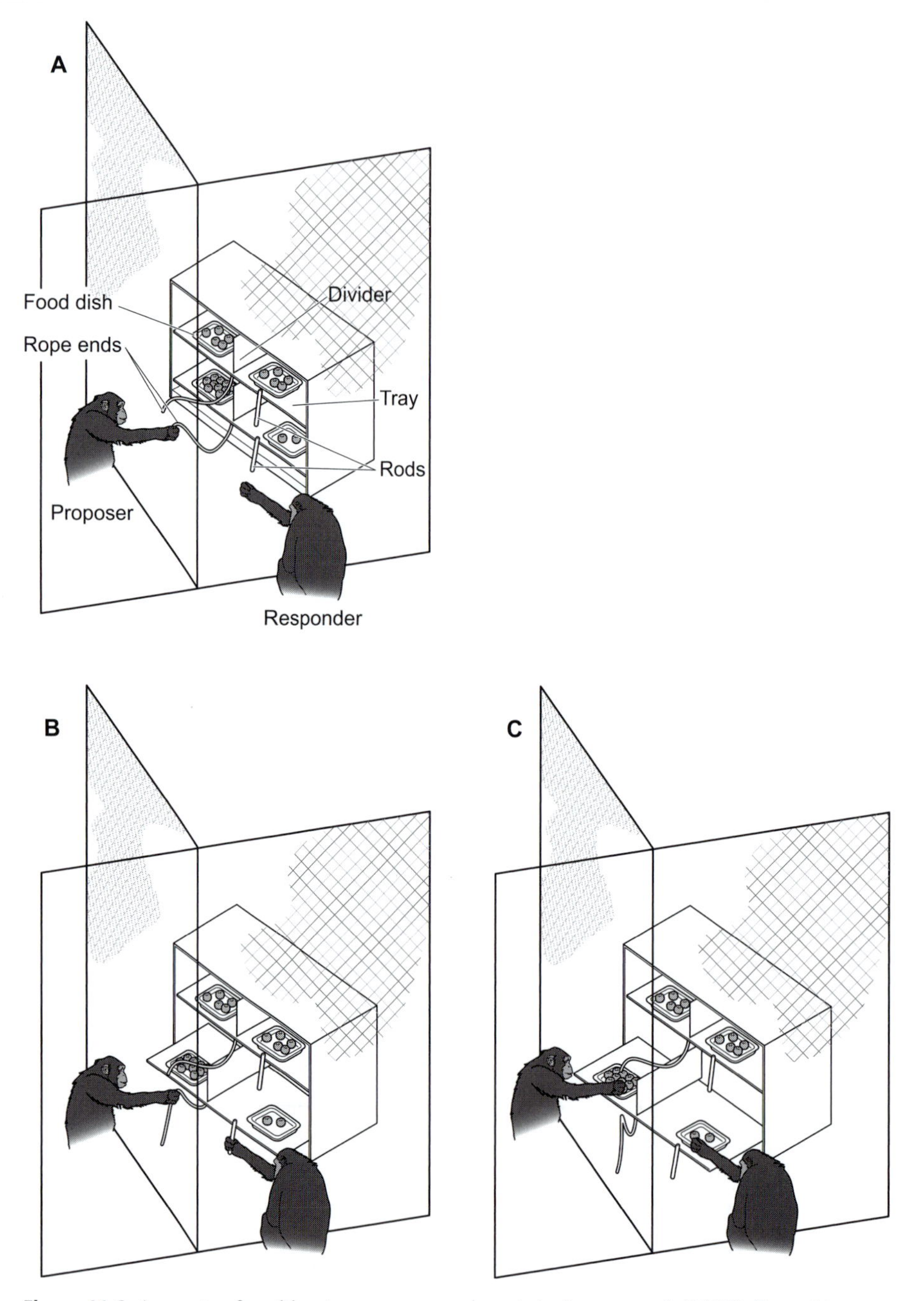

Figure 11.6 Apparatus for ultimatum game experiments in Jensen *et al.* (2007). Two chimpanzees could play the game: one took the role of the proposer and chose between two ropes that could pull in a shelf containing food, and one played the role of the responder and accepted or rejected the proposer's choice.

Researcher profile: **Dr. Joan Silk**

Figure 11.7
Dr. Joan Silk.

The interdisciplinary field of comparative cognition can be greatly enriched by the research and theories of anthropologists. For example, Joan Silk's work on prosocial behavior, kinship, friendship, and reconciliation in primates has focused on the evolutionary bases of these behaviors but also delves into the underlying cognitive mechanisms. At the heart of this work is Silk's interest in the primate origins of human social behaviors.

Silk's fascination with animal behavior began in childhood, and as an undergraduate student, observations of wild monkeys during an archaeological expedition in Africa provided a direction for her burgeoning research interests. After earning a PhD at the University of California, Davis, and spending time working as a postdoctoral researcher and instructor, Silk joined the Anthropology Department at the University of California, Los Angeles, where she worked for many years. She is now at the School of Human Evolution and Social Change at Arizona State University.

Silk has engaged in both field and laboratory work to examine how social behavior in primates has been shaped by natural selection. For example, Silk studies the adaptive benefits of close female bonds among chacma baboons (*Papio ursinus*), the sex ratio of offspring in anubis baboons (*Papio anubis*), and particularly relevant to this chapter, cooperation and food sharing in chimpanzees. This work has led Silk to propose that "despite [the] intriguing parallels in the patterns of cooperation and the correlates of social bonds among humans and other primates, there are also important differences in the scope of cooperation" (Silk & House, 2011, p. 10912). For Silk, this is just the start: "It is important to continue efforts to chart the size and dimensions of the gap between humans and other primates if we want to understand the evolutionary forces that have shaped human social preferences" (p. 10976).

Realistically, Silk and her colleagues recognize the difficulty in this task and conclude, "It seems unlikely that we will ever settle on a single account of how we became such an unusual species. But we now have a much richer body of theory and comparative data that allow us to develop more complete and compelling hypotheses that we can critically evaluate" (Kappeler *et al.*, 2010, p. 12).

Brosnan and de Waal (2003) suggested that the results supported the proposal that the capuchin monkeys were demonstrating inequity aversion; the equal work of trading a token was not met with equal pay in terms of quality of food. However, several arguments have been made against this interpretation. For example, the observing capuchins' responses may have been due to general frustration at not being able to obtain a desired food, regardless of the fact that the other capuchin was receiving that food (e.g. Dubreuil *et al.*, 2006). Even if one does accept an interpretation based on inequity aversion, however, it must be noted that the aversion only seems to occur when receiving the lesser food; capuchins did not turn down the preferred food when they saw their group mates receive the cucumber (Henrich, 2004; Brosnan & de Waal, 2004). These data, thus, fall short of demonstrating a clear sense of fairness in capuchins, though it has been proposed that perhaps the tendency to reject offers that are comparably disadvantageous to oneself is an initial stage in the evolution of inequity aversion (Brosnan *et al.*, 2005; Silk, 2007a). After all, in order to recognize fairness, one must be able to compare the value of commodities in context (e.g. when there is only cucumber available, its value is high, but the value decreases if grapes are present).

Taken together, it is clear that there is still much unknown about the frequency and mechanisms underlying sharing behavior in animals. In the wild, there is ample observation of food sharing among relatives in cooperatively breeding species. When the study of sharing is moved into the laboratory and primate species such as chimpanzees and capuchin monkeys are tested, the results are mixed. Direct sharing is not seen in dictator games, punishment for not sharing in a fair manner is not found with regularity, and inequity aversion is, at best, limited to situations of unfairness to oneself. These findings from laboratory studies form a strong contrast with those from a large body of research with humans. Thus, both the ultimate and proximate causes of sharing behavior, across species, are still the focus of active research.

11.5 Comforting

Thus far, this chapter has discussed prosocial behaviors that occur in conjunction with others' instrumental needs (helping) and material needs (sharing). One can also, however, act prosocially in response to others' emotional needs, which in common language can be thought of as comforting someone who is in emotional distress. In human behavior, one typical comforting behavior is a hug, perhaps directed toward an individual who is crying. Yet, comforting behaviors might also take the form of repairing a broken toy for someone who is saddened by the potential loss of a favorite plaything or pretending to slay the monster under the bed for a frightened child (e.g. Dunfield & Kuhlmeier, 2013).

In this section, research on comforting behavior across species will be presented. Of all of the prosocial behaviors discussed, comforting might be one of the hardest to interpret in animals. With humans, one can discuss such emotions as sadness and emotional distress, but with animals, it is unclear what emotional state, if any, underlies a particular vocalization and how it is experienced by the animal. Likewise, the receiver of the distress signal may or may not attribute the underlying emotional state to the signaler (i.e. responses may be simply elicited by a particular behavioral signal). Thus, similar to the discussion of helping behavior in Section 11.3, this section will consider behaviors that *function* to assist others who are exhibiting behaviors that human observers code as distress. Later in the section, however, the possible role of empathy as a proximate cause of comforting will be discussed.

11.5.1 Response to separation distress calls

Many young mammals emit a vocalization when separated from the nest or their mothers. These vocalizations are often called **distress calls**, and in many cases, result in maternal intervention (e.g. retrieval of the infant). This behavioral interaction between infant and mother has been very well studied in rats, and one common interpretation is the following: the pup is separated from the nest, and as its body starts to cool rapidly, it emits an ultrasonic call which the mother responds to with a retrieving behavior. Indeed, this interpretation has been the basis for considering the rat mother/infant interaction as a model for separation anxiety in human infants (Brain *et al.*, 1991; also see Box 11.3). That said, there is also suggestion that the infant rat's call is not the outcome of separation anxiety *per se*, but is a vocal byproduct of compression of the abdomen due to cold exposure, much as the sound of a sneeze is a byproduct of a non-communicative physiological function (Blumberg & Sokoloff, 2001).

Regardless of whether the infant rat intends to signal distress (and the pup is very likely in distress, at least at a physiological level), or whether the mother attributes distress to the pup, the mother's response functions to reduce the negative state by retrieving the pup and bringing it back to

Box 11.3 Attachment and expectations for comforting

Many human infants seek out and accept comfort from their caregivers, but other infants do not. In the laboratory, these individual differences are often observed in the **Strange Situation Test**, created by developmental psychologist Ainsworth (e.g. Ainsworth *et al.*, 1978). During the test, infants interact with a primary caregiver who, over time, leaves and returns to the room. Behaviors of particular interest include infant reactions to the caregiver's absence and reappearance (e.g. seeks out contact, is comforted by the contact, etc.). Infants who readily seek and accept comfort from caregivers are considered securely attached, whereas infants who are reluctant to seek comfort or have difficulty accepting the comfort are considered insecure in their attachments.

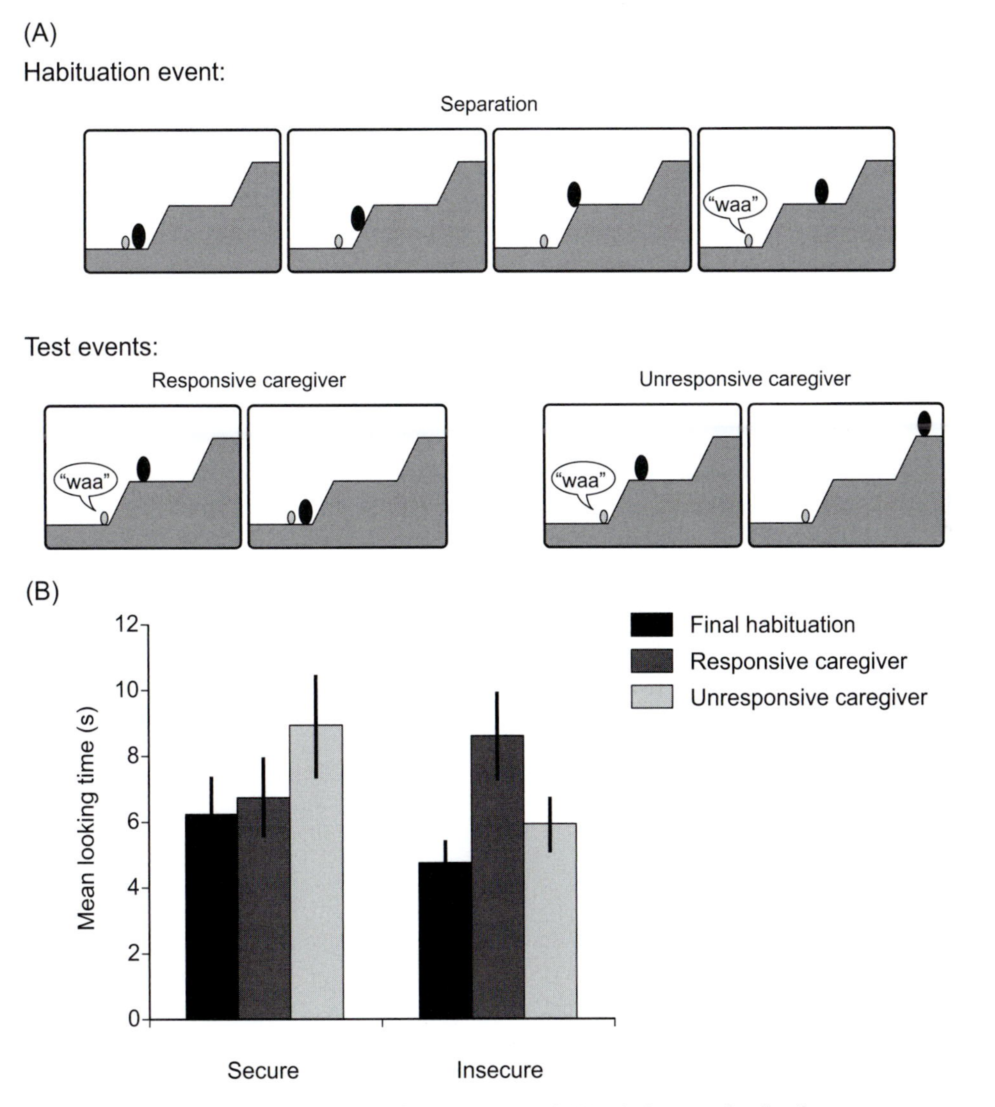

Figure 11.8 Stimuli (A) and results (B) for Johnson *et al.* (2010). See text for details.

Ainsworth was influenced by the theories of Bowlby, who in turn was inspired by the ethologist Lorenz, and the psychologist Harlow. Lorenz's work on imprinting (described in Chapter 1) encouraged Bowlby to think of attachment to the caregiver as an adaptive behavior for survival, and Harlow had presented work with rhesus monkeys (*Macaca mulatta*) that emphasized the importance of the 'creature comfort' and secure base that a caregiver can provide (e.g. Harlow, 1958). Bowlby subsequently proposed that human infants have a behavioral system that responds to feelings of fear or threat by encouraging proximity to the caregiver, in turn promoting survival. Further, Bowlby (1969) suggested that this behavioral system is based on cognitive representations of social interactions, or **internal working models of attachment**. These models are 'working models' in the sense that they are constantly being revised due to experience, but they contain beliefs that the infants hold about themselves and their caregivers.

Ainsworth's Strange Situation Task was designed to tap into these models, and more recently, another experimental task has shed light on these models of attachment. For example, Johnson and colleagues (e.g. 2010) repeatedly showed infants the following scene until they habituated: a small circle (the 'child') and a large circle (the 'caregiver') were separated on a hillside (Figure 11.8A) at which time the small circle emitted a cry. Infants then watched two test events (Figure 11.8B), and infants who had been previously categorized as securely attached via the Strange Situation Task looked longer at the test event in which the caregiver circle moved away from the distressed child circle, suggesting that this action was unexpected given their model of attachment. Further, insecurely attached infants showed evidence of a different working model; they looked longer when the caregiver approached the child.

The possibility of cross-species attachment has also been studied, particularly within the dog–human relationship. Adult dogs show selective attachment behavior toward their owner during the Strange Situation Test (e.g. Topál *et al.*, 1998), and dog puppies are also more responsive to their owners than to unfamiliar humans. In fact, these attachment behavior patterns of the dog puppies were not observed for socialized, hand-reared wolves, suggesting that selective processes during domestication may have led to genetic changes related to the attachment system of the dog (Topál *et al.*, 2005).

the nest. Given the genetic relationship between the individuals, this prosocial behavior fits models of kin selection, but what remains unclear are the cognitive processes and other mechanisms that form the proximate causes of the behavior.

11.5.2 Consolation

Comforting behavior has also been observed between unrelated individuals, and in particular, researchers like de Waal and his colleagues have discussed instances of consolation among captive chimpanzees and bonobos (*Pan paniscus*). Consolation behavior refers to instances in which bystanders approach, often to groom or embrace, a victim of aggression (de Waal & van Roosmalen, 1979). Supporting the hypothesis that the actions are a reaction to emotional distress, de Waal's research indicates that those who engage in consoling behavior contact victims of aggression more often than they contact aggressors. Further, these bystanders contact victims of serious aggression more than victims of mild aggression.

Figure 11.9 Bill twining in rooks. This behavior will occur after a bird has an aggressive encounter with a third individual.

Subsequent studies documented these behaviors in wild chimpanzees (Kutsukake & Castles, 2004; Wittig & Boesch, 2003) and in birds (Seed *et al.*, 2007). For example, rooks (*Corvus frugilegus*), birds that are members of the corvid family, will engage in bill twining (i.e. interlocking the mandibles of their beaks) with a partner after that partner has had an aggressive encounter with another rook (Figure 11.9). Interestingly, although present in great apes and corvids, there is no evidence for consolation behaviors in monkey species, thus raising the possibility of convergent evolution of the behavior between apes and some corvids (Silk, 2007b).

However, some have warned that further examination of this possible convergence should proceed with caution as researchers cannot be sure that the consolation behavior is truly consoling in the sense of reducing stress and providing comfort. A study by Koski and Sterck (2007), for example, found no evidence that affiliative interactions such as embracing after conflicts serve to reduce stress in captive chimpanzees. Specifically, in that study, the interactions, even with close kin, did not decrease scratching behavior (a common behavior associated with stress). Instead of consolation, the authors suggest that the behavior might serve to strengthen social bonds and signal that the pair is allied. In sum, with consolation behaviors, researchers are still searching for answers to two questions: (1) do those who console actually intend to comfort? and (2) do the actions function to decrease distress or do they serve another function?

11.5.3 The role of empathy

For some, a particularly important reason to continue to study comforting (and indeed, all prosocial behavior) in animals is the examination of **empathy** as a cause of the behavior. Empathy is a term commonly used to refer to the ability to understand the feelings of others, and it is distinguished from the concept of **sympathy**, which refers to feelings of concern for others stemming from empathy (e.g. Silk, 2007b). By considering the role of empathy in animal comforting behavior, researchers aim to address the first of the two questions above regarding the 'motive' behind an animal's behavior.

Because empathy refers to the ability to understand the feelings of others, researchers also have to be clear as to what they mean by 'understanding' feelings. In the field of developmental psychology, Hoffman (e.g. 1981) has proposed levels of empathy development, and researchers focused on animal behavior, such as Frans de Waal (2008), have proposed similar levels in order to consider the underlying cognitive mechanisms in animals. At the first of de Waal's three levels of empathy, and foundational to the other levels, is the ability to match another's emotional state, referred to as **emotional contagion** (Hatfield *et al.*, 1993). For example, human infants, rats, and pigeons display distress in response to perceived distress in a conspecific (Church, 1959; Sagi & Hoffman, 1976;

Watanabe & Ono, 1986). Similarly, chimpanzees show similar changes in brain and peripheral skin temperature in response to emotionally charged images of fellow chimpanzees who were darted with a tranquilizer gun as to an image of the gun itself (Parr, 2001).

Emotional contagion may serve to motivate prosocial behavior such as comforting and consoling because by dampening the emotional distress signals of others, the individual who comforts will also experience a decrease in their own vicarious emotional distress (de Waal, 2008). An explanation such as this may, for example, serve as an interpretation for the helping behavior observed in rats that was described above. An outstanding question remains, however, as to whether all responses to others' distress might be motivated solely by the drive to extinguish one's own emotional distress that has been brought about by emotional contagion. As one may assume, there remains much debate as to the correct answer to this question. de Waal proposes that some animal behaviors may be based on his second level of empathy, **sympathetic concern**, in which emotional contagion is combined with appraisal of the current context as well as attempts to understand the underlying cause of the other's distress. Sympathy is not simply personal distress, but a concern for the other individual, and this element of concern means that one acts in an other-oriented, unselfish manner. Sympathetic concern, for example, may underlie the actions of pets such as dogs when they comfort adults who have been instructed to feign sadness; dogs often put their heads on the lap of the distressed person (Zahn-Waxler *et al.*, 1984; de Waal, 2008). However, as some have argued (e.g. Custance & Mayer, 2012), it is more parsimonious at this point to describe such behaviors as emotional contagion that is combined with a previous learning history in which there has been reward for approaching distressed humans (e.g. affection or decreased arousal due to the calming of the human).

If finding evidence for sympathetic concern in animals is difficult, one may surely imagine that evidence for de Waal's third level of empathy, **empathetic perspective-taking**, would be even harder to find (also see Box 11.4). At this level, an individual "understand[s] another's specific situation and needs separate from one's own – combined with vicarious emotional arousal" (de Waal, 2008, p. 285). The previous two levels, emotional contagion and sympathetic concern, are present here, but the additional element is a consideration of the goals of others and the means by which one can intervene to enable those goals. It is possible to propose, as de Waal (2008) does, that helping behavior may constitute empathetic perspective-taking because the goal of the individual who needs help must be attributed and accompanied by a concern that motivates the helper's intervention. The nature of the 'perspective-taking' in terms of goal attribution is difficult to know at this point, though, as goals may be attributed via a cognitive system that reasons about behavior or one that reasons about the mental states that cause behavior (see Chapter 10).

Box 11.4 Contagious yawning and empathy

Observing another person yawn, or even thinking or reading about yawning, leads to yawning in about 40–60% of human adults (Platek *et al.*, 2003; Provine, 1986, 1989). Various interpretations of contagious yawning have been proposed, including those that consider the behavior to be a fixed action pattern in response to a yawn stimulus (Provine, 1986). However, more relevant to the present chapter, others have suggested that contagious yawning is based in empathy because the frequency of the behavior is related to scores on a test of empathy (e.g. Platek *et al.*, 2003). Further, children with autism spectrum disorder, a developmental disorder that has been linked to limitations in empathic responding, show less contagious yawning than typically developing children even though their rates of spontaneous yawning do not differ (Senju *et al.*, 2007).

Given the research with humans thus far, however, caution should be taken in considering the connection between contagious yawning and empathy. First, studies with individuals with autism will need to address the fact that these individuals tend to fixate more on mouth regions than on eyes (e.g. Klin *et al.*, 2002), and yet the eyes of yawning people are particularly relevant stimuli for contagious yawning (Provine, 1989). Second, the underlying cognitive and neural mechanisms of contagious yawning have been hard to pinpoint, and thus clear connections with theories of empathy (and proposed levels of empathy as described earlier) have been impossible to make.

Still, research on contagious yawning behavior continues, including work with animal species. These studies have often used procedures that measure yawn frequency when subjects are watching another individual engage in an actual yawn (experimental condition) and when subjects are watching an individual engage in a mouth movement unrelated to a yawn (control condition), and these frequencies are statistically compared. Chimpanzees, for example, engage in more yawning after viewing a video or a computer animation of another chimp yawning than after viewing another chimp engaged in other species-typical mouth movements (Anderson *et al.*, 2004; Campbell *et al.*, 2009). Data from dogs are more discrepant. One study showed that 21 of 29 dogs yawned during a 5-minute interval in which an unfamiliar human engaged in multiple yawns in front of them, whereas no dogs yawned in a control condition (Joly-Mascheroni *et al.*, 2008). However, in a study that used video stimuli of dogs or humans yawning, the frequency of contagious yawning in dogs was much lower (Harr *et al.*, 2009).

In sum, the function and underlying mechanisms of contagious yawning are still unknown. It remains possible that some form of empathy underlies the behavior; the representation of the observed action may modulate the observer's behavior such that a matching autonomic response is made, akin to emotional contagion. Alternatively, the contagious yawning may be a response to a stressful situation (Joly-Mascheroni *et al.*, 2008). When stumptail macaques (*Macaca arctoides*) were tested, for example, contagious yawns were often accompanied by self-scratching, which is thought to be a sign of tension (Paukner & Anderson, 2006; but see Palagi *et al.*, 2009).

11.6 Cooperation

As described in Section 11.2, this chapter refers to 'cooperation' as behavior in which two or more individuals work together to achieve a shared goal. For example, humans engage in cooperative interactions when they work together to remove a fallen tree from a highway, play a symphony, or build an irrigation system. Cooperation is mutually beneficial behavior in which there are immediate gains for all actors, and the risk of cheaters or defection is relatively low since individuals gain more by working together than by acting alone. However, the risk of cheaters increases when the group size increases, particularly if the outcome of the cooperation is publicly available in the sense that individuals who did not take part in the cooperative activities can access the benefits (often referred to by economists as the '**tragedy of the commons**').

Instances of cooperative behavior can be found across many species, but questions remain as to the underlying cognitive mechanisms that support cooperation and limit the risk of cheaters. These questions, in turn, raise a further question: to what extent, if any, is human cooperative behavior distinct from animals'? In this section, a wide variety of cooperative behaviors will be presented, both from observation in the wild and experimental manipulations in the laboratory.

11.6.1 Hunting

Hunting sometimes involves the activity of many individuals, particularly if the prey is large or difficult to subdue by any one individual acting alone. Group hunting has been well studied in social carnivores such as lions (*Panthera leo*; e.g. Scheel & Packer, 1991), hyenas (*Crocuta crocutae*; e.g. Kruuk, 1972), and wolves (*Canis lupus*; e.g. Mech, 1970). Primates also engage in group hunting; for example, researchers like Boesch and Stanford have examined the group hunting behaviors of chimpanzees, whose prey are often monkey species like the red colobus monkey (*Colobus badius*). Though chimpanzees are much larger and stronger than red colobus monkeys, a chimpanzee's large size can actually make it difficult for one individual to catch a monkey because a red colobus can stay out of reach by moving with great agility through high, thin branches in the forest canopy. One option that chimpanzees have is to hunt individually by attempting to drive a monkey to areas in which the canopy is interrupted, effectively cornering the prey. Alternatively, chimpanzees may hunt in groups so that even in the forest canopy, they can be successful (Boesch, 2002).

Boesch has documented chimpanzee group hunting behavior, noting that during a hunt, different roles may be played by individual chimpanzees. A 'driver', for example, may follow prey without actually trying to catch it. A 'blocker' may deter the progression of prey, whereas a 'chaser' tries to actually catch up with the prey. Finally, an 'ambusher' is a hunter who places himself in a position at which the prey is ultimately likely to arrive. Among the hunts that Boesch observed, sometimes males played multiple roles in the same hunt, and in total, all of the males performed each of the roles at least once over the observed hunts (Boesch, 2002). Each of these roles likely involves different cognitive processing, though direct study is difficult in the wild. The ambusher, for example, appears to reason about the behavior of the prey, even in its absence, so as to predict likely locations for its appearance. As detailed in Chapter 10, though, it would be possible to do so without also reasoning about the mental states underlying the prey's behavior; skilled hunting takes about 20 years to develop, which may allow for the formation of representations of behavioral tendencies that lead to predictions of likely prey behaviors. An additional consideration is that research from populations of chimpanzees in other regions suggests that hunting is not truly a cooperative effort but instead consists of individual, independent actions by multiple chimpanzees (Stanford, 1998). Thus, the study of the cognitive processes underlying hunting may require further observational work to determine whether, and under what circumstances, cooperative roles are taken during a hunt.

When the prey has been captured, new considerations exist. Cooperation may not thrive as a behavioral system if others, bystanders for example, also benefitted to the same extent from the hunt; if that were the case, the most logical behavior would be not to hunt but to share in the spoils of others' work. Further, if at the end of the hunt, only one chimpanzee is holding the prey, cooperation may not thrive unless the food is shared. As mentioned in Section 11.4, passive sharing behavior has been documented for wild chimpanzees (Muller & Mitani, 2005), and this behavior is primarily seen after the hunt. However, the sharing is not equal across all members of the troop. By one account, chimpanzees allow others to obtain meat according to their role in the hunt. Boesch and Boesch (1989) report that hunters obtain more food than bystanders who did not engage in the hunt. There are at least two possibilities regarding the cognitive processes underlying the food distribution: (1) chimpanzees keep track of others' contributions, or more simply (2) the individuals who hunted are also better at securing the most food for themselves (e.g. size, strength, dominance, etc., Melis & Seemann, 2010).

Future research on the cognitive processes that underlie group hunting will help to elucidate these mechanisms. Yet, research on cooperation in the wild can be quite difficult. The chimpanzee hunt, for example, occurs high in the forest canopy and occurs at very high speed, both reasons why documentation by human observers is difficult. For these reasons, among others, researchers have also examined cooperation in the laboratory, which is the subject of the next section.

11.6.2 Cooperation in the laboratory

When the study of cooperation is moved to the laboratory, the tasks do not typically involve hunting; they involve other forms of cooperative problem-solving that lead to benefits for all participants. With adult humans, for example, experimenters have used a public goods game in which four players have to decide, simultaneously, whether (and how much) to contribute to a pool of money. The pool is then doubled and divided among the players equally. Thus, cheating by not contributing always yields better gains for an individual since there is no initial loss of money. Typically, there is a high level of cooperation at the start, but it ultimately wanes when the game is played repeatedly (Melis & Seemann, 2010; Milinski *et al.*, 2002). However, since humans are capable of keeping track of others' contributions, cheating can be minimized by behaviors such as punishment or ostracism, ensuring that cooperation survives.

Keeping track of others' contributions is one mechanism that supports cooperation, and laboratory studies of primate species have pointed to others. Laboratory tasks often create situations in which pairs of animals must work together to achieve food using an apparatus like that in Figure 11.10. In this task, two participants (typically apes, monkeys, or human children, though also hyenas, rooks, and elephants) must each pull an end of the rope to bring the platform close to obtain the food. One participant alone cannot complete the task: the ropes are too far away to be pulled together at once, and further, if only one rope end is pulled, the participant will simply end up with a rope, not food (e.g., Melis *et al.*, 2006). Other variants of this task are analogous, but instead of a loose rope, these tasks use two ropes that need to be pulled by two participants because the object being pulled in is too heavy for one individual (e.g. Crawford, 1937).

In a series of studies using this task with capuchin monkeys, the monkey pairs acted simultaneously and obtained the food (e.g. Mendres & de Waal, 2000; Visalberghi *et al.*, 2000). However, debate exists as to the most accurate interpretation of this behavior. Visalberghi and colleagues have

Figure 11.10 Apparatus for studying cooperation in the laboratory. Here, two participants must each pull an end of the rope to obtain food (e.g. Melis *et al.*, 2006).

concluded that success on their task was due to simultaneous pulling by chance and that participants were not taking into account the presence of the partner. Under this interpretation, no cooperation was truly occurring, and pullers were not considering the role of the partner. In contrast, Mendres and de Waal propose that since participants showed improved performance when they could clearly see each other (and thus communicate), there is evidence of coordinated activity. In turn, coordination (here, and likely in many other contexts) may be the outcome of using the partner's behavior as a discriminative stimulus for one's own behavior (e.g. learning that 'pulling when they pull leads to reward'; Shettleworth, 2010).

In addition to learning mechanisms, other factors may impact the frequency of cooperative behavior. Melis and colleagues, for example, found that tolerance can increase cooperation on a rope-pulling task. Pairs of chimpanzees that showed more ability to share food outside of the testing session were also more likely to work together to solve the task (Melis *et al.*, 2006). Additional evidence comes from bonobos, a species that tends to show greater tolerance in relation to food than chimpanzees. When tested in the rope-pulling task, bonobos engaged in greater cooperation than chimpanzees, particularly when the food outcome could be monopolizable because it was clumped centrally on the platform (Hare *et al.*, 2007). That is, when pulling resulted in two, divided clumps of food, both species cooperated. In contrast, the more tolerant bonobos also cooperated when the ultimate food reward would have to be shared.

Box 11.5 Oxytocin and prosocial behavior

Oxytocin has become the focus of attention for researchers interested in the biological mechanisms of social cognition, and specific to this chapter, prosocial behavior. Oxytocin is a hormone that regulates contractions during childbirth as well as lactation, but oxytocin also acts in the brain. It is released by neurons (thus, a **neurohormone**), and can greatly influence social behavior. In rodents, for example, oxytocin is released in pathways that are responsible for maternal behaviors such as building nests, licking and protecting pups, and retrieving pups back to the nest. Thus, in general, it can be said that oxytocin is relevant to kin-directed prosocial behavior.

However, in humans, oxytocin appears to affect sharing behavior toward strangers as well. Studies often rely on a procedure in which oxytocin is administered to participants via a nasal spray. (Control groups are given a placebo spray.) Those who receive oxytocin tend to give more money in two economic games described earlier in this chapter: the ultimatum game (Zak *et al.*, 2007) and the trust game (Kosfeld *et al.*, 2005). One limitation of studies that administer oxytocin and examine behavioral effects is that they cannot speak to the underlying mechanisms and functions of oxytocin. Recently, however, researchers have begun to look to the actual receptor for answers.

Neurohormones like oxytocin have specific receptors on cells that allow them to interact with the cells. Oxytocin has one receptor (OXTR), but variations in genes for OXTR can increase or reduce the effects of oxytocin. Indeed, one nucleotide can differ among individuals such that one person can have an adenine (A) where another person has a guanine (G), and those with a G instead of an A seem to be those who engage in greater maternal care and have a prosocial temperament (see Poulin, 2012, for a summary).

As research grows in this area, new theories are being presented. For example, a study by Poulin *et al.* (2012) has further examined the way in which OXTR might influence prosocial behavior. First, a large sample of adults completed surveys that covered topics such as donating blood or giving money to charities, the perceived duty to make sacrifices for society, and opinions about whether people are generally benevolent or untrustworthy. Participants also provided saliva samples for DNA analysis, and thus each participant was determined to have either the G or the A variant of OXTR. Notably, whereas those with the A variant showed less sharing behavior if they assumed others were generally malevolent, those with the G variant did not decrease sharing behavior even if they viewed others as untrustworthy. Poulin (2012) suggested that this variant of OXTR may lead to diminished concerns over others' likelihood to 'repay' by reciprocating. Further, he proposes that perhaps oxytocin can make humans think of strangers as they would kin, thus creating a situation in which actual reciprocity is less of a consideration for engaging in prosocial behavior. Of course, further work is necessary, including work that examines these theories with animal species.

11.6.3 Conclusions

Thus, laboratory studies are beginning to discover important underlying factors for cooperation. Some cognitive components appear to be available to many species, such as learning to coordinate actions. Other aspects may vary greatly among species, such as a level of tolerance that allows for the sharing of a public resource. Indeed, it is here that some of the greatest differences with humans, a relatively tolerant species, can be seen. For example, 3-year-old children tested with the rope-pulling task readily cooperated even when the food (or sticker) reward was clumped in the middle of the platform. Even more telling, most of the time, children subsequently shared the clumped reward equally (Warneken *et al.*, 2011).

More generally, tolerance may play a role in all types of prosocial behavior, though as discussed throughout this chapter, a wide range of cognitive processes that are shared across many species seem to support prosocial actions, including: memory, individual associative learning, social learning, theory of mind and/or representations of behavioral patterns, and kin recognition, among others. As has been seen, there may be unique characteristics of human prosocial behavior – perhaps based on unique cognitive processes – though there is no simple answer to what these unique aspects are.

Chapter Summary

- Biologists, psychologists, mathematicians, and economists have all examined how it can be that behaviors that lead to decreased survival and reproductive success for the actor, yet increase the survival and reproduction of others, are behaviors that will evolve.
- Kin selection as well as reciprocity can explain the evolution of prosocial behavior. The latter has been extensively modeled by economic games such as the Prisoner's Dilemma.
- Different behaviors fit under the heading of prosocial behavior, such as helping, sharing, comforting, and cooperation. Examining these behaviors separately may allow for a more careful consideration of the behaviors and their underlying cognitive mechanisms.
- Helping behaviors function to assist others with certain instrumental needs. A variety of helping behaviors have been observed across animal species, including aiding conspecifics to have access to mates, defending the nest, alerting others to a threat, and even picking up an object that a human experimenter has dropped.
- Sharing consists of giving some of one's own material resources to benefit another individual who is in need. In the wild, food sharing is most clearly seen in cooperative breeders. In the laboratory, sharing is often modeled by the Dictator Game, and performance on this game varies between humans and other animals. Humans appear to more readily share, and we also seem to place a greater value on fairness than other species.
- Comforting is a prosocial act in response to others' emotional needs. Examples include responding to distress calls and consolation, and of current debate is whether empathy can be said to underlie these behaviors.
- Cooperation describes behavior in which two or more individuals work together to achieve a shared goal. A paradigmatic case from observations in the wild is group hunting, though the cognitive processes that underlie this behavior are still being examined. In the laboratory, the behavior of apes and monkeys on cooperative tasks has led to proposals that a certain level of tolerance is necessary to cooperate effectively when rewards must ultimately be divided.

Questions

1. Why was the existence of prosocial behavior a challenge for evolutionary theorists, and what major theories now serve to explain its existence?
2. How do humans typically respond in the Ultimatum Game, and what aspect of this behavior does this game model? How has this been tested in chimpanzees?
3. How can helping behavior be tested in the laboratory with chimpanzees? What cognitive processing might underlie this behavior?
4. What are some reasons why humans and other species might respond differently in the Dictator Game?
5. What are evidence-based arguments for and against the existence of inequity aversion in capuchin monkeys?
6. What is an important distinction between empathy and sympathy?

7. Imagine that occasionally, when its owner is sad, a cat approaches the owner and lays its head on her lap. What level of empathy may most parsimoniously describe the cat's behavior?
8. In what two situations is cooperation particularly at risk of cheaters?
9. In the text, cognitive mechanisms that may underlie the behavior of the 'ambusher' in a chimpanzee group hunt were proposed. What might be the mechanisms that underlie the behavior of the 'chaser'?
10. What might the comparative study of prosocial behavior ultimately tell us about potentially unique aspects of human prosociality?

FURTHER READING

The original papers and books on prosocial behavior by evolutionary theorists provide a rich understanding of the history of the field. Some examples include:

Dawkins, R. (1976). *The Selfish Gene*. Oxford: Oxford University Press.
Hamilton, W. D. (1963). The evolution of altruistic behavior. *Am Naturalist*, **97**, 354–356.
Trivers, R. L. (1971). The evolution of reciprocal altruism. *Q Rev Biol*, **46**, 35–57.

The following books and papers present views on empathy and its role in prosocial behavior:

de Waal, F. B. M. (2008). Putting the altruism back in altruism: The evolution of empathy. *Ann Rev Psychol*, **59**, 279–300.
Hoffman, M. L. (2000). *Empathy and Moral Development: Implications for Caring and Justice*. New York, NY: Cambridge University Press.
Silk, J. B. (2007). Empathy, sympathy, and prosocial preferences in primates. In R. I. M. Dunbar and L. Barrett (eds.). *The Oxford Handbook of Evolutionary Psychology* (pp. 115–126). Oxford: Oxford University Press.

The following papers provide a discussion of the role of oxytocin in social behaviors including prosocial behavior:

Campbell, A. (2010). Oxytocin and human social behavior. *Pers Soc Psychol Rev*, **14**, 281–295.
Poulin, M. (2012). Our genes want to be altruists. *Observer*, **25**. Association for Psychological Science.

For recent theories on the uniqueness of human prosocial behavior, see:

Silk, J. B., & House, B. R. (2011). Evolutionary foundations of human prosocial sentiments. *Proc Natl Acad Sci USA*, **108**, 10910–10917.
Tomasello, M. (2009). *Why We Cooperate*. Cambridge, MA: MIT Press/Boston Review.

12 Communication

Background

From across the beach, the female came running at full speed. When she was 10 feet from the male, she stopped, and they both stood still until, suddenly, he dropped into a bow by laying his front legs on the ground. After a pause, they burst toward each other and proceeded to chase, bite, and bark, stopping only to bow again and then resume. The movements were fast and powerful, a flurry of fur and paws. When both were tired, they trotted over to a bowl of water and shared it, drinking side by side.

The intensity of dog play behavior can be impressive, and at times, it can easily be mistaken for fighting by human observers. Socialized dogs (and other members of the family *Canidae* such as wolves and coyotes) appear to recognize the difference between play and aggression, and normal dominance hierarchies and 'friendships' appear to resume after play bouts. Play is made possible through communication such as body postures; even Darwin used the example of bowing behavior in *The Expression of the Emotions in Man and Animals* when discussing the evolution of communication (Darwin, 1872). Animal communication can take as many forms as perception allows and is integral to the survival of social species. It has been one of the primary areas of interest in comparative cognition, with general lines of inquiry that ask questions at the proximate level regarding whether communicative behaviors truly refer to objects in the world (e.g. food, predators, etc.) and if communicators aim to modify others' behavior with their own. At the ultimate level, research considers the evolution of communication, and more specifically, the evolution of human language.

Chapter plan

This chapter begins with a broad consideration of the topic of communication, defining terms and themes, but also details the special aspects of human language as a form of communication. With these fundamentals in place, some particularly well-studied natural communication systems are examined, including the 'dances' of honeybees, alarm and food calls, and play signals. Then, a survey of the attempts to teach human language to apes will be presented. As will be seen, it has been quite common to consider animal communication systems in relation to human language, owing to a more general question of whether language actually marks a qualitative difference between humans and animals.

12.1 Features of communication

Communication is occurring almost continuously. On a short walk through a suburban neighborhood, for example, one can encounter a line of ants following each other through chemoreception, a dog smelling the urine of a dog that marked a tree hours earlier, the sound of bird song, the vivid coloration of butterflies, and the daily newspaper. The topic of communication encompasses a broad range of behaviors, transmission across varying media, and perception via different sensory modalities. To study communicative systems, then, it is important to define the general features of information transmission. This section will begin by considering communication as an instance of behavior that influences others' behavior and will define relevant terms. Then, because the field of comparative cognition has traditionally been interested in what, if any, aspects of human language are found in animal communication systems, some key features of language (as identified by Hockett, 1960, and Fitch, 2005) will be presented.

12.1.1 Signals, senders, and receivers

The transmission of information between individuals occurs through communicative **signals** produced by a **sender** to a **receiver**. Not every behavior is a signal, though clear differentiation between signals and non-signals is not always easy. The work of early ethologists often examined patterns of behavior that appeared to be selected for specific interactions between members of a species (Shettleworth, 2010). These were often stereotyped and repeated behaviors that occurred in a particular context, such as 'courtship dances' performed by males in the presence of females during mating season. Another example is the display behavior of chimpanzees (*Pan troglodytes*), which occurs in an arousing situation and consists of an individual (typically male) swaying or moving about with hair standing on end and vocalizing or otherwise violently making noise using objects in the environment (Figure 12.1A). The display is quite impressive and intimidating, even for human observers.

Behavioral patterns such as these are classified as communication, but information can also be transmitted in a passive manner. A classic example of passive communication is **aposematism**, in which an unpalatable species has coloration, odors, or warning sounds that signal their danger (e.g. toxic or causing illness) to potential predators (Figure 12.1B). These signals typically are conspicuous, and predators may learn them relatively quickly, particularly if the punishment is strong (Ruxton *et al.*, 2004). Predators also appear to generalize the learning of signals to other, similar-looking species, a topic expanded upon in Section 12.2.

For successful communication to occur, the receiver requires an appropriate perceptual system to process the signal. Indeed, human observers can be unaware of other species' communication systems due to our limited perception. For example, until recently, it was not known that elephants (e.g. *Elphas maximus*) send, receive, and respond to low frequency vocalizations that are out of human perceptual range (e.g. Heffner & Heffner, 1982; O'Connell-Rodwell *et al.*, 2000; Payne *et al.*, 1986; Poole *et al.*, 1988). Within a species, though, the perceptual match of signal to receiver can also allow for **eavesdropping** behavior, or passive listening by other members of a group (Johnstone, 2001; McGregor, 1993). During conflict, two aggressors may exchange signals that are also detectable by individuals that are not engaged in the fight. These signals can influence the eavesdropper's subsequent behavior toward the fighters (e.g. submission or avoidance of fight).

Figure 12.1 (A) Chimpanzee display behavior. (B) A striped skunk (*Mephitis mephitis*) with coloration exhibiting aposematism.

12.1.2 Signal set

Human communication is characterized by an unlimited signal set. Humans use words in many contexts, on a broad range of topics spanning past, present, and future. When a new word is needed, one is created; language is open-ended. In contrast, based on observations thus far, there is reason to believe that most animals have a **limited signal set**. That is, animals tend to communicate with a relatively small number of signals in a small number of contexts (e.g. food, aggression/competition, sex, and danger from predators). Some signals might be graded, like an increase in pitch or volume with increased excitement (see, for example, the later discussion of chimpanzee food barks), but the number of signals remains small.

The unlimited signal set of human language is enhanced by combinatorial rules that allow for new meaning to come from different arrangements of words. In English, "*The dog bit the man*" has a much different meaning than "*The man bit the dog*." Further, language has the property of **recursion**, which, when applied to language, refers to the unlimited extension that is made possible by embedding clauses within clauses (e.g. Chomsky, 1965; see Fitch, 2010, for other definitions). The recursive ability can create meaningful sentences like "*Mary saw John eat the peach that she had purchased this morning*" from embedded simple sentences like "*Mary saw John*," "*John ate the peach*," and "*Mary purchased a peach this morning*." By one proposal, recursion is the one ability that is unique to human language. In this way, it is the structure of language – and the ability for limitless expression though recombination of units – that makes it unique among communication systems (e.g. Hauser *et al.*, 2002). Alternative proposals have been made which claim that there is not one specific ability that distinguishes human language; instead, it is proposed that some abilities that are shared with other species have, for humans, been co-opted for our form of communication (e.g. Pinker & Jackendoff, 2005; see also Fitch *et al.*, 2005).

12.1.3 Intention to inform

In humans, the sender typically uses language with the *intention* to provide information to the receiver. Behavior is thus altered given the situation and the characteristics of the receiver. If, for example, a potential receiver is likely already aware of the information, a speaker might not communicate or might alter the words used. In this way, human language use intersects with our theory of mind abilities. But, just as it has proven difficult to achieve consensus on the social cognitive mechanisms of animals (Chapter 10), it is also difficult to determine whether their signals are produced with an underlying intention to inform.

To approach this question, researchers have examined whether animal communication shows an **audience effect** such that signaling behavior is altered by the presence or absence of receivers as well as the receivers' response. Alarm calls will be more fully discussed in Section 12.4, and they are a logical communicative signal in which to look for audience effects. Calling in the presence of a threat potentially puts the signaler at increased risk of predation, and thus calling in the absence of conspecifics may hold no benefit for the signaler. Roosters' (*Gallus gallus domesticus*) alarm calls do seem to be modulated by the presence of others; in the laboratory, alarm calling increased when roosters could see the image of a videotaped hen (Evans & Marler, 1992). Studies such as this, and those with other species (see Gyger, 1990), show evidence of conditional control of the senders' behavior by the receivers, but as Shettleworth (2010) points out, there is little evidence to support an interpretation based on the senders' understanding of receivers' mental states.

Human communication also shows sensitivity on the part of the *receiver* to the mental states of the *sender*. Children, for example, often use a speaker's gaze as a clue to the referent of a new word (Baldwin, 1993). For adults, the on-line use of theory of mind during communication has been examined in a task that involves a 'director' issuing commands to a participant to move objects from shelf to shelf in a 4×4 shelving array (e.g. Keysar *et al.*, 2003). The director and participant sit on opposite sides of the set of shelves, and because some of the objects are occluded from the director's view, the participant must consider what the director can and cannot see to understand what he or she intends to refer to. For example, if the participant can see that two shelves hold tennis balls, but the director can see only one of them, a command to "move the tennis ball up one shelf" should lead the participant to move the ball that is observable by the director. Adults actually have some difficulty with this task and occasionally err by answering based on an egocentric viewpoint, especially when the number of possible objects increases (Dumontheil *et al.*, 2010; Keysar *et al.*, 2003). Yet, the task is often solved correctly and appears to activate areas of the brain associated with theory of mind reasoning (Dumontheil *et al.*, 2010). Later sections of this chapter will consider whether the response of recipients in animal communication are modulated by characteristics of the sender, but again, there is little evidence to date that animal communication systems incorporate sophisticated consideration of the mental states of those involved.

12.1.4 Functional reference

A signal can encode information about the sender's motivational state and identity as well as stimuli in the environment such as food or predators (e.g. Evans, 1997; Marler, 1967; Box 12.2). If the signals provide information specific to a particular object or event and are not simply signals that convey, say, generalized arousal, and if the receivers of the signal respond to the signal even in the absence of the stimulus that elicited it, the signal is considered to be **functionally referential** (Evans, 1997). These signals provide receivers with sufficient information to predict events in the environment (e.g. the presence of a predator) even when the receivers cannot directly perceive the event themselves (e.g. Hauser, 1996). By including the word 'functionally,' though, researchers are acknowledging that many of the cognitive processes underling signal production (e.g. the intention to inform, etc.) are either poorly understood or presumed to be absent; thus, signals are said to at least *function* as referents.

Later in the chapter, evidence will be presented that some signals in animal communication fit the criteria for functional reference. In human language, however, a speaker can also refer to objects, people, or events that have occurred in the past, are occurring now, or will occur in the future. The **situational freedom** inherent to human language means that we do not have to be directly experiencing a stimulus to signal about it. Whether or not some animal signals exhibit situational freedom is still a point of discussion (Roitblat, 1987; Shettleworth, 2010). There is likely an important difference between our ability to refer to past, future, and even fictional events, and the ability of many animal species to signal about predators or food sources that were just observed.

12.2 Evolution of communication

In the study of communication, two main lines of inquiry dominate. The first is the search within animal communication for the features of language that were introduced in the previous section, and the second is the evolution of communication, which will be introduced in this section. Both lines will continue through this chapter, and both intersect with the study of human language. The evolution of communication has a rich literature, with seminal theories originating in ethology and considered in psychology and linguistics. The traditional ethological view focused on communication as a prosocial behavior (e.g. Marler, 1968; Tinbergen, 1964), which was discussed in Chapter 11 when considering information sharing (via alarm calls) as an act of helping. Broadly, communication can benefit both signaler and receiver in environments in which kin altruism or reciprocity is possible. These original ideas have been elaborated to consider implications of individual selection (e.g. Dawkins & Krebs, 1978), emphasizing that signalers can benefit from manipulating others. Yet, it has also been proposed that in some cases, honest signaling should evolve. Both of these seemingly contradictory accounts will be considered below.

12.2.1 Manipulation and deception

Sometimes, what is in the best interest of the signaler is not in the best interest of the receiver. A chimpanzee's display behavior is costly and seemingly not of benefit for others, but potentially benefits the signaler by giving the impression of being bigger and healthier than he actually is. As Maynard Smith and Harper (2003) discuss, signals are not necessarily truthful indicators of the quality of the sender but can benefit the sender by manipulating the receiver. As dishonest signals are selected for though, so too may be the receiver's ability to distinguish between honest and dishonest signals. Thus, an evolutionary 'arms race' occurs in which the signals that better manipulate the receiver are selected, and the detection system of the receiver that better filters this signal is selected (Krebs & Dawkins, 1984).

It is worth considering at this point that, although the topic of dishonest signaling in animal communication is laden with terms such as 'deception' and 'manipulation,' there is no assumption that the signalers are intending to deceive. In contrast, it is readily accepted that human communication can be used to intentionally mislead or deliberately create a false belief through reasoning about the mental states of others. The ethological theory simply implies that the signals of animals can function to deceive the receiver to the benefit of the sender.

In fact, even a passive signal can be considered deceptive. In the case of **Batesian mimicry**, a palatable species has similar physical features to an aposematic species, in essence dishonestly using another species' honest warning signal. The viceroy butterfly (*Limenitis archippus*), for example, does not have the toxicity of the monarch butterfly (*Danaus plexippus*), yet has similar coloration and is avoided by the predators that learn to avoid the monarch. Batesian mimicry can also be seen with auditory warning signals. By using high-speed, infrared cameras to observe bat–moth interactions, Barber and Conner (2007) examined red and big brown bats' (*Lasiurus borealis* and *Eptesicus fuscus*, respectively) reactions to sound-producing, noxious tiger moths (*Cycnia tenera*). As expected, the bats quickly learned to avoid the tiger moths. The bats also then avoided another sound-producing species (*Syntomeida epilais*) even though it lacked the chemical protection of the tiger moths. Interestingly, a subset of the red bats subsequently learned that the Batesian mimic was palatable, suggesting that in this interaction, the bats exert a strong selective force on mimetic resemblance. That is, in Batesian mimicry, the duped receiver may come to recognize subtle physical (or auditory) differences between the noxious and palatable species, exerting selective pressure for more accurate mimicry.

12.2.2 Honest signaling

In some cases, the nature of a signal would suggest that it is too costly to fake, or simply could not be faked, and thus honest signals can evolve (e.g. Grafen, 1990; Zahavi, 1975). Zahavi's **handicap principle**, for example, suggests that a costly trait (like that of a peacock's tail) can be used as an honest signal because only those individuals who can develop the tail and survive the increased predation risk will, in fact, have the trait. Other proposed examples of honest signaling include the stotting behavior of gazelles in which an individual springs into the air, lifting all four legs off of the ground (Figure 12.2), and 'conspicuous consumption' in humans in which some individuals may purchase and clearly display valuable goods (Johnstone, 1998; Veblen, 1899). Even aposematic species can be said to be engaging in honest signaling to potential predators, advertising their very real danger. Initially, Zahavi's handicap principle and the general proposal of honest signaling was the topic of much debate; however, recently developed mathematical models have lent support (e.g. Grafen, 1990; but see Getty, 1998), and the ideas are now generally accepted (e.g. Maynard Smith & Harper, 2003).

Figure 12.2 A gazelle exhibiting stotting behavior, proposed as an example of honest signaling.

Box 12.1 *FOXP2* and the genetics of language

Speech and language disorders have long been thought to involve the mutation of genes, particularly because certain disorders run in families. For example, across three generations of the KE family (researchers use coded names to protect privacy), approximately half of the members have a verbal dyspraxia, an impaired ability to complete the coordinated movements necessary for speech. The disorder affects expression and articulation, but comprehension is preserved. After studying the difficulties faced by affected members and using neuroimaging and genetic techniques, researchers identified mutated gene *FOXP2* (Lai *et al.*, 2001). This gene codes for a protein that controls expression of other genes, serving as a gene regulator in the brain and other organs (e.g. Vargha-Khadem *et al.*, 2005).

There has been great interest in the evolution of *FOXP2*, and comparisons of the nucleotide and amino acid sequences of human *FOXP2* and animal *FoxP2* (nomenclature varies slightly for the animal versions) have been completed. Although a number of nucleotide changes have accumulated since the human and mouse lineages diverged approximately 70 million years ago, only three amino acids have changed in the human sequence. Two of these three changes are present in humans but not the great apes, suggesting that the changes are relatively new, occurring only 4 to 6 million years ago (see Vargha-Khadem *et al.*, 2005, for summary). Some have estimated that this represents a selective 'sweep' in which the advantageous genotype spread quickly through the human population. There is indication based on DNA extraction from fossils that Neanderthals may have already had the changes (Krause *et al.*, 2007), though this does not imply that Neanderthals spoke as modern humans do since there are, of course, other genes involved in language and speech. The findings also do not indicate that the KE family members represent a reversion to an ancestral state, as the mutation is fundamentally different (Enard *et al.*, 2002).

The animal *FoxP2* has also been the focus of much study. Songbirds, for example, use social learning mechanisms when developing species-typical songs (see Chapter 13), and the pattern of *FoxP2* expression in the brain of one species of songbird, the zebra finch (*Taeniopygia guttata*), appears to be similar to that of the mammalian brain. When individuals are engaged in song learning, *FoxP2* appears to be upregulated in the same regions of the brain that had previously been assumed to be associated with song learning, and further, interfering with *FoxP2* expression results in incomplete and inaccurate learning (e.g. Haesler *et al.*, 2007).

More work is necessary to fully determine the behavioral role of *FOXP2* across species. In humans, it is possible that the skilled, coordinated movements related to *FOXP2* expression have been co-opted for vocal communication, and like in songbirds, *FOXP2* may be implicated in the development of circuits that are important in learning of vocal sequences (see Fisher & Marcus, 2006, for review). Importantly, though, *FOXP2* is simply one piece of a much larger puzzle of the evolution of communication, but it hints at the promising future of interdisciplinary, comparative research in this field.

12.3 Bee dance

With foundational terminology and evolutionary theory pertaining to animal communication in place, the next three sections will focus on natural communication systems. The research presented here has been completed in the wild and the laboratory, with the goals of documenting a given communication system, considering the features of communication (e.g. audience effects, functional reference, etc.), and addressing theory related to the evolution of communication. It is fitting to start this discussion with the dancing of honeybees (Genus *Apis*), because it is one of the most studied (and initially controversial) animal communication systems. The original research by von Frisch (e.g. 1953, 1967) was integral to the growth of the field of ethology, and along with the work of Tinbergen and Lorenz, earned the trio a Nobel Prize. In this section, the dance behavior will be detailed, and then, studies designed to specifically examine the information embedded in this signal will be discussed.

12.3.1 The waggle dance: initial observations and experiments

von Frisch carefully detailed the behavior of bees when they returned to the hive after finding nectar at far away locations. On the vertical side of the honeycomb, returning bees performed a **waggle dance** consisting of a straight 'run' during which the abdomen is 'waggled' back and forth, and ending with a turn to the right or left to circle back around and start the run again. After each run, the dancing bee alternates right and left turns such that the dance forms a figure-eight pattern (Figure 12.3). Other bees at the hive encircle the dancing bee, making contact with their antennae. After observing the dance, many other bees are observed to leave the hive and find food at the location previously visited by the dancer, leading to the assumption that the dance is a signal to the location of the food.

Given that the food source can be over 100 meters away, it seems amazing that such a simple dance can convey distinct information regarding a particular location. Different elements of the dance, however, appear to contain information relevant to the distance and direction of the food source. The distance to the food is conveyed by the length of the run and the duration of the dance. The angle of the dance 'run' to a hypothetical vertical line drawn on the honeycomb corresponds to the angle of the food location to the sun's azimuth angle (labeled in Figure 12.3). Other cues to the food are conveyed by the odors of the food remaining on her body.

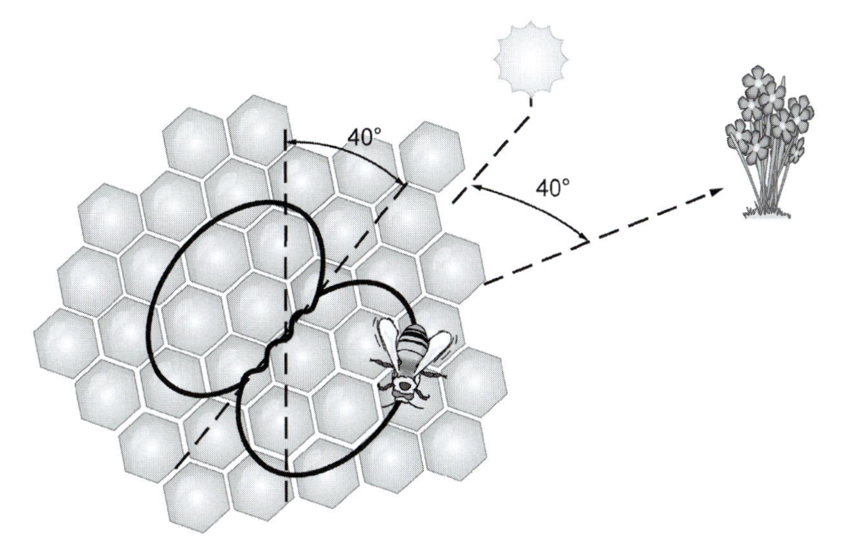

Figure 12.3 A bee performing a waggle dance. During the run, the abdomen is moved back and forth, ending with a turn to the right or left to circle back around and starting the run again. The angle of the dance in relation to the vertical corresponds to the angle of the food location to the sun's azimuth angle (labeled as 40° in this figure).

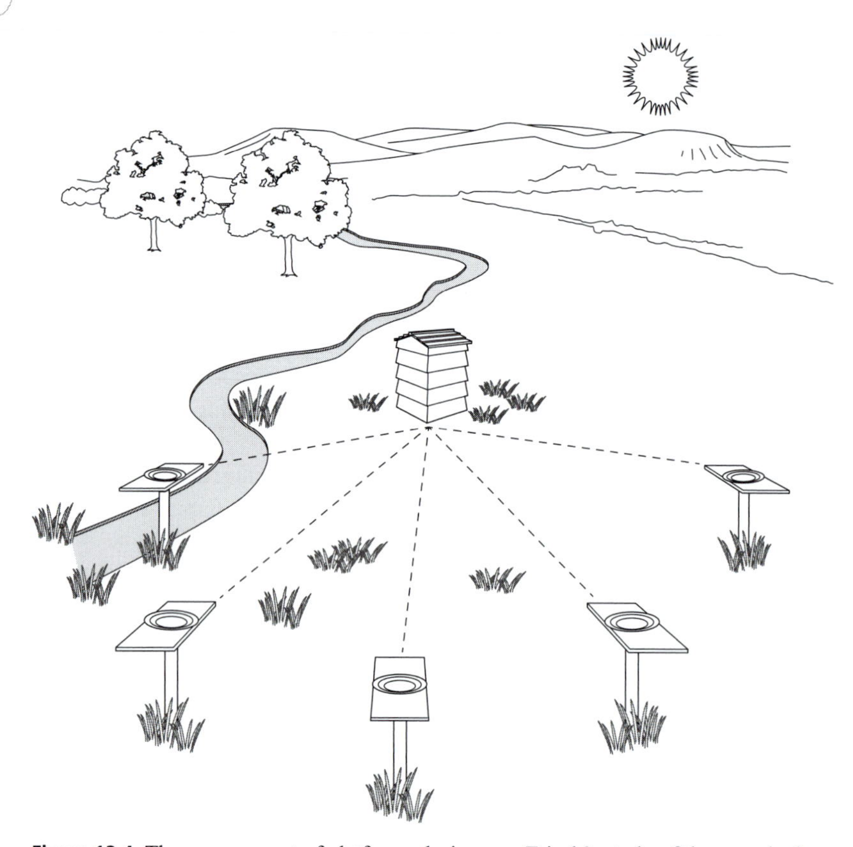

Figure 12.4 The arrangement of platforms during von Frisch's study of the waggle dance. The dances appeared to contain information about the direction of a food source; recruited bees were more likely to visit the baited platform than control platforms.

The initial research by von Frisch included experiments designed to examine whether bees that observed the dance truly used the information to find the food location (Figure 12.4). To test whether receivers (often called 'recruits') use information regarding direction, certain bees were marked and trained to find food at a feeding platform. During training days, the food was a weakly sugared solution that did not elicit dancing back at the hive, but on testing days, the solution was made much stronger. One platform was baited, but control platforms were also placed in the environment, each equidistant (but set at different angles) from the hive. After the marked bees visited the strong solution at the training feeder, they returned to the hive, danced, and the recruits subsequently arrived at the correct location, suggesting that they had learned information about direction from the dance. To test for distance information, a similar experiment was conducted, but now, the platforms were set up in a line from the hive such that the angle was identical but the distances differed. Again, though, recruits looked for food at the correct platform.

Although von Frisch's experiments seem to be compelling evidence that bees use information from the dance to indicate the location of food, controversy and debate abounded. When bees dance, the motion is often stopped, and the dancer regurgitates small amounts of the food. This, and the odor cues from the dancer's body may be all that the recruits are using to find the food outside of the hive, a possibility supported by Wenner *et al.* (1967). Wenner (1967) further proposed that the sounds made by the movement of the dancer's wings carry the information regarding the distance to the food. The alternatives to von Frisch's account of the honeybee 'dance language' were not without merit, but as will be seen in the following section, subsequent studies have shown conclusively that bees do use information from the dance to find food, though they might not rely exclusively on it.

12.3.2 The waggle dance: subsequent experimental support

Strong support for von Frisch's claims came from Gould (1975), who incorporated a novel experimental design that, in effect, caused dancing bees to misinform the recruits. The overarching logic was that if recruited bees show up to a location that was indicated by a dance, yet no food is actually there, then it cannot be said that recruits solely use environmental cues such as odor to lead them to a location. Gould based the experiment on the fact that bees in dark hives dance according to the vertical (as described above), but bees in bright hives dance in relation to the source of light. Further, after covering some bees' light-sensitive ocelli, they would behave as if they were in the dark. That is, bees with covered ocelli would dance according to the vertical, but the potential recruits in the bright hive with normal, uncovered ocelli would interpret the dance as angled relative to the light. These recruits subsequently searched for food at the misinterpreted direction, indicating that the dance itself is indeed used as communicative information.

Further evidence comes from work by Kirchner and Towne (1994) that utilized an artificial, model bee. Because experimenters could manipulate the model bee, they could systematically observe which aspects of the dance event were used by recruits. When the artificial bee danced, dispensed small amounts of the food, and vibrated its wings, recruits left the hive and searched at the indicated location. When any of the three elements were missing, fewer bees were recruited to search for the location. Thus, all aspects of the dance may be important: features like the sound and vibration might encourage bees to attend to the dance and help them determine the dance orientation (e.g. Tautz *et al.*, 2001), the dance itself conveys distance and direction information, and the odors may influence recruiters as they near the location.

12.3.3 The waggle dance: cognitive mechanisms

The cognitive mechanisms underlying dance behavior appear to include aspects of spatial cognition. Dancing bees provide straight-line information in relation to the food, yet their own search was likely circuitous. Thus, dancers may use path integration to encode and report the location (Menzel *et al.*, 2006). Receivers appear to 'read' the dance, but there is also suggestion that their prior experience in the environment can override dance information. Given prior experience, receivers can match odor from the dancer to memory of the source location in the environment (Grüter *et al.*, 2008). Note that this does not discount theories of the importance of bee communicative dance; it simply suggests that bees may not always need to rely on the information.

The dance behavior of bees has often been referred to as a 'language,' though it may be more appropriate to refer to it broadly as a communication system. When compared to the features of human language, it could be argued that bee dance has situational freedom in that the dance relates to a location that is not currently perceived (Roitblat, 1987). Yet, as noted earlier, situational freedom in human language is not limited to just-experienced events or locations. Because recruits respond to the signal in the absence of the stimulus that elicited it, there is evidence of functional reference; however, the reference appears to be a location as dances can recruit observers to visit food sites or even potential nest sites (Seeley, 1995). In contrast, studies of alarm calls

Box 12.2 Dolphin signature whistles

In 1965, Caldwell and Caldwell observed that individual bottlenose dolphins (*Tursiops truncatus*) will often produce their own specific whistle (a **signature whistle**) (Caldwell & Caldwell, 1965). Since the initial report, signature whistles have been observed for approximately 200 captive and wild bottlenose dolphins as well as for members of other dolphin species (Sayigh *et al*., 2007). Some researchers have questioned the findings that dolphins produce individually specific whistles (McCowan & Reiss, 2001), but even these researchers have since acknowledged their existence (e.g. Marino *et al*., 2007; see also Sayigh *et al*., 2007). Signature whistles are thought to broadcast the identity of an individual; when dolphins meet they often exchange signature whistles before joining each other (Quick & Janik, 2012). Dolphins live in a **fission–fusion society**, with individuals merging and separating with larger and smaller groups, and vocal signals of identity may have been selected for in this dynamic social environment.

Evidence that these whistles play a role in individual recognition comes from experiments in which researchers play back prerecorded whistles through underwater speakers. In one such study, dolphins responded more strongly to whistles of related individuals than those of familiar, yet unrelated individuals (Janik *et al*., 2006). Additionally, dolphins can learn vocalizations throughout their lives, and the signature whistle of one individual can become part of the whistle repertoire of another individual, though it is used only rarely (e.g. Tyack & Sayigh, 1997). It is possible that the copied whistles are used to address the 'owner' of the signature whistle, and in fact, individuals do appear to recognize their own whistle. For example, King and Janik (2013) played back recorded signature whistles as well as other whistle sounds in the bottlenose dolphin repertoire and signature whistles from other populations. Individuals only responded to their own signature whistles, and they did so by sounding back with the same signature whistle.

Is a signature whistle, then, equivalent to a name? The fact that signature whistles are individually distinct does make them unlike other animals' use of species-general vocalizations when isolated or even the use of group-specific signals for cohesion (Boughman & Moss, 2003). Signature whistles are not solely genetically specified; they are learned by calves and often modeled after existing whistles in the group (e.g. Fripp *et al*., 2005). Some have hypothesized that the whistles are used with the intention to contact a specific individual, though alternatively, individuals may learn that producing the whistles is associated with a desirable result (e.g. King & Janik, 2013). More information is needed to determine the underlying cognitive mechanisms allowing for the production and learning of signature whistles, and specifically, knowledge of the development and creation of the whistles may shed light on these processes.

(to be discussed in the next section) have found evidence that different signals can be used with specificity for different predator species.

12.4 Alarm and food calls

The relatively limited signal set that characterizes most animal communication includes calls that are elicited by predators and food sources. These calls tend to be species specific, and at least in the case of alarm calls, do not appear to show regional dialects like bird song (see Chapter 13). The

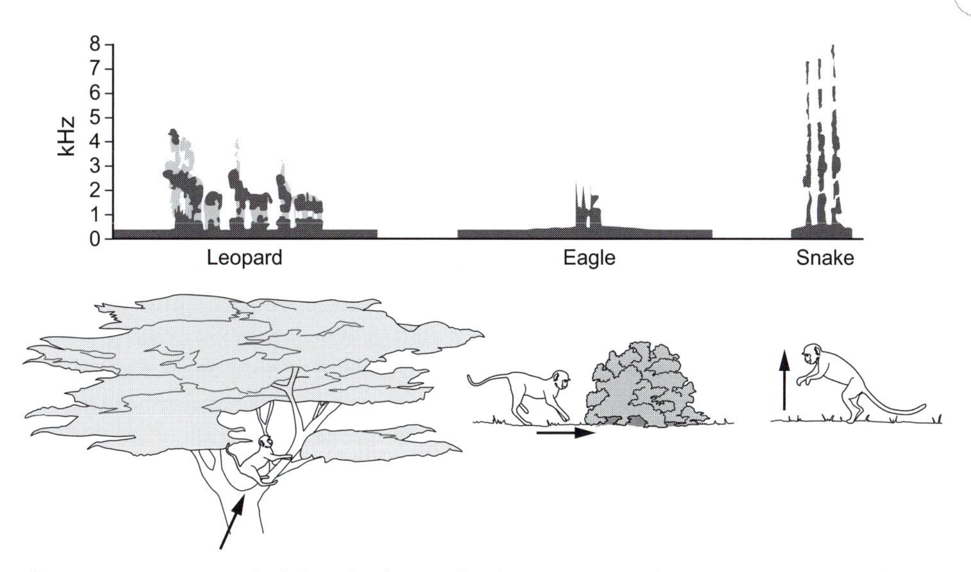

Figure 12.5 Sonograms depicting the alarm calls of one vervet monkey. Behavioral responses of recievers differ according to the call type.

learning and context of these calls has been the focus of research both in the wild and in laboratory settings. Of particular interest has been whether the calls may be considered functionally referential, how the calls are learned, how specific the calls may be to particular referents, and relatedly, how the receiver represents the call (e.g. do calls simply elicit an appropriate behavior, or do they elicit a representation of a particular type of food or predator?). Many species engage in food and alarm calling, but here, focus will be placed on a few particularly well-studied species.

12.4.1 Alarm calling in vervet monkeys

Vervet monkeys (*Chlorocebus pygerythrus*) live in troops in savannah areas of southern and eastern Africa. They give three different sounding alarm calls to three types of predator: leopards, eagles, and snakes (e.g. Seyfarth *et al.*, 1980; Figure 12.5). To do so, adult vervets must make fine distinctions among bird species, as calls are not given to non-predators like storks and vultures. Though young vervets begin to make the calls at a very young age, the ability to discriminate among predators appears to develop over the first 4 years of life (e.g. Seyfarth & Cheney, 1986, 2003). Infants, for example, may give eagle alarm calls to harmless objects in the air. Yet, these young vervets already show some discrimination; the eagle calls are seldom given in response to terrestrial predators. The first years of life, thus, appear to provide learning experiences that allow the alarm call system to be tuned to appropriate threats. It is likely that observing others use the calls provides experience for learning (e.g. via observational conditioning, Chapter 13), though there is no evidence for direct teaching by older members of the group (Cheney & Seyfarth, 1990).

The behavioral responses to each of the three calls differ too (Seyfarth *et al.*, 1980). After a leopard call, for example, receivers climb up trees, but after snake calls, they look around on the ground and, at times, mob the snake. Receivers respond to eagle calls by looking up and often running for cover. To test if these responses were truly elicited by the calls and not personal observation of the predator, Seyfarth, Cheney, and Marler conducted a field experiment in which prerecorded calls were played back from concealed speakers (a **playback experiment**). Because calls were played back when there were no actual predators present, any responses would be based on the call itself. For each type of call, the response was appropriate, suggesting that these calls show functional reference.

However, playback experiments alone cannot determine why the calls are effective. Do the calls simply elicit particular responses, or do they have meaning and elicit the representation of a particular predator? Both are likely supported by associative learning mechanisms, though distinguishing between the possibilities helps to elucidate the content of signals like alarm calls. Seyfarth and Cheney (2003) suggest that the alarm calls activate representations of the predator, a proposal based on a series of studies (e.g. Cheney & Seyfarth, 1988; Zuberbühler *et al.*, 1999). In Zuberbühler *et al.* (1999), for example, one finding was that monkeys did not alarm call in response to the sound of a predator if they previously heard a conspecific alarm calling for that same type of predator. It is possible the alarm call elicited a representation of the predator such that the subsequent sound of the predator required no new calls. In contrast, when the sound of the predator occurred after a conspecific alarm call for a different type of predator, vervets did alarm call after hearing the predator.

12.4.2 Attention to other species' alarm calls

When one species lives in close proximity to another, there is the potential to learn associations between another species' calls and the events (e.g. presence of predator) that elicited the call. Diana monkeys (*Cercopithecus diana*) for example, respond with alarm calls to the alarm screams of chimpanzees with whom they share a common predator (leopards), but they do not respond to the social screams of chimpanzees (Zuberbühler, 2003). Learning to distinguish between the chimpanzees' screams is adaptive for diana monkeys; the monkeys can benefit from the chimpanzee alarm screams even if no conspecifics have seen the predator. Further, remaining quiet when chimpanzees are engaging in social screaming makes the diana monkeys less conspicuous to the chimpanzees, which are also one of their predators.

Attention to other species' alarm calls can also lead to deceptive signaling. In particular, when a species is capable of vocal mimicry, it can learn to produce the alarm calls of another species and subsequently also learn that production of the mimicked call affects the behavior of members of that species. For example, fork-tailed drongos (*Dicrurus adsimilis*), a bird species, mimic the alarm calls of other species (e.g. meerkats, *Suricata suricatta*) when they are handling food, which causes the other animals to flee and leave the food unattended (Figure 12.6). The drongos are then able to 'steal' the food (Flower, 2011). Even actual drongo alarm calls (i.e. not mimicry) can elicit fleeing in sympatric species if members of those species have learned the drongo calls. Drongos appear to take advantage of this by also producing their own calls (i.e. false alarms) before stealing food.

It is likely that deceptive alarm calls are effective because the cost of ignoring real alarms outweighs the cost of responding to false alarms (Munn, 1986). It is also the case, though, that deceptive signals can become ineffective via the process of extinction if they are used too frequently in relation to real calls. Pied babblers (*Turdoides bicolor*), for example, did not abandon food after hearing playbacks of drongo alarms (a call also frequently used in false alarms). Yet, the pied babblers did abandon food after playback of the drongos mimicking another species' alarm (Flower, 2011). It is possible that vocal mimicry is adaptive in this case

Figure 12.6 A fork-tailed drongo. Drongos mimic the alarm calls of other species as well as producing their own species-specific false alarms, which allows them to 'steal' the food of fleeing members of other species.

because it allows species to have variety in false alarm calls, which in turn may weaken the process of extinction.

12.4.3 Food calls

Calls made in association with food have been reported for many primate and bird species (see Bugnyar *et al.*, 2001, and Clay *et al.*, 2012, for reviews). The costs of potentially having to share or give up a food source to conspecifics who hear the food call are thought to be outweighed by various benefits, including: the social recruitment of group members so as to lower predation risk or take over defended food (e.g. Elgar, 1986), attraction of potential mates (Marler *et al.*, 1986), the enhancement of inclusive fitness by attracting kin (Hauser & Marler, 1993), or the display of high status (e.g. Heinrich & Marzluff, 1991). As with alarm calls, researchers have been interested in determining whether food calls can be classified as functionally referential.

Some of the strongest evidence for functional reference in food calls comes from chickens. After discovering food in the presence of a hen, roosters emit a specific vocalization that has an acoustically distinct structure. When these calls are played back to hens, they look to the ground as if searching for food (e.g. Evans & Marler, 1994). Further, if the hen hears the call just after eating a few kernels of corn, she engages in significantly less searching behavior, suggesting that the calls do elicit a representation of food, yet provided no new information since the hens were already aware of the food (Evans & Evans, 1999). A logical concern is that the hens were simply satiated, but too little food was provided for this to explain the results. (Readers of Chapter 4 may find the logic of this experiment similar to that of devaluation experiments that distinguish between S-S and S-R associations.)

Researcher profile: Dr. Dorothy Cheney

Figure 12.7 Dr. Dorothy Cheney.

In 1977, Dorothy Cheney received her PhD from the University of Cambridge, focusing on theory and method in the study of animal behavior. While there, she studied with one of the most seminal figures in ethology, Dr. Robert Hinde. After graduate training, Cheney and her longtime collaborator and partner, Dr. Robert Seyfarth, began research on vervet monkeys as postdoctoral fellows at Rockefeller University, with support from their mentor, Dr. Peter Marler. Marler, a world-renowned expert in avian communication, taught Cheney about vocal communication and field experiments.

Cheney, now a professor of biology at the University of Pennsylvania, has become a seminal figure in her own right; her work with Seyfarth has increased our understanding of primate communication as well as social knowledge (detailed in Chapter 10). Much of the research detailed in this chapter was completed in Amboseli National Park, where Cheney and her colleagues engaged in playback experiments with vervet monkeys that greatly advanced our knowledge of the content of alarm calls. This work not only provided a detailed account of a natural communication system, but it provided researchers with a new set of questions to ask about animal communication in relation to its evolution, development, and cognitive underpinnings. Detailed review of the research was reported in the first of two popular books written by Cheney and Seyfarth, *How Monkeys See the World: Inside the Mind of Another Species* (1990), and numerous journal articles.

In 1992, Cheney and her colleagues began their study of a baboon group in the Okavango Delta of Botswana, and research continued through 2008. There, by studying social interactions, vocalizations, and even feces, they have increased our understanding of the baboon's complex social structure, offering a window into the evolution of human cognition. This work has appeared in scientific journals, as well as the 2007 book, *Baboon Metaphysics: The Evolution of a Social Mind.*

The work is characteristic of the diversity of research paradigms and interdisciplinary frameworks in the field of comparative cognition. According to Cheney and Seyfarth, studies of primates in the wild "allow us to examine how an individual's behavior affects her survival and reproduction. They also allow us to study social cognition in the absence of human training, in the social and ecological contexts in which it evolved." (p. 10, 2007).

Other strong candidates for functional reference are chimpanzee and bonobo (*Pan paniscus*) food barks, which are grunt sounds that can vary in pitch. In playback experiments in a captive setting, chimpanzee and bonobo participants spent more time looking for food at a feeding site after hearing food barks than when no sounds were played (Clay & Zuberbühler, 2011; Slocombe & Zuberbühler, 2005). More specifically, if they heard barks that were produced in relation to a high-quality food, they searched at a location where high-quality food had previously been found. In contrast, they searched at a low-quality food location after hearing barks produced in relation to low-quality food. However, Clay *et al.* (2012) note that since the subjects were originally trained to find high- or low-quality food at these locations, it is possible that the detection of caller arousal led to search behaviors at locations of matching arousal for the receivers, not the representation of a particular type of food. In fact, these authors suggest that many studies of food calls may not provide conclusive evidence regarding functional reference because calls may simply work to recruit others based on general excitement, and not an expectation of food. Systematic studies with other species that are based on the work of Evans and Marler with chickens, however, may ultimately provide more information about the nature of food-associated calls.

12.5 Communication in play behavior

Play is a behavior that many animal species engage in but has proven difficult to define. One oft-cited definition is the following by Bekoff and Byers (1981):

> Play is a motor activity performed postnatally that appears to be purposeless, in which motor patterns from other contexts may often be used in modified forms and altered temporal sequencing. If the activity is directed toward another living being it is called social play. (p. 300–301)

Yet the authors of the definition and other researchers studying play acknowledge its limitations. Two primary concerns are that the definition could arguably include such behaviors as pacing and that the function of play is not considered (e.g. Dugatkin, 2004). What appears to be accepted, though, is that play is an important behavior during development and the behavior includes multiple play-specific communicative signals. In this section, the function and signals of interactive, social play will be discussed.

12.5.1 The function of play

It is commonly thought that play provides training in physical and social skills that are needed for survival and reproduction as an adult. In bighorn sheep (*Ovis canadensis*), for example, adult males compete aggressively for females. As juveniles, male bighorn sheep spend more time in play than females, and this play consists of contact like head-butting and pushing, which is preceded by stereotyped movements (Berger, 1980). In chimpanzees, play has been hypothesized to provide skills that will be relevant for later coalition forming (Mendoza-Granados & Sommer, 1995). In humans, play has been studied extensively, and similar to proposed functions of play for animals, it is thought that play is integral to cognitive and social development. (As the educator Maria Montessori famously said, "Play is the child's work.")

A particularly well-studied form of social play is referred to as 'rough-and-tumble' play in humans, and in animals, it is typically called 'play fighting.' Juvenile rats that have been deprived of interactions involving play fighting with peers have been found to develop various deficits in social competence, even if other types of social interaction were possible during development (Von Frijtag *et al.*, 2002). Further, Panksepp and colleagues have found that play fighting in rats elevates levels of brain-derived neurotrophic factor (a growth factor) in the amygdala and dorsolateral frontal cortex, suggesting that play might 'program' brain regions involved in emotional behaviors (Gordon *et al.*, 2003). In humans, boys engage in more vigorous, rough-and-tumble play than girls, a difference that appears to be universal across human cultures (Pellegrini, 2011). Girls do engage in some rough-and-tumble play, and higher levels are related to levels of prenatal testosterone in both boys and girls (Auyeung *et al.*, 2009). The period of development in which human rough-and-tumble play between children and fathers is most common appears to correspond with the development of frontal lobe functioning, which supports the regulation of behavior and emotion. It remains unclear whether the developing frontal lobes make rough-and-tumble play possible or if play provides experiences relevant to brain maturation, but further comparative neuroscience studies appear warranted.

12.5.2 The signals of play

Play fighting appears to be important to the development of social species, but an important question is how an animal might signal that aggressive-looking behaviors are, in fact, play. Research by Bekoff and colleagues has suggested three manners in which a play session may be distinguished, but the one most relevant to the study of communication is production of **play markers**, specific and highly stereotyped behaviors that can serve to initiate and maintain the play context (Bekoff, 1995).

Among dogs (*Canis familiaris*), juvenile wolves (*Canis lupus*), and juvenile coyotes (*Canis latrans*), a bowing motion is often associated with bouts of play. A bow occurs when an individual bends on its forelimbs but remains standing on its hind legs, and may be accompanied by a tail wag and bark. To examine whether bowing serves as a play marker, Bekoff analyzed the temporal placement of bows within recorded play sessions. The bows did not occur at random, but instead were used immediately before or after an action that is typically used in dominance or predatory interactions (e.g. biting with side-to-side shaking of the head). That is, the bows appeared to be primarily used to mark instances in which another behavior might be misinterpreted.

The cognitive mechanisms underlying play communication are not well studied. According to Bekoff, play markers may serve to communicate a message akin to "what follows is play" or "this is still play," yet while this description may characterize the function of the signal, the underlying representational content (if any) is unknown. Other means of distinguishing play from aggressive contexts appear to be available, as many species that engage in play fighting do so without any discernable play markers *per se* (see Pellis & Pellis, 1996, for review). Facial expressions such as open mouths with teeth covered by the top lip are present in play in many primates, for example, and may also serve to communicate a play context. Play signals in general, thus, may provide 'meta-communication' by altering the significance of other signals that would otherwise be associated with fighting and dominance (e.g. Altmann, 1962; Bateson, 1955). Rooney (1999), for example, observed that the outcome of some games like tug-of-war does not impact real-world dominance relationships (here, between a human and dog), suggesting that play signals can serve to keep the results of play fighting context specific.

Box 12.3 Lip-smacking, vocalizing, and the evolution of speech

Human speech represents a unique motoric and vocal feat. Box 12.1 considered the role of the *FOXP2* gene in the production and evolution of language, and there has also been a growing interest in the possible evolutionary links between human speech and non-human primate facial communication. Ghazanfar *et al.* (2012), for example, have proposed that primate lip-smacking (i.e. rhythmic movement of the lips as they open and close, in the absence of vocalizations) may be a precursor to speech. Support for this proposal comes from the fact that lip-smacking has a 3–8 Hz rhythm which is similar to the periodicity of integral aspects of speech in many human languages. Primate vocalizations, in contrast, do not share this feature with speech (the lips and tongue remain relatively motionless), nor do other primate facial movements like chewing. Until recently, there was no known example of a rhythmic primate vocalization that was produced while lip-smacking.

Geladas (*Theropithecus gelada*) are Old World monkeys closely related to baboons (Genus *Papio*). Males produce a vocalized lip-smack (a 'wobble') during affiliative interactions with females, though females occasionally produce the sound as well (Gustison *et al.*, 2012). Bergman (2013) examined recordings of wobbles from six different males in the wild and found that the rhythm of the wobbles closely matched that of human speech. That is, the vocalized lip-smack produced sounds that were structurally similar to speech. (The original paper links to a movie file in which wobbles can be heard.)

The wobble of the gelada suggests that a combination of lip-smacking and vocalizations can produce a sound with strong similarities to human speech. In turn, the original lip-smacking hypothesis of Ghazanfar and colleagues has gained some additional support. As Bergman (2013) notes, however, it is still unknown whether human speech actually did evolve in this manner, and further questions remain, such as how these sounds came to represent complex meaning and ultimately form language.

12.6 Teaching language to animals

In the second half of the 1900s there was a series of attempts to teach human language to great apes. In hindsight, the studies seem inevitable: they occurred at a time when research in animal learning was synthesizing with interests in cognition and ethology, when field studies by Goodall and others were revealing the complex social and tool behavior of chimpanzees, and when linguists like Chomsky (1968) were proposing that humans had a species-unique, innate language module that allowed for Universal Grammar. A logical question to ask, therefore, was if any animals could learn language. At the risk of giving away the conclusion of this section, the answer to this question is now accepted to be "no."

Yet, some of the main studies will be detailed here because they are revealing in terms of what animals did not learn (Fitch, 2005). The studies also played an important role in stimulating research and new methodology in the study of language development in children, which in turn further pointed to the differences between the 'language trained' apes and children. Even if chimpanzees, for example, are exposed to human, species-typical language environments, they do not develop in the same way as a child. Some of this was already known earlier in the 1900s, before the main thrust of the ape language studies. In the 1930s and 1940s, two husband and wife teams (the Hayeses and the Kelloggs) each raised a young chimpanzee in their home (Figure 12.8). Both of the chimpanzees, Viki and Gua, came to use some human artifacts (e.g. spoons), but the endeavor to get them to

Figure 12.8 Raising infant chimpanzees in the home. Early studies of ape language ability attempted to teach language within the setting of a human home. As chimpanzees develop into larger juveniles and adults, they are unable to live safely in a setting such as this.

produce speech sounds failed (also see Box 12.1 and see Box 12.3 for further consideration of speech). We now know that chimpanzees lack the anatomical substrates for speech, and this encouraged the first phase of ape language research to attempt to try again by teaching language through a non-vocal medium.

12.6.1 Sign language

In the late 1960s, another team of researchers, the Gardners, attempted to teach American Sign Language (ASL) to a young female chimpanzee named Washoe. The researchers working with Washoe communicated in ASL, and Washoe also learned by explicit operant conditioning. By 5 years of age, she was reported to know over 100 signs, but also of interest was the possibility that she would combine signs in meaningful ways. According to one anecdote, Washoe was seen to sign "water bird" upon seeing a swan. Other words were combined such as "you tickle me" (Gardner & Gardner, 1969; Gardner *et al.*, 1989). The reports were met with great excitement until the publishing of a deflationary paper 10 years later (Terrace *et al.*, 1979).

In that paper, work with another chimpanzee, Nim Chimpsky (a play on the name of linguist Noam Chomsky), was reported. Nim was also trained with ASL signs and was observed to produce them in combinations, though many of the strings consisted of repeated words and the mean 'utterance' length remained small relative to that of developing children. The critical finding reported in Terrace *et al.* (1979), however, was that many of Nim's signs (and those of Washoe's after reanalysis) were repetitions of signs that the trainer had just made. There was little evidence of the style of conversation that even young children can have; there was no real turn-taking, and most sign usage appeared similar to learned operant responses. A more recent review of the records from the Nim and Washoe studies came to a similar conclusion (Rivas, 2005).

12.6.2 Shapes as words

Concurrent with the sign language studies, two other language studies used a system of visual symbols to represent words. The advantages of this method are that the trainer cannot inadvertently cue the chimpanzee or engage in non-standard signs, and further, the behavior of the chimpanzees can be more objectively coded. Disadvantages include the fact that the learning environment is stripped of the social pragmatics of human conversation and that the procedures become more frequently based on operant conditioning (Shettleworth, 2010). In work by Premack and colleagues, the chimpanzee Sarah was trained to associate plastic shapes with objects (Figure 12.9). With these associations in place, Sarah was able to respond to questions that examined her ability to understand concepts like 'same/different' or 'color of.' For example, if shown symbols for 'color of' and a red apple, Sarah would pick a blue triangle shape, which had previously been associated with the color red. The focus of this work became to examine abilities such as analogical reasoning, which Premack proposed was an ability promoted by the initial training with the symbol system (Gillan *et al.*, 1981; Premack, 1971; Premack & Premack, 1983).

Figure 12.9 Depiction of a chimpanzee in a language study by Premack and colleagues. A chimpanzee (e.g. Sarah) would respond to questions displayed through plastic shapes.

Figure 12.10 Depiction of Kanzi pressing a button on his communication system in work by Savage-Rumbaugh and colleagues. During interactions such as these, his human caretakers would frequently use English words to respond and to describe the environment.

Another chimpanzee, Lana, was trained with a communication system consisting of a set of keys labeled with various shapes that was connected to a computer that would register all key presses (Rumbaugh, 1977). Not unlike the sign language trained chimpanzees, Lana used the system to press a series of keys that led to a desired food. For example, she might press a series that translated to "please machine give drink." Years later, even some of the primary researchers involved suggested that this paradigm held little similarity to actual language learning and could best be explained by operant conditioning, though they noted that chimpanzees may still be capable of language learning (Rumbaugh *et al.*, 1994).

12.6.3 Understanding English sentences

In part, the acknowledgement that earlier attempts to teach language had methodological flaws was influenced by the apparent language learning of a male bonobo, Kanzi, that occurred under different conditions. Kanzi began to learn the computer system by observing the training of his foster mother (Savage-Rumbaugh *et al.*, 1986). After showing evidence of learning associations between the computer keys and their real-world referents through observation, Kanzi spend years interacting with Savage-Rumbaugh and her colleagues via a portable computer system while the human caretakers used English words to describe objects and events (Figure 12.10). In one experiment, Kanzi's understanding of English was examined by presenting him with verbal instructions from an experimenter who was unable to see him (and thus was unable to cue his actions). The instructions consisted of novel series of actions with a wide variety of objects, such as "Put the ball in the refrigerator," and Kanzi performed correctly on most trials (Savage-Rumbaugh *et al.*, 1993). Compared to any other project that attempted to teach language to an animal, it is much harder to describe Kanzi's behavior as stemming solely from operant conditioning. It would also seem unlikely that bonobos have an evolved language ability that is not present in chimpanzees, though at this point, no studies have directly tested this (see also Rumbaugh & Savage-Rumbaugh, 1994). Psychologists and linguists are still learning about human language development, but it is important to note that even with the extensive training and experience Kanzi received, he has not achieved the language skills that most children have by school age (see also Box 12.4 for research with dogs). In comparison, his productions are limited and consist primarily of requests rather than comments.

Box 12.4 Do dogs use fast mapping when learning words?

As discussed in Chapter 10, the domestication process of dogs may have selected for reduced aggression and fear toward humans that led to increased attention to behavior, subsequently allowing dogs to respond to human social cues such as pointing. Dogs also seem to learn to connect human sounds (words) with objects in the environment, and judging by the performance of a border collie named Rico, they seem to be able to make these connections very quickly.

Human children can learn words through a process of **fast mapping**. With minimal experience (as simple as being shown a novel object and hearing, "This is a blicket"), toddlers can learn a new word. Fast mapping can also take place when two objects are shown, one that is known and one that is novel, and an experimenter asks, "Which one is the blicket?". With this minimal experience, children learn not only the association between a new sound and a new object, but they are able to understand and use it in new sentences; that is, they learn a *word*. Fast mapping is typically considered to be an integral aspect of language development (e.g. Bloom, 2004; Markson & Bloom, 1997). Thus, the attention of many was piqued when researchers suggested that Rico learned words via fast mapping (Kaminski *et al.*, 2004).

Rico was able to choose a novel item among familiar items when told to "Fetch [novel word]." That is, when presented with a novel word, he ignored known objects and returned with the novel one. Weeks later, there was evidence that Rico remembered these new word–object pairings (Kaminski *et al.*, 2004). Although Rico's learning is undeniably impressive, concerns were presented regarding the claim that Rico had learned *words* by fast mapping (e.g. Bloom, 2004; Markman & Abelev, 2004). For instance, it is unclear from this study whether Rico learned that the novel word referred to the object itself or whether the sound was encoded as part of a command to fetch. The former would allow for understanding of the word in other contexts that do not involve fetching. As Bloom (2004) notes, children "appreciate that a word [e.g. 'sock'] can refer to a category, and thereby be used to request a sock, or point out a sock, or comment on the absence of one" (p. 1605).

In a study inspired by research with Rico and the subsequent commentary, another border collie, Chaser, was trained by her owner to respond to over 1000 words, and then specific aspects of Bloom's concerns were examined (Pilley & Reid, 2011). For example, because Chaser was able to produce appropriate behavior when three different commands were randomly paired with three different objects, Pilley and Reid suggested that Chaser understood the names independent of the behavior directed toward the object (e.g. understood "sock" rather than as a command to "fetch sock"). Still, important differences exist between the way dogs and children learn words. Children learn words from overheard speech and can use words outside of contexts in which actions are performed on objects (and learn words for concepts such as love or happiness). Although the work with Rico and Chaser is challenging the way researchers think of word learning, there is still good reason to think that children understand reference in a different way, and it is the task of the fields of comparative cognition and developmental psychology to formulate exactly what these differences are.

12.6.4 Conclusions

Even if the most parsimonious interpretation of the ape language studies is that the individuals tested were basing responses on learned associations, as previous chapters of this book have noted, this learning mechanism can lead to representations of the world that affect behavior. The underlying cognitive processes may still be quite complex, even if they do not completely overlap with the processes that allow human language development. Arguably, though, the study of natural communication systems has increased our understanding of cognition and evolution as much if not more than studies aimed at directly teaching language to animals. Again, further research is required to fully understand the nature of the representations that are generated from, for example, alarm calls or food calls, or even how one signal might influence the interpretation of subsequent ones as may occur with play signals. Existing theory has been developed from research that effectively combined observations of behavior in the wild with field and laboratory experiments, and further work using this comparative approach will bring us closer to understanding the cognitive processes of communication, and at an ultimate level, the evolution of communication and human language.

Chapter Summary

- The topic of communication encompasses a broad range of behaviors, transmission across varying media, and perception via different sensory modalities. The transmission of information between individuals occurs through communicative signals produced by a sender to a receiver.
- Signals can be active, such as alarm calls, or passive, like the coloration of poisonous tree frogs. Human communication is characterized by an unlimited signal set, but it is thought that most animals have a relatively limited signal set.
- If the signals provide information specific to a particular object or event and are not simply signals that convey, say, generalized arousal, and if the receivers of the signal respond to the signal even in the absence of the stimulus that elicited it, the signal is considered to be functionally referential. These signals provide receivers with sufficient information to predict events in the environment when the receivers cannot directly perceive the event themselves.
- Signals are not necessarily truthful indicators of the quality of the sender, but can benefit the sender by manipulating the receiver. In some cases, though, the nature of a signal would suggest that it is too costly to fake, or simply could not be faked, and thus honest signals can evolve.
- von Frisch carefully detailed the behavior of bees when they returned to the hive after finding nectar at far away locations. On the vertical side of the honeycomb, returning bees performed a waggle dance. Different elements of the dance appear to contain information relevant to the distance and direction of the food source. There is evidence of functional reference; however, the reference appears to be a location as dances can recruit observers to visit food sites or even potential nest sites.
- The relatively limited signal set that characterizes most animal communication includes calls that are elicited by predators and food sources. Vervet monkeys, for example, give three different sounding alarm calls to three types of predator: leopards, eagles, and snakes. Playback experiments suggest that these calls are functionally referential.
- Calls made in association with food have been reported for many primate and bird species. Some of the strongest evidence for functional reference in food calls comes from chickens. After discovering food in the presence of a hen, roosters emit a specific vocalization that has an acoustically distinct structure. When these calls are played back to hens, they look to the ground as if searching for food.
- It is commonly thought that play provides training in physical and social skills that are needed for survival and reproduction as an adult. Play markers, specific and highly stereotyped behaviors like the bowing motion of many canids, can serve to initiate and maintain the play context.
- In the second half of the 1900s there was a series of attempts to teach human language to great apes. The studies also played an important role in stimulating research and new methodology in the study of language development in children, which in turn pointed to the differences between the 'language trained' apes and children.
- Existing theory has been developed from research that effectively combined observations of behavior in the wild with field and laboratory experiments, and further work using this comparative approach will bring us closer to understanding the cognitive processes of communication, and at an ultimate level, the evolution of communication and human language.

Questions

1. Why is animal communication thought to be characterized by a limited signal set? What are the typical contexts in which communication is utilized by animal species?
2. Human signalers often consider what receivers know or do not know when communicating. Given what is currently known about animals' understanding of others' mental states (Chapter 10), how might animal communication differ? How might it be the same?
3. What role might a predatory species play in the evolution of accurate, Batesian mimicry?
4. At first glance, it might seem contradictory to say that both honest and dishonest signals can evolve. What are the differences between both types of signals and in what ways might they be advantageous to the signalers?
5. Do animal communication systems show evidence of situational freedom? In what ways might the situational freedom inherent to human language differ?
6. Meerkats have acoustically different alarm calls for snakes, aerial predators, and jackals. Design a field experiment to examine whether these calls show functional reference.
7. Taking into account our current understanding of animal social cognition (Chapter 10), is it accurate to say that drongos truly intend to deceive (by considering others' mental states) when mimicking alarm calls of other species? If not, how might the mimicking behavior develop?
8. Why might some animal species emit calls when encountering food?
9. What is the basis for the current consensus that past attempts to teach language to apes have not shown full language learning that is analogous to that of humans?
10. In your opinion, is human language qualitatively or quantitatively different from animal communication systems?

FURTHER READING

Key readings on the evolution of communication from psychology, neuroscience, and ethology include:

Fitch, W. T. (2005). The evolution of language: A comparative review. *Biol Philos*, **20**, 193–230.
Ghazanfar, A. A., & Cohen, Y. E. (2008). Primate communication: Evolution and neurobiology. In L. Squire *et al.* (eds.). *The Encyclopedia of Neuroscience*. Oxford, UK: Elsevier Press.
Maynard Smith, J., & Harper, D. (2003). *Animal Signals*. Oxford, UK: Oxford University Press.

Recent research on the genetic and neuronal mechanisms of human language have been summarized in papers such as the following:

Fisher, S. E., & Marcus, G. F. (2006). The eloquent ape: genes, brains, and the evolution of language. *Nat Rev Genet*, **7**, 9–20.
Vargha-Khadem, F., Gadian, D. G., Copp, A., & Mishkin, M. (2005). FOXP2 and the neuroanatomy of speech and language. *Nat Rev Neurosci*, **6**, 131–138.

Thorough reviews of natural communication systems, which could only be selectively surveyed in this chapter, can be found in:

Bradbury, J. W., & Vehrenkamp, S. L. (1998). *Principles of Animal Communication*. Sunderland, MA: Sinauer Associates.
Cheney, D., & Seyfarth, R. M. (1990). *How Monkeys See the World: Inside the Mind of Another Species*. Chicago, IL: University of Chicago Press.
Cheney, D., & Seyfarth, R. M. (2007). *Baboon Metaphysics: The Evolution of a Social Mind*. Chicago, IL: University of Chicago Press.
Marler, P., & Slabbekoorn, H. (eds.) (2004). *Nature's Music*. San Diego, CA: Elsevier Academic Press.

Reviews of projects that attempted to teach language to animals tend to take strong stances supporting one side or the other. Readers are encouraged to read from both sides of the argument:

Pepperberg, I. M. (1999). *The Alex Studies: Cognitive and Communicative Abilities of Grey Parrots*. Cambridge, MA: Harvard University Press.

Savage-Rumbaugh, E. S., & Lewin, R. (1984). *Kanzi, the Ape at the Brink of the Human Mind*. New York, NY: John Wiley & Sons, Inc.

Wallman, J. (1992). *Aping Language*. Cambridge: Cambridge University Press.

13 Learning from others

Background

In the 1950s, the behavior of a hungry Japanese macaque (*Macaca fuscata*) on Koshima Island set off a flurry of discussion regarding whether and how animals learn from each other and whether animals can be said to have culture. The hungry macaque was a young female named Imo who had started to wash the sweet potatoes that researchers were placing on her beach. The food on the beach would become covered with sand, and wading out into the water and washing off the potatoes likely allowed for a more easily consumed treat (and potentially a desirable salty flavoring). Researchers soon observed that other macaques in the troop were washing their potatoes too (e.g. Kawai, 1965; Kawamura, 1959), and Imo's actions were interpreted as an innovation that was imitated by others in her group.

In the subsequent years, various other interpretations were made. Some researchers suggested the spread of the behavior could be the result of individual (asocial) learning instead of individuals imitating each other. For example, Visalberghi and Fragaszy (1990) provided captive tufted capuchins (*Sapajus paella*) with sandy fruit and a tub of water; many of the monkeys started to wash their fruit within hours, but there was no evidence of imitation. Others pointed out that imitation is only one form of social learning, and perhaps other mechanisms could account for the washing behavior (e.g. encountering wet potatoes left by others could create conditions that enhanced individual learning). Researchers may never be certain what caused the sweet potato washing behavior among the Japanese macaques, but there has been much learned about social learning in the subsequent years.

Chapter plan

This chapter will examine how animals and humans learn from conspecifics, what is commonly referred to as 'social learning.' Attention will first be paid to theories regarding the evolution of social learning and the mechanisms by which one can learn from others. Then, social learning in four contexts will be examined: food and mate preferences, communication systems, and object use. Finally, the larger question of whether animals can be said to have culture will be explored, continuing a theme that will carry throughout the chapter regarding whether human social learning has species-unique qualities.

13.1 Evolution and selectivity of social learning

Learning can occur through an individual's own, asocial experiences with the world (e.g. though operant conditioning) or through observation of others' actions. Both asocial and social learning have been described as forms of **phenotypic plasticity**; they allow animals to change behavior to fit changing environments (Boyd & Richerson, 1988). There are important differences between these two learning processes, though. Individual learning is not passed on from generation to generation; the learning mechanisms may be inherited genetically, but the learning occurs through an individual's own interaction with the environment. In contrast, behaviors learned through social learning can be spread from one generation to another. In fact, it is typically assumed that social learning can be more efficient than individual learning because learning from others serves as a shortcut relative to iterations of trial-and-error learning.

Still, individual learning may be necessary to provide an innovative behavior that others can observe and learn about. Also, in some situations, individual learning may be a preferred strategy over social learning. For example, nine-spined sticklebacks (*Pungitius pungitius*) tend to ignore social information about the value of food patches when their own personal experience has provided up-to-date information. However, these same fish will switch to learning about food patches from others' behavior if their own individual information is unreliable (Van Bergen *et al.*, 2004). That is, when one's own information is trustworthy, it is not beneficial to learn (potentially inaccurate) information from others, yet in some instances, information provided by others may be better than one's own outdated information. Thus, the reliance on social versus individual learning depends on the relative costs and benefits of each.

Many researchers have suggested that indiscriminately engaging in social learning is not adaptive and instead predict that animals will be *selective* in terms of when to rely on social learning and whom to learn from (e.g. Boyd & Richerson, 1985; Galef, 1995; Giraldeau *et al.*, 2002; Laland, 2004). By these accounts, natural selection in animals capable of social learning should favor strategies that allow for discriminate learning. Importantly, these accounts are not suggesting that animals are aware of the strategies or understand why the strategy works (Laland, 2004). Two proposed strategies will be detailed below: being discriminative about when to learn from others and being discriminative about who to learn from.

13.1.1 When to engage in social learning

Laland proposed three "when" strategies that describe the circumstances in which individuals may start to learn from others' behavior (e.g. Laland, 2004). The first of these strategies is relatively simple: if one's established behavior becomes unproductive, switch to copy the behavior of others. This strategy characterizes the behavior of pigeons (*Columba livia*) in a study by Lefebvre and Palameta (1988) in which birds pecked at a carton to obtain seeds. Pigeons took food from other birds that were opening the cartons unless this scrounging behavior led to too little food, which occurred when there were too few birds opening boxes. Unproductive scroungers then began to open the cartons in a similar manner to other birds. Importantly, the researchers had previously demonstrated that the carton-opening behavior was acquired by social learning; scroungers apparently learned from others to open the boxes when scrounging proved to be unproductive.

A second strategy is to engage in social learning when the costs of asocial learning are high, an idea described in great detail by Boyd and Richerson (1988). Support for this strategy comes from the behavior of the nine-spined stickleback fish mentioned above. These fish will engage in social learning about food patches, but the closely related three-spined stickleback (*Gasterosteus aculeatus*) does not tend to use information gained from others' choice of food patch (Coolen *et al.*, 2003). Coolen and colleagues propose that individually learning about food patches is more costly for the nine-spined sticklebacks than the three-spined species because the former has weaker defenses; the three-spined sticklebacks have larger spines than the nine-spined fish, as well as armored body plates. For the weaker nine-spined species, asocial personal sampling of food patches may be relatively costly and outweighed by the benefits of feeding where others were observed to feed successfully.

Lastly, social learning should occur more when the potential demonstrators share the same environment as the potential learner (Boyd & Richerson, 1988; Galef & Laland, 2005). The logic is thus: it is better to learn from another individual who has learned behavior that is effective in the same environment as one is currently in. If the environment rapidly changes, asocial learning may be favored. Some evidence of this strategy comes from rats (*Rattus norvegicus*) in a study by Galef and Whiskin (2004). Rats who were maintained in a frequently changing environment copied the food preferences of other rats less often than those living in a more stable environment.

13.1.2 Whom to learn from

In 1995, Coussi-Korbel and Fragaszy proposed that in some cases of social learning, the identity of the demonstrator increases his or her salience for an observing individual. The authors called this **directed social learning**, separating it from instances of non-specific social learning in which any one individual's behavior could influence many other individuals. The difference between these two instances of social learning, they argued, is not in terms of the type of information being learned or even the underlying cognitive processes. Instead, characteristics of the demonstrator in relation to the observer make the former's actions more likely to be salient to the latter (Coussi-Korbel & Fragaszy, 1995).

The effectiveness of learning from others will depend to some extent on whether the individual who is being observed is actually successful and productive. Thus, an obvious strategy to use when deciding which individuals to learn from would be to focus on individuals who have cues to being successful (Boyd & Richerson, 1985). One cue, for example, might be that an individual has obtained food. Wilkinson (1992) found that in a colony of bats (*Nycticeius humeralis*), previously unsuccessful females would follow previously successful females as they left the colony to forage. Other cues to being successful may include one's status in the social hierarchy in general, with high-status individuals being copied more than low-status individuals (Galef & Laland, 2005; Horner *et al.*, 2010; Nicol & Pope, 1999). Relatedly, Henrich and Gil-White (2001) have proposed that there may be a preference to learn from whomever others are learning from, a prestige-bias, that in turn makes skilled individuals highly valued (and perhaps highly ranked) in a group. Three- and four-year-old human children, for example, more frequently copied the tool-use actions of an adult demonstrator who had previously been attended to by others than the actions of an adult who was ignored. Interestingly, this directed social learning was specific to the actions observed; the food preferences of prestigious tool-using models were not copied (Chudek *et al.*, 2012).

Another proposed strategy when engaging in directed social learning is to do what the majority of individuals are doing, sometimes described as **conformity** (Boyd & Richerson, 1985). In guppies (*Poecilia reticulata*), rats, and pigeons, for example, there is evidence that the probability of copying a behavior increases in relation to the proportion of other individuals doing the behavior. This is only weak evidence of conformity, however, as other strategies may better

Figure 13.1 Procedure for study of conformity by Haun *et al.* (2012). Chimpanzees observed three different chimpanzees drop a ball into the same box to receive a reward and also watched one chimpanzee place a ball into a different box for reward, three times.

describe these instances of social learning (e.g. Laland, 2004). Research by Haun and colleagues, however, suggests that chimpanzees (*Pan troglodytes*) and human children are more likely to copy an action performed by three individuals than the action of one individual who repeats himself three times. Participants watched three different conspecifics put a ball into the same box to receive a reward and also watched one conspecific put a ball into a different box for reward, three times (Figure 13.1). The chimpanzees and human children copied the box choice of both the single individual and that of the group of three (the majority), but the box used by the majority was copied more often (Haun *et al.*, 2012). This work is preliminary, however, as the authors warn that an alternative strategy may be at work: an observer may simply be choosing one individual at random to copy, and it is statistically more likely that this individual would be in the group than be the single model.

The evolution and selectivity of social learning is currently under study, and the major questions are still being answered. Indeed, most research in social learning has focused on what types of social learning animals engage in, not the contexts under which social learning occurs. What is clear for now is that social learning is not necessarily better than asocial learning; reliance on one or the other depends on their relative costs and benefits. It is likely that animals and humans are discriminative in terms of when to engage in social learning as well as selective in relation to whom to copy.

13.2 Types of social learning

Until now, this chapter has been using the word 'copying' to broadly refer to an animal engaging in behavior that matches the behavior of other individuals. In reality, the corpus of social learning research, particularly research that has focused on the cognitive mechanisms underlying social learning, uses a vast number of terms. Some have even characterized the number of terms, and their often overlapping meanings, as 'bewildering' (Heyes, 1993; Shettleworth, 2010). The cause of this abundance of terminology is actually good scientific practice; experimentalists and theoreticians aim to carefully characterize and distinguish among the processes that drive social learning (e.g. Want & Harris, 2002; Whiten & Ham, 1992). In this section, the goal is to present the most contemporary distinctions. Figure 13.2 serves to summarize the distinctions between these terms using the example of chimpanzee termite fishing discussed in Chapter 8.

13.2.1 Enhancement

At the beginning of Chapter 1, the bottle-opening behavior of blue tits (*Cyanistes caeruleus*) was described; these birds quickly came to open the milk bottles delivered to houses in Great Britain in order to drink the cream off of the top. Because the behavior occurred in only a few isolated areas, Hinde and Fisher (1951) proposed that the behavior was being spread socially, but they also pointed out that it may have been the products of others' behavior that were actually the salient aspects to the observing birds. That is, pecking at a food source comes naturally to many bird species, but it was the direction of this behavior onto the bottles that was learned. When this was tested directly in the lab with another bird species, black-capped chickadees (*Poecile atricapillus*), the product of others' actions (i.e. opened tubs) was enough to lead to increased opening of sealed containers by naïve birds (Sherry & Galef, 1984).

Learning can thus be facilitated by having one's attention drawn to a locale or stimulus, a process called **local enhancement** or **stimulus enhancement** (Roberts, 1941; Spence, 1937; Thorpe, 1956). Locales (places) and stimuli (e.g. a particular object that is being acted on) can be indistinguishable, but are not always the same, so the two processes are often considered separately. An example of stimulus enhancement comes from a study by Zentall and Levine (1972) in which rats (*Rattus norvegicus*) watched another rat press a lever for water. The observing rats learned to press their own levers for water faster than rats that had observed another rat receiving water without pressing a lever. It is likely that the demonstrator's activity at its own lever gained the attention of the observers, and stimulus generalization may have occurred between the two levers (Zentall, 2011).

A natural question to ask is whether stimulus and local enhancement should be considered a type of social learning *per se*. It typically is, because in these situations it is the other individuals in the group who are creating the conditions for learning. Yet, after an individual is drawn to an enhanced stimulus or location, individual trial-and-error learning is likely taking place. As will be seen below, claims regarding the presence of imitation, another type of social learning, are often pitted against lower level (and thus presumably more parsimonious) explanations based on enhancement.

13.2.2 Observational conditioning

When observing the actions of a conspecific, attention may be drawn to the location or an object (e.g. a lever), but also, an association might be formed between that location or object and reinforcement (e.g. presentation of food). This higher order conditioning (the observer does not personally experience the unconditioned stimulus) is typically called **observational conditioning** (e.g. Whiten & Ham, 1992; Zentall, 2006). It is also possible for the affective state that is aroused in

Type of Social Learning	Description
Enhancement	The attention of an observing chimpanzee is drawn to the termite mound or the fishing stick. Subsequently, the chimpanzee may independently learn that engaging in certain activities with the stick at the mound leads to food.
Observational Conditioning	Hearing the food barks of a chimpanzee at a termite mound causes an observing chimpanzee to engage in food barking. Subsequently, the mound is associated with food.
Emulation	
affordance learning	An observing chimpanzee learns that a stick can push through a termite mound and/or that termites will attach to the stick.
object movement re-enactment	An observing chimpanzee learns how a stick can be moved to obtain food, though may not subsequently move its hand in the same manner as the other chimpanzee.
end-state emulation	An observing chimpanzee learns that termites can be retrieved and subsequently acts to achieve this by any means.
Mimicry	An observing chimpanzee learns the motions but not the purpose of the other chimpanzee's actions and thus may subsequently poke sticks into other piles of earth, etc.
Imitation	An observing chimpanzee recognizes the end-state of the actions (to obtain termites) and uses the same actions as the observed chimpanzee.

Figure 13.2 Distinctions among the types of social learning discussed in this chapter. Here, the example of one chimpanzee observing another chimpanzee engage in termite fishing is used to illustrate the social learning processes (adapted from Want & Harris, 2002).

an observer from seeing the affective state of a demonstrator (via, for example, emotional contagion) to itself be associated with whatever stimulus is present.

Learning to recognize predators, for example, is an example of observational conditioning. In many small birds, the response to a predator is to engage in mobbing behavior; birds in a group approach the predator while calling in a particular manner. Naïve birds, however, must learn about predators from others in their group, and a study by Curio *et al.* (1978) suggests that observational conditioning is a likely mechanism. In that study, a demonstrating European blackbird (*Turdus merula*) and an observer blackbird were in two separate cages such that the observer could not see what the demonstrator could see. The demonstrator, a bird who had already learned to recognize predators, was presented with a stuffed version of a common predator and began to engage in species-typical mobbing behavior. The observer could see and hear this mobbing, and through contagion began to engage in mobbing behavior as well. However, in front of the observer was a non-predator (another type of bird, or even a bottle of the same size). The observer, when presented later with the harmless bird or bottle, engaged in mobbing behavior. It appears that the demonstrator's mobbing behavior elicited mobbing behavior in the observer, and the observer learned an association between this behavior and the non-predator object (also see Box 13.1). Shettleworth

Box 13.1 Social influence

In 1964, Tolman noted that a satiated chicken will start to eat again if it is introduced to a hungry chicken that begins to eat (Tolman, 1964). Humans and some other animals (see Box 11.4) will often yawn soon after observing a yawn. Some bird species engage in anti-predator mobbing behavior upon seeing others doing this behavior. **Contagion** is a term used to describe these reflexive, species-typical responses to the behavior of others (Zentall, 2011). The behavior is considered to be unlearned; it is a behavior that is 'released' by the behavior of others (e.g. Thorpe, 1963).

Contagious behaviors are social in the sense that the behavior of others influences an individual's behavior, but they are not considered a type of social learning. That said, an individual animal may learn to associate its current, socially elicited behavioral and emotional state (e.g. the behaviors and arousal elicited by observing a conspecific engage in anti-predator behavior) with a stimulus in the environment. Thus, contagion may sometimes be a 'first step' on the way to social learning (see Section 13.2.2).

Other social influence factors may affect the behavior of an individual, and one of the most studied examples is that of **social facilitation** (also called 'social enhancement'). Sometimes, the mere presence of a conspecific can increase or decrease an individual's arousal, regardless of the conspecific's behavior. Changes in arousal, in turn, can affect an individual's task performance, exploratory behavior, fear, and other behaviors (Zajonc, 1965). In general, if a task is very easy or well-learned (e.g. a relatively easy maze for which an individual has previous experience), the presence of others who are either watching or also engaged in the task seems to enhance the individual's performance. In contrast, the presence of others seems to decrease performance on difficult tasks.

Zajonc (1965) proposed that social facilitation effects were due to an activation of the endocrine system and elevation of cortisol levels that occurred when others were present, but even today the mechanisms underlying social facilitation are not well known. Some researchers have noted that when examining social learning, social facilitation effects can be controlled for by including a condition in which observers are in the presence of a conspecific, yet this conspecific does not perform the target behavior. By doing so, the experimental and control groups both presumably experience similar levels of social facilitation (Klein & Zentall, 2003).

(2010) points out that this learning can even occur after the observer has been habituated to the safe object, suggesting that latent inhibition may not be strong in this situation.

Another example of the social transmission of predator learning via observational conditioning comes from the manner in which captive rhesus monkeys (*Macaca mulatta*) learn to fear snakes. Monkeys born in the laboratory do not express a fear of snakes, though in the wild, this same species will react with a strong fear response including vocalization and withdrawing. If laboratory monkeys are shown a video of wild monkeys responding to a snake, they too engage in the fear behaviors. This also occurs later when the videotaped monkey is no longer present, yet a snake is near (Mineka & Cook, 1989). Importantly, observational conditioning that allows for the social transmission of predator recognition also appears to be selective. In a follow-up experiment, Cook and Mineka (1990) found that observers did not learn to fear flowers after seeing video events that had been edited to show a conspecific reacting in fear to flowers. Subsequent work with human infants, who also do not naturally show a fear of snakes, has found that infants pay more attention to images of snakes if these images are paired with fearful adult voices, but this does not occur with other types of images (DeLoache & LoBue, 2009).

13.2.3 Emulation

Perhaps one of the most examined types of social learning has been emulation, and likely not by coincidence, the term 'emulation' has historically been used in different ways (Want & Harris, 2002). Here, three variants of emulation are described, using contemporary definitions. Further examples of each will appear in Section 13.5, as emulation tends to apply best to situations in which observers are learning about objects.

Affordance learning

In one of its earliest uses in comparative cognition, emulation referred to the process of learning about the properties of objects or relationships between objects and thus was referred to as **affordance learning**. Through a demonstrator's actions, an observer may learn, for example, that a lock holds a lid down (Byrne, 1998), but is not learning about the specific actions of the demonstrator. Whiten (2006) conceptualizes this form of learning as 'learning that' instead of 'learning how.' Note that this is different from stimulus enhancement, since actual properties of the stimulus are being learned, instead of simply having greater attention paid to the stimulus. It is a little harder to distinguish affordance learning from observational conditioning (Byrne, 1998). However, Zentall (2012) suggests that in observational conditioning the learned response is closely related to the unconditioned response to the reward (e.g. approaching a lever that has been associated with food), but in affordance learning, the response is more arbitrary, reflecting learning about features (e.g. pushing a door to obtain food).

Object movement reenactment

Tomasello (1998) explained that emulation consists of "learning about the environment, not about behavior" (p. 704). **Object movement reenactment** refers to instances in which, through the actions of others, an individual has learned how objects can move. For example, object movement reenactment can refer to a situation in which an individual has learned how a bolt must move before a lid can be opened, but affordance learning is when an individual has learned that bolts can be moved and lids can be opened. Both are processes in which the specific actions of the demonstrator are not necessarily encoded; instead, what is learned is information about the environment (the object).

End-state emulation

In 1992, Whiten and Ham identified a different form of emulation, **end-state emulation**. Again, in this form of emulation, the actions of the demonstrator are not learned, but in end-state emulation, it is the fact that a particular end state can be achieved that is learned. For example, an observer might replicate the end result of a demonstrator's actions, but do so using completely different means. The term is used regardless of whether the observer actually understands the mental states underlying the goal-directed behavior (see Chapter 10), though when this is assumed, the phrase 'goal emulation' is favored (e.g. Whiten, 2006).

13.2.4 Mimicry

All the variants of emulation described above share the common meaning of *attending to the environment or end states instead of the actions themselves*. In contrast, **mimicry** involves learning about a demonstrator's actions without understanding the objects involved or even the end state of the action (Tomasello *et al.*, 1993; Want & Harris, 2002). For example, the copying of human speech by psittacine birds (parrots) is an instance of mimicry; although these birds can map the sounds they hear to their own vocal behavior, they are likely not understanding the purpose or meaning of the sounds. A human child, however, appears to understand something of the communicative intent of the sounds others are making when he or she learns and repeats words. That said, other human behaviors do fit this definition of mimicry. Bjorklund and Blasi (2012) provide an example of a 2-year-old child who, after watching his father step on a scale and look down at the numbers, would do the same behavior, even without understanding the purpose of scales or what the numbers meant. In adult humans, many instances of unconscious mimicry have also been studied (Box 13.2).

Box 13.2 Unconscious mimicry in humans

When one of the authors (V. Kuhlmeier) moved to Canada, her American friends and relatives jokingly worried that she would soon be speaking like a Canadian, adding "eh?" to the end of sentences and pronouncing the word "about" as "aboot" instead of using the typical American accent. Though obviously nothing to worry about (particularly since these are just stereotypes and do not reflect every Canadian's speaking patterns!), their prediction *was* based on an accurate assumption about people's propensity to blend in with their social environments. Humans mimic all kinds of behaviors without realizing it and do not notice when others are subtly mimicking them (Chartrand & Bargh, 1999; DePaulo & Friedman, 1998; Hatfield *et al.*, 1994).

The current research on unconscious mimicry (often from social psychology laboratories) tends to focus on the reproduction of postures, gestures, and mannerisms. These studies suggest that when people are interacting, they unknowingly mimic body movements such as crossing of legs and touching the face, a phenomenon termed the **chameleon effect** (Chartrand & Bargh, 1999). Additionally, when compared to participants in a control group, participants who have been modestly mimicked by a confederate during an interaction later report greater willingness to help the confederate (van Baaren *et al.*, 2004) and give increased positive ratings of the interaction (Chartrand & Bargh, 1999) even without recognizing that they were mimicked.

There appear to be certain contexts and motives that influence whether people mimic. For example, adults mimic less frequently in competitive interactions and in interactions with individuals who have just ostracized them. In contrast, they are more likely to mimic when they have the goal of affiliating or a desire to create rapport (Chartrand & Dalton, 2009). Thus, there are situations in which our human 'chameleon' nature is more or less likely to appear, and in general this suggests that the function of this type of mimicry may be to help solidify social bonds.

In a related line of work, Celia Heyes and her colleagues have found that humans will learn the actions of others without any conscious intention to imitate or awareness that they have learned from others, a process called **automatic imitation**. In a series of experiments, observers watched a demonstrator complete a sequence of actions using a computer keyboard. When a subsequent task required the observers to complete a similar sequence of actions, responses were faster than in other tasks in which they had to complete a dissimilar sequence, suggesting they had learned the sequence through observation of the demonstrator. Importantly, this occurred even though the observers were not told to learn the sequence and in most instances, the observers were not consciously aware of what they had learned (Bird & Heyes, 2005; Bird *et al.*, 2005).

13.2.5 Imitation

Often considered the peak of high-fidelity copying, the term **imitation** is reserved for instances in which the observer recognizes the end state (sometimes referred to as 'goal') and reproduces the specific actions that were used to bring about that state (e.g. Tomasello, 1990; Want & Harris, 2002). Others have noted that imitation is the term commonly used for behavior that matches that of the demonstrator, but cannot be described by any of the other types of social learning just described (e.g. Zentall, 2011). One further consideration is the extent to which affordance learning is also an aspect of imitation. Want and Harris (2002) note that a human or animal that imitates an action on an object can faithfully reproduce both the actions and the goal but not actually understand the affordances of the object, a type of imitation they call 'blind imitation' (e.g. many people know how to use a car key and why to use it, but not how it works).

As will be seen in Section 13.5, studies with the aim of examining imitation in a wide range of species have gone to great pains to design experimental methodologies that can rule out other social learning processes. Additionally, studies often incorporate actions that are novel for the species and thus unlikely to emerge spontaneously. Still, it is difficult to determine whether an action is truly novel, so often 'improbable' suffices. Regardless of the study design, the distinction between imitation and one alternative social learning process, goal emulation, can be hard to make; it can be difficult for a researcher to decide whether the actions that an observer produces match those of the demonstrator to a sufficient extent. For this reason, some researchers have suggested that a more productive research question may be how the observer decides what aspects of an action to reproduce, not whether the reproduction is actually imitation (e.g. Csibra, 2007; Williamson & Markman, 2006).

13.3 Learning food and mate preferences

Research on social learning has typically either examined naturally occurring behaviors (in the wild or in semi-naturalistic laboratory settings) or examined the transmission of novel behaviors, often with artificial apparatus, in the laboratory. In this section and in Section 13.4, the former will be discussed, with special focus on the learning of food preferences (learning what to eat), mate preferences (learning who to mate with), and later, communicative behaviors (learning species-typical communication). In much of this work, the focus has been to examine whether and under what circumstances social learning has occurred. The question of what *type* of social learning is implicated has only been a secondary aim.

13.3.1 Learning what to eat

For those in the rat extermination business, the difficulties in using poison to remove a colony are well known. Repeatedly injecting the same poison into permanent baiting stations in infested areas tends to be successful initially, but over time, the rats stop eating the bait and the colony can come back to full size. The initial culling of the colony is due to death resulting from poison ingestion, but some rats that eat poison may survive if they only ate a non-lethal dose. These survivors do not typically eat the poison again (an example of conditioned taste aversion, Chapter 4), and their young avoid the types of food that adults avoid, without ever tasting it for themselves (Galef & Laland, 2005; Steiniger, 1950). Initially, it was thought that the adult survivors were communicating food avoidance to their young via warning pheromones left on the poisoned bait, but through extensive laboratory experiments, Galef and his colleagues have detailed a much different social learning process between members of rat colonies in which young rats learn from older rats what foods to eat, not what foods to avoid.

The learning of food preferences in Norway rats starts early in development, a likely important adaptation for an omnivorous species that will have many foods to learn about. Hepper (1988) found that if pregnant rats were fed garlic, their pups would prefer the taste of garlic to onion 12 days after birth, even if these pups were being reared by foster dams in the interim. A similar process seems to exist in humans. In one study, pregnant women drank carrot juice 4 days a week for 3 weeks during their third trimester. At 5 months of age, their infants showed a greater positive reaction to cereal prepared with carrot juice than infants whose mothers did not drink carrot juice (Mennella *et al.*, 2001). Thus, prenatal experience appears to influence the food choices that are made later in development.

Newborn rats (and humans) continue to learn about food from their mothers through experience with flavors in milk. Flavors consumed by the lactating mother will also be evident in her milk and can lead to preferences for certain foods upon weaning. In one experiment, rat pups showed an increased preference for a flavored food that had been eaten by a female from whom they had previously nursed. This preference did not exist for a group of rat pups that had been exposed to a female who had eaten the flavored food and had acted maternally yet did not give milk (Galef & Sherry, 1973). Additionally, when young rats begin to leave the nest, they tend to forage where adult rats are foraging and thus consume similar foods. It appears, then, that early in development, rats have three ways of experiencing the foods that the adults in their colony are safely eating, and they come to prefer these foods themselves. Some have suggested that in addition to this social learning system, there may be a general preference for familiar over novel flavors which effectively ensures that young rats come to eat the same foods as others (Shettleworth, 2010).

In the aforementioned studies, the young rats had direct experience with the flavors of the food; however, rats will also come to choose certain foods based on exposure to the breath of a conspecific who has recently eaten. In a study by Galef and Wigmore (1983), pairs of rats were housed together and given ordinary laboratory chow. One of the rats (the demonstrator) was then removed and given food flavored with either cinnamon or cocoa. The demonstrator was then placed back with its companion (the observer), and they interacted without food for a short time. After, the observer and

Figure 13.3 Social learning of food preferences by rats. Rats can detect the combination of food and carbon disulfide on the breath of a conspecific when engaging in mouth sniffing behavior.

demonstrator were separated again, and the observer was given two bowls of food, one with cinnamon flavor and one with cocoa flavor. Over the testing session, the observer ate significantly more cinnamon flavored food if their companion had eaten cinnamon, but more cocoa if their companion had eaten cocoa. Further, the short interaction with the demonstrator could affect food choices over a month later, and the socially learned preference could even overcome a species-typical aversion to the taste of pepper (Galef & Whiskin, 2003).

Thus, during an interaction with a conspecific who has just eaten, rats appear to gain information that subsequently affects food choices. It is not as simple as detecting small amounts of food on the partner's whiskers, though. Instead, upon encountering each other, rats smell the combination of food and carbon disulfide (a natural constituent of breath) when they engage in natural mouth sniffing behavior (Figure 13.3). There are two important points to make here. First, it is the combination of food odor and natural elements of breath that is necessary for learning to occur; the food odor presented alone does not influence later food choices. This effectively ensures that the rat is learning about food that has actually been consumed. Second, the mouth sniffing behavior is an example of a natural behavior that ensures subsequent social learning about safe foods.

As mentioned in the introduction to this section, characterization of the *type* of social learning implicated in food preference learning has been only a secondary aim of the research. That said, the rich body of work on the topic allows for some evidence-based proposals. Hoppitt and Laland (2008), for example, propose that while the learning of food preferences may have some similarity to stimulus enhancement, there are some important differences. First, exposure to the stimulus (food odor) alone does not lead to increased eating of that food; there must also be cues to the presence of other rats. This is in contrast to one of the paradigmatic examples of stimulus enhancement described above: chickadees will start to open tubs of cream after being exposed solely to opened tubs. Second, the fact that exposure to other rats' breath can overcome an observer's original aversion to a flavor would suggest that the exposure did more than simply sensitize the observer to the food. Thus, these researchers and others have set aside this behavior into its own category: the social enhancement of food preferences.

Examining species differences in social learning

When examining the social learning of food preferences from a comparative perspective, it may be tempting to predict that a relatively solitary species will differ in its reliance on social learning as compared to a species that lives in a large colony. Yet, in most mammal and bird species there is interaction between adults and offspring, regardless of whether adults spend time with other adults, and this interaction is occurring at a time in the juveniles' lives when social learning may be of utmost importance. Thus, it may be expected that social learning exists even in relatively solitary species as an **ontogenetic adaptation**, a characteristic that serves an adaptive function at a certain time in development (Galef & Laland, 2005).

Supporting evidence for this prediction comes from a study by Lupfer *et al.* (2003). Young golden hamster pups (*Mesocricetus auratus*), like the Norway rats discussed above, showed a preference for the types of foods that were eaten by their mother. However, as adults, members of this relatively solitary species do not learn food preferences from each other. Thus, at least in the social learning of food preferences, predictions as to which species may or may not engage in this form of learning must also consider how the reliance on social learning may change across development.

13.3.2 Learning about potential mates

In many species, females are choosier than males in terms of deciding whom to mate with. Indeed, the outcomes of mating, offspring, can be costlier to females than males, and females may not have as many chances at reproduction. Sexual selection has often resulted in male traits that convey good mate quality, but in situations in which these traits are not immediately clear and taking time to assess is costly, copying the mate choice of other females may be beneficial.

In the 1990s, Dugatkin and his collaborators examined **mate-choice copying** in Trinidadian guppies (*Poecilia reticulata*). In the wild, males and females of this fish species will often gather together in shoals, which allows guppies to observe others' mating behavior. To mimic this environment in the laboratory, Dugatkin created a rectangular aquarium system in which sections could be divided off by clear plexiglass walls. One male was placed into each of two end sections. A female (the observer) was placed into a clear plexiglass canister in the middle section of the aquarium while another female (the demonstrator) was free to swim around this middle section and approach the clear divider near either of the two males. Thus, the observer could see which of the two males the other female spent more time engaging in courtship behaviors with. Later, the demonstrator was removed, and the observer was freed from her canister, allowing her to approach either of the two males. Females tended to approach whichever male the demonstrator female had chosen (e.g. Dugatkin, 1992). Control experiments suggested that the observer female's behavior was not simply due to choosing an area that recently contained a larger number of fish (a possibility since this is a schooling fish species).

The findings with guppies have been difficult to replicate, but studies of other fish have provided further data supporting mate-choice copying (Brown & Laland, 2003). Research with birds has also found similar results. In Japanese quail (*Coturnix japonica*), for example, females will tend to affiliate more with a male who was previously observed to be courting and mating with another female (e.g. Galef & White, 1998). Further experiments suggested that if a female observed a courting male who had a particular physical characteristic, the observing female not only later preferred this male, but also preferred other males with that same characteristic (White & Galef, 2000). Thus, social learning in quail may be able to affect the sexual selection of traits in males.

Mate-choice copying is likely an instance of social learning via stimulus enhancement (Hoppitt & Laland, 2008; Zentall, 2011). Indeed, many other forms of social learning can be ruled out. For example, the preferences learned during mate-choice copying are not location-specific, and thus local enhancement is not likely. Additionally, learning and replicating another female's actions (e.g. mimicry) does not seem to characterize the behavior; in experiments with quail, placing a fake, stuffed female demonstrator near a male yields the same results even though this 'female' does not move (Akins *et al.*, 2002). Instead, the presence of a female near a male seems to enhance that male for observer females, which results in later interaction with him.

In sum, two behaviors that are particularly important for survival and reproduction, choosing edible food and choosing high-quality mates, appear to occur in part through social learning processes. In any particular species, the use of social learning may serve an adaptive function only at a certain time in development (e.g. when golden hamster pups are learning what to eat, or when females are of reproductive age). Alternatively, in some species, social learning of certain behaviors may continue to reap benefits over and above asocial learning (e.g. when adult rats learn what others have recently eaten). In the next section, the social learning of communicative behaviors will be examined, and again, the existent research has primarily focused on whether, and under what circumstances, social learning has occurred.

13.4 Social learning and communication: vocal learning

Chapter 12 presented research and findings on natural communication, but one aspect of communication, the vocal learning that results in behaviors such as birdsong, fits squarely in a chapter on social learning. Song learning by birds is a well-studied example of how the development of a communication system can critically depend on social interaction. The first part of this section will consider song development. Later, the question of how to characterize this form of social learning will be addressed.

13.4.1 Bird song acquisition

Though many bird species produce sounds, not all sounds are songs. **Song** refers to species-typical, musical vocalizations that are primarily produced during the mating season by males of the order *Passeriformes*. Songs can play an important role in territory defense and attracting females, and for many species of songbird, we now know that the development of species-typical song depends on social interaction with conspecifics. In some of the most studied species, this interaction begins soon after hatching in the spring or summer, a time during which mature males are producing song.

A good proportion of what is currently known about bird song learning stems from the work of Thorpe, Marler, and their colleagues, and much attention has been paid to one particular species, the white-crowned sparrow (*Zonotrichia leucophrys*). Male white-crowned sparrows have a single song, and that song is shared with neighbors. However, across geographical areas, the songs will differ, meaning that there are local dialects, which in turn was an early clue that some form of social learning was at work (e.g. Marler & Tamura, 1964). In the laboratory, white-crowned sparrows raised in isolation will come to produce songs; these songs start off disorganized and varied, but then begin to have some elements of the species-typical song. They are, however, abnormal. In contrast, if isolated birds hear a recorded playback of species-typical song during a sensitive period (10–50 days of age), they will develop a song similar to the one they experienced (Marler, 1970). This occurs even if the song to which they are exposed comes from a different dialect area than the one in which their parents lived.

Importantly, young white-crowned sparrows will not learn the songs of other species of songbird. There appears to be something special about how the songs of one's own species are processed; it is even the case that isolated birds who have never heard their own species' song will show increased heart rate and begging behavior when first exposed to a recording of the song (Dooling & Searcy, 1980; Nelson & Marler, 1993). Thus, the songbirds may be predisposed to discriminate their own species' song. This finding, combined with the fact that even isolated birds produce song that, though abnormal, has some similarity to a typical song, suggests that the social learning of birdsong builds from an existing sensory template that may be unlearned (e.g. Konishi, 1965).

With exposure to species-typical song during the sensitive period, young birds develop song production in phases. First, they produce variable sounds, a type of 'subsong' that has been likened

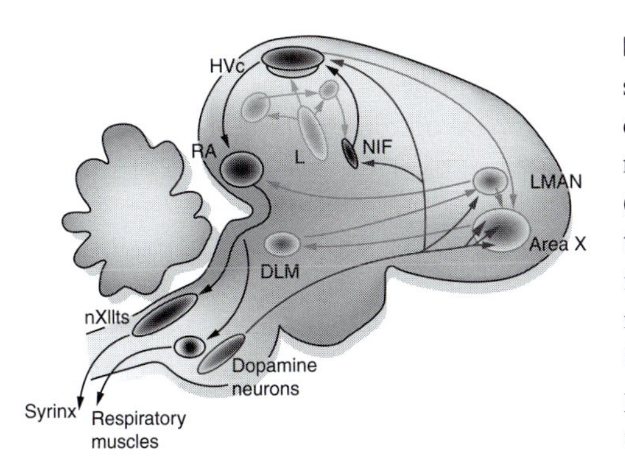

Figure 13.4 The 'song system' of the songbird brain. While the specifics of the system go beyond the focus of this chapter, some details are provided here. The pathway that includes the HVc, the robust nucleus of the archistriatum (RA), and the hypoglossal nucleus (nXIIts, which projects to the syrinx, the vocal organ) is involved in the production of song throughout life. The rostral pathway that includes the HVc, the lateral magnocellular nucleus of the anterior neostriatum (LMAN), Area X, and the dorsolateral thalamus (DLM) is thought to have a role in song learning. The Field L complex is the primary forebrain auditory area, which in turn projects to other auditory areas.

to human babbling. Later, more structured 'plastic songs' are produced which contain elements of the song they are being exposed to. Finally, the song becomes 'crystallized' and will be similar to the song of the tutor (demonstrator) to which they had the most exposure. This social learning process appears to be supplemented by experience in hearing one's own song production; birds deafened after exposure to a tutor's song produce aberrant songs (Konishi, 1965).

The neurocircuitry of birdsong learning is arguably the most well-defined system in the field of social learning. The 'song system' has been proposed to include neural pathways for the motoric behavior of sound production, the perceptual learning of sounds, and the comparison of heard sounds and the internal template (see Brainard & Doupe, 2002, for review). Research in this area is ongoing, and includes single-cell recordings that indicate the equivalent of mirror neurons (see Chapter 10), cells that fire when a bird is currently singing or listening to a bird sing the same song (Prather *et al.*, 2008). Figure 13.4 depicts the neural substrates for song learning.

13.4.2 Changes in bird song over time

Over the last 30 years, savannah sparrows (*Passerculus sandwichiensis*) have been studied on Kent Island in the Bay of Fundy (New Brunswick, Canada). Individual birds are known and many generations of song production have been recorded. During this period of time, researchers have found changes in the species-typical song (e.g. Williams *et al.*, 2013). The savannah sparrow song has multiple segments, and all but one of these segments has undergone change. For example, one part of the song, a trill at the end that differs slightly across males, appears to be particularly important for females as they choose mates. What this means is that the mate-choice behavior of females can actually have some effect on the song styles learned by young males and thus can influence changes in the song over time.

Specifically, female savannah sparrows tend to prefer males who produce a shorter trill at the end of the song, and these males show higher reproductive success. Additionally, some researchers have suggested that young males who are learning song will preferentially learn the song patterns of reproductively successful, neighboring males (e.g. West *et al.*, 2003; Williams, 2008). In time, then, the most common song form in a region can begin to change, spearheaded by the mate-choice behavior of the females.

In contrast, one segment of the savannah sparrow song has remained unchanged over the decades. A 'buzz' segment that occurs before the trill still retains the same features as it did when observation began. Williams and her colleagues suggest that this segment may serve to define the species or the local dialect of the island population. In fact, similar unchanged segments occur in white-crowned sparrow song (Nelson *et al.*, 2004). It is possible that, along with a bias to copy song elements of successful males, another learning bias exists for segments that do not differentiate successful and unsuccessful males: copy the most common form that is heard (Williams *et al.*, 2013). Thus, changes in birdsong over time may provide examples of directed social learning and conformity.

13.4.3 Is vocal learning a special type of social learning?

There is no dispute that birdsong learning is an example of vocal learning that is dependent on social experience, but how to characterize the learning process is still a matter of discussion. Some researchers will refer to the behavior as mimicry or imitation; however, many are quick to point out the differences between vocal learning and visually learned behaviors that are traditionally described using these terms.

For example, Zentall (2011) refers to song learning as a "special case of social learning" (p. 243), pointing out two important differences. First, as detailed above, birdsong learning is built upon a foundation of species-typical behavior, and thus, the social learning occurs in a very constrained manner. When compared to the types of behaviors learned from conspecifics in relation to object use (Section 13.5), this constrained learning appears quite different. Second, vocal learning is unique in that a learner can hear its own vocalization and compare it to the memory of the demonstrator's vocalization. That is, the stimulus produced by the demonstrator and the stimulus produced by the observer can be closely matched, which is not always the case in visual imitation and may create an easier perceptual or cognitive process that is specific to vocal learning (Shettleworth, 2010; Fitch, 2000; though also see Box 13.3).

Box 13.3 Learning human speech sounds: perceptual narrowing of phonemes

There are several reports of human children who were deprived of linguistic input (neither human vocal speech nor sign language) and subsequently failed to develop language. One famous child is Victor, who was named the "Wild Child" after being found abandoned, living in the woods near Aveyron, France, in 1800. When discovered at around 12 years of age, Victor could make various sounds, but had no language. After many years of socialization, Victor only gained a few words (Lane, 1976). His story and that of others (e.g. Genie, a more modern "wild child") dramatically emphasize the important role of social learning in language development.

Language development has been extensively studied by developmental psychologists, and, of course, current knowledge goes well beyond the scope of this book. One aspect, however, points to an interesting process underlying the social learning of speech sounds, or **phonemes**. Learning human speech requires the ability to distinguish among phonemes, an ability that is present in infancy (e.g. Eimas *et al.*, 1971). However, a difference between infants' and adults' discrimination of speech sounds is that young infants are actually better at it. Consider that any one language uses only a subset of the large number of phonemes that can exist. Adults do not perceive differences among some speech sounds that do not appear in their native language. Yet, using various experimental measures like looking-time duration, researchers have found that young infants can distinguish among all phonemes that have been tested (Jusczyk, 1997). At this point, then, the early discrimination of speech sounds appears to not rely on social learning; young infants can discriminate sounds that they have never heard before.

However, this discrimination ability does not last past approximately 9 to 10 months of age. In a clever set of experiments, Werker and her colleagues demonstrated the loss of discrimination of certain phonemes that are part of the Hindi language if infants were being raised in exclusively English-speaking homes. Infants were tested in a conditioning procedure in which they learned that a change in a series of sounds would correspond with the appearance of an interesting toy to their right. Predictive head turns would thus indicate that a phoneme distinction was made. Only young infants appeared to perceive the distinction among certain phonemes in Hindi; older infants (and adults) had great difficulty in noticing a difference (e.g. Werker & Tees, 1984).

What appears to be happening is that infants begin to focus their social learning to the most common speech sounds in their environment. This **perceptual narrowing** seems to result in greater expertise in the predominant language (Kuhl *et al.*, 2006). Thus, for humans, social learning of speech sounds may be enhanced by a tuning of perceptual systems to the most frequent (and thus potentially the most relevant) sounds occurring in the environment.

Others suggest that song learning may be an example of end-state emulation (e.g. Byrne, 2002); birds that are listening to others' songs may be, in effect, copying the final result. A full understanding of the social learning processes that characterize song learning will require further testing, but it may be the case that increased knowledge of the neuronal bases of both vocal and visual forms of social learning will shed light on the similarities and differences of the underlying processing.

13.5 Learning to use objects

In 1911, Thorndike's experiments with puzzle boxes revealed that cats could learn, by trial-and-error, how to open a latch to escape a box (Chapter 4). In other experiments, Thorndike allowed cats to observe other cats opening the box before being tested themselves. He reasoned that if imitation had occurred, the observing cats would have escaped faster than cats with no prior observation, but no such facilitation was evident. Even if observing cats had shown an advantage, we now know that one of many possible social learning mechanisms could account for the behavior, including stimulus enhancement of the latch.

At the end of the twentieth century and to the present, the study of social learning has continued to focus on the question of whether 'true imitation' is evident in animals and human children. Now, however, researchers are armed with a set of carefully controlled laboratory methodologies and a battery of social learning mechanisms to consider as they examine what type of learning might best characterize the development of certain behaviors. Much of this work studies the manner in which animals and human children learn to manipulate objects, often inspired by the evidence for social learning of tool use in the wild (Chapter 8). From this research, new insight has been gained into the conditions under which imitation appears to occur as well as the potential abilities that are unique to humans and may serve to support culture.

13.5.1 Testing for imitation: bidirectional control and two-action procedures

Examining imitation depends on carefully controlled experiments, yet it has been notoriously difficult to find a way to rule out the various non-imitative forms of social learning. One attempt has been the **bidirectional control procedure** in which an observer watches a demonstrator move

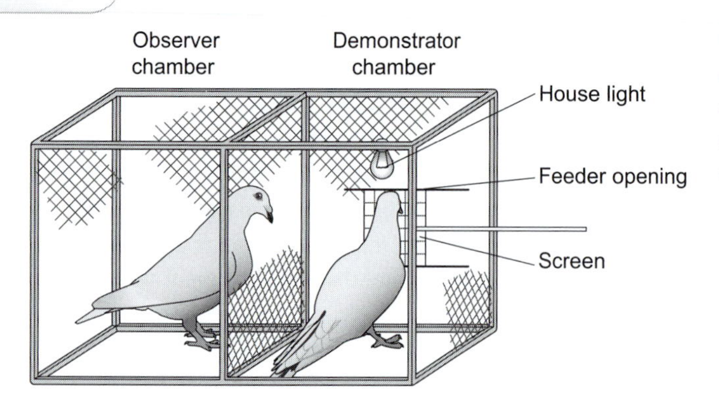

Figure 13.5 Testing arena for Klein & Zentall (2003). The observing pigeons pushed a sliding door (the 'screen') to the left or to the right to obtain food, depending on the direction a demonstrating bird had previously pushed.

an object in one of two possible directions to obtain a goal. Heyes and Dawson (1990), for example, trained one group of rats to serve as demonstrators; these rats learned to move a pole either to the left or to the right. After watching a demonstrator, observers subsequently pushed the pole in the direction they had observed. At first glance, the observers would appear to be engaging in imitation; both the end state and the specific actions that were used to bring about that state were replicated. However, it is possible that rats were engaging in emulation in the form of affordance learning. That is, they may have learned that that pole could be moved in a specific way.

To control for affordance learning, a follow-up study moved the pole's location to a perpendicular wall in the testing chamber such that now, to push in the correct direction, rats would have to have encoded the direction of the push relative to the body. Rats in this condition continued to move the pole in the direction that had been observed, and the authors interpreted the results as indicative of imitative learning. Further study, however, revealed that the rats' behavior may have had more to do with attraction to odor cues present on one side of the pole than actual imitation (Mitchell *et al.*, 1999).

Despite this rocky start, the bidirectional control procedure has been used in recent years with novel controls and with a species for which odor cues are less important, namely the pigeon. Klein and Zentall (2003), for example, have found that pigeons will push a sliding door to the left or to the right, depending on the direction a demonstrating bird has previously pushed (Figure 13.5). A control for affordance learning was also included in which the door slid in one of the two directions independently (via a hidden experimenter). In this 'ghost condition,' affordance learning was still possible; however, matching the push direction could not be said to occur via imitation. Two other conditions controlled for social facilitation (by having another bird present while the door moved on its own) and odor and sound cues (by having the actions blocked from view). Pigeons were most likely to push in the observed direction after observing a conspecific push, providing evidence for imitation. Interestingly, though, pigeons also pushed in the matching direction in the ghost condition when no other pigeon was present, yet to a lesser extent. One interpretation of this pattern of results is that pigeons can imitate, but will also engage in affordance learning in the absence of a demonstrator.

A second testing procedure has been developed that may control for emulation more effectively than the bidirectional control procedure. In the **two-action test**, the object moves in the same manner, but the action used to cause the movement differs across demonstrators. That is, in contrast to the bidirectional procedure, the affordances remain the same, but observers will differ in terms of what specific actions they witness. For example, Akins and Zentall (1996) trained Japanese quail demonstrators to use either a stepping or pecking motion to move a treadle to obtain food. Observers

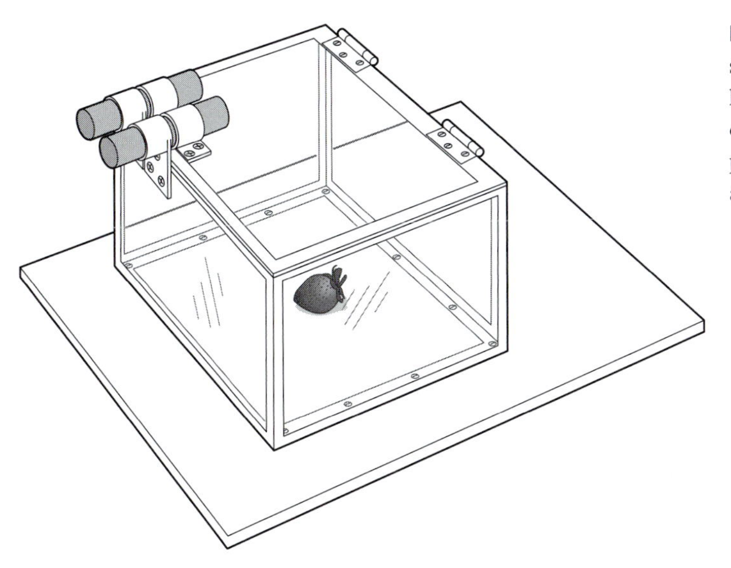

Figure 13.6 'Artificial fruit' used to examine social learning in chimpanzees and young human children. In this version, participants observed a human demonstrator either pull or poke out 'bolts' in order to unlock the box and obtain food (e.g. Whiten *et al.*, 1996).

then watched a demonstrator obtain food. If an observer watched stepping motion, it also engaged in stepping motion, but if pecking was observed, the observer subsequently pecked. Similar results were found with pigeons (Zentall *et al.*, 1996).

A two-action task such as this can rule out emulation by affordance learning because the object's movement was exactly the same across conditions (i.e. the treadle moved in the same way). Thus, the birds appear to have encoded the demonstrator's actions. Additionally, Zentall (2012) points out that there is little visual similarity between what the observer witnesses during the demonstration phase and what is seen when they have the opportunity to manipulate the treadle (e.g. they cannot necessarily see their foot touch the treadle given the position of the eyes). For this reason, it is unlikely that the results are due to simply moving in such a way as to match two visual stimuli.

Chimpanzees, children, and 'artificial fruit'

Because human children are such good imitators, it has been tempting to assume that ape species, our closest living relatives, would also be good at aping. Yet, results from variants on the two-action test just described have indicated some intriguing similarities and differences. Whiten and colleagues have pioneered studies using what they have called 'artificial fruit,' transparent boxes that can be opened in various manners to obtain fruit (Figure 13.6). In one version, for example, two bolts held the box closed, and chimpanzees or children (ages 2 to 4 years) individually watched a human adult demonstrator either pull or poke the bolts out and open the box. The participants were then given a newly baited box and allowed to try to open it.

Children copied the actions of their demonstrator with the most fidelity, and while the chimpanzees were also more likely to use the observed action, they showed less exact copying (Whiten *et al.*, 1996). One obvious question is whether chimpanzees would show greater fidelity if they had observed a conspecific initially demonstrate the box, but even in this case, results remained the same (Whiten *et al.*, 2004). Thus, human children may be more predisposed to closely copy the actions of demonstrators, a finding that will be elaborated on in the next section and Box 13.4. As for the chimpanzees, some have suggested that the lower fidelity copying is

Box 13.4 Overimitation

Though often selective in terms of whom to imitate, children appear to be less discriminating about what to imitate. As described in Section 13.5, children are more likely than chimpanzees to imitate the extraneous actions of a demonstrator (e.g. Horner and Whiten, 2005). This behavior, often called **overimitation**, has been shown across cultures (Nielsen & Tomaselli, 2010), and it increases with age over childhood (McGuigan *et al.*, 2011). Overimitation exists for such actions as waving a stick over an apparatus or lightly tapping a cover, even when children are left alone and motivated to work as fast as possible (Lyons *et al.*, 2007).

Several theories have been posited as to why overimitation occurs, although so far none has satisfactorily explained all the factors. Carpenter (2006), Meltzoff (2007), and Nielsen (2006) suggest that children overimitate the actions of adult demonstrators in order to promote affiliation with the demonstrator. Lyons *et al.* (2007) suggest that even in a causally transparent system, children may come to believe that the non-causal actions are somehow necessary to achieve the final goal. Others, in contrast, have shown that children are able to differentiate between necessary and irrelevant actions, suggesting that they do not believe the non-causal actions truly serve a mechanistic purpose (Kenward *et al.*, 2011). Instead, these researchers suggest that overimitation occurs because children interpret the demonstrated actions as social conventions (or norms) that should be copied.

Indeed, human behavior is strongly influenced by cultural practices that have conventions (specifying what one typically does) and norms (specifying what one ought to do). This allows us to act collectively, for example, to create artifacts that we all use in certain ways and to have a common framework for what constitutes appropriate action (e.g. Rakoczy *et al.*, 2008; Searle, 1995). Preschool age children are already aware of this, heartily protesting ("No, it does not go like this!") when others act in direct contrast with the rules of a game (Rakoczy *et al.*, 2008).

It is thus possible that some instances of social learning in humans may be constrained so as to place importance on the manner in which a goal was achieved, even if this sometimes leads to inefficient behavior. The potential benefit of this cognitive system, however, is the learning of actions that are conventional and normative in one's society.

actually indicative of emulation, perhaps via object movement reenactment. That is, they may not have learned about the actions but instead learned that the bolts could move in a certain way. The behavior of capuchin monkeys (*Cebus apella*) in a similar task addresses this interpretation. Custance *et al.* (1999) found that monkeys copied the direction of movement of the bolts, but not the manner in which they were handled. For example, they would sometimes pull the bolts out from the back instead of poking them out. However, chimpanzees were not observed to engage in this type of behavior.

The conflicting interpretations of the results from artificial fruit tasks likely stem from the fact that it can be difficult for a researcher to decide whether the actions that an observer produces match those of the demonstrator to a sufficient extent. As mentioned earlier, some researchers have suggested that a more productive research question may be how the observer decides what aspects of an action to reproduce, not whether the reproduction is actually imitation. In the next section, this question will be considered, again in studies comparing the behavior of chimpanzees and children.

Researcher profile: **Dr. Thomas Zentall**

Figure 13.7 Dr. Thomas Zentall.

Thomas Zentall has spent over 40 years examining some of the foundational questions in comparative cognition. "The approach my students and I use is to define a cognitive behavior that is characteristic of humans in a way that clearly distinguishes it from simple associative learning and then to examine the conditions under which it can be found in animals" (https://psychology.as.uky.edu/users/zentall). This approach has been applied successfully through the years to advance our knowledge of memory, concept formation, and social learning.

In 1963, Zentall completed bachelor's degrees in both psychology and electrical engineering at Union College, Schenectady, New York. He then headed west to complete a PhD at the University of California, Berkeley, under the mentorship of Dr. Donald A. Riley. After 6 years at the University of Pittsburgh, Zentall moved to the University of Kentucky in 1976 and started the Comparative Cognition Laboratory. Of the large body of work to come out of his lab, perhaps the most relevant to this chapter is a 1988 book that Zentall co-edited with Bennett Galef, Jr. The book, *Social Learning: Psychological and Biological Perspectives*, was an interdisciplinary synthesis bringing together field and laboratory researchers who were studying social learning from separate but overlapping frameworks. The volume remains seminal to this day.

Research from Zentall's lab has carefully considered the variety of social learning mechanisms that exist and offered critical experimental designs that help to tease apart the contexts in which animals may or may not engage in imitation. Outside of the lab, Zentall has taken on various leadership roles in the field, serving as the president of the Comparative Cognition Society (2004–2006) and the editor of the journal *Comparative Cognition and Behavior Reviews* (2011–2014). An esteemed mentor (his current and former students affectionately call him 'Z'), he has supervised the undergraduate and graduate training of many current researchers in the field of comparative cognition.

Like most researchers in this field, Zentall believes that the comparative approach expands our understanding of both cognitive mechanisms and simpler learning processes in humans and other animals. "The cognitive approach to animal behavior can have value whether or not one can actually find evidence for cognitive processes in animals. I suggest that adopting a cognitive approach has great heuristic value because it will lead one to carry out research that otherwise would not have been conducted." (Zentall, 1993, p. 3).

13.5.2 Switching between imitation and emulation

In some instances, copying the exact actions of a model can result in inefficient behavior, for example when the model is demonstrating actions that are extraneous or not causally relevant. Thus, one could argue that as long as a species has an understanding of the object physics involved, causally irrelevant actions may be disregarded, which in turn would lead to end-state emulation instead of imitation. In a seminal study on this topic, Horner and Whiten (2005) tested young chimpanzees and 4-year-old children with two different boxes containing reward (food for the chimpanzees, stickers for the children). One box was made of clear plastic, and the other was

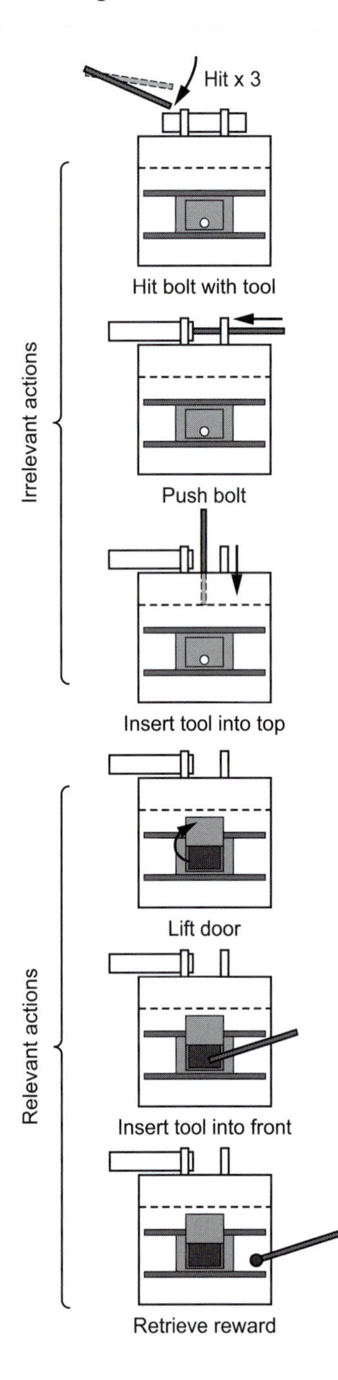

Figure 13.8 Apparatus used in Horner & Whiten (2005). A human demonstrator was observed to engage in an irrelevant action (e.g. touching the tool to the interior ledge) prior to inserting the tool into the front of the apparatus and retrieving a reward. When the box was clear, the irrelevant and relevant actions are apparent, but with the opaque box, the causal relevance of the action at the top of the box (touching the ledge) is not immediately apparent.

opaque, but in all other ways they were identical (Figure 13.8). The human demonstrator used a stick to reveal a hole at the top of the box, and then pushed the stick into this hole, tapping on a plastic ledge three times. The stick was then removed and inserted into a second hole, retrieving the reward. When the box is clear, it is apparent that the actions in the first hole are causally unrelated to obtaining the reward; the ledge that the stick taps is separate from the area housing the reward. With the opaque box, however, the causal relevance of this action is not immediately apparent.

When the opaque box was presented first, both children and chimpanzees replicated all of the demonstrator's actions, showing evidence of imitation. However, with the clear box, only the children repeated the extraneous action, even though an understanding of barriers and solidity is assumed to be present by this age. When the clear box was presented first, chimpanzees skipped the irrelevant action and continued to skip it with the opaque box. Children, in contrast, always repeated all the actions. Horner and Whiten (2005) suggest that chimpanzees thus attend to different aspects of the demonstration depending on the situation: when the causal structure of a task is clear, emulation may occur more readily than imitation. In contrast, children may be more strongly biased to imitate, a proposal discussed in detail in Box 13.4.

13.5.3 Attending to a model's intention

Even though human children often imitate even the extraneous actions of others, they are not complete slaves to imitation. Toddlers, for example, imitate intentional actions, but do not repeat superficially similar, accidental acts (Carpenter *et al.*, 1998). Further, young children appear to imitate actions based on the rationality of the action given the goal and the current situational constraints. When young children watched a demonstrator turn on a light by using her forehead while her hands were occupied holding a blanket, they subsequently turned on the light with their hands (Gergely *et al.*, 2002). Compared to this condition, after witnessing a demonstrator whose hands were not holding anything – yet used her head – children were more likely to use their own heads to turn on the light. These findings suggest that children interpreted the use of the head action, even when hands were available, as indicating that the purpose of the action was *to turn on the light with the head*, instead of just *to turn on the light*. Thus, when learning from others, children may take into account the likely goal of a demonstrator given the current constraints and either engage in a direct copy of the actions (e.g. use their head) or choose an alternative means (e.g. use their hands, a form of goal emulation).

Similar results to Gergely *et al.* (2002) have been found for chimpanzees using an analogous task. Chimpanzees observed a human demonstrator operating six objects using an unusual body part (e.g. pressing with the forehead). Again, in two conditions, the demonstrator would either have free or occupied hands. Like the young children, chimpanzees repeated the modeled action using the unusual body part more often when the demonstrator's hands were free (Buttelmann *et al.*, 2007). However, two aspects of these findings suggest that further research is necessary. First, using the unusual body part is actually quite infrequent, though it does occur more in the free hand conditions as predicted. Second, it is still unclear as to the manner in which chimpanzees interpret the goals of others (e.g. by representing behavior or representing the mental states that underlie behavior; see Chapter 10). In sum, though, it is possible that at least for children and chimpanzees, decisions regarding what aspects of observed actions should be copied may be influenced by consideration of the underlying goal.

13.6 Culture

When considering human society, 'culture' refers to population-specific shared traditions, beliefs, and behaviors. Human culture is shared from one human generation to another by a combination of social learning processes, language, and teaching (see Box 13.5). Thus, the field of comparative cognition has been hesitant to use the term 'culture,' though examination of the foundations of

Box 13.5 Do animals teach?

In social learning research, the term 'demonstrator' is commonly used to denote an individual whose behavior is being observed, regardless of whether this individual recognizes that he or she is being observed or alters behavior to better impart information. But, can any animal species be said to demonstrate in the sense of *teaching*? In 1992, Caro and Hauser created a functional definition of teaching that has been widely accepted today (Caro & Hauser, 1992). To be considered teaching, an individual must: (1) modify behavior in front of naïve individuals in a manner that increases the chance of learning, (2) incur immediate cost to oneself, and (3) cause the student to learn something more rapidly than would have occurred otherwise.

Currently, the most systematic study suggesting teaching in an animal species focused on meerkats (*Suricata suricatta*; Thornton & McAuliffe, 2006). Meerkats are mammals found in Southern Africa that engage in cooperative breeding. Pups are reared by many members of the group, and when adults leave to search for food, the young follow and make begging calls. These calls encourage the adults to bring them prey, including scorpions, which can be difficult to kill. The manner in which the prey is presented varies with the age (and thus experience) of the pup; young pups are given dead or disabled prey, while older pups are given intact prey. Additionally, the time spent with a pup after providing prey is dependent on the pup's age, with younger pups being accompanied longer. A follow-up experiment found that provisioning of disabled prey does facilitate learning relative to an unprovisioned control group.

Thus, all three of Caro and Hauser's requirements for teaching are met. The adults vary the type of prey provided based on the age of the pup, a cost is likely incurred due to loss of foraging time (and giving away food), and the pups appear to learn a critical behavior faster than they would without the interaction. Still, many have argued that a functional definition of teaching such as this does not capture many key components of human teaching.

Gergely Csibra, for example, suggests that while the meerkat behavior is a good example of scaffolding (modifying the environment in such a way as to facilitate individual learning), human teaching is much more than this (Csibra, 2007). Human teaching often entails demonstrating information about the hidden properties of objects (e.g. their labels, their functions, their palatability), and this information is generalizable to different contexts. While one can say that the meerkats have learned the generalizable skill of how to kill, they likely learned this individually after being presented with prey. Human teaching, however, often includes description of how to engage in certain behaviors so that this is learned socially, not individually. Thus, although some examples of animal teaching may fit the current operational definition of teaching, this definition may need to be modified to more accurately define what we know about human teaching.

human culture in non-human primates is an active research endeavor incorporating strict operational definitions. Much of this work has been conducted with chimpanzees, both in the wild and in captivity.

13.6.1 Chimpanzee tool use cultures

Chimpanzees exist in Africa in geographically separated populations, and each has been the subject of long-term observation. In 1999, Whiten and eight field researchers collaborated to examine whether certain behaviors might be best described as cultural variants that differed across the populations. For each population, the frequency of occurrence of 65 behaviors was estimated on a continuum from customary to absent. Among these behaviors were tool-use behaviors described in Chapter 8, including ant-fishing, nut-cracking, and water-sponging, as well as grooming postures that included holding hands. If a behavior was not observed in a population, researchers considered whether an ecological reason could be given for its absence (e.g. were there nuts available to crack?). Of the 65 behaviors considered, 39 were common in some populations, but absent in others, even when the required ecological conditions were present.

These behaviors are thus candidates for cultural variants, yet two other criteria were considered essential. First, there should be no reason to think that genetic differences could account for the group differences. Indeed, some behaviors, like ant-fishing, occurred in different manners (e.g. swiping the ants off a stick with the hand versus putting the stick directly into the mouth) in nearby communities with no known genetic differences. As a second criterion, the behavior should be considered to be the outcome of social learning where information is transmitted across generations, and consistent with this, these behaviors are considered to be learned by juveniles from observing adults (Whiten *et al.*, 1999).

Given these criteria for culture, some behavioral variations in wild chimpanzee populations have been described as being cultural. This characterization, however, is debated. One hesitancy was alluded to above: the term 'culture' has a nuanced meaning in the fields of anthropology, archaeology, and psychology, and a short list of criteria fails to capture the full meaning. A second cause for concern is whether the observation of behavioral differences in the wild really can exclude ecological causes or be truly said to result from social learning (e.g. Galef, 2004; Laland & Janik, 2006; Shettleworth, 2010)

13.6.2 Evidence for traditions

Given the difficulties inherent in studying social transmission in wild populations, the study of culture in animals has started to take a new tack. Researchers are able to examine the transmission of tool-using behavior within captive populations where the precise spread of a behavior can be tracked and the ecological variance can be minimized. Additionally, some researchers have opted to use the word 'tradition' to characterize a population-wide variation in behavior.

These studies have often used boxes such as the artificial fruit apparatus used by Whiten *et al.* (1996), effectively creating a two-action test. A chimpanzee demonstrator in one group is trained to open the box using one of two actions, while a second demonstrator in another group uses the other action. When a demonstrator is introduced to a group, the group tends to learn the action demonstrated (Whiten *et al.*, 2005). In one study using this methodology, a transmission chain was documented such that an observer learned the box opening from the trained chimpanzee, and then another observer learned from the first observer, and so on (Horner *et al.*, 2006).

Though these data support the claim that a novel behavior can be spread through a group of chimpanzees via social learning, and that other groups may develop other novel behaviors in the same manner, the question remains as to whether chimpanzees can be said to have culture in any meaningful way. The traditions that are seen with chimpanzees may be best understood as the building blocks of the *cumulative* culture of human populations (Zentall, 2012). Human culture is considered unique because traditions learned by imitation are modified and improved by individual learning and teaching, effectively 'ratcheting up' over generations (e.g. Tennie *et al.*, 2009). Species that are limited to learning from others primarily through emulation or other forms of social learning must spend time relearning much of what others have already learned (Richerson & Boyd, 2005; Shettleworth, 2010). Thus, consideration of the evolution of culture points to some possible outcomes of species differences in social learning processes. As was seen in this chapter, however, there is still much to learn about these differences.

Chapter Summary

- Learning can occur through an individual's own, asocial experiences with the world or through observation of others' actions. Individual learning is not passed on from generation to generation; the learning mechanisms may be inherited genetically, but the learning occurs through an individual's own interaction with the environment. In contrast, behaviors learned through social learning can be spread from one generation to another.
- By some accounts, natural selection in animals capable of social learning should favor 'strategies' that allow for discriminate learning. Two proposed strategies entail being discriminative about when to learn from others, and being discriminative about whom to learn from.
- Research that has focused on the cognitive mechanisms underlying social learning has characterized at least five types of learning, some of which contain various subtypes: enhancement, observational conditioning, emulation, mimicry, and imitation.
- Many species appear to learn food preferences, mate preferences, and aspects of communicative behaviors from others. In much of this work, the focus has been to examine whether, and under what circumstances, social learning has occurred.
- At the end of the twentieth century and to the present, the study of social learning has continued to focus on the question of whether 'true imitation' is evident in animals and human children. From this research, new insight has been gained into the conditions under which imitation appears to occur as well as the potential abilities that are unique to humans.
- Consideration of the evolution of culture points to some possible outcomes of species differences in social learning processes. Without imitation or teaching, it is possible that a species cannot have a human-like cumulative culture.

Questions

1. What are some of the proposed characteristics of individuals from whom it would be most beneficial to learn? Using examples from this chapter, provide supporting evidence for these proposals.
2. What are the key differences between stimulus enhancement and observational conditioning? Provide examples of each.
3. One morning, after observing his father apply shaving cream, a 3-year-old child carefully applies some cream onto his own face. What type of social learning would likely underlie this behavior?
4. What evidence is there that rats begin to learn what to eat even while in utero?
5. Explain what is meant by the following statement: It may be expected that social learning exists in even relatively solitary species as an ontogenetic adaptation.
6. What type of learning might best characterize mate-choice copying? Why?
7. What are the typical stages of bird song learning in white-capped chickadees? Is there evidence for a sensitive period of development?
8. What can the songs of isolated birds that have never heard species-typical song tell us about the unlearned foundations of song learning?

9. A chimpanzee watches a demonstrator open a box by first sliding a door to the left to reveal a bolt, then twisting and pulling to remove the bolt. When given the box, what is the chimpanzee likely to do? How might this compare with the actions of a 4-year-old child?
10. How might the human propensity for imitation and teaching have affected the evolution of culture?

FURTHER READING

Social learning has been the focus of interdisciplinary study for many decades. Volumes and review papers that document the major findings and theoretical frameworks include the following:

Byrne, R. W., & Russon, A. E. (1998). Learning by imitation: A hierarchical approach. *Behav Brain Sci*, **21**, 667–684.

Heyes, C. M., & Galef, B. G., Jr. (1996). *Social Learning in Animals: The Roots of Culture.* San Diego, CA: Academic Press.

Rogers, S. J., & Williams, J. H. G. (2006). *Imitation and the Social Mind: Autism and Typical Development.* New York, NY: Guilford Press.

A special issue of the *Journal of Comparative Psychology* focused on social learning in humans and animals: Volume **126**, No. 2, 2012.

Detailed characterization of the types or mechanisms of social learning can be found in the following papers:

Want, S. C., & Harris, P. L. (2002). Do children ape? Applying concepts from the study of nonhuman primates to the developmental study of "imitation" in children. *Dev Sci*, **5**, 1–13. (Including commentaries.)

Zentall, T. R. (2012). Perspectives on observational learning in animals. *J Comp Psychol*, **126**, 114–128.

Learning about food and mates has had a rich research history, often led by the work of Bennett Galef. Some reviews include:

Galef, B. G., Jr. (1976). Social transmission of acquired behavior: a discussion of tradition and social learning in vertebrates. *Adv Study Behav*, **6**, 77–100.

Galef, B. G., Jr., & Laland, K. N. (2005). Social learning in animals: empirical studies and theoretical models. *BioScience*, **55**, 489–499.

Birdsong learning has been the focus of decades of research. Many excellent books and review papers examine behavioral and neurobiological aspects:

Brainard, M. S., & Doupe, A. J. (2002). What songbirds teach us about learning. *Nature*, **417**, 351–358.

Marler, P., & Slabbekoorn, H. (2004). *Nature's Music: The Science of Birdsong*. San Diego, CA: Elsevier Academic Press.

The question of animal culture has been discussed in books and seminal articles:

de Waal, F. (2001). *The Ape and the Sushi Master: Cultural Reflections of a Primatologist.* New York, NY: Basic Books.

Laland, K. N., & Janik, V. M. (2006). The animal cultures debate. *Trends Ecol Evol*, **21**, 542–547.

Whiten, A., Goodall, J., McGew, W. C., *et al.* (1999). Cultures in chimpanzees, *Nature*, **399**, 682–685.

GLOSSARY

absolute pitch (AP): the ability to identify or produce the pitch of a musical note without any external reference; also called perfect pitch.

action potential: transmission of an electrical signal from the cell body of a neuron, along the axon, to the presynaptic terminal, resulting in the release of neurotransmitter.

adaptation: an evolved solution to a problem of survival and reproduction.

adaptive specialization: the relative ease with which animals acquire certain associations, based on their evolutionary history.

affordance learning: a social learning process in which observers learn the properties of objects or relationships between objects.

algorithm: a specific procedure or set of instructions that will yield the correct solution.

analog magnitude representation: a 'noisy' representation of a set of items that is proportional to the number of items being represented.

aposematism: a form of passive communication in which an unpalatable species has coloration, odors, or warning sounds that signal danger (e.g. toxic or causing illness) to potential predators.

arbitrary causality: a relationship in which the generative processes of the causal interaction are not readily apparent (e.g. pecking a key causes food to appear; see **natural causality** for contrast).

attention: the mental process that selects sensory information for further processing; a focusing on a particular stimulus or event.

audience effect: occurs when signaling behavior is altered by the presence or absence of receivers as well as the receivers' response.

autism spectrum disorders: developmental disorders characterized by communicative and social deficits, among others.

automatic imitation: a process describing how humans will learn the actions of others without any conscious intention to imitate or awareness that they have learned from others.

availability heuristic: a cognitive bias to judge more memorable events as occurring at higher frequencies.

axon: output fiber of a neuron that ends in presynaptic terminals.

Batesian mimicry: describes instances in which a palatable species has similar physical features to an aposematic species, in essence dishonestly using another species' honest warning signal.

Bayesian inference: a form of inference that considers prior probabilities and updates based on new observations.

beacon: a stimulus that marks a goal location.

behavioral ecology: an offshoot field of ethology that is characterized by increased attention to the ultimate causes of behavior; the scientific study of interactions between organisms and their environment, and how these interactions result in differential survival and reproduction.

behavioral economics: a field merging behavioral analysis of choice with economics; a description of choice behavior in terms of costs and gains.

behavioral neuroscience: the study of brain–behavior relationships in all animals.

behaviorism: a subdiscipline of psychology, associated with John B. Watson, that adheres to the principle that behavior (not thoughts or ideas or any other cognitive process) is the only justifiable object of study in psychology.

bias: in choice experiments, preference for one behavioral option that is independent of reinforcement payoff.

bidirectional control procedure: a procedure used to study social learning in which an observer watches a demonstrator move an object in one of two possible directions to obtain a goal.

binocular vision: overlap of the two visual fields so that visual information reaching the two eyes is similar.

biological motion: the movement of animals, particularly vertebrates, is in most cases non-rigid, yet it is constrained by the configuration of the body and gravity.

bioluminescence: chemical light produced when organic compounds are mixed; occurs in the organs of deep-sea animals.

blindsight: a phenomenon in which individuals with cortical blindness are able to locate a visual stimulus; residual visual function with no awareness of this function.

blocking: a phenomenon in which an association between one stimulus and a US disrupts subsequent conditioning to a second stimulus when the two are presented in compound conditioning trials.

camouflage: structural adaptations that allow a species to blend in with its environment.

categorization: the cognitive process of classifying items or events into groups based on one or more common features.

category-specific semantic deficit: the inability to access information about one semantic category while leaving others intact.

causal Bayes nets: causal structures of the world are depicted with graphical models that formalize inferences based on conditional probabilities and consequences of interventions.

central place forager: an animal that carries food back to a home base, such as a nest or burrow, to horde or consume it.

chameleon effect: a term used to describe the finding that humans have a propensity to unknowingly mimic body movements of people they interact with.

choice: the outcome of decision making, measured as a motor action or verbal response.

chunking: a process that increases encoding capacity by reorganizing discrete elements of sensory input into larger units.

circadian rhythm: a rhythm of activity or physiological state that appears to be the result of an internal clock with an approximately 24-hour cycle.

classical conditioning: a type of non-declarative memory; the process whereby stimuli that do not elicit a response initially acquire this ability through association with a motivationally significant event.

cognition: knowledge or thinking; the acquisition, storage, and processing of mental information in humans and animals.

common adaptation: a process whereby individual species develop different strategies for dealing with the same problem; convergence on the same solution through independent evolutionary change.

compensatory plasticity hypothesis: the idea that a loss or deficit in one sense leads to a heightened capacity in another.

computer science: a branch of science that deals with the design of computers and theories of computation.

concept: the abstract set of rules that define membership in a category.

concept formation: the cognitive process of establishing and updating the abstract set of rules for category membership.

concurrent schedule of reinforcement: operant conditioning in which two or more schedules of reinforcement are in effect simultaneously.

conditioned avoidance: an operant paradigm in which animals learn to avoid a stimulus associated with an aversive event.

conditioned response (CR): in classical conditioning, a response that is elicited by the CS following conditioning.

conditioned stimulus (CS): in classical conditioning, a stimulus that acquires motivational significance through pairing with the US.

conditioned taste aversion (CTA): an aversive reaction to the smell and taste of a particular food caused by a prior food–nausea association.

conformity: basing one's behavior on the behavior of the majority of individuals in one's group.

CONLEARN: a proposed visual system that allows for further learning about the facial characteristics of conspecifics.

consolidation: the second stage of memory processing in which encoded representations are modified to become more stable over time.

CONSPEC: a proposed visual system that orients the newborn to face-like patterns.

conspecific recognition: a form of social categorization that involves recognition by one member of a species of another member of the same species; intra-species recognition.

contagion: a term used to describe reflexive, species-typical matching responses to the behavior of others.

continuity hypothesis: the idea that trait differences between animals and humans will be quantitative, not qualitative.

continuous reinforcement: (CRF) a reinforcement schedule in which every response produces a reinforcer; equivalent to an FR1 schedule.

cooperative breeding: situations in which individuals – who are not parents themselves – are actively engaged in raising others' offspring.

dark adaptation: a reduction in the threshold for detecting light that occurs under reduced illumination.

decision making: the cognitive process of selecting one course of action from a variety of options.

declarative memory: a subcategory of reference memory; a knowledge-based system that is expressed through explicit statements and depends on conscious awareness.

delay discounting: see temporal discounting.

dendrites: tiny fibers that branch from the cell body of neurons. Synaptic receptors located along dendritic surfaces allow chemical signals to be received from other neurons.

devaluation: a three-stage experiment: 1. CS–US pairings; 2. devaluation of the US through association with an aversive stimulus; 3. testing conditioned properties of CS.

developmental psychology: the study of systematic changes that occur across the lifespan in humans.

Dictator Game: an economic game in which a player must choose how much money to give to a second player.

directed social learning: in some cases of social learning, the identity of the demonstrator increases his or her salience for an observing individual.

discrimination: the cognitive process of distinguishing items or events based on one or more distinct features.

disinhibition: the recovery of a CR following extinction when a novel stimulus is presented.

dispersal: movement away from a parent or conspecific population due to a declining habitat, overcrowding, or mate competition.

distress calls: vocalizations emitted by many young mammals when separated from the nest or their mothers.

divided attention: the ability to process, simultaneously, sensory input from more than one source.

eavesdropping: passive listening by other members of a group.

ecological psychology: a subdiscipline of psychology, associated with James J. Gibson, promoting the idea that behavior (and cognition) can only be understood in the environmental context in which it took place.

EEG/ERP: electroencephalography (EEG) is a neuroscience technique in which brain activity is measured along the scalp by multiple electrodes. EEG can measure event-related potentials (ERP), which are brain responses to specific sensory, cognitive, or motor events.

elaboration: the process of adding meaning, images, or other complex information to incoming sensory input such that encoding is enhanced.

elastic demand: changes in the price of an item are inversely related to demand for that item.

emotional contagion: matching another individual's emotional state.

empathetic perspective-taking: de Waal's third level of empathy in which an individual understands others' needs and experiences vicarious emotional arousal.

empathy: the ability to understand the feelings of others.

encoding: the first stage of memory processing in which incoming information is converted into neural signals that are used for later processing.

end-state emulation: a type of social learning in which the actions of the demonstrator are not learned, but the fact that a particular end state can be achieved is learned.

entrainment: the process by which the internal circadian rhythm falls into synchrony with the external day/night cycle.

episodic memory: a subcategory of declarative memory describing knowledge for events in a personal past.

equipotentiality: the idea that associations between different stimuli, responses, and reinforcers could be formed with equal ease.

equivalence classes: sets of stimuli that have been grouped together via experience such that if one member of an equivalence class becomes discriminative for a given behavior, the other members become discriminative for the same behavior.

ethology: the scientific study of the causes and adaptive value of animal behavior; a subtopic of zoology and biology.

exemplar: a distinct item within a category.

extinction: removal of a US that leads to a reduction in responding.

extractive foraging: physically extracting a food source from the environment.

false beliefs: the ability to represent the inaccurate belief states of others is often considered a hallmark of mental state attribution as these beliefs diverge from one's own belief states.

fast mapping: in the study of language development, the ability to connect a new sound and object (e.g. so as to learn a word) with minimal experience.

feature integration theory: a modern version of structuralism which posits that the elements of sensory input are combined to produce perceptual wholes.

fission–fusion society: a social organization in which individuals merge and separate with larger and smaller groups.

fixed action pattern (FAP): stereotyped, instinctive behaviors that occur in a rigid order and are triggered by a 'sign' stimulus in the external environment.

fixed interval (FI): a reinforcement schedule in which reinforcement is delivered following the first responses that occur after a set period of time.

fixed ratio (FR): a reinforcement schedule in which a set number of responses produces the reinforcer.

fMRI: functional magnetic resonance imaging.

free-running rhythm: the rhythm of the internal clock when light/dark input from the environment is removed in experimental studies of the circadian rhythm.

frequency coding: a principle of sensory detection describing the relationship between stimulus intensity and firing rate of sensory neurons; as stimulus intensity increases, so does the rate of action potentials.

function: a term in the study of tool use that refers to what an object or tool was made for.

functional categorization: the process of grouping stimuli together based on a shared meaning or association; associative categorization.

functional fixedness: a term used to describe situations in which humans have difficulty ignoring an artifact's original function so as to use it to solve a new problem.

functionally referential: a characteristic of signals that provide information specific to a particular object or event and are responded to by receivers even in the absence of the object or event.

game theory: the study of cooperation and competition among individuals using simulations of interactions.

genotype: an organism's genetic code.

glia: non-neuronal cells in the central nervous system. The functions of glia include supporting and nourishing neurons, guiding neuronal growth during development, and removing debris after neuronal injury or death.

green-beard gene: a gene that codes for mechanisms that ultimately allow organisms to 'recognize' a fellow carrier of certain genes.

habit learning (S–R learning): responses that are elicited automatically by environmental stimuli and are relatively insensitive to changes in the value of the reinforcer.

habituation: decreased responding to repeated presentation of a stimulus that is often discussed in terms of reflex modification; a type of non-declarative memory.

handicap principle: some costly traits may serve as honest signals of biological fitness.

handling: the time and energy required to extract a consumable food item from its source.

heuristic: in decision making, a cognitive process that ignores or discounts a portion of available information, thereby saving time and mental resources; an informal 'rule of thumb.'

homing: a type of large-scale navigation that involves returning home after displacement, often from an unfamiliar location.

hyperthymesia: enhanced autobiographical memory.

ideal free distribution model: a theory explaining how conspecifics divide themselves during group foraging; the number of animals aggregating at a particular food patch is proportional to the amount of resources available at that patch.

imitation: a type of social learning in which the observer recognizes the end state and reproduces the specific actions that were used to bring about that state.

imprinting: a particular type of learning in which exposure to specific stimuli or events, usually at a young age, alters behavioral traits of the animal.

impulsivity: preference for a small immediate reward over a large delayed reward.

inclusive fitness: a measure of an individual's fitness that takes into account his or her own offspring (direct fitness) and the offspring of relatives (indirect fitness).

indifference point: the point at which immediate and delayed rewards are equally valued.

individual recognition: a form of social categorization that involves recognition of a specific member of the same species, based on individual features and not familiarity.

inelastic demand: changes in the price of an item have relatively little impact on demand for the item.

inequity aversion: a negative reaction to situations in which there is 'unequal pay for equal work.'

innate: developmentally fixed; not the product of conditioning or learning.

instinct: a behavioral pattern that appears in full form the first time it is displayed, even if the organism has no previous experience with the stimuli that elicits this behavior.

internal working models of attachment: cognitive representations of social interactions that are constantly being revised due to experience through development.

interval timing: keeping track of and remembering durations on the order of seconds or minutes.

just noticeable difference (JND): the amount by which two sensory stimuli must differ in order for the difference to be detected.

kinesis: non-directional movement in response to a stimulus.

kin recognition: a form of social categorization that involves recognition of genetic relations.

landmark: a stimulus that has a fixed spatial relationship to a goal location.

latent inhibition: a phenomenon in which prior exposure to a stimulus blocks or retards subsequent conditioning to this stimulus; also called CS pre-exposure effect.

lateral geniculate nucleus: a group of neurons in the thalamus that receives a major projection from visual receptors; the first relay station in visual processing beyond the retina.

launching events: physical events (like those of colliding billiard balls) have a certain motion pattern that yields the perception of causality.

law of effect: Thorndike's idea that responses that produce a satisfying outcome will be repeated whereas those that produce unsatisfying outcomes will not.

learning: the acquisition of information; measured as the modification of behavior based on past experience.

learning curve: changes in behavior that occur across conditioning trials (either classical or operant), characterized by smaller and smaller increments as the trials progress.

limited signal set: a term used to characterize most animal communication, a relatively small number of signals used in a small number of contexts (e.g. food, aggression/competition, sex, and danger from predators).

linguistics: the scientific study of the nature and structure of human language.

local enhancement: a type of social learning in which one's attention is drawn to a locale, thus facilitating learning.

long-term potentiation (LTP): persistent increase in synaptic strength following high-frequency firing of the presynaptic neurons.

marginal value: the point at which foraging in one patch yields a net energy gain that is lower than the average gain of the habitat.

Mark Test: an experimental design used to study mirror self recognition by examining whether participants notice and remove a mark on their body when looking in a mirror.

matching law: a rule that describes choice behavior; the relative rate of responding on one alternative matches the relative rate of reinforcement on that alternative.

mate-choice copying: an instance of social learning in which an individual chooses to mate with a conspecific with whom another member of the group has mated.

memory: the mental processes of acquiring and retaining information for later retrieval; measured as the modification of behavior based on past experience.

mental states: feelings, desires, emotions, beliefs, and intentions that are the assumed causes of the behavior.

methodological behaviorism: an experimental approach that involves quantifiable measures of behavioral output; involves tight control of extraneous variables that may affect behavioral measures.

migration: a type of large-scale navigation that involves seasonal movement between spatially distinct habitats.

mimicry: a type of social learning in which an observer learns about a demonstrator's actions without understanding the objects involved or the end state of the action.

mirror neurons: a system of neurons in the premotor cortex that respond to observations of action as well as the production of the same action.

Morgan's canon: a term coined by the nineteenth century British psychologist C. Lloyd Morgan; the principle that animal behavior should not be interpreted in terms of higher cognitive processes when simpler explanations are possible.

mutualism: a term used to describe situations in which unrelated individuals simultaneously benefit from an interaction.

natal homing: large-scale navigation that involves returning to the birthplace to reproduce.

natal philopatry: a behavioral trait of remaining in the vicinity, or returning to one's birthplace, to breed and rear young.

natural causality: physical causal events in which the generative processes are relatively apparent (e.g. launching events; see **arbitrary causality** for contrast).

navigation: the process of identifying a specific location, regardless of one's current position. Divided into small-scale navigation (short-distance travel within familiar territory) and large-scale navigation (long-distance travel often through unfamiliar territory).

negative reinforcement: a positive relationship between a response and an aversive stimulus; removal of the reinforcer increases responding.

neuroeconomics: the study of how the brain interacts with the environment to enable decision making; an interdisciplinary endeavor aimed at understanding the neurological underpinnings of decision making.

neurohormone: a hormone secreted into the circulation for systemic effect that can also have a role as a neurotransmitter (e.g. oxytocin).

neuron: a nerve cell; the primary function of neurons is to receive and transmit information in the central nervous system.

nominal scale: a qualitative level of measurement that can contain numbers that serve to differentiate items or that are used as labels.

non-declarative memory: a subcategory of reference memory that does not depend on awareness or explicit knowledge to be expressed.

object displacement tasks: a methodological technique used to examine object permanence and the ability to track objects through occlusion.

object movement re-enactment: a type of emulation characterizing instances in which, through the actions of others, an individual has learned how objects can move.

object permanence: recognizing that objects exist even when they are out of view (or imperceptible via scent or sound).

object physics: the regularities in the displacements of objects and their interactions with other objects.

object tracking system: a proposed mechanism that likely underlies small number discrimination and entails a small number of mental 'indexes' for individual objects.

observational conditioning: a type of social learning in which attention may be drawn to a location or an object (e.g. a lever), and an association might be formed between that location or object and reinforcement (e.g. presentation of food).

omission: a negative relationship between a response and an aversive stimulus; removal of the reinforcer decreases responding.

ontogenetic adaptation: a characteristic that serves an adaptive function at a certain time in development.

operant conditioning: a change in behavior that occurs because the behavior produces a consequence (positive or negative).

operations: procedures on numbers such as addition or subtraction.

optic flow: the movement of elements in a visual scene, relative to the observer; a critical component of ecological theories of perception.

ordinal scale: a qualitative level of measurement that contains numbers that are ordered in a rank based on their relative magnitudes.

orientation: an angle measured with respect to a reference direction; the process of taking up a particular bearing with respect to a current position, regardless of the ultimate destination.

oscillators: mechanisms that fluctuate in a rhythmic, sinusoidal manner.

overimitation: faithful imitation of a demonstrator's actions, including extraneous, non-causal actions.

overmatching: a deviation from the matching law in which subjects are more sensitive to reinforcement payoffs than predicted; in concurrent choice paradigms, the relative rate of responding on the lower payoff alternative is lower than the relative payoff provided by that alternative.

overshadowing: a phenomenon in which one stimulus acquires stronger conditioning than a second stimulus when the two are presented in compound conditioning trials.

pacemaker: a hypothetical generator of pulses that are sent to an accumulator within the information processing theory of timing and number.

patches: discrete areas within a habitat containing food that are separated by areas with little or no food.

path integration: a mechanism of small-scale navigation that allows animals to return to a starting location (usually their home) by keeping track of the combined distance and direction travelled on a single journey. Also called dead reckoning.

peak procedure: a procedure for examining timing in animals that involves exposing subjects to many trials in which food can be accessed a fixed time after the onset of a stimulus; with experience, target behavior peaks around the fixed time.

perception: the interpretation of sensations; involves processing, organization and filtering of sensory information.

perceptual categorization: the process of grouping stimuli together based on shared physical features.

perceptual narrowing: a process by which experiences with certain stimuli results in a system that is tuned to those features, allowing for increased recognition and distinctions within those stimuli as compared to other stimuli.

perceptual priming: facilitated identification of a stimulus as a consequence of prior exposure to the stimulus; a type of non-declarative memory.

periodic timing: a primary type of timing for cyclical activity that occurs during a 24-hour period.

phenotype: an organism's observable traits, both physiological and behavioral.

phenotypic matching: a possible mechanism of kin recognition in which an individual compares a conspecific's phenotype with its own.

phenotypic plasticity: the ability of animals to change behavior to fit changing environments.

pheromone: a chemical released by one animal that is used as a scent signal by another animal, usually a conspecific.

phonemes: a basic unit of speech sound that can be combined with other phonemes to create meaningful units (e.g. morphemes).

place cells: pyramidal neurons in the hippocampus that fire when an animal is in a particular location.

place fields: the spatial area that produces the greatest response in a place cell.

playback experiment: an experimental technique in which pre-recorded vocal signals are played back from concealed speakers.

play markers: specific and highly stereotyped behaviors that can serve to initiate and maintain the play context.

point-light displays: visual displays depicting the biological movements of individual joints with single lights, effectively producing an animation of movement.

population coding: a principle of sensory detection describing the relationship between stimulus intensity and the number of sensory neurons firing action potentials; as the stimulus intensity increases, so does the number of firing neurons.

positive reinforcement: a positive relationship between a response and an appetitive stimulus; presentation of the reinforcer increases responding.

precursor behaviors: behaviors observed early in the development of tool use that contain elements of mature tool use.

presynaptic terminal: the swelled tip of axons where neurotransmitter is released when an action potential arrives.

the Prisoner's Dilemma: game analyzed in game theory that has been used to model the evolution of prosocial behavior.

procedural memory: the process underlying a gradual change in behavior based on feedback; a type of non-declarative memory.

prosimians: primates that split from the common ancestor to monkeys, apes, and humans.

prosopagnosia: a deficit in face recognition with no loss in the ability to identify other objects; may be congenital or acquired.

prototype: the typical or most representative member of a category.

proximate cause: a description of a trait or behavior in terms of its development or its mechanisms.

pseudocategory: a random collection of stimuli that have no obvious cohesive feature.

psychophysics: the study of the relationship between sensations and perceptions; the relationship between the physical properties of a stimulus and the interpretation of that stimulus.

punishment: a negative relationship between a response and an appetitive stimulus; presentation of the reinforcer decreases responding.

ratio strain: cessation of operant responding with high FR schedules.

receiver: the recipient of a signal in a communicative interaction.

recursion: when applied to language, refers to the unlimited extension that is made possible by embedding clauses within clauses.

redirected aggression: after being threatened by one individual, an individual may redirect his own threatening behaviors to an individual related to the one who originally threatened him.

reference memory: long-term memory; involves encoding, consolidation, and retrieval of information in long-term store.

reinforcement schedule: in operant conditioning, the relationship between responding and the rate of reinforcement delivery.

relational categorization: the process of grouping stimuli together based on the same connection between two or more items; also called rule-based learning.

relative numerosity: an approximate quantity in relation to another quantity.

releaser: see **sign stimulus**.

reproductive fitness: the contribution to the gene pool that is made by an individual (see also **inclusive fitness**).

response renewal (reacquisition): recovery of a CR following extinction when extinction is conducted in a novel environment.

retrieval: the third stage of memory processing in which stored information is accessed.

risk sensitivity: variable responses to choice situations involving unpredictable outcomes across time.

rotational bias: a tendency to use geometric relationships in small-scale navigation that leads to systematic errors in locating a goal when environmental cues are rotated.

scalar property: a term used to describe situations in which the variability of responses is proportional to the length of the interval being timed.

search image: a mental representation of a target; foraging animals scan the environment for stimuli that match the search image.

selective attention: the ability to attend to a limited range of sensory information while actively inhibiting competing input.

self-control: preference for a large delayed reward over a small immediate reward.

semantic memory: a subcategory of declarative memory describing general knowledge of the world.

sender: the producer of a signal in a communicative interaction.

sensation: the activation of sensory receptors by physical stimuli in the environment.

sensitization: increased responding to repeated presentation of a stimulus that is often discussed in terms of reflex modification; a type of non-declarative memory.

sensitive period: a period in which experience-dependent manipulations can have profound effects on development.

sensory bias: enhanced responding to sensory stimuli that are not part of an animal's natural environment, and were not part of the natural environment of their ancestors.

sensory drive hypothesis: a theory to explain ecological divergence within a species. When populations occupy new habitats with different sensory environments, natural selection favors adaptations that maximize the effectiveness of intra- and inter-species communication.

sensory exploitation: an explanation for sensory bias; a sensory signal that has an evolutionary advantage in one context has been co-opted in another.

sensory preconditioning: a three-stage experiment: 1. Two neutral stimuli (A and B) are presented together; 2. one stimulus (A) is paired with a US; 3. the other stimulus (B) is tested for conditioned responses.

sensory threshold: the minimum level of stimulation that can be detected by each sensory system.

set size signature: a characteristic of small number discrimination such that performance is determined by the size of the arrays.

sexual dimorphism: a systematic difference in a physical trait or behavior between males and females of the same species.

signals: information sent by a sender to a receiver in a communicative interaction.
signature whistle: individual-specific whistles of dolphins.
sign stimulus: the essential feature of a stimulus that is necessary to elicit a response; sometimes called a releaser.
situational freedom: inherent to human language, a speaker can refer to objects, people, or events that have occurred in the past, are occurring now, or will occur in the future.
social categorization: the process of classifying organisms into groups that include conspecifics versus heterospecifics, kin versus non-kin, and familiar versus unfamiliar conspecifics.
social facilitation: occurs when the mere presence of a conspecific can increase or decrease an individual's arousal, regardless of the conspecific's behavior.
social knowledge: the recognition of individuals and their relationships to one another.
social tolerance: the ability to accept the presence of conspecifics in, for example, a feeding situation.
soma: cell body that contains the nucleus and other structures that aid in cellular metabolism.
somatic marker hypothesis: the idea that decision making is informed by bodily reactions that are triggered by emotions.
song: species-typical, musical vocalizations that are primarily produced during the mating season by males of the order *Passeriformes*.
speciation: the process by which new species evolve, usually as the result of isolation from the main population.
spontaneous recovery: the reappearance of a CR following extinction.
stimulus enhancement: a type of social learning in which one's attention is drawn to an object or event as a result of the actions (past or present) of others.
stimulus filtering: the process of separating and extracting meaningful (i.e. biologically relevant) information from the abundance and diversity of sensory cues in the environment.
strabismus: a condition in which the visual axes of the two eyes are misaligned; disrupts depth perception and binocular vision.
the Strange Situation Test: a procedure used to test attachment between an infant and a caregiver.
structuralism: the idea that perceptions are built from the elements of sensations.
supernormal stimuli: stimuli with exaggerated naturalistic features that elicit greater responses than the natural stimuli on which they are modeled.
suppression ratio: the dependent measure in a classical conditioning test of suppression of lever pressing; calculated as (lever presses during the CS)/(lever presses during the CS plus lever presses during an equal period of time preceding the CS).
sustained attention: the ability to focus on one aspect of the environment for extended periods of time.
sympathetic concern: de Waal's second level of empathy in which emotional contagion is combined with appraisal of the current context as well as attempts to understand the underlying cause of the other's distress.
sympathy: feelings of concern for others stemming from empathy.
synapse: a narrow gap between the presynaptic terminal and the cell surface of another neuron.
taxis: directional movement in response to a stimulus.
temporal discounting: the tendency for future rewards to decline in value.
temporal generalization: when responses continue for durations that are different yet close to a trained, target duration.
teratology: the study of abnormal development, specifically defects that are present at birth; from the Greek roots teratos (monster) and ology (to study).
theory of mind: the ability to recognize that those around you have minds and mental states, like intentions and beliefs, which guide their behavior.
third-party relationship: the relationship between other individuals, in which the self is not a principal party.
tool manufacture: modifying objects in the environment to produce tools.
tragedy of the commons: a term referring to situations in which the outcome of the cooperation is publicly available in the sense that individuals who did not take part in the cooperative activities can access the benefits.
transduction: the transfer of physical events in the environment (sound or light waves, pressure on the skin, etc.) to electrical signals in the central nervous system; a build up of electrical signals causes an action potential in the sensory nerves.
trap-tube task: an experimental procedure used to examine the ability to understand the causal relationships among a probing tool, a tube, and the food contained within the tube.
triangulation: a term used to characterize experimental procedures that examine transfer of understanding from one task to another, implying some abstraction of the similar causal structure of the two tasks.
two-action test: an experimental procedure developed to study imitation while controlling for emulation by having demonstrators use one of two different actions to achieve an outcome and examining the subsequent action used by their observers.

Ultimatum Game: an economic game in which a player must choose how much money to give to a second player, who can then choose to accept or reject the offer.

unconditioned response (UR): in classical conditioning, a response that is elicited prior to conditioning.

unconditioned stimulus (US): in classical conditioning, a stimulus that has motivational significance prior to conditioning.

undermatching: a deviation from the matching law in which subjects are less sensitive to reinforcement payoffs than predicted; in concurrent choice paradigms, the relative rate of responding on the lower payoff alternative is higher than the relative payoff provided by that alternative.

ultimate cause: a description of a trait or behavior in terms of its adaptive value.

utility: the subjective value of an outcome.

variable interval (VI): a reinforcement schedule in which reinforcement is delivered following the first responses that occur after an average time interval has elapsed.

variable ratio (VR): a reinforcement schedule in which an average number of responses produces the reinforcer.

vestibular system: sense organs in the inner ear that monitor body rotation and movement in relation to gravity.

vigilance task: a measure of sustained attention in which subjects are required to monitor a particular location and respond when a stimulus is presented in this area.

waggle dance: a movement performed by bees returning to the hive after finding nectar; the movement conveys information as to the direction and distance to the food source.

working memory: the process of maintaining information in short-term store so that it can be used in other cognitive processes.

Zugunruhe: migratory restlessness.

REFERENCES

Abbott, K. R., & Sherratt, T. N. (2011). The evolution of superstition through optimal use of incomplete information. *Anim Behav*, **82**, 85–92.

Ackerman, P. L., Beier, M. E., & Boyle, M. O. (2005). Working memory and intelligence: The same or different constructs? *Psychol Bull*, **131**, 30–60.

Adams, C. D., & Dickinson, A. (1981a). Actions and habits: Variations in associative representations during instrumental learning. In N. E. Spear & R. R. Miller (eds.). *Information Processing in Animals: Memory Mechanisms* (pp. 143–165). Hillsdale, NJ: Lawrence Erlbaum Associates.

Adams, C. D., & Dickinson, A. (1981b). Instrumental responding following reinforcer devaluation. *Q J Exp Psychol*, **34B**, 77–98.

Ader, R., & Cohen, N. (1975). Behaviorally conditioned immunosuppression. *Psychosom Med*, **37**, 333–340.

Ader, R., Felten, D., & Cohen, N. (1990). Interactions between the brain and the immune system. *Ann Rev Pharmacol Toxicol*, **30**, 561–602.

Ainslie, G. (1974). Impulse control in pigeons. *J Exp Anal Behav*, **21**, 485–489.

Ainsworth, M. D., Blehar, M. C., Waters, E., & Wall, S. (1978). *Patterns of Attachment: A Psychological Study of the Strange Situation*. Oxford, UK: Lawrence Erlbaum.

Akins, C. K. (2004). The role of Pavlovian conditioning in sexual behavior: A comparative analysis of human and nonhuman animals. *Int J Comp Psychol*, **17**, 241–262.

Akins, C. K., Levens, N., & Bakondy, H. (2002). The role of static features of males in the mate choice behavior of female Japanese quail (*Coturnix japonica*). *Behav Process*, **58**, 97–103.

Akins, C. K., & Zentall, T. R. (1996). Imitative learning in male Japanese quail (*Coturnix japonica*) using the two-action method. *J Comp Psychol*, **110**, 316–320.

Alcock, J. (1972). The evolution of the use of tools by feeding animals. *Evolution*, **26**, 464–473.

Alerstam, T. (2006). Conflicting evidence about long-distance animal navigation. *Science*, **313**, 791–794.

Alexander, B. K., Coambs, R. B., & Hadaway, P. F. (1978). The effect of housing and gender on morphine self-administration in rats. *Psychopharmacology*, **58**, 175–179.

Allan, L. G. (1980). A note on measurement of contingency between two binary variables in judgment tasks. *Bull Psychonom Soc*, **15**, 147–149

Allan, L.G. (2005). Learning of contingent relationships [Special issue]. *Anim Learn Behav*, **33**, 127–129.

Almeida, L. G., Ricardo-Garcell, J., Prado, H., Barajas, L., Fernández-Bouzas, A., Avila, D., & Martinez, R. B. (2010). Reduced right frontal cortical thickness in children, adolescents and adults with ADHD and its correlation to clinical variables: A cross-sectional study. *J Psychiatr Res*, **44**, 1214–1223.

Altmann, S. A. (1962). Social behavior of anthropoid primates: Analysis of recent concepts. In E. L. Bliss (ed.). *Roots of Behavior* (pp. 277–285). New York, NY: Harper.

Alvarez-Borda, B., Ramírez-Amaya, V., Pérez-Montfort, R., & Bermúdez-Rattoni, F. (1995). Enhancement of antibody production by a learning paradigm. *Neurobiol Learn Mem*, **64**, 103–105.

Alvernhe, A., Sargollini, F., & Poucet, B. (2012). Rats build and update topographical representations through exploration. *Anim Cogn*, **15**, 359–368.

Amos, B., Twiss, S., Pomeroy, P., & Anderson, S. (1995). Evidence for mate fidelity in the gray seal. *Science*, **268**, 1897–1899.

Anderson, J. R., Myowa-Yamakoshi, M., & Matsuzawa, T. (2004). Contagious yawning in chimpanzees. *Proc R Soc Lond B Biol Sci*, **271**, 468–470.

Anderson, J. R., & Schooler, L. J. (1991). Reflections of the environment in memory. *Psychol Sci*, **2**, 396–408.

Ando, J., Ono, Y., & Wright, M. J. (2001). Genetic structure of spatial and verbal working memory. *Behav Genet*, **31**, 615–624.

Angie, A. D., Connelly, S., Waples, E. P., & Kligyte, V. (2013). The influence of discrete emotions on judgement and decision-making: A meta-analytic review. *Cogn Emotion*, **25**, 1393–1422.

Apperly, I. A., & Butterfill, S. A. (2009). Do humans have two systems to track beliefs and belief-like states? *Psychol Rev*, **116**, 953–970.

Arkes, H. R., & Ayton, P. (1999). The sunk cost and Concorde effects: Are humans less rational than lower animals? *Psychol Bull*, **15**, 591–600.

Ashcraft, M. H., & Klein, R. (2010). *Cognition*. Toronto: Pearson Education Canada.

Atkinson, A. P., & Adolphs, R. (2011). The neuropsychology of face perception: Beyond simple dissociations and functional selectivity. *Phil Trans R Soc Lond B*, **366**, 1726–1738.

Atkinson, R. C., & Shiffrin R. M. (1968). Human memory: A proposed system and its control processes. In W. K. Spence & J. T. Spence (eds.). *The Psychology of Learning and Motivation: Advances in Research and Theory* (Vol. 2) (pp. 89–195). New York, NY: Academic Press.

Auyeung, B., Baron-Cohen, S., Ashwin, E., Knickmeyer, R., Taylor, K., Hackett, G., & Hines, M. (2009). Fetal testosterone predicts sexually differentiated childhood behavior in girls and in boys. *Psychol Sci*, **20**, 144–148.

Avarguès-Weber, A., Deisig, N., & Giurfa, M. (2011). Visual cognition in social insects. *Ann Rev Entomol*, **56**, 423–443.

Baddley, A. D., & Hitch, G. (1974). Working memory. In G. H. Bower (ed.). *The Psychology of Learning and Motivation* (Vol. 8) (pp. 47–89). New York, NY: Academic Press.

Baharloo, S., Johnston, P., Service, S., Gitschier, J., & Freimer, N. (1998). Absolute pitch: An approach for identification of genetic and nongenetic components. *Am J Hum Genet*, **62**, 224–231.

Bailey, K., West, R., & Anderson, C. A. (2010). A negative association between video game experience and proactive cognitive control. *Psychophysiology*, **47**, 34–42.

Baillargeon, R. (1995). Physical reasoning in infancy. In M. S. Gazzaniga (ed.). *The Cognitive Neurosciences* (pp. 181–204). Cambridge, MA: MIT Press.

Baillargeon, R., Kotovsky, L., & Needham, A. (1995). The acquisition of physical knowledge in infancy. In A. J. Premack, D. Premack, & D. Sperber (eds.). *Causal Cognition: A Multidisciplinary Debate* (pp. 79–116). Oxford, UK: Clarendon Press.

Balda, R. P., & Kamil, A. C. (1989). A comparative study of cache recovery by three corvid species. *Anim Behav*, **38**, 486–495.

Balda, R. P., & Kamil, A. C. (2006). The ecology and life history of seed caching corvids. In M. F. Brown & R. G. Cook (eds.). *Animal Spatial Cognition: Comparative, Neural, and Computational Approaches* [On-line]. Available at: www.pigeon.psy.tufts.edu/asc/balda/

Baldwin, D. A. (1993). Early referential understanding: Infants' ability to understand referential acts for what they are. *Dev Psychol*, **29**, 832–843.

Baldwin, D. A., Baird, J. A., Saylor, M. M., & Clark, M. A. (2001). Infants parse dynamic action. *Child Dev*, **72**, 708–717.

Balleine, B. W., Delgado, M. R., & Hikosaka, O. (2007). The role of the dorsal striatum in reward and decision-making. *J Neurosci*, **27**, 8161–8165.

Balleine, B. W., & Ostlund, S. B. (2007). Still at the choice point: Action selection and initiation in instrumental conditioning. *Ann N Y Acad Sci*, **1104**, 147–171.

Bannerman, D. M., Good, M. A., Butcher, S. P., Ramsay, M., & Morris, R. G. M. (1995). Distinct components of spatial learning revealed by prior training and NMDA receptor blockade. *Nature*, **9**, 182–186.

Barber, J. R., & Conner, W. E. (2007). Acoustic mimicry in a predator–prey interaction. *Proc Natl Acad Sci USA*, **104**, 9331–9334.

Barnard, C. (2004). *Animal Behaviour: Mechanisms, Development, Function and Evolution*. Harlow: Pearson Education Limited.

Barner, D., Wood, J., Hauser, M., & Carey, S. (2008). Evidence for a non-linguistic distinction between singular and plural sets in rhesus monkeys. *Cognition*, **107**, 603–622.

Baron-Cohen, S. (1985). *Mindblindness*. Cambridge, MA: MIT Press.

Bartal, I.B., Decety, J., & Mason, P. (2011). Empathy and prosocial behavior in rats. *Science*, **334**, 1427–1430.

Barth, J., & Call, J. (2006). Tracking the displacement of objects: A series of tasks with great apes (*Pan troglodytes*, *Pan paniscus*, *Gorilla gorilla*, and *Pongo pygmaeus*) and young children (*Homo sapiens*). *J Exp Psychol Anim Behav Process*, **32**, 239–252.

Barth, J., Reaux, J. E., & Povinelli, D. J. (2005). Chimpanzees' (*Pan troglodytes*) use of gaze cues in object-choice tasks: Different methods yield different results. *Anim Cogn*, **8**, 84–92.

Bartlett, F. C. (1932). *Remembering: A Study in Experimental and Social Psychology*. London, UK: Cambridge University Press.

Basolo, A. L. (1990). Female preference predates the evolution of the sword in swordtail fish. *Science*, **250**, 808–810.

Bateson, G. (1955). A theory of play and fantasy. *Psychiatr Res Rep*, **2**, 39–51.

Baumard, N., André, J. B., & Sperber, D. (2013). A mutualistic approach to morality: The evolution of fairness by partner choice. *Behav Brain Sci*, **36**, 59–122.

Baxter, D. A., & Byrne, J. H. (2006). Feeding behavior of *Aplysia*: A model system for comparing cellular mechanisms of classical and operant conditioning. *Learn Mem*, **13**, 669–680.

Beatty, W. W., & Shavalia, D. A. (1980). Spatial memory in rats: Time course of working memory and effects of anesthetics. *Behav Neural Biol*, **28**, 454–462.

Beck, B. B. (1980). *Animal Tool Behavior: The Use and Manufacture of Tools by Animals*. New York, NY: Garland STPM Press.

Bee, M. A., & Gerhardt, H. C. (2002). Individual voice recognition in a territorial frog (*Rana atesbeiana*). *Proc R Soc Lond B*, **269**, 1443–1448.

Behrmann, M., & Avidan, G. (2005). Congenital prosopagnosia: Face-blind from birth. *Trends Cogn Neurosci*, **9**, 180–187.

Bekoff, M. (1995). Play signals as punctuation: The structure of social play in canids. *Behaviour*, **132**, 5–6.

Bekoff, M., & Byers, J. A. (1981). A critical reanalysis of the ontogeny and phylogeny of mammalian social and locomotor play: An ethological hornet's nest. In K. Immelmann, G. W. Barlow, L. Petrinovich, & M. Main (eds.). *Behavioral Development* (pp. 296–337). London, UK: Cambridge University Press.

Benhamou, S. (2001). Orientation and movement patterns of the wood mouse (*Apodemus sylvaticus*) in its home range are not altered by olfactory or visual deprivation. *J Comp Physiol A*, **187**, 243–248.

Benhamou, S. (2010). Orientation and navigation. *Encyclopedia Behav Neurosci*, **2**, 497–503.

Benhamou, S., Sudre, J., Bourjea, J., Ciccione, S., De Santis, A., & Luschi, P. (2011). The role of geomagnetic cues in green turtle open sea navigation. *PLoS One*, **6**, e26672.

Benzer, S. (1973). Genetic dissection of behavior. *Sci Am*, **229**, 24–37.

Beran, M.J. (2001). Summation and numerousness judgments of sequentially presented sets of items by chimpanzees (*Pan troglodytes*). *J Comp Psychol*, **115**, 181–191.

Beran, M. J., & Beran, M. M. (2004). Chimpanzees remember the results of one-by-one addition of food items to sets over extended time periods. *Psychol Sci*, **15**, 94–99.

Beran, M. J., & Minahan, M. F. (2000). Monitoring spatial transpositions by bonobos (*Pan paniscus*) and chimpanzees (*P. troglodytes*). *Int J Comp Psychol*, **13**, 1–15.

Berger, J. (1980). The ecology, structure and function of social play in Bighorn sheep (*Ovis canadensis*). *J Zool*, **192**, 531–542.

Bergerud, A. T., Luttich, S. N., & Camps, L. (2012). *The Return of the Caribou to Ungava*. Montreal, CA: McGill-Queen's University Press.

Bergman, T.J. (2013). Speech-like vocalized lip-smacking in geladas. *Curr Biol*, **23**, R268-R269.

Bertenthal, B. I., Proffitt, D. R., & Cutting, J. E. (1984). Infant sensitivity to figural coherence in biomechanical motions. *J Exp Child Psychol*, **37**, 213–230.

Berthold, P. (1998). Bird migration: Genetic programs with high adaptability. *Zoology*, **101**, 235–245.

Berthold, P., Helbig, A. J., Mohr, G., & Querner, U. (1992). Rapid microevolution of migratory behaviour in a wild bird species. *Nature*, **360**, 668–670.

Berthold, P., & Querner, U. (1981). Genetic basis of migratory behavior in European warblers. *Science*, **212**, 77–79.

Bertram, B. C. R. (1980). Vigilance and group size in ostriches. *Anim Behav*, **28**, 278–286.

Bettis, T. J., & Jacobs, L. F. (2009). Sex-specific strategies in spatial orientation in C57BL/6J mice. *Behav Process*, **82**, 249–255.

Bhatt, R. S., Wasserman, E. A., Reynolds, W. F., & Knauss, K. S. (1988). Conceptual behavior in pigeons: Categorization of both familiar and novel examples from four classes of natural and artificial stimuli. *J Exp Psychol Anim Behav Process*, **14**, 219–234.

Biebach, H., Falk, H., & Krebs, J. R. (1991). The effect of constant light and phase shifts on a learned time-place association in garden warblers (*Sylvia borin*): Hourglass or circadian clock? *J Biol Rhythms*, **6**, 353–365.

Biebach, H., Gordijn, M., & Krebs, J. R. (1989). Time-and-place learning by garden warblers, *Sylvia borin*. *Anim Behav*, **37**, 353–360.

Billington, E. J., & DiTommaso, N. M. (2003). Demonstrations and applications of the matching law in education. *J Behav Education*, **12**, 91–104.

Bingman, V. P., Bagnoli, P., Ioalé, P., & Casini, G. (1984). Homing behavior of pigeons after telencephalic ablations. *Brain Behav Evol*, **24**, 94–106.

Bingman V. P., Jechura, T., & Kahn, M. C. (2006). Behavioral and neural mechanisms of homing and migration in birds. In M. F. Brown & R. G. Cook (eds.). *Animal Spatial Cognition: Comparative, Neural, and Computational Approaches*. [On-line]. Available at: www.pigeon.psy.tufts.edu/asc/bingman/

Bingman, V. P., & Sharp, P. E. (2006). Neuronal implementation of hippocampal-mediated spatial behavior: A comparative evolutionary perspective. *Behav Cogn Neurosci Rev*, **5**, 80–91.

Bird, G., & Heyes, C. (2005). Effector-dependent learning by observation of a finger movement sequence. *J Exp Psychol Hum Percept Perform*, **31**, 262–275.

Bird, G., Osman, M., Saggerson, A., & Heyes, C. (2005). Sequence learning by action, observation and action observation. *Br J Psychol*, **96**, 371–388.

Biro, D., Sumpter, D. J., Meade, J., & Guilford, T. (2006). From compromise to leadership in pigeon homing. *Curr Biol*, **16**, 2123–2128.

Bíró, S., Koós, O., Gergely, G., & Csibra, G. (1997). Understanding rational action in infancy. *Psychol Lang Commun*, **1**, 29–37.

Bitterman, M. E. (1965). Phyletic differences in learning. *Am Psychol*, **20**, 396–410.

Bitterman, M. E. (1996). Comparative analysis of learning in honeybees. *Anim Learn Behav*, **24**, 123–141.

Bjorklund, D. F., & Blasi, C. H. (2012). *Child and Adolescent Development: An Integrated Approach*. Belmont, CA: Wadsworth, Cengage Learning.

Blaisdell, A. P., Sawa, K., Leising, K. J., & Waldmann, M. R. (2006). Causal reasoning in rats. *Science*, **311**, 1020–1022.

Blakemore, C. (1976). The conditions required for the maintenance of binocularity in the kitten's visual cortex. *J Physiol (Lond)*, **261**, 423–444.

Blakemore, S.-J., & Robbins, T. W. (2012). Decision-making in the adolescent brain. *Nat Neurosci*, **15**, 1184–1191.

Blanchard, R. J., & Blanchard, D. C. (1969). Crouching as an index of fear. *J Comp Physiol Psychol*, **67**, 370–375.

Blanchard, R. J., Blanchard, D. C., Weiss, S. M., & Meyer, S. (1990). The effects of ethanol and diazepam on reactions to predatory odors. *Pharmacol Biochem Behav*, **35**, 775–780.

Bliss, T. V. P., & Lomo, T. (1973). Long-lasting potentiation of synaptic transmission in the dentate area of the anesthetized rabbit following stimulation of the perforant path. *J Physiol*, **232**, 331–356.

Bloom, P. (2004). Can a dog learn a word? *Science*, **304**, 1605–1606.

Blough, D. S. (1956). Dark adaptation in the pigeon. *J Comp Physiol Psychol*, **49**, 425–430.

Bluff, L. A., Weir, A. A., Rutz, C., Wimpenny, J. H., & Kacelnik, A. (2007). Tool-related cognition in new Caledonian crows. *Comp Cogn Behav Rev*, **2**, 1–25.

Blumberg, M.S., & Sokoloff, G. (2001). Do infant rats cry? *Psychol Rev*, **108**, 83–95.

Boccia, M. M., Blake, M. G., Acosta, G. B., & Baratti, C. M. (2005). Memory consolidation and reconsolidation of an inhibitory avoidance task in mice: Effects of a new different learning task. *Neuroscience*, **135**, 19–29.

Boesch, C. (2002). Cooperative hunting roles among Tai chimpanzees. *Hum Nature*, **13**, 27–46.

Boesch, C., & Boesch, H. (1989). Hunting behavior of wild chimpanzees in the Tai National Park. *Am J Phys Anthropol*, **78**, 547–573.

Boinski, S., Kauffman, L., Westoll, A., Stickler, C. M., Cropp, S., & Ehmke, E. (2003). Are vigilance, risk from avian predators and group size consequences of habitat structure? A comparison of three species of squirrel monkey (*Saimiri oerstedii*, *S. boliviensis*, and *S. sciureus*). *Behaviour*, **140**, 1421–1467.

Boisvert, M. J., & Sherry, D. F. (2006). Interval timing by an invertebrate, the bumble bee *Bombus impatiens*. *Curr Biol*, **16**, 1636–1640.

Bolhuis, J. J., & Giraldeau, L. A. (2005). *The Behavior of Animals: Mechanisms, Function, and Evolution*. Oxford: Blackwell Publishing Ltd.

Boles, L. C., & Lohmann, K. J. (2003). True navigation and magnetic maps in spiny lobsters. *Nature*, **421**, 60–63.

Bolles, R. C. (1970). Species-specific defense reactions and avoidance learning. *Psychol Rev*, **71**, 32–48.

Bonabeau, E., Cabibbo, N., Candelier, R., *et al*. (2008). Interaction ruling collective animal behavior depends on topological rather than metric distance: Evidence from a field study. *Proc Nat Acad Sci USA*, **105**, 1232–1237.

Bonadonna, F., & Nevitt, G. A. (2004). Partner-specific odor recognition in an Antarctic seabird. *Science*, **306**, 835.

Borrero, J. C., Crisolol, S. S., Tu, Q., *et al*. (2007). An application of the matching law to social dynamics. *J Appl Behavr Anal*, **40**, 589–601.

Boughman, J. W. (2002). How sensory drive can promote speciation. *Trends Ecol Evol*, **17**, 571–577.

Boughman, J.W., & Moss, C. F. (2003). Social sounds: Vocal learning and development of mammal and bird calls. In A.M. Simmons, R.R. Fay, & A.N. Popper (eds.). *Springer Handbook of Auditory Research: Acoustic Communication* (Vol. 16) (pp. 138–224). New York, NY: Springer.

Bouton, M.E. (1994). Conditioning, remembering, and forgetting. *J Exp Psychol Anim Behav Process*, **20**, 219–231.

Bouton, M.E., & Moody, E.W. (2004). Memory processes in classical conditioning. *Neurosci Biobehav Rev*, **28**, 663–674.

Bovet, D., & Vauclair, J. (1998). Functional categorization of objects and their pictures in baboons (*Papio anubis*). *Learn Motiv*, **29**, 309–322.

Bowlby, J. (1969). *Attachment and loss: Attachment* (Vol. 1). London, UK: Hogarth.

Boyd, R., & Richerson, P. J. (1985). *Culture and the Evolutionary Process*. Chicago, IL: University of Chicago Press.

Boyd, R, & Richerson, P. J. (1988). The evolution of reciprocity in sizable groups. *J Theor Biol*, **132**, 337–356.

Boysen, S.T. (1993). Counting in chimpanzees: Nonhuman principles and emergent properties of number. In S.T. Boysen & E.J. Capaldi (eds.). *The Development of Numerical Competence: Animal and Human Models. Comparative Cognition and Neuroscience* (pp. 39–59). Hillsdale, NJ: Lawrence Erlbaum Associates, Inc.

Boysen, S.T., & Berntson, G. G. (1989). Numerical competence in a chimpanzee (*Pan troglodytes*). *J Comp Psychol*, **103**, 23–31.

Boysen, S.T., & Berntson, G.G. (1995). Responses to quantity: Perceptual versus cognitive mechanisms in chimpanzees (*Pan troglodytes*). *J Exp Psychol Anim Behav Process*, **21**, 82–86.

Boysen, S. T., & Capaldi, E. J. (1993). *The Development of Numerical Competence: Animal and Human Models*. Hillsdale, NJ: Erlbaum.

Bradbury, J. W., & Gibson, R. M. (1983). Leks and mate choice. In P. Bateson (ed.). *Mate Choice*. Cambridge, UK: Cambridge University Press.

Bradbury, J.W., & Vehrenkamp, S.L. (1998). *Principles of Animal Communication*. Sunderland, MA: Sinauer Associates.

Brain, P. F., Kusumorini, N., & Benton, D. (1991). 'Anxiety' in laboratory rodents: A brief review of some recent behavioural developments. *Behav Process*, **25**, 71–80.

Brainard, M. S., & Doupe, A. J. (2002). What songbirds teach us about learning. *Nature*, **417**, 351–358.

Brannon, E.M., Abbott, S., & Lutz, D. J. (2004). Number bias for the discrimination of large visual sets in infancy. *Cognition*, **93**, B59-B68.

Brannon, E.M., Suanda, S., & Libertus, K. (2007). Temporal discrimination increases in precision over development and parallels the development of numerosity discrimination. *Dev Sci*, **10**, 770–777.

Brannon, E.M., & Terrace, H.S. (1998). Ordering of the numerosities 1 to 9 by monkeys. *Science*, **282**, 746–749.

Brannon, E.M., & Terrace, H.S. (2000). Representation of the numerosities 1–9 by rhesus macaques (*Macaca mulatta*). *J Exp Psychol Anim Behav Process*, **26**, 31–49.

Brannon, E.M., Wusthoff, C.J., Gallistel, C.R., & Gibbon, J. (2001). Numerical subtraction in the pigeon: Evidence for a linear subjective number scale. *Psychol Sci*, **12**, 238–243.

Brasil-Neto, J.P., Pascual-Leone, A., Valls-Solé, J., Cohen, L.G., & Hallett, M. (1992). Focal transcranial magnetic stimulation and response bias in a forced-choice task. *J Neurol Neurosurg Psychiatry*, **55**, 964–966.

Breen, R.B., & Zuckerman, M. (1999). 'Chasing' in gambling behaviour: Personality and cognitive determinants. *Pers Indiv Diff*, **27**, 1097–1111.

Breiter, H.C., Aharon, I., Kahneman, D., Dale, A., & Shizgal, P. (2001). Functional imaging of neural responses to expectancy and experience of monetary gains and losses. *Neuron*, **30**, 619–639.

Breland, K., & Breland, M. (1961). The misbehavior of organisms. *Am Psychol*, **16**, 681–175.

Brembs, B. (2011). Towards a scientific concept of free will as a biological trait: Spontaneous actions and decision-making in invertebrates. *Proc Roy Soc Lond B*, **278**, 930–939.

Brembs, B., & Plendl, W. (2008). Double dissociation of PKC and AC manipulations on operant and classical conditioning in *Aplysia*. *Curr Biol*, **18**, 1168–1171.

Brodbeck, D.R. (1994). Memory for spatial and local cues: A comparison of a storing and a nonstoring species. *Anim Learn Behav*, **22**, 119–133.

Brodbeck, D.R. (1997). Picture fragment completion: Priming in the pigeon. *J Exp Psychol Anim Behav Process*, **23**, 461–468.

Broesch, T., Callaghan, T., Henrich, J., Murphy, C., & Rochat, P. (2011). Cultural variations in children's mirror self-recognition. *J Cross Cult Psychol*, **42**, 1018–1029.

Brosch, T., Scherer, K.R., Grandjean, D., & Sander, D. (2013). The impact of emotion on perception, attention, memory, and decision-making. *Swiss Med Weekly*, **143**, w13786.

Brosnan, S. F., & de Waal, F. B. M. (2003). Monkeys reject unequal pay. *Nature*, **425**, 297–299.

Brosnan, S. F., & de Waal, F. B. M. (2004). Animal behaviour: Fair refusal by capuchin monkeys. *Nature*, **428**, 140.

Brosnan, S. F., Schiff, H. C., & de Waal, F. B. M. (2005). Tolerance for inequity may increase with social closeness in chimpanzees. *Proc R Soc B Biol Sci*, **272**, 253–258.

Brown, A.A., Spetch, M.L., & Hurd, P.L. (2007). Growing in circles: Rearing environment alters spatial navigation in fish. *Psychol Sci*, **18**, 569–573.

Brown, C., & Laland, K. N. (2003). Social learning in fishes: A review. *Fish Fisheries*, **4**, 280–288.

Brown, G.E., Brown, J.A., & Crosbie, A.M. (1993). Phenotypic matching in juvenile rainbow trout. *Anim Behav*, **46**, 1223–1225.

Brown, M. F., & Demas, G. E. (1994). Evidence for spatial working memory in honeybees (*Apis mellifera*). *J Comp Psychol*, **108**, 344–352.

Brown, R., & Kulik, J. (1977). Flashbulb memories. *Cognition*, **5**, 73–99.

Bruner, J.S., Goodnow, J. J, & Austin, G.A. (1956). *A Study of Thinking*. New York, NY: Wiley.

Buckwalter, J.G., Stanczyk, F.Z., McCleary, C.A., *et al.* (1999). Pregnancy, the postpartum, and steroid hormones: Effects on cognition and mood. *Psychoneuroendocrinology*, **24**, 69–84.

Bugnyar, T., & Heinrich, B. (2005). Ravens, *Corvus corax*, differentiate between knowledgeable and ignorant competitors. *Proc R Soc B Biol Sci*, **272**, 1641–1646.

Bugnyar, T., Kijne, M., & Kotrschal, K. (2001). Food calling in ravens: Are yells referential signals? *Anim Behav*, **61**, 949–958.

Buhusi, C. V., & Meck, W. H. (2005). What makes us tick? Functional and neural mechanisms of interval timing. *Nat Rev Neurosci*, **6**, 755–765.

Bulow, P. J., & Meller, P. J. (1998). Predicting teenage girls' sexual activity and contraception use: An application of the matching law. *J Comm Psychol*, **26**, 581–596.

Burgess, N. (2006). Spatial memory: How egocentric and allocentric combine. *Trends Cogn Sci*, **10**, 551–557.

Burkart, J. M., Fehr, E., Efferson, C., & van Schaik, C. P. (2007). Other-regarding preferences in a non-human primate: Common marmosets provision food altruistically. *Proc Natl Acad Sci USA*, **104**, 19762–19766.

Burley, N. T., & Symanski, R. (1998). "A taste for the beautiful": Latent aesthetic mate preferences for white crests in two species of Australian grassfinches. *Am Nat*, **152**, 792–802.

Bushnell, I. W. R., Sai, R., & Mulin, J. T. (2011). Neonatal recognition of the mother's face. *Dev Psychol*, **7**, 3–15.

Buske-Kirschbaum, A., Kirschbaum, C., Stierle, H., Jabaij, L., & Hellhammer, D. (1994). Conditioned manipulation of natural killer (NK) cells in humans using a discriminative learning protocol. *Biol Psychol*, **38**, 143–155.

Buss, D. M. (2004). *Evolutionary Psychology.* Boston, MA: Pearson.

Buttelmann, D., Carpenter, M., Call, J., & Tomasello, M. (2007). Enculturated chimpanzees imitate rationally. *Dev Sci*, **10**, F31–F38.

Butterworth, B. (2005). The development of arithmetical abilities. *J Child Psychol Psychiatry*, **46**, 3–18.

Buzsáki, G., & Draguhn, A. (2004). Neuronal oscillations in cortical networks. *Science*, **304**, 1926–1929.

Byrne, R.W. (1995). *The Thinking Ape: Evolutionary Origins of Intelligence.* Oxford, UK: Oxford University Press.

Byrne, R. W. (1998). Comment on chimpanzee and human cultures. *Curr Anthropol*, **39**, 591–614.

Byrne, R. W. (2002). Imitation of novel complex actions: What does the evidence from animals mean? *Adv Study Behav*, **31**, 77–105.

Cabrera, F., Sanabria, F., Shelley, D., & Killeen, P. R. (2009). The "lunching" effect: Pigeons track motion towards food more than motion away from it. *Behav Process*, **82**, 229–235.

Cacchione, T., & Burkart, J. M. (2012). Dissociation between seeing and acting: insights from common marmosets (*Callithrix jacchus*). *Behav Process*, **89**, 52–60.

Cacchione, T., & Call, J. (2010). Intuitions about gravity and solidity in great apes: The tubes task. *Dev Sci*, **13**, 320–330.

Cacchione, T., Call, J., & Zingg, R. (2009). Gravity and solidity in four great ape species (*Gorilla gorilla*, *Pongo pygmaeus*, *Pan troglodytes*, *Pan paniscus*): Vertical and horizontal variations of the table task. *J Comp Psychol*, **123**, 168–180.

Cacchione, T., & Krist, H. (2004). Recognizing impossible object relations: Intuitions about support in chimpanzees (*Pan troglodytes*). *J Comp Psychol*, **118**, 140–148.

Cahill, L., Uncapher, M., Kilpatrick, L., Alkire, M. T., & Turner, J. (2004). Sex-related hemispheric lateralization of amygdala function in emotionally influenced memory: An fMRI investigation. *Learn Mem*, **11**, 261–266.

Cain, S. W., McDonald, R. J., & Ralph, M. R. (2008). Time stamp in conditioning avoidance can be set to different circadian phases. *Neurobiol Learn Mem*, **89**, 591–594.

Caldwell, M.C., & Caldwell, D.K. (1965). Individualized whistle contours in bottlenose dolphins (*Tursiops truncatus*). *Nature*, **207**, 434–435.

Caldwell, R. (1992). Recognition, signaling and reduced aggression between former mates in a stomatopod. *Anim Behav*, **44**, 11–19.

Call, J., Hare, B., Carpenter, M., & Tomasello, M. (2004). 'Unwilling' versus 'unable': Chimpanzees' understanding of intentional human action. *Dev Sci*, **7**, 488–498.

Call, J., & Tomasello, M. (1998). Distinguishing intentional from accidental actions in orangutans (*Pongo pygmaeus*), chimpanzees (*Pan troglodytes*) and human children (*Homo sapiens*). *J Comp Psychol*, **112**, 192–206.

Call, J., & Tomasello, M. (2008). Does the chimpanzee have a theory of mind? 30 years later. *Trends Cogn Sci*, **12**, 187–192.

Campbell, M. W., Carter, J. D., Proctor, D., Eisenberg, M. L., & de Waal, F. B. (2009). Computer animations stimulate contagious yawning in chimpanzees. *Proc R Soc B Biol Sci*, **276**, 4255–4259.

Candland, D. K. (1993). *Feral Children & Clever Animals: Reflections on Human Nature.* New York, NY: Oxford University Press.

Cantlon, J. F. (2012). Math, monkeys, and the developing brain. *Proc Nat Acad Sci USA*, **109**, 10725–10732.

Cantlon, J. F., & Brannon, E. M. (2007). How much does number matter to a monkey (*Macaca mulatta*)? *J Exp Psychol Anim Behav Process*, **33**, 32–41.

Cantlon, J. F. & Brannon, E. M. (2010). Animal arithmetic. In N. Clayton (ed.). *Encyclopedia of Animal Behavior.* Oxford, UK: Elsevier Press.

Cantlon, J. F., Platt, M. L., & Brannon, E. M. (2009). Beyond the number domain. *Trends Cogn Sci*, **13**, 83–91.

Caramazza, A., & Shelton, J. R. (1998). Domain specific knowledge systems in the brain: The animate-inanimate distinction. *J Cogn Neurosci*, **10**, 1–34.

Carey, S. (2004). Bootstrapping & the origin of concepts. *Daedalus*, **133**, 59–68.

Carlson, S. M., Davis, A. C., & Leach, J. G. (2005). Less is more: executive function and symbolic representation in preschool children. *Psychol Sci*, **16**, 609–616.

Caro, T. M., & Hauser, M. D. (1992). Is there teaching in nonhuman animals? *Q Rev Biol*, **67**, 151–174.

Carpenter, M. (2006). Instrumental, social, and shared goals and intentions in imitation. In S. J. Rogers & J. H. G Williams (eds.). *Imitation and the Social Mind: Autism and Typical Development* (pp. 48–70). New York, NY: The Guilford Press.

Carpenter, M., Akhtar, N., & Tomasello, M. (1998). Fourteen-through 18-month-old infants differentially imitate intentional and accidental actions. *Infant Behav Dev*, **21**, 315–330.

Carr, D., Wilkinson, K. M., Blackman, D., & McIlvane, W. J. (2000). Equivalence classes in individuals with minimal verbal repertoires. *J Exp Anal Behav*, **74**, 101–114.

Cartwright, B.A., & Collett, T. S. (1982). How honey bees use landmarks to guide their return to a food source. *Nature*, **295**, 560–564.

Casler, K., & Kelemen, D. (2005). Young children's rapid learning about artifacts. *Dev Sci*, **8**, 472–480.
Casler, K., & Kelemen, D. (2007). Reasoning about artifacts at 24 months: The developing teleo-functional stance. *Cognition*, **103**, 120–130.
Castellanos, F.X., Giedd, J.N., Marsh, W.L., *et al.* (1996). Quantitative brain magnetic resonance imaging in attention-deficit hyperactivity disorder. *Arch Gen Psychiatry*, **53**, 607–616.
Catania, A. C. (1970). Reinforcement schedules and psychophysical judgments: A study of some temporal properties of behavior. In W.N. Schoenfeld (ed.). *The Theory of Reinforcement Schedules* (pp. 1–42). New York, NY: Appleton-Century-Crofts.
Célérier, A., Bon, C., Malapert, A., Palmas, P., & Bonadonna, F. (2011). Chemical kin labels in seabirds. *Biol Lett*, **7**, 807–810.
Chai, X.J., & Jacobs, L.F. (2010). Effects of cue types on sex differences in spatial memory. *Behav Brain Res*, **208**, 336–342.
Chapius, N., & Scardigli, P. (1993). Shortcut ability in hamsters (*Mesocricetus auratus*): The role of environmental and kinesthetic information. *Anim Learn Behav*, **29**, 336–353.
Chapman, B., Morrell, L., Tosh, C., & Krause, J. (2010). Behavioural consequences of sensory plasticity in guppies. *Proc R Soc B Biol Sci*, **277**, 1395–1401.
Chappell, J. (2006). Avian cognition: Understanding tool use. *Curr Biol*, **16**, R244–R245.
Chaput, H.H., & Cohen, L. B. (2001). A model of infant causal perception and its development. In J.D. Moore & K. Stenning (eds.). *Proceedings of the Twenty-Third Annual Conference of the Cognitive Science Society* (pp. 182–187). Mahwah, NJ: Lawrence Erlbaum.
Charmantier, A., Keyser, A. J., & Promislow D. E. L. (2007) First evidence for heritable variation in cooperative breeding behaviour. *Proc R Soc B Biol Sci*, **274**, 1757–1761.
Charnov, E.L. (1976). Optimal foraging: Attack strategy of a mantid. *Am Nat*, **110**, 141–151.
Charnov, E.L. (1976). Optimal foraging, the marginal value theorem. *Theor and Popul Biol*, **9**, 129–136.
Charpentier, M. J. E., Deubel, D., & Peignot, P. (2008). Relatedness and social behaviors in *Cercopithecus solatus*. *Int J Primatol*, **29**, 487–495.
Chartrand, T. L., & Bargh, J. A. (1999). The chameleon effect: The perception-behavior link and social interaction. *J Pers Social Psychol*, **76**, 893–910.
Chartrand, T. L., & Dalton, A. N. (2009). Mimicry: Its ubiquity, importance, and functionality. In E. Morsella, J. A. Bargh, & P. M Gollwitzer (eds.). *Oxford Handbook of Human Action: Social Cognition and Social Neuroscience* (pp. 458–483). New York, NY: Oxford University Press.
Cheney, D. L., & Seyfarth, R. M. (1986). The recognition of social alliances by vervet monkeys. *Anim Behav*, **34**, 1722–1731.
Cheney, D. L., & Seyfarth, R. M. (1988). Assessment of meaning and the detection of unreliable signals by vervet monkeys. *Anim Behav*, **36**, 477–486.
Cheney, D. L., & Seyfarth, R. M. (1989). Redirected aggression and reconciliation among vervet monkeys, *Cercopithecus aethiops*. *Behaviour*, **110**, 258–275.
Cheney, D. L., & Seyfarth, R. M. (1990a). *How Monkeys See the World: Inside the Mind of Another Species*. Chicago, IL: University of Chicago Press.
Cheney, D. L., & Seyfarth, R. M. (1990b). The representation of social relations by monkeys. *Cognition*, **37**, 167–196.
Cheney, D. L., & Seyfarth, R. M. (2007). *Baboon Metaphysics: The Evolution of a Social Mind*. Chicago, IL: University of Chicago Press.
Cheng, K. (1986). A purely geometrical module in the rat's spatial representation. *Cognition*, **23**, 149–178.
Cheng, K., Pena, J., Porter, M.A., & Irwin, J.D. (2002). Self-control in honeybees. *Psychon Bull*, **9**, 259–263.
Cheng, K., Shettleworth, S.J., Huttenlocher, J., & Rieser, J.J. (2007). Bayesian integration of spatial information. *Psychol Bull*, **133**, 625–637.
Cheng, K., Spetch, M.L., Kelly, D.M., & Bingman, V.P. (2006). Small-scale spatial cognition in pigeons. *Behav Processes*, **72**, 115–127.
Cheng, P.W. (1997). From covariation to causation: A causal power theory. *Psychol Rev*, **104**, 367–405.
Chiappori, P., Levitt, S., & Groseclose, T. (2002). Testing mixed-strategy equilibria when players are heterogeneous: The case of penalty kicks in soccer. *Am Econ Rev*, **92**, 1138–1151.
Chiou, T.H., Kleinlogel, S., Cronin, T., *et al.* (2008). Circular polarization vision in a stomatopod crustacean. *Curr Biol*, **18**, 429–434.
Chittka, L., Geiger, K., & Kunze, J. (1995). The influence of landmark use on distance estimation of honey bees. *Anim Behav*, **50**, 23–31.
Chomsky, N. (1959). A review of Skinner's verbal behavior. *Language*, **35**, 26–38.
Chomsky, N. (1965). *Aspects of the Theory of Syntax* (Vol. 11). Cambridge, MA: The MIT press.
Chomsky, N. (1968). *Language and Mind*. New York, NY: Harcourt, Brace, and World, Inc.
Christianson, S. (1989). Flashbulb memories: Special, but not so special. *Memory and Cognition*, **17**, 435–443.
Chudek, M., Heller, S., Birch, S., & Henrich, J. (2012). Prestige-biased cultural learning: Bystander's differential attention to potential models influences children's learning. *Evol Hum Behav*, **33**, 46–56.
Church, R.M. (1959). Emotional reactions of rats to the pain of others. *J Comp Physiol Psychol*, **52**, 132.
Church, R. M., & Broadbent, H. A. (1990). Alternative representations of time, number, and rate. *Cognition*, **37**, 55–81.
Church, R.M., & Gibbon, J. (1982). Temporal generalization. *J Exp Psychol Anim Behav Process*, **8**, 165–186.
Clay, Z., & Zuberbühler, K. (2011). Bonobos extract meaning from call sequences. *PloS One*, **6**, e18786.
Clay, Z., Smith, C.L., & Blumstein, D.T. (2012). Food-associated vocalizations in mammals and birds: What do these calls really mean? *Anim Behav*, **83**, 323–330.
Clayton, N.S., Bussey, T.S., & Dickinson, A. (2003). Can animals recall the past and plan for the future? *Nat Rev Neurosci*, **4**, 685–691.
Clayton, N.S., & Dickinson, A. (1998). Episodic-like memory during cache recovery by scrub jays. *Nature*, **395**, 272–274.

Clayton, N. J., & Krebs, J. R. (1994). One-trial associate memory: Comparison of food-storing and non-storing species of birds. *Anim Learn Behav*, **22**, 336–372.

Clearfield, M. W., & Mix, K. S. (1999). Number versus contour length in infants' discrimination of small visual sets. *Psychol Sci*, **10**, 408–411.

Clutton-Brock, T. H., O'Riain, M. J., Brotherton, P. N. M., *et al.* (1999). Selfish sentinels in cooperative mammals. *Science*, **284**, 1640–1644.

Cockburn, A. (2006). Prevalence of different modes of parental care in birds. *Proc R Soc B Biol Sci*, **273**, 1375–1383.

Coffin, H. R., Watters, J. R., & Mateo, J. M. (2011). Odor-based recognition of familiar and related conspecifics: A first test conducted on captive Humbold penguins (*Sphenicus humboldti*). *PLoS ONE*, **6**, e25002.

Cohen, L. B., & Oakes, L. M. (1993). How infants perceive a simple causal event. *Dev Psychol*, **29**, 421–433.

Colombo, J., & Mitchell, D. W. (2009). Infant visual adaptation. *Neurobiol Learn Mem*, **92**, 225–234.

Collett, T. S. (1995). Making learning easy: The acquisition of visual information during orientation flight of social wasps. *J Comp Physiol A*, **177**, 737–747.

Collett, T. S., & Baron, J. (1994). Biological compasses and the coordinate frame of landmark memories in honeybees. *Nature*, **368**, 137–140.

Collier, G., Johnson, D. F., & Morgan, C. (1992). The magnitude-of-reinforcement function in closed and open economies. *J Exp Anal Behav*, **57**, 81–89.

Conrod, P. J., O'Leary-Barrett, M., Newton, N., *et al.* (2013). Effectiveness of a selective, personality-targeted prevention program for adolescent alcohol use and misuse: A cluster randomized controlled trial. *J Am Med Assoc Psychiatry*, **70**, 334–342.

Cook, M., & Mineka, S. (1990). Selective associations in the observational conditioning of fear in rhesus monkeys. *J Exp Psychol Anim Behav Process*, **16**, 372–389.

Cook, R. G. (1992). Acquisition and transfer of visual texture discriminations by pigeons. *J Exp Psychol Anim Behav Process*, **18**, 341–353.

Cook, R. G., Katz, J. S., & Cavoto, B. R. (1997). Pigeon same-different concept learning with multiple stimulus classes. *J Exp Psychol Anim Behav Process*, **23**, 417–433.

Coolen, I., Van Bergen, Y., Day, R. L., & Laland, K. N. (2003). Species difference in adaptive use of public information in sticklebacks. *Proc Roy Soc Lond B Biol Sci*, **270**, 2413–2419.

Cordes, S., & Brannon, E. M. (2008). The difficulties of representing continuous extent in infancy: Using number is just easier. *Child Dev*, **79**, 476–489.

Cordes, S., & Brannon, E. M. (2009a). The relative salience of discrete and continuous quantity in young infants. *Dev Sci*, **12**, 453–463.

Cordes, S., & Brannon, E. M. (2009b). Crossing the divide: Infants discriminate small from large numerosities. *Dev Psychol*, **45**, 1583–1594.

Cordes, S., Gelman, R., Gallistel, C. R., & Whalen, J. (2001). Variability signatures distinguish verbal from nonverbal counting for both large and small numbers. *Psychon Bull Rev*, **8**, 698–707.

Cordes, S., Williams, C. L., & Meck, W. H. (2007). Common representations of abstract quantities. *Current Directions in Psychological Science*, **16**, 156–161.

Cosmides, L., & Tooby, J. (2013). Evolutionary psychology: New perspectives on cognition and motivation. *Ann Rev Psychol*, **64**, 201–229.

Courtney, S. M., Petit, L., Maisog, J. M., Underleider, L. G., & Haxby, J. V. (1998). An area specialized for spatial working memory in human frontal cortex. *Science*, **279**, 1347–1351.

Coussi-Korbel, S., & Fragaszy, D. M. (1995). On the relation between social dynamics and social learning. *Anim Behav*, **50**, 1441–1453.

Coutlee, C. G., & Huettel, S. A. (2012). The functional neuroanatomy of decision making: Prefrontal control of thought and action. *Brain Res*, **1428**, 3–12.

Couvillon, P. A., Arakaki, L., & Bitterman, M. E. (1997). Intramodal blocking in honeybees. *Anim Learn Behav*, **25**, 277–282.

Couvillon, P. A., Campos, A. C., Bass, T. D., & Bitterman, M. E. (2001). Intermodal blocking in honeybees. *Q J Expl Psychol B*, **4**, 369–381.

Couzin, I. D. (2009). Collective cognition in animal groups. *Trends Cogn Sci*, **13**, 36–43.

Couzin, I. D., Krause, J., Franks, N. R., & Levin, S. A. (2005). Effective leadership and decision-making in animal groups on the move. *Nature*, **433**, 513–516.

Cowie, R. J. (1977). Optimal foraging in great tits (*Parus major*). *Nature*, **268**, 137–139.

Craik, F. I. M., & Lockhart, R. S. (1972). Levels of processing: A framework for memory research. *J Verb Learn Verb Behav*, **12**, 599–607.

Crawford, M. P. (1937). *The Cooperative Solving of Problems by Young Chimpanzees*. Baltimore, MD: Johns Hopkins Press.

Cresswell, W., Hilton, G. M., & Ruxton, G. D. (2000). Evidence for a rule governing the avoidance of superfluous escape flights. *Proc Roy Soc Lond B*, **267**, 733–737.

Cronin, A. L., Molet, M., Doums, C., Monnin, T., & Peeters, C. (2013). Recurrent evolution of dependent colony foundation across eusocial insects. *Ann Rev Entomol*, **58**, 37–55.

Cronin, K. A., Schroeder, K. K., Rothwell, E. S., Silk, J. B., & Snowdon, C. T. (2009). Cooperatively breeding cottontop tamarins (*Saguinus oedipus*) do not donate rewards to their long-term mates. *J Comp Psychol*, **123**, 231–241.

Croze, H. (1970). Search image in carrion crows. *Z Tierpsychol*, **Supplement 5**, 1–85.

Csibra, G. (2007). Teachers in the wild. *Trends Cogni Sci*, **11**, 95–96.

Csibra, G., & Gergely, G. (1998). The teleological origins of mentalistic action explanations: A developmental hypothesis. *Dev Sci*, **1**, 255–259.

Csibra, G., Gergely, G., Bíró, S., Koós, O., & Brockbank, M. (1999). Goal attribution without agency cues: The perception of 'pure reason' in infancy. *Cognition*, **72**, 237–267.

Curio, E., Ernst, U., & Vieth, W. (1978). Cultural transmission of enemy recognition: One function of mobbing. *Science*, **202**, 899–901.

Curtiss, S. (1977). *Genie: A Psycholinguistic Study of a Modern-Day 'Wild Child'*. New York, NY: Academic Press.

Cusato, B., & Domjan, M. (1998). Special efficacy of sexual conditioned stimuli that include species typical cues: Tests with a conditioned stimulus preexposure design. *Learn Motiv*, **29**, 152–167.

Custance, D., & Mayer, J. (2012). Empathic-like responding by domestic dogs (*Canis familiaris*) to distress in humans: An exploratory study. *Anim Cogn*, **15**, 851–859.

Custance, D., Whiten, A., & Fredman, T. (1999). Social learning of an artificial fruit task in capuchin monkeys (*Cebus apella*). *J Comp Psychol*, **113**, 13–23.

Dallal, N. L., & Meck, W. H. (1990). Hierarchical structures: Chunking by food type facilitates spatial memory. *J Exp Psychol Anim Behav Process*, **16**, 69–84.

Dally, J. M., Clayton, N. S., & Emery, N. J. (2006a). The behaviour and evolution of cache protection and pilferage. *Anim Behav*, **72**, 13–23.

Dally, J. M., Emery, N. J., & Clayton, N. S. (2006b). Food-caching western scrub-jays keep track of who was watching when. *Science*, **312**, 1662–1665

Damasio, A., Grabowski, T. J., Tranel, D., Hichwa, R. D., & Damasio, A. R. (1996). A neural basis for lexical retrieval. *Nature*, **380**, 499–505.

Damasio, H. (1994). *Descartes' Error: Emotion, Reason, and the Human Brain*. New York, NY: G. P. Putnam's Sons.

Damasio, H., Grabowski, T., Frank, R., Galaburda, A. M., & Damasio, A. R. (1994). The return of Phineas Gage: Clues about the brain from the skull of a famous patient. *Science*, **264**, 1102–1105.

D'Amato, M. R., & Colombo, M. (1988). Representation of serial order in monkeys (*Cebus appella). J Exp Psychol*, **14**, 131–139.

D'Amato, M. R., & Van Sant, P. (1988). The person concept in monkeys (*Cebus paella). J Exp Psychol Anim Behav Process*, **14**, 43–56.

Danchin, È., Giraldeau, L.-A., & Cézilly, F. (eds.). (2008). *Behavioural Ecology*. Dunod, Paris: Oxford University Press.

Darwin, C. (1859). *On the Origin of Species by Means of Natural Selection*. London, UK: Murray.

Darwin, C. (1871). *The Descent of Man and Selection in Relation to Sex*. London, UK: John Murray.

Darwin, C. (1872). *The Expression of the Emotions in Man and Animals*. Chicago, IL: University of Chicago Press.

Dasser, V. (1987). Slides of group members as representations of the real animals (*Macaca fascicularis*). *Ethology*, **76**, 65–73.

Dasser, V. (1988). A social concept in Java monkeys. *Anim Behav*, **36**, 225–230.

Davies, N. B., & Brooke, M. d. L. (1988). Cuckoos versus red warblers: Adaptations and counteradaptations. *Anim Behav*, **36**, 262–284.

Davies, N. B., Brooke, M. d. L., & Kacelnik, A. (1996). Recognition errors and probability of parasitism determine whether reed warblers should accept or reject mimetic cuckoo eggs. *Proc Roy Soc Lond B*, **263**, 925–931.

Dawkins, R. (1976). *The Selfish Gene*. Oxford: Oxford University Press.

Dawkins, R. (2006). A user's guide to animal welfare science. *Trends Ecol Evol*, **21**, 77–82.

Dawkins, R., & Krebs, J. R. (1978). Animal signals: Information or manipulation? In J. R. Krebs & N. B. Davies (eds.). *Behavioural Ecology: An Evolutionary Approach* (pp. 282–309). Oxford, UK: Blackwell.

de Blois, S. T., Novak, M. A., & Bond, M. (1998). Object permanence in orangutans (*Pongo pygmaeus*) and squirrel monkeys (*Saimiri sciureus*). *J Comp Psychol*, **112**, 137–152.

Defeyter, M. A., & German, T. P. (2003). Acquiring an understanding of design: Evidence from children's insight problem solving. *Cognition*, **89**, 133–155.

De Fraja, G., Oliveira, T., & Zanchi, L. (2010). Must try harder: Evaluating the role of effort in educational attainment. *Rev Econ Stat*, **92**, 577–597.

Dehaene, S. (2009). Origins of mathematical intuitions. *Ann NY Acad Sci*, **1156**, 232–259.

Dehaene, S. & Brannon, E. M. (eds.). (2011). *Space, Time, and Number in the Brain: Searching for the Foundations of Mathematical Thought*. Oxford, UK: Elsevier Press.

De Houwer, J., Vandorpe, S., & Beckers, T. (2005). Evidence for the role of higher order reasoning processes in cue competition and other learning phenomena. *Learn Behav*, **33**, 239–249.

Deipolyi, A., Santos, L., & Hauser, M. D. (2001). The role of landmarks in cotton-top tamarin spatial foraging: Evidence for geometric and non-geometric features. *Anim Cogn*, **4**, 99–108.

DeLoache, J. S., & LoBue, V. (2009). The narrow fellow in the grass: Human infants associate snakes and fear. *Dev Sci*, **12**, 201–207.

Dennett, D. (2003). *Freedom Evolves*. New York, NY: Viking Press.

Denniston, J. C., Savastano, H. I., Blaisdell, A. P., & Miller, R. R. (2003). Cue competition as a retrieval deficit. *Learn Motiv*, **34**, 1–31.

DePaulo, B. M., & Friedman, H. S. (1998). Nonverbal communication. In D. T. Gilbert, S. T. Fiske, & G. Lindzey (eds.). *The Handbook of Social Psychology*, Vols. 1 and 2 (4th Edn.) (pp. 3–40). New York, NY: McGraw-Hill.

Depy, D., Fagot, J., & Vauclair, J. (1997). Categorization of three-dimensional stimuli by humans and baboons: Search for prototype effects. *Behav Proc*, **39**, 299–306.

de Quervain, D.J., Fischbacher, U., Treyer, V., *et al.* (2004). The neural basis of altruistic punishment. *Science*, **305**, 1254–1258.

de Resenda, B.D., Ottoni, E.B., & Fragaszy, D.M. (2008). Ontogeny of manipulative behaviour and nut-cracking in young tufted capuchin monkeys (*Cebus apella*): A perception-action perspective. *Dev Sci*, **11**, 828–840.

D'Ettorre, P, & Heinze, J. (2005). Individual recognition in queen ants. *Curr Biol*, **15**, 2170–2174.

de Waal, F. B. M. (2008). Putting the altruism back in to altruism: The evolution of empathy. *Ann Rev Psychol*, **59**, 279–300.

de Waal, F. B. M., & van Roosmalen, A. (1979). Reconciliation and consolation among chimpanzees. *Behav Ecol Sociobiol*, **5**, 55–66.

DeWind, N. K., & Brannon, E. M. (2012). Malleability of the approximate number system: Effects of feedback and training. *Front Hum Neurosci*, **6**, 1–10.

de Wit, S., & Dickinson, A. (2009). Associative theories of goal-directed behavior: A case for animal-human translational models. *Psychol Res*, **73**, 463–476.

Dewsbury, D. A. (2000). Comparative cognition in the 1930s. *Psychon Bull Rev*, **7**, 267–283.

Dickinson, A. (1980). *Contemporary Animal Learning Theory.* Cambridge, UK: Cambridge University Press.

Dickinson, A. (2001). The 28th Bartlett memorial lecture causal learning: An associative analysis. *Q J Exp Psychol B*, **54**, 3–25.

Dickinson, A., & Balleine, B. W. (1994). Motivational control of goal-directed action. *Anim Learn Behav*, **22**, 1–18.

Dickinson, A., & Balleine, B. (2000). Causal cognition and goal-directed action. In C. M. Hayes and L. Huber (eds.). *The Evolution of Cognition* (pp. 185–204). Cambridge, MA: MIT Press.

Dittman, A. H., & Quinn, T. P. (1996). Homing in Pacific salmon: Mechanisms and ecological basis. *J Exp Biol*, **199**, 83–91.

Dittman, A. H., Quinn, T. P., & Nevitt, G. A. (1996). Timing of imprinting to natural and artificial odors by coho salmon (*Oncorhynchus kisutch*). *Can J Fish Aquatic Sci*, **53**, 434–442.

Domjan, M. (2002). Cognitive modulation of sexual behavior. In M. Bekoff, C. Allen, & G. M. Burghard (eds.). *The Cognitive Animal: Empirical and Theoretical Perspectives on Animal Cognition* (pp. 89–96). Cambridge, MA: MIT Press.

Domjan, M. (2010). *The Principles of Learning and Behavior* (6th Edn.) Belmont, CA: Wadsworth, Cage Learning.

Domjan, M., Blebois, E., & Williams, J. (1998). The adaptive significance of sexual conditioning: Pavlovian control of sperm release. *Psychol Sci*, **9**, 411–415.

Dooling, R., & Searcy, M. (1980). Early perceptual selectivity in the swamp sparrow. *Dev Psychobiol*, **13**, 499–506.

Downes, S. (2001). Trading heat and food for safety: Costs of predator avoidance in a lizard. *Ecology*, **82**, 2870–2881.

Dreisig, H. (1995). Ideal free distributions of nectar foraging bumblebees. *Oikos*, **72**, 161–172.

Dringenberg, H. C., Kornelsen, R. A., & Vanderwolf, C. H. (1994). Food carrying in rats is blocked by the putative anxiolytic agent buspirone. *Pharmacol Biochem Behav*, **49**, 741–746.

Druzhkova, A. S., Thalmann, O., Trifonov, V. A., *et al.* (2013) Ancient DNA analysis affirms the Canid from Altai as a primitive dog. *PLoS ONE*, **8**, e57754.

Dubreuil, D., Gentile, M.S., & Visalberghi, E. (2006). Are capuchin monkeys (*Cebus apella*) inequity averse? *Proc R Soc B: Biol Sci*, **273**, 1223–1228.

Duchaine, B., Germine, L., & Nakayama, K. (2007). Family resemblance: Ten family members with prosopagnosia and within-class object agnosia. *Cogn Neuropsychol*, **24**, 419–430.

Dudai, Y. (2004). The neurobiology of consolidations, or, how stable is the engram? *Ann Rev Psychol*, **55**, 51–86.

Dudai, Y., Jan, Y. N., Byers, D., Quinn, W. G., & Benzer, S. (1976). Dunce: A mutant in *Drosophila* deficient in learning. *Proc Natl Acad Sci USA*, **73**, 1684–1688.

Dugatkin, L. A. (1992). Sexual selection and imitation: Females copy the mate choice of others. *Am Nat*, **139**, 1384–1389.

Dugatkin, L. A. (2004). *Principles of Animal Behavior.* New York, NY: W. W. Norton & Co Inc.

Dukas, R., & Bernays, E. A. (2000). Learning improves growth rate in grasshoppers. *Proc Natl Acad Sci USA*, **97**, 2637–2640.

Dukas, R., & Kamil, A. C. (2000). The cost of limited attention in blue jays. *Behav Ecol*, **11**, 502–506.

Dumontheil, I., Küster, O., Apperly, I. A., & Blakemore, S. J. (2010). Taking perspective into account in a communicative task. *Neuroimage*, **52**, 1574–1583.

Duncan, C. P. (1949). The retroactive effect of electroconvulsive shock. *J Comp Physiol Psychol*, **42**, 32–44.

Duncker, K. (1945). On problem-solving. *Psychol Monogr*, **58**, 1–113.

Dunfield, K. A., & Kuhlmeier, V. A. (2013). Classifying prosocial behavior: Children's responses to instrumental need, emotional distress, and material desire. *Child Dev*, **84**, 1766–1776.

Dunfield, K., Kuhlmeier, V. A., O'Connell, L., & Kelley, E. (2011). Examining the diversity of prosocial behavior: Helping, sharing, and comforting in infancy. *Infancy*, **16**, 227–247.

Dungl, E., Schratter, D., & Huber, L. (2008). Discrimination of face-like patterns in the giant panda (*Ailuropoda melanoleuca*). *J Comp Psychol*, **122**, 335–343.

Dyer, F. C. (1998). Cognitive ecology of navigation. In R. Dukas (ed.). *Cognitive Ecology: The Evolutionary Ecology of Information Processing and Decision Making* (pp. 201–260). Chicago, IL: University of Chicago Press.

Dyer, F. C., & Dickinson, J. A. (1994). Development of sun compensation by honeybees: How partially experienced bees estimate the sun's course. *Proc Natl Acad Sci USA*, **91**, 4471–4474.

Dyer, J. R. G., Johansson, A., Helbing, D., Couzin, I. D., & Krause, J. (2009). Leadership, consensus decision making and collective behaviour in humans. *Philos Trans Roy Soc Lond B*, **364**, 781–789.

Ebbinghaus, H. (1885). *Memory: A Contribution to Experimental Psychology.* Translated by H. A. Ruger & C. E. Bussenius. New York, NY: Dover.

Edwards, C. A., Jagielo, J. A., & Zentall, T. R. (1983). Same/different symbol use by pigeons. *Anim Learn Behav*, **11**, 349–355.

Egri, A., Blaho, M., Kriska, G., *et al.* (2012). Poarotactic tabanids find striped patterns with brightness and/or polarization modulation least attractive: An advantage of zebra stripes. *J Exp Biol*, **215**, 736–745.

Eichenbaum, H., Dudchenko, P., Wood, E., Shapiro, M., & Tanila, M. (1999). The hippocampus, memory, and place cells: Is it spatial memory or a memory space? *Neuron*, **23**, 209–226.

Eilam, D. (2005). Die hard: A blend of freezing and fleeing as a dynamic defense: implications for the control of defensive behavior. *Neurosci Biobehav Rev*, **29**, 1181–1191.

Eimas, P. D. (1994). Categorization in early infancy and the continuity of development. *Cognition*, **50**, 83–93.

Eimas, P. D., Siqueland, E. R., Jusczyk, P., & Vigorito, J. (1971). Speech perception in infants. *Science*, **171**, 303–306.

Eisenberg, D. T., MacKillop, J., Modi, M., *et al.* (2007). Examining impulsivity as an endophenotype using a behavioral approach: A DRD2 TaqI A and DRD4 48-bp VNTR association study. *Behav Brain Funct*, **3**, 2.

Eisenberger, R., & Adornetto, M. (1986). Generalized self-control of delay and effort. *J Pers Soc Psychol*, **51**, 1020–1031.

Elgar, M. A. (1986). House sparrows establish foraging flocks by giving chirrup calls if the resources are divisible. *Anim Behav*, **34**, 169–174.

Elner, R. W., & Hughes, R. N. (1978). Energy maximization in the diet of the shore crab *Carcinus maenas*. *J Anim Ecol*, **47**, 103–116.

Emery, N. J., & Clayton, N. S. (2001). Effects of experience and social context on prospective catching strategies by scrub jays. *Nature*, **414**, 443–446.

Emlen, J. M. (1966). The role of time and energy in food preferences. *Am Nat*, **100**, 611–617.

Emlen, S. T. (1970). Celestial rotation: Its importance in the development of migratory orientation. *Science*, **170**, 1198–1201.

Emlen, S. T., & Emlen, J. T. (1966). A technique for recording migratory orientation of captive birds. *Auk*, **83**, 361–367.

Enard, W., Khaitovich, P., Klose, J., *et al.* (2002). Intra- and interspecific variation in primate gene expression patterns. *Science*, **296**, 340–343.

Endler, J. A. (1992). Signals, signal conditions, and the direction of evolution. *Am Nat*, **139**, 125–153.

Endler, J. A., Basolo, A., Glowacki, S., & Zerr, J. (2001). Variation in response to artificial selection for light sensitivity in guppies (*Poecilia reticulata*). *Am Nat*, **158**, 36–48.

Engh, A. L., Siebert, E. R., Greenberg, D. A., & Holekamp, K. E. (2005). Patterns of alliance formation and postconflict aggression indicate spotted hyenas recognize third-party relationships. *Anim Behav*, **69**, 209–217.

Evans, C. S. (1997). Referential signals. In D. H. Owings, M. D. Beecher, & N. S. Thompson (eds.). *Perspectives in Ethology: Communication* (Vol. 12) (pp. 99–143). New York, NY: Plenum Press.

Evans, C. S., & Evans, L. (1999). Chicken food calls are functionally referential. *Anim Behav*, **58**, 307–319.

Evans, C. S., & Marler, P. (1992). Female appearance as a factor in the responsiveness of male chickens during anti-predator behaviour and courtship. *Anim Behav*, **43**, 137–145.

Evans, C. S., & Marler, P. (1994). Food calling and audience effects in male chickens, *Gallus gallus*: Their relationships to food availability, courtship and social facilitation. *Anim Behav*, **47**, 1159–1170.

Everett, D. (2005). Cultural constraints on grammar and cognition in Pirahã: Another look at the design features of human language. *Curr Anthropol*, **46**, 621–646.

Fabre-Thorpe, M., Richard, G., & Thorpe, S. J. (1998). Rapid categorization of natural images by rhesus monkeys. *Neuroreport*, **9**, 303–308.

Falkenberg, G., Fleissner, G., Schurchardt, K., *et al.* (2010). Avian magnetoreception: Elaborate iron mineral containing dendrites in the upper beak seem to be a common feature of birds. *PLoS One*, **5**, e9231.

Falls, J. B., & Brooks, J. R. (1975). Individual recognition by song in white-throated sparrows. II. Effects of location. *Can J Zool*, **53**, 1412–1420.

Fehr, E., & Fischbacher, U. (2004). Third-party punishment and social norms. *Evol Hum Behav*, **25**, 63–87.

Feigenson, L., & Carey, S. (2003). Tracking individuals via object-files: Evidence from infants' manual search. *Dev Sci*, **6**, 568–584.

Feigenson, L., Carey, S., & Hauser, M. D. (2002). The representation underlying infants' choice of more: Object files versus analog magnitude. *Psychol Sci*, **13**, 150–156.

Feigenson, L., Carey, S., & Spelke, E. (2002). Infants' discrimination of number vs. continuous extent. *Cogn Psychol*, **44**, 33–66.

Feigenson, L., Dehaene, S., & Spelke, E. (2004). Core systems of number. *Trends Cogn Sci*, **8**, 307–314.

Feigenson, L., Libertus, M. E., & Halberda, J. (2013). Links between the intuitive sense of number and formal mathematics ability. *Child Dev Perspect*, **7**, 74–79.

Feistner, A. T. C., & Chamove, A. S. (1986) High motivation toward food increases food-sharing in cotton-top tamarins. *Dev Psychobiol*, **19**, 439–452.

Festinger, L. (1957). *A Theory of Cognitive Dissonance*. Stanford, CA: Stanford University Press.

Fias, W., Lammertyn, J., Reynvoet, B., Dupont, P., & Orban, G. A. (2003). Parietal representation of symbolic and nonsymbolic magnitude. *J Cogn Neurosci*, **15**, 47–56.

Fiset, S., Beaulieu, C., & Landry, F. (2003). Duration of dogs' (*Canis familiaris*) working memory in search for disappearing objects. *Anim Cogn*, **6**, 1–10.

Fisher, J., & Hinde, R. A. (1949). The opening of milk bottles by birds. *Br Birds*, **42**, 437–457.

Fisher, S. E., & Marcus, G. F. (2006). The eloquent ape: Genes, brains and the evolution of language. *Nat Rev Genet*, **7**, 9–20.

Fitch, W. T. (2000). The evolution of speech: A comparative review. *Trends Cogn Sci*, **4**, 258–267.

Fitch, W. T. (2005). The evolution of language: A comparative review. *Biol Philos*, **20**, 193–230.

Fitch, W. T. (2010). *The Evolution of Language*. New York, NY: Cambridge University Press.

Fitch, W. T., Hauser, M. D., & Chomsky, N. (2005). The evolution of the language faculty: Clarifications and implications. *Cognition*, **97**, 179–210.

Flemming, T.M., Beran, M. J., & Washburn, D. A. (2007). Disconnect in concept learning by rhesus monkeys: Judgment of relations and relations-between-relations. *J Exp Psychol Anim Behav Process*, **33**, 55–63.

Flemming, T. M., & Kennedy, E. H. (2011). Chimpanzee (*Pan troglodytes*) relational matching: Playing by their own (analogical) rules. *J Comp Psychol*, **125**, 207–215.

Flexner, L. B., Flexner, J. B., & Stellar, E. (1965). Memory and cerebral protein synthesis in mice as affected by graded amounts of puromycin. *Exp Neurol*, **13**, 264–272.

Flombaum, J. I., Junge, J. A., & Hauser, M. D. (2005). Rhesus monkeys (*Macaca mulatta*) spontaneously compute addition operations over large numbers. *Cognition*, **97**, 315–325.

Flombaum, J. I., & Santos, L. R. (2005). Rhesus monkeys attribute perceptions to others. *Curr Biol*, **15**, 447–452.

Floresco, S. B., St Onge, J. R., Ghods-Sharifi, S., & Winstanley, C. A. (2008). Cortico-limbic-striatal circuits subserving different forms of cost-benefit decision making. *Cogn Affect Behav Neurosci*, **8**, 375–389.

Flower, T. (2011). Fork-tailed drongos use deceptive mimicked alarm calls to steal food. *Proc Roy Soc B*, **278**, 1548–1555.

Fong, D.W., Kane, T. C., & Culver, D. C. (1995). Vestigialization and loss of nonfunctional characters. *Ann Rev Ecol Syst*, **26**, 249–268.

Foster, K. R., & Kokko, H. (2009). The evolution of superstitious and superstition-like behaviour. *Proc R Soc B*, **276**, 31–37.

Fox, R., & McDaniel, C. (1982). The perception of biological motion by human infants. *Science*, **218**, 486–487.

Foxall, G. R., James, V. K., Olivereira-Castro, J. M., & Ribier, S. (2010). Product substitutability and the matching law. *Psychol Record*, **60**, 185–216.

Fraenkel, G. S., & Gunn, D. L. (1961). *The Orientation of Animals*. New York, NY: Drover.

Frank, M. C., Everett, D.L., Fedorenko, E., & Gibson, E. (2008). Number as a cognitive technology: Evidence from Piraha language and cognition. *Cognition*, **108**, 819–824.

Freake, M. J., Muheim, R., & Phillips, J. B. (2006). Magnetic maps in animals: A theory comes of age? *Q Rev Biol*, **81**, 327–347.

Freedman, D. J., & Miller, E. K. (2008). Neural mechanisms of visual categorization: Insights from neurophysiology. *Neurosci Biobehav Rev*, **32**, 311–329.

Fretwell, S. D., & Lucas, H. L. (1970). On territorial behavior and other factors influencing habitat distribution in birds. I. Theoretical development. *Acta Biotheoretica*, **19**, 16–36.

Fripp, D., Owen, C., Quintana-Rizzo, E., *et al.* (2005). Bottlenose dolphin (*Tursiops truncatus*) calves appear to model their signature whistles on the signature whistles of community members. *Anim Cogn*, **8**, 17–26.

Frost, B. J., & Mouritsen, H. (2006). The neural mechanisms of long distance animal navigation. *Curr Opin Neurobiol*, **16**, 481–488.

Froy, O., Grotter, A. L., Casselman, A. L., & Repper, S. M. (2003). Illuminating the circadian clock in monarch butterfly migration. *Science*, **300**, 1303–1305.

Fujita, K., Kuroshima, H., & Asai, S. (2003). How do tufted capuchin monkeys (*Cebus apella*) understand causality involved in tool use? *J Exp Psychol Anim Behav Process*, **29**, 233–242.

Fullard, J. H., & Yack, J. E. (1993). The evolutionary biology of insect hearing. *Trends Ecol Evol*, **8**, 248–252.

Funahashi, S., Bruce, C. J., & Goldman-Rakic, P. S. (1989). Mnemonic coding of visual space in the monkey's dorsolateral prefrontal cortex. *J Neurophysiol*, **61**, 1–19.

Funder, D., & Sneed, C. (1993). Behavioral manifestations of personality: An ecological approach to judgmental accuracy. *J Pers Soc Psychol*, **64**, 479–490.

Fuster, J. M. (1985). Temporal organization of behavior. *Hum Neurobiol*, **4**, 57–60.

Fuster, J. M., & Alexander, G. E. (1971). Neuron activity related to short-term memory. *Science*, **173**, 652–654.

Fyhn, N., Molden, S., Witter, M. P., *et al.* (2004). Spatial representation in the entorhinal cortex. *Science*, **305**, 1258–1264.

Galea, L. A. M., Kavaliers, M., & Ossenkopp, K. P. (1996). Sexually dimorphic spatial learning in meadow voles *Microtus pennsylvanicus* and deer mice *Peromyscus maniculatus*. *J Exp Biol*, **199**, 195–200.

Galef, B.G. (1995). Why behaviour patterns that animals learn socially are locally adaptive. *Anim Behav*, **49**, 1325–1334.

Galef, B. G. (2004). Approaches to the study of traditional behaviors of free-living animals. *Anim Learn Behav*, **32**, 53–61.

Galef, B. G. (2008). Social influences on the mate choices of male and female Japanese quail. *Comp Cogn Behav Rev*, **3**, 1–12.

Galef, B. G. (2009). A most unlikely animal behaviorist. In L. Drickamer & D. Dewsbury (eds.). *Leaders in Animal Behavior: The Second Generation* (pp. 279–308). Cambridge, UK: Cambridge University Press.

Galef, B. G., & Clark, M. M. (1971). Social factors in the poison avoidance and feeding behavior of wild and domesticated rat pups. *J Comp Physiol Psychol*, **75**, 341–357.

Galef, B. G., & Laland, K. N. (2005). Social learning in animals: Empirical studies and theoretical models. *BioScience*, **55**, 489–499.

Galef, B. G., & Sherry, D. F. (1973). Mother's milk: A medium for transmission of cues reflecting the flavor of mother's diet. *J Comp Physiol Psychol*, **83**, 374–378.

Galef, B. G., & Whiskin, E. E. (2003). Socially transmitted food preferences can be used to study long-term memory in rats. *Anim Learn Behav*, **31**, 160–164.

Galef, B. G., & Whiskin, E. E. (2004). Effects of environmental stability and demonstrator age on social learning of food preferences by young Norway rats. *Anim Behav*, **68**, 897–902.

Galef, B. G., & White, D. J. (1998). Mate-choice copying in Japanese quail, *Coturnix coturnix japonica*. *Anim Behav*, **55**, 545–552.

Galef, B. G., & Wigmore, S. W. (1983). Transfer of information concerning distant foods: A laboratory investigation of the 'information-centre' hypothesis. *Anim Behav*, **31**, 748–758.

Gallese, V., & Goldman, A. (1998). Mirror neurons and the simulation theory of mind-reading. *Trends Cogn Sci*, **2**, 493–501.

Gallese, V., Keysers, C., & Rizzolatti, G. (2004). A unifying view of the basis of social cognition. *Trends Cogn Sci*, **8**, 396–403.

Gallistel, C. R. (1989). Animal cognition: The representation of space, time and number. *Ann Rev Psychol*, **40**, 155–189.

Gallistel, C. R. (1990). *The Organization of Learning*. Cambridge, MA: The MIT Press.

Gallistel, C. R. (1993). A conceptual framework for the study of numerical estimation and arithmetic reasoning in animals. In S. T. Boysen & E. J. Capaldi (eds.). *The Development of Numerical Competence: Animal and Human Models* (pp. 211–223). Hillsdale, NJ: Lawrence Erlbaum Associates, Inc.

Gallistel, C. R. (2007). Commentary on Le Corre & Carey. *Cognition*, **105**, 439–445.

Gallistel, C. R., & Cramer, A. E. (1996). Computations on metric maps in mammals: Getting oriented and choosing a multi-destination route. *J Exp Biol*, **199**, 211–217.

Gallistel, C. R., & Gelman, R. (1992). Preverbal and verbal counting and computation. *Cognition*, **44**, 43–74.

Gallistel, C. R., & Gelman, R. (2000). Non-verbal numerical cognition: From reals to integers. *Trends Cogn Sci*, **4**, 59–65.

Gallup, G. G. (1970). Chimpanzees: Self-recognition. *Science*, **167**, 86–87.

Garcia, J., & Koelling, R. A. (1966). Relation of cue to consequence in avoidance learning. *Psychon Sci*, **4**, 123–124.

Gardner, R. A., & Gardner, B. T. (1969). Teaching sign language to a chimpanzee. *Science*, **165**, 664–672.

Gardner, R. A., Gardner, B. T., & Van Cantfort, T. E. (eds.) (1989). *Teaching Sign Language to Chimpanzees*. Albany, NY: State University of New York Press.

Garland, H. (1990). Throwing good money after bad: The effect of sunk costs on the decision to escalate commitment to an ongoing project. *J Appl Psychol*, **75**, 728–731.

Gass, J. T., Osborne, M. P., Watson, M. L., Brown, J. L., & Olive, M. F. (2009). mGluR antagonism attenuates methamphetamine reinforcement and prevents reinstatement of methamphetamine-seeking behavior in rats. *Neuropsychopharmacology*, **34**, 820–833.

Gaulin, S. J. C., & Fitzgerald, R. W. (1986). Sex differences in spatial ability: An evolutionary hypothesis and test. *Am Nat*, **127**, 74–88.

Gelman, R., & Butterworth, B. (2005). Number and language: How are they related? *Trends Cogn Sci*, **9**, 6–10.

Gelman, R., & Gallistel, C. R. (1978). *The Child's Understanding of Number*. Cambridge, MA: Harvard University Press.

Gemberling, G. A., & Domjam, M. (1982). Selective association in one-day-old rats: Taste-toxicosis and texture shock aversion learning. *J Comp Physiol Psychol*, **96**, 105–113.

Gentner, D., & Rattermann, M. J. (1991). Language and the career of similarity. In S. Gelman & J. P. Byrnes (eds.). *Perspectives on Thought and Language: Interrelations in Development* (pp. 225–277). London, UK: Cambridge University Press.

Gergely, G., Bekkering, H., & Király, I. (2002). Developmental psychology: Rational imitation in preverbal infants. *Nature*, **415**, 755.

Gerlach, G., Hodgins-Davis, A., Alolio, C., & Schunter, C. (2008). Kin recognition in zebrafish: A 24-hour window for olfactory imprinting. *Proc Roy Soc Lond B*, **275**, 2165–2170.

German, T. P., & Barrett, H. C. (2005). Functional fixedness in a technologically sparse culture. *Psychol Sci*, **16**, 1–5.

German, T. P., & Defeyter, M. A. (2000). Immunity to functional fixedness in young children. *Psychon Bull Rev*, **7**, 707–712.

Getty, T. (1998). Handicap signalling: When fecundity and viability do not add up. *Anim Behav*, **56**, 127–130.

Ghazanfar, A.A., & Cohen, Y.E. (2008). Primate communication: Evolution and neurobiology. In L. Squire *et al.* (eds.). *The Encyclopedia of Neuroscience*. Oxford, UK: Elsevier Press.

Ghazanfar, A. A., Takahashi, D. Y., Mathur, N., & Fitch, W. (2012). Cineradiography of monkey lip-smacking reveals putative precursors of speech dynamics. *Curr Biol*, **22**, 1176–1182.

Gibbon, J. (1977). Scalar expectancy theory and Weber's law in animal timing. *Psychol Rev*, **84**, 279–325.

Gibbon, J., & Balsam, P. (1981). Spreading association in time. In C. M. Locurto, H. S. Terrace, & J. Gibbon (eds.). *Autoshaping and Conditioning Theory* (pp. 219–253). New York, NY: Academic Press.

Gibbon, J., Church, R. M., & Meck, W. H. (1984). Scalar timing in memory. *Ann N Y Acad Sci*, **423**, 52–77.

Gibson, J. J. (1979). *The Ecological Approach to Visual Perception*. Boston, MA: Houghton Miffin.

Gigerenzer, G., & Brighton, H. (2009). Homo heuristicus: Why biased minds make better inferences. *Top Cogn Sci*, **1**, 107–143.

Gigerenzer, G., & Gaissmaier, W. (2011). Heuristic decision making. *Ann Rev Psychol*, **62**, 451–482.

Gill, R. E., Piersma, T., Hufford, G., Servranckx, R., & Riegen, A. (2005). Crossing the ultimate ecological barrier: Evidence for an 11000-km-long nonstop flight from Alaska to New Zealand and eastern Australia by bar-tailed godwits. *Condor*, **107**, 1–20.

Gillan, D. J., Premack, D., & Woodruff, G. (1981). Reasoning in the chimpanzee: I. Analogical reasoning. *J Exp Psychol Anim Behav Process*, **7**, 1–17.

Giraldeau, L.-A., & Kramer, D. L. (1982). The marginal value theorem: A quantitative test using load size variation in a central place forager, the eastern chipmunk, *Tamias striatus*. *Anim Behav*, **30**, 1036–1042.

Giraldeau, L.-A., Valone, T. J., & Templeton, J. J. (2002). Potential disadvantages of using socially acquired information. *Philos Trans Roy Soc Lond B Biol Sci*, **357**, 1559–1566.

Giurfa, M., Zhang, S., Jenett, A., Menzel, R., & Srinivisan, M. W. (2001). The concept of 'sameness' and 'difference' in an insect. *Nature*, **410**, 930–933.

Glimcher, P. W., Camerer, C. F., Fehr, E., & Poldrack, R. A. (2009). Introduction: A brief history of neuroeconomics. In: P. W. Glimcher, E. Fehr, C. F. Camerer, & R. A. Poldrack (eds.). *Neuroeconomics: Decision Making and the Brain* (pp. 1–12). London, UK: Academic Press.

Gmelch, G., & Felson, R. (1980). Can a lucky charm get you through organic chemistry? *Psychol Today*, **14**, 75–78.

Godard, R. (1991). Long-term memory of individual neighbors in a migratory songbird. *Nature*, **350**, 228–229.

Godin, J-G., & Keenleyside, M. H. A. (1984). Foraging on patchily distributed prey by a cichlid fish (*Teleostei cichlidae*): A test of the ideal free distribution theory. *Anim Behav*, **32**, 120–131.

Gopnik, A., Sobel, D. M., Schulz, L. E., & Glymour, C. (2001). Causal learning mechanisms in very young children: Two-, three-, and four-year-olds infer causal relations from patterns of variation and covariation. *Dev Psychol*, **37**, 620–629.

Gopnik, A., & Wellman, H. M. (2012). Reconstructing constructivism: Causal models, Bayesian learning mechanisms, and the theory theory. *Psychol Bull*, **138**, 1085–1108.

Gordon, N. S., Burke, S., Akil, H., Watson, S. J., & Panksepp, J. (2003). Socially-induced brain 'fertilization': Play promotes brain derived neurotrophic factor transcription in the amygdala and dorsolateral frontal cortex in juvenile rats. *Neurosci Lett*, **341**, 17–20.

Gordon, P. (2004). Numerical cognition without words: Evidence from Amazonia. *Science*, **306**, 496–499.

Gould, J. L. (1975). Honey bee recruitment: The dance-language controversy. *Science*, **189**, 685–693.

Gould, J. L., & Gould, C. G. (1988). *The Honey Bee*. New York, NY: Freeman Press.

Gould, J. L., & Gould, C. G. (1994). *The Animal Mind*. New York, NY: Scientific American Library.

Gould, K. L., Kelly, D. M., & Kamil, A. C. (2010). What scatter-hording animals have taught us about small-scale navigation. *Philos Trans Roy Soc Lond B*, **365**, 901–914.

Gordon, W. C., & Spear, N. E. (1973). Effect of reactivation of a previously acquired memory on the interaction between memories in the rat. *J Exp Psychol*, **99**, 349–355.

Gouzoules, S. (1984). Primate mating systems, kin associations and cooperative behavior: Evidence for kin recognition. *Yearb Phys Anthropol*, **27**, 99–134.

Grabenhorst, F., & Rolls, E.T. (2012). Value, pleasure and choice in the ventral prefrontal cortex. *Trends Cogn Sci* **15**, 56–67.

Graef, S., Schonknecht, P., Sabri, O., & Hergl, U. (2011). Cholinergic receptor subtypes and their role in cognition, emotion, and vigilance control: An overview of preclinical and clinical findings. *Psychopharmacology*, **215**, 205–229.

Grafen, A. (1990). Biological signals as handicaps. *J Theor Biol*, **144**, 517–546.

Graham, J.M., & Desjardins, C. (1980). Pavlovian conditioning: Induction of luteinizing hormone and testosterone secretion in anticipation of sexual activity. *Science*, **210**, 1039–1041.

Grand, T. (1997). Foraging site selection by juvenile coho salmon: Ideal free distributions of unequal competitors. *Anim Behav*, **53**, 185–196.

Griffin, D.R. (1944). Echolocation by blind men, bats and radar. *Science*, **100**, 589–590.

Griffin, D.R. (1992). *Animal Minds*. Chicago, IL: University of Chicago Press.

Griffin, D.R. (2001). *Animal Minds: Beyond Cognition to Consciousness*. Chicago, IL: University of Chicago Press.

Grinnell, J. (1931). Some angles in the problem of bird migration. *The Auk*, **48**, 22–32.

Grüter, C., Balbuena, M.S., & Farina, W.M. (2008). Informational conflicts created by the waggle dance. *Proc Roy Soc B Biol Sci*, **275**, 1321–1327.

Gustison, M.L., le Roux, A., & Bergman, T.J. (2012). Derived vocalizations of geladas (*Theropithecus gelada*) and the evolution of vocal complexity in primates. *Philos Trans Roy Soc B Biol Sci*, **367**, 1847–1859.

Gutierrez, G., & Domjan, M. (1996). Learning and male-male sexual competition in Japanese quail (*Coturnix japonica*). *J Comp Psychol*, **110**, 170–175.

Guttman, N., & Kalish, H.I. (1956). Discriminability and stimulus generalization. *J Exp Psychol*, **51**, 79–88.

Gwinner, E. (1977). Circannual rhythms in bird migration. *Ann Rev Ecol Syst*, **8**, 381–405.

Gwinner, E. (2003). Circannual rhythm in bird migration. *Curr Opin Neurobiol*, **13**, 770–778.

Gyger, M. (1990). Audience effects on alarm calling. *Ethol Ecol Evol*, **2**, 227–232.

Haesler, S., Rochefort, C., Georgi, B., Licznerski, P., Osten, P., & Scharff, C. (2007). Incomplete and inaccurate vocal imitation after knockdown of FoxP2 in songbird basal ganglia nucleus Area X. *PLoS Biol*, **5**, e321.

Hafting, T., Fyhn, M., Molden, S., Moser, M.-B., & Moser, E.I. (2005). Microstructure of a spatial map in the entorhinal cortex. *Nature*, **436**, 801–806.

Hailman, J.P. (1967). The ontogeny of an instinct. *Behav Suppl*, **15**, 1–159.

Halberda, J., Mazzocco, M.M., & Feigenson, L. (2008). Individual differences in non-verbal number acuity correlate with maths achievement. *Nature*, **455**, 665–668.

Hall, J.K., Hutton, S.B., & Morgan, M.J. (2010). Sex differences in scanning faces: Does attention to the eyes explain *female superiority* in facial expression recognition? *Cogn Emot*, **24**, 629–637.

Hamann, S.B., & Squire, L.R. (1995). On the acquisition of new declarative knowledge in amnesia. *Behav Neurosci*, **109**, 1027–1044.

Hamilton, W.D. (1964). The genetical evolution of social behaviour I and II. *J Theor Biol*, **7**, 1–52.

Hammerstein, P., & Stevens, J.R. (eds) (2012). *Evolution and the Mechanisms of Decision Making*. Boston, MA: MIT Press.

Hampton, R.R., Sherry, D.F., Shettleworth, S.J., Khurgel, M., & Ivy, G. (1995). Hippocampal volume and food-storing behavior are related in Parids. *Brain Behav and Evol*, **45**, 54–61.

Hare, B., Brown, M., Williamson, C., & Tomasello, M. (2002). The domestication of social cognition in dogs. *Science*, **298**, 1634–1636.

Hare, B., Call, J., Agnetta, B., & Tomasello, M. (2000). Chimpanzees know what conspecifics do and do not see. *Anim Behav*, **59**, 771–785.

Hare, B., Call, J., & Tomasello, M. (2001). Do chimpanzees know what conspecifics know? *Anim Behav*, **61**, 139–151.

Hare, B., Melis, A.P., Woods, V., Hastings, S., & Wrangham, R. (2007). Tolerance allows bonobos to outperform chimpanzees on a cooperative task. *Curr Biol*, **17**, 619–623.

Hare, B., Plyusnina, I., Ignacio, N., *et al.* (2005). Social cognitive evolution in captive foxes is a correlated by-product of experimental domestication. *Curr Biol*, **15**, 226–230.

Hare, B., & Tomasello, M. (2005). Human-like social skills in dogs? *Trends Cogn Sci*, **9**, 439–444.

Harlow, H.F. (1953). Mice, men, monkeys, and motives. *Psychol Rev*, **60**, 23–32.

Harlow, H.F. (1958). The nature of love. *Am Psychol*, **13**, 673–685.

Harlow, J.M. (1868). Recovery from the passage of an iron bar through the head. *Publications of the Massachusetts Medical Society*, **2**, 327–347.

Harper, D.G.C. (1982). Competitive foraging in mallards: 'Ideal free' ducks. *Anim Behav*, **30**, 575–584.

Harr, A.L., Gilbert, V.R., & Phillips, K.A. (2009). Do dogs (*Canis familiaris*) show contagious yawning? *Anim Cogn*, **12**, 833–837.

Hartley, C.A., & Phelps, E.A. (2012). Anxiety and decision-making. *Biol Psychiatry*, **72**, 113–118.

Hatfield, E., Cacioppo, J.T., & Rapson, R.L. (1993). Emotional contagion. *Curr Dir Psychol Sci*, **2**, 96–99.

Hatfield, E., Cacioppo, J.T., & Rapson, R.L. (1994). *Emotional Contagion: Studies in Emotion and Social Interaction*. New York, NY: Cambridge University Press.

Haun, D., Rekers, Y., & Tomasello, M. (2012). Majority-biased transmission in chimpanzees and human children, but not orangutans. *Curr Biol*, **22**, 727–731.

Hauser, M.D. (1996). *The Evolution of Communication*. Cambridge, MA: The MIT Press.

Hauser, M.D. (1997). Artifactual kinds and functional design features: What a primate understands without language. *Cognition*, **64**, 285–308.

Hauser, M.D., & Carey, S. (2003). Spontaneous representations of small numbers of objects by rhesus macaques: Examinations of content and format. *Cogn Psychol*, **47**, 367–401.

Hauser, M. D., Carey, S., & Hauser, L. B. (2000). Spontaneous number representation in semi-free-ranging rhesus monkeys. *Proc Roy Soc Lond B Biol Sci*, **267**, 829–833.

Hauser, M. D., Chomsky, N., & Fitch, W. T. (2002). The faculty of language: What is it, who has it, and how did it evolve? *Science*, **298**, 1569–1579.

Hauser, M. D., & Marler, P. (1993). Food-associated calls in rhesus macaques (*Macaca mulatta*): I and II. *Behav Ecol*, **4**, 194–212.

Hauser, M. D., Williams, T., Kralik, J. D., & Moskovitz, D. (2001). What guides a search for food that has disappeared? Experiments on cotton-top tamarins (*Saguinus oedipus*). *J Comp Psychol*, **115**, 140–151.

Hawkins, R. D., Abrams, T. W., Carew, T. J., & Kandel, E. R. (1983). A cellular mechanism of classical conditioning in *Aplysia*: Activity-dependent amplification of presynaptic facilitation. *Science*, **219**, 400–406.

Hawkins, R. D., Kandel, E. R., & Bailey, C. H. (2006). Molecular mechanisms of memory storage in *Aplysia*. *Biol Bull*, **210**, 174–191.

Hayden, B. Y., Pearson, J. M., & Platt, M. L. (2011). Neuronal basis of sequential foraging decisions in a patchy environment. *Nat Neurosci*, **14**, 933–939.

Hayton, S. J., Lovett-Barron, M., Dumont, E. C., & Olmstead, M.C. (2010). Target-specific encoding of response inhibition: Increased contribution of AMPA to NMDA receptors at excitatory synapses in the prefrontal cortex. *J Neurosci*, **30**, 11493–11500.

Healey, T. J. (2013). *History of Psychology: From Antiquity to Modernity*. Upper Saddle River, NJ: Pearson Education.

Healy, S. D. (2006). An adaptationist's view of spatial cognition. In M. F. Brown & R. G. Cook (eds.). *Animal Spatial Cognition: Comparative, Neural, and Computational Approaches*. [On-line]. Available at: www.pigeon.psy.tufts.edu/asc/balda/

Healy, S. D., Ginner, E., & Krebs, J. R. (1996). Hippocampal volume in migratory and non-migratory warblers: Effects of age and experience. *Behav Brain Res*, **81**, 61–68.

Healy, S. D., & Hurly, T. A. (1995). Spatial memory in rufous hummingbirds (*Selasphorus rufus*): A field test. *Anim Learn Behav*, **23**, 63–68.

Healy, S. D., & Krebs, J. R. (1992). Food storing and the hippocampus in corvids: Amounts and volume are correlated. *Proc Roy Soc Lond B*, **248**, 241–245.

Hebb, D. O. (1949). *The Organization of Behavior*. New York, NY: John Wiley.

Heffner, R. S., & Heffner, H. E. (1982). Hearing in the elephant (*Elephas maximus*): Absolute sensitivity, frequency discrimination, and sound localization. *J Comp Physiol Psychol*, **96**, 926–944.

Heider, F., & Simmel, M. (1944). An experimental study of apparent behaviour. *Am J Psychol*, **57**, 243–259.

Heinrich, B., & Marzluff, J. M. (1991). Do common ravens yell because they want to attract others? *Behav Ecol Sociobiol*, **28**, 13–21.

Helbig, A. J. (1991). Inheritance of migratory direction in a bird species: A cross-breeding experiment with SE- and SW-migratory blackcaps (*Slyvia atricapilla*). *Behav Ecol Sociobiol*, **28**, 9–12.

Henderson, J., Hurly, T. A., Bateson, M., & Healy, S. D. (2006). Timing in free-living rufous hummingbirds, *Selasphorus rufus*. *Curr Biol*, **16**, 512–515.

Henrich, J. (2004). Inequity aversion in capuchins? *Nature*, **428**, 139.

Henrich, J., & Gil-White, F. J. (2001). The evolution of prestige: Freely conferred deference as a mechanism for enhancing the benefits of cultural transmission. *Evol Hum Behav*, **22**, 165–196.

Henrich, J., McElreath, R., Barr, A., *et al.* (2006). Costly punishment across human societies. *Science*, **312**, 1767–1770.

Hepper, P. G. (1988). Adaptive fetal learning: Prenatal exposure to garlic affects postnatal preferences. *Anim Behav*, **36**, 935–936.

Hermer, L., & Spelke, E. (1996). Modularity and development: The case of spatial orientation. *Cognition*, **61**, 57–59.

Hernik, M., & Csibra, G. (2009). Functional understanding facilitates learning about tools in human children. *Curr Opin Neurobiol*, **19**, 34–38.

Herrnstein, R. J. (1961). Relative and absolute strength of response as a function of frequency of reinforcement. *J Exp Anal Behav*, **4**, 267–272.

Herrnstein, R. J., & Loveland, D. H. (1964). Complex visual concept in the pigeon. *Science*, **146**, 549–551.

Herrnstein, R. J., Loveland, D. H., & Cable, C. (1976). Natural concepts in pigeons. *J Exp Psychol Anim Behav Process*, **2**, 285–301.

Heron-Delaney, M., Wirth, S., & Pascalis, O. (2011). Infants' knowledge of their own species. *Philos Trans Roy Soc Lond B*, **366**, 1753–1763.

Hershberger, W. A. (1986). An approach through the looking glass. *Anim Learn Behav*, **14**, 443–451.

Heyes, C. M. (1993). Anecdotes, training, trapping and triangulation: Do animals attribute mental states? *Anim Behav*, **46**, 177–188.

Heyes, C. M. (1994). Reflections on self-recognition in primates. *Anim Behav*, **47**, 909–919.

Heyes, C. M., & Dawson, G. R. (1990). A demonstration of observational learning in rats using a bidirectional control. *Q J Exp Psychol*, **42**, 59–71.

Hilgard, E. R. (1936). The nature of the conditioned response: I. The case for and against stimulus-substitution. *Psychol Rev*, **43**, 366–385.

Hillman, K. L., & Bilkey, D. K. (2012). Neural encoding of competitive effort in the anterior cingulate cortex. *Nat Neurosci*, **15**, 1290–1297.

Hinde, R. A. (1970). *Animal Behavior: A Synthesis of Ethology and Comparative Psychology* (2nd Edn.). New York, NY: McGraw-Hill.

Hinde, R. A., & Fisher, J. (1951). Further observations on the opening of milk bottles by birds. *Br Birds*, **44**, 393–396.

Hirsch, S. M., & Bolles, R. C. (1980). On the ability of prey to recognize predators. *Z Tierpsychol*, **54**, 71–84.

Hockett, C. F. (1960). Logical considerations in the study of animal communication. In W. E. Lanyon & W. N. Tavolga (eds.). *Animal Sounds and Communication* (pp. 392–430). Washington, DC: American Institute of Biological Sciences.

Hodos, W. (1961). Progressive ratio as a measure of reward strength. *Science*, **134**, 934–944.

Hoffman, M. L. (1981). The development of empathy. In J. P. Rushton & R. M. Sorrentino (eds.). *Altruism and Helping Behavior: Social, Personality, and Developmental Perspectives* (pp. 41–63). Hillsdale, NJ: Laurence Erlbaum Associates.

Holland, P. C. (2008). Cognitive versus stimulus-response theories of learning. *Learn Behav*, **36**, 227–241.

Holland, R. A., Thorup, K., Vonhof, J. M., Cochran, W. W., & Wikelski, M. (2006). Navigation: Bat orientation using Earth's magnetic field. *Nature*, **444**, 702.

Holland, S. M., & Smulders, T. V. (2011). Do humans use episodic memory to solve a *What-Where-When* memory task? *Anim Cogn*, **14**, 95–102.

Hollerman, J. R., & Schultz, W. (1998). Dopamine neurons report an error in the temporal prediction of reward during learning. *Nat Neurosci*, **1**, 304–309.

Hollis, K. L. (1984). The biological function of Pavlovian conditioning: The best defense is a good offence. *J Exp Psychol Anim Behav Process*, **10**, 413–425.

Hollis, K. L. (1997). Contemporary research on Pavlovian conditioning: A "new" functional analysis. *Am Psychol*, **52**, 956–965.

Hollis, K. L., Pharr, V. L., Dumas, M. J., Britton, G. B., & Field, J. (1997). Classical conditioning provides paternity advantage for territorial blue gouramis (*Trichogaster trichopterus*). *J Comp Psychol*, **111**, 219–225.

Holloway, I. D., & Ansari, D. (2009). Mapping numerical magnitudes onto symbols: The numerical distance effect and individual differences in children's mathematics achievement. *J Exp Child Psychol*, **103**, 17–29.

Holmes, W. G., & Sherman, P. W. (1982). The ontogeny of kin recognition in two species of group squirrels. *Am Zool*, **22**, 491–517.

Honig, W. K., & Stewart, K. E. (1989). Discrimination of relative numerosity by pigeons. *Anim Learn Behav*, **17**, 134–146.

Hood, B. M. (1995). Gravity rules for 2–4-year-olds? *Cogn Dev*, **10**, 577–598.

Hood, B. M., Hauser, M. D., Anderson, L., & Santos, L. (1999). Gravity biases in a non-human primate? *Dev Sci*, **2**, 35–41.

Hood, B. M., Santos, L., & Fieselman, S. (2000). Two-year-olds' naïve predictions for horizontal trajectories. *Dev Sci*, **3**, 328–332.

Hooper, J. (2002). *Of Moths and Men: An Evolutionary Tale: The Untold Story of Science and the Peppered Moth*. New York, NY: W. W. Norton and Co.

Hoppitt, W., & Laland, K. N. (2008). Social processes influencing learning in animals: A review of the evidence. *Adv Study Behav*, **38**, 105–165.

Horner, V., Devyn Carter, J., Suchak, M., & de Waal, F. B. M. (2011). Spontaneous prosocial choice by chimpanzees. *Proc Natl Acad Sci USA*, **108**, 13847–13851.

Horner, V., Proctor, D., Bonnie, K. E., Whiten, A., & de Waal, F. B. M. (2010). Prestige affects cultural learning in chimpanzees. *PloS One*, **5**, e10625.

Horner, V., & Whiten, A. (2005). Causal knowledge and imitation/emulation switching in chimpanzees (*Pan troglodytes*) and children (*Homo sapiens*). *Anim Cogn*, **8**, 164–181.

Horner, V., Whiten, A., Flynn, E., & de Waal, F. B. M. (2006). Faithful replication of foraging techniques along cultural transmission chains by chimpanzees and children. *Proc Natl Acad Sci USA*, **103**, 13878–13883.

Horton, T. W., Holdaway, R. N., Zerbini, A. N., *et al.* (2011). Straight as an arrow: Humpback whales swim constant course tracks during long-distance migration. *Biol Lett*, **7**, 674–679.

Howard, R. W., & Blomquist, G. J. (2005). Ecological, behavioral, and biochemical aspects of insect hydrocarbons. *Ann Rev Entomol*, **50**, 371–393.

Hsu, C. Y., & Li, C. W. (1994). Magnetoreception in honeybees. *Science*, **265**, 95–97.

Hubel, D., & Wiesel, T. (1977). Functional architecture of the macaque monkey visual cortex (Ferrier lecture). *Proc Roy Soc Lond B*, **198**, 1–59.

Huber, L., & Aust, U. (2006). A modified feature theory as an account of pigeon visual categorization. In E. A. Wasserman & T. R. Zentall (eds.). *Comparative Cognition: Experimental Explorations of Animal Intelligence* (pp. 325–342). New York, NY: Oxford University Press.

Huber, L., & Lenz, R. (1993). A test of the linear feature model of polymorphous concept discrimination with pigeons. *Q J Exp Psychol*, **46B**, 1–18.

Hull, C. L. (1930). Knowledge and purpose as habit mechanisms. *Psychol Rev*, **30**, 511–525.

Hume, D. (1748/1977). *An Enquiry Concerning Human Understanding*. Indianapolis, IN: Hackett Publishing.

Hunt, G. R. (1996). Manufacture and use of hook-tools by New Caledonian crows. *Nature*, **379**, 249–251.

Hunt, G. R., Corballis, M. C., & Gray, R. D. (2001). Laterality in tool manufacture by crows. *Nature*, **414**, 707.

Hunt, G. R., & Gray, R. D. (2003). Diversification and cumulative evolution in New Caledonian crow tool manufacture. *Proc Roy Soc Lond B Biol Sci*, **270**, 867–874.

Huntingford, F. A., & Wright, P. J. (1992). Inherited population differences in avoidance conditioning in three-spined stickleback, *Gasterosteus aculeatus*. *Behaviour*, **122**, 264–273.

Hursh, S. R. (1980). Economic concepts for the analysis of behavior. *J Exp Anal Behav*, **34**, 219–238.

Hutchinson, J. M. C., & Gigerenzer, G. (2005). Simple heuristics and rules of thumb: where psychologists and behavioural biologists might meet. *Behav Process*, **69**, 87–124

Hyman, I. E., & Pentland, J. (1996). The role of mental imagery in the creation of false childhood memories. *J Mem Lang*, **35**, 101–117.

Inoue, T., Hasegawa, T., Takara, S., Lukats, B., Mizuno, M., & Aou, S. (2008). Categorization of biologically significant objects, food and gender in rhesus monkeys: I. Behavioral study. *Neurosc Res*, **61**, 70–78.

Iriki, A., Tanaka, M., & Iwamura, Y. (1996). Coding of modified body schema during tool use by macaque postcentral neurones. *Neuroreport*, **7**, 2325–2330.

Ito, M., Takatsuru, S., & Saeki, D. (2000). Choice between constant and variable alternatives by rats: Effects of different reinforcer amounts and energy budgets. *J Exp Anal Behav*, **73**, 79–92.

Jaakkola, K., Guarino, E., Rodriguez, M., Erb, L., & Trone, M. (2010). What do dolphins (*Tursiops truncatus*) understand about hidden objects? *Anim Cogn*, **13**, 103–120.

Jackendoff, R. (1987). *Consciousness and the Computational Mind*. Cambridge, MA: MIT Press.

Jacobs, L. F., Gaulin, S. J. C., Sherry, D. F., & Hoffman, G. E. (1990). Evolution of spatial cognition: Sex-specific patterns of spatial behavior predict hippocampal size. *Proc Natl Acad Sci USA*, **87**, 6349–6352.

Jacobs, L. F., & Schenk, F. (2003). Unpacking the cognitive map: The parallel map theory of hippocampal function. *Psychol Rev*, **110**, 285–315.

Jacobs, L. F., & Spencer, W. D. (1994). Natural space-use patterns and hippocampal size in kangaroo rats. *Brain Behav Evol*, **44**, 125–132.

Jacobsen, C. F. (1935). Functions of frontal association areas in primates. *Arch Neurol Psychiatry*, **33**, 558–569.

Jaeggi, A.V., Burkart, J.M., & Van Schaik, C.P. (2010). On the psychology of cooperation in humans and other primates: Combining the natural history and experimental evidence of prosociality. *Philos Trans Roy Soc B Biol Sci*, **365**, 2723–2735.

Janik, V. M., Sayigh, L. S., & Wells, R. S. (2006). Signature whistle shape conveys identity information to bottlenose dolphins. *Proc Natl Acad Sci USA*, **103**, 8293–8297.

Jeanson, R., Dussutour, A., & Fourcassié, E. (2012). Key factors for the emergence of collective decision in invertebrates. *Front Neurosci*, **6**, 1–15.

Jeneson, A., Mauldin, K. N., & Squire, L. R. (2010). Intact working memory for relational information after medial temporal lobe damage. *J Neurosci*, **30**, 13624–13629.

Jenkins, H. H., & Harrison, R. H. (1962). Generalization gradients of inhibition following auditory discrimination learning. *J Exp Anal Behav*, **5**, 435–441.

Jenkins, H. M., & Moore, B. R. (1973). The form of the autoshaped response with food or water reinforcers. *J Exp Anal Behav*, **20**, 163–181.

Jensen, K., Call, J., & Tomasello, M. (2007). Chimpanzees are rational maximizers in an ultimatum game. *Science*, **318**, 107–109.

Jensen, K., Hare, B., Call, J., & Tomasello, M. (2006). What's in it for me? Self-regard precludes altruism and spite in chimpanzees. *Proc Roy Soc B Biol Sci*, **273**, 1013–1021.

Jentsch, J. D., & Taylor, J. R. (1999). Impulsivity resulting from frontostriatal dysfunction in drug abuse: Implications for the control of behavior by reward-related stimuli. *Psychopharmacology*, **146**, 373–390.

Jitsumori, M. (1994). Discrimination of artificial polymorphous categories by rhesus monkeys (*Macaca mulatta*). *Q J Exp Psychol*, **47B**, 371–386.

Jitsumori, M., Ohkita, M., & Ushitani, T. (2011). The learning of basic level categories by pigeons: The prototype effect, attention, and the effects of categorization. *Learn Behav*, **39**, 271–287.

Johansen, J. P., Cain, C. K., Ostroff, L. E., & LeDoux, J. E. (2012). Molecular mechanisms of fear learning and memory. *Cell*, **147**, 509–524.

Johansson, G. (1973). Visual perception of biological motion and a model for its analysis. *Atten Percept Psychophys*, **14**, 201–211.

Johansson, G. (1976). Spatio-temporal differentiation and integration in visual motion perception. *Psychol Res*, **38**, 379–393.

Johnson, M. H. (2005). Sub-cortical face processing. *Nat Rev Neurosci*, **6**, 766–774.

Johnson, M. H., Dziurawiec, S., Ellis, H., & Morton, J. (1991). Newborns' preferential tracking of face-like stimuli and its subsequent decline. *Cognition*, **40**, 1–19.

Johnson, S. C., Dweck, C. S., Chen, F. S., *et al.* (2010). At the intersection of social and cognitive development: Internal working models of attachment in infancy. *Cogn Sci*, **34**, 807–825.

Johnstone, R. A. (1998). Conspiratorial whispers and conspicuous displays: Games of signal detection. *Evolution*, **52**, 1554–1563.

Johnstone, R. A. (2001). Eavesdropping and animal conflict. *Proc Natl Acad Sci USA*, **98**, 9177–9180.

Joly-Mascheroni, R. M., Senju, A., & Shepherd, A. J. (2008). Dogs catch human yawns. *Biol Lett*, **4**, 446–448.

Jones, F. R. H. (1955). Photo-kinesis in the ammocoete larva of the brook lamprey. *J Exp Biol*, **34**, 492–503.

Jones, J. E., Antoniadis, E., Shettleworth, S. J., & Kamil, A. C. (2002). A comparative study of geometric rule learning by nutcrackers (*Nucifraga columiana*), pigeons (*Columba livia*), and jackdaws (*Corvus monedula*). *J Comp Psychol*, **116**, 350–356.

Josselyn, S. A., Kida, S., & Silva, A. J. (2004). Inducible repression of CREB function disrupts amygdala-dependent memory. *Neurobiol Learn Mem*, **82**, 159–163.

Jozefowiez, J., & Staddon, J. E. R. (2008). Operant behavior. In R. Menzel (ed.). *Learning Theory and Behavior* (pp. 75–102). Oxford, UK: Elsevier Press.

Jupp, B., Caprioli, D., & Dalley, J. W. (2013). Highly impulsive rats: Modelling an endophenotype to determine the neurobiological, genetic and environmental mechanisms of addiction. *Dis Model Mech*, **6**, 302–311.

Jusczyk, P. W. (1997). *The Discovery of Spoken Language*. Cambridge, MA: Massachusetts Institute of Technology.

Kacelnik, A. (1984). Central place foraging in starlings (*Sturnus vulgaris*). I. Patch residence time. *J Anim Ecol*, **53**, 283–299.

Kacelnik, A., & Bateson, M. (1996). Risky theories: The effect of foraging on risky decisions. *Am Zool*, **36**, 420–434.

Kacelnik, A., & Bateson, M. (1997). Risk-sensitivity: Crossroads for the theories of decision-making. *Trends Cogn Sci*, **1**, 304–309.

Kacelnik, A. & Marsh, B. (2002). Cost can increase preference in starlings. *Anim Behav*, **63**, 245–250.

Kahneman, D., Treisman, A., & Gibbs, B. J. (1992). The reviewing of object files: Object-specific integration of information. *Cogn Psychol*, **24**, 175–219.

Kamil, A. C., & Cheng, K. (2001). Way-finding and landmarks: The multiple bearing hypothesis. *J Exp Biol*, **204**, 103–113.

Kamin, L. J. (1969). Predictability, surprise, attention and conditioning. In B. A. Campbell & R. M. Church (eds.). *Punishment and Aversive Behavior* (pp. 279–296). New York, NY: Appleton-Century-Crofts.

Kaminski, J., Call, J., & Fischer, J. (2004). Word learning in a domestic dog: Evidence for "fast mapping". *Science*, **304**, 1682–1683.

Kaminski, J., Riedel, J., Call, J., & Tomasello, M. (2005). Domestic goats, *Capra hircus*, follow gaze direction and use social cues in an object choice task. *Anim Behav*, **69**, 11–18.

Kandel, E. R. (2001). The molecular biology of memory storage: A dialogue between genes and synapses. *Science*, **294**, 1030–1038.

Kappeler, P. M., Silk, J. S., Burkart, J. M., & van Schaik, C. P. (2010). Primate behavior and human universals: Exploring the gap. In P. M. Kappeler & J. B. Silk (eds.). *Mind the Gap: Tracing the Origins of Human Universals* (pp. 3–15). Berlin: Springer.

Karavanich, C., & Atema, J. (1998). Individual recognition and memory in lobster dominance. *Anim Behav*, **56**, 1553–1560.

Kass, J. H. (2008). The evolution of complex sensory and motor systems of the human brain. *Brain Res Bull*, **18**, 384–390.

Kastak, C. R., Schusterman, R. J., & Kastak, D. (2001). Equivalence classification by California sea lions using class-specific reinforcers. *J Exp Anal Behav*, **76**, 131–158.

Katz, J. S., Cook, R. G., & Magnotti, J. F. (2010). Toward a framework for evaluating feature binding in pigeons. *Behav Process*, **85**, 215–225.

Kawai, M. (1965). Newly-acquired pre-cultural behavior of the natural troop of Japanese monkeys on Koshima Islet. *Primates*, **6**, 1–30.

Kawamura, S. (1959). The process of sub-culture propagation among Japanese macaques. *Primates*, **2**, 43–60.

Kelly, D. M., & Spetch, M. L. (2001). Pigeons encode relative geometry. *J Exp Psychol Anim Behav Process*, **27**, 417–422.

Kennedy, E. H., & Fragaszy, D. M. (2008). Analogical reasoning in a capuchin monkey (*Cebus apella*). *J Comp Psychol*, **122**, 167–175.

Kennedy, M., & Gray, R. D. (1993). Can ecological theory predict the distribution of foraging animals? A critical evaluation of experiments on the ideal free distribution. *Oikos*, **68**, 158–166.

Kennedy, P. J., & Shapiro, M. L. (2009). Motivational states activate distinct hippocampal representations to guide goal-directed behaviors. *Proc Natl Acad Sci USA*, **106**, 10805–10810.

Kenward, B., Karlsson, M., & Persson, J. (2011). Over-imitation is better explained by norm learning than by distorted causal learning. *Proc Roy Soc B Biol Sci*, **278**, 1239–1246.

Kenward, B., Rutz, C., Weir, A. A., & Kacelnik, A. (2006). Development of tool use in new Caledonian crows: Inherited action patterns and social influences. *Anim Behav*, **72**, 1329–1343.

Kenward, B., Weir, A. A., Rutz, C., & Kacelnik, A. (2005). Behavioural ecology: Tool manufacture by naive juvenile crows. *Nature*, **433**, 121.

Kenward, R. E. (1978). Hawks and doves: Factors affecting success and selection in goshawk attacks on wood-pigeons. *J Anim Ecol*, **47**, 449–460.

Kettlewell, H. B. D. (1955). Selection experiments on industrial melanism in the Lepidoptera. *Heredity*, **9**, 323–343.

Keysar, B., Lin, S., & Barr, D. J. (2003). Limits on theory of mind use in adults. *Cognition*, **89**, 25–41.

Kimble, G. A. (1967). The definition of learning and some useful distinctions. In G. A. Kimble (ed.). *Foundations of Conditioning and Learning* (pp. 82–99). New York, NY: Appleton, Century, Crofts.

King, S. L., & Janik, V. M. (2013). Bottlenose dolphins can use learned vocal labels to address each other. *Proc Natl Acad Sci USA*, **110**, 13216–13221.

Kirby, K. N., & Petry, N. M. (2004). Heroin and cocaine abusers have higher discount rates for delayed rewards than alcoholics or non-drug controls. *Addiction*, **99**, 461–471.

Kirchner, W. H., & Towne, W. F. (1994). The sensory basis of the honeybee's dance language. *Sci Am*, **270**, 74–80.

Kisilevsky, B. S., Hains, S. M., Lee, K., *et al.* (2003). Effects of experience on fetal voice recognition. *Psychol Sci*, **14**, 220–224.

Klein, E. D., Bhatt, R. S., & Zentall, T. R. (2005). Contrast and the justification of effort. *Psychon Bull Rev*, **12**, 335–339.

Klein, E. D., & Zentall, T.R. (2003). Affordance learning by pigeons (*Columba livia*). *J Comp Cogn*, **117**, 414–419.

Klein, S. B., Cosmides, L., Tooby, J., & Chance, S. (2002). Decisions and evolution of memory: Multiple systems, multiple functions. *Psychol Rev*, **109**, 306–329.

Klin, A. (2000). Attributing social meaning to ambiguous visual stimuli in higher-functioning autism and Asperger syndrome: The social attribution task. *J Child Psychol Psychiatry*, **41**, 831–846.

Klin, A., Jones, W., Schultz, R., Volkmar, F., & Cohen, D. (2002). Visual fixation patterns during viewing of naturalistic social situations as predictors of social competence in individuals with autism. *Arch Gen Psychiatry*, **59**, 809–816.

Knierim, J. J., Skaggs, W. E., Kudrimoti, H. S., & McNaughton, B. L. (1996). Vestibular and visual cues in navigation: A tale of two cities. *Ann N Y Acad Sci*, **781**, 399–406.

Knowlton, B. J., Mangels, J. A., & Squire, L. R. (1996). A neostriatal habit learning system in humans. *Science*, **256**, 675–677.

Knowlton, B. J., & Squire, L. R. (1993). The learning of categories: Parallel brain systems for item memory and category knowledge. *Science*, **262**, 1747–1749.

Kobayashi, T., Hiraki, K., Mugitani, R., & Hasegawa, T. (2004). Baby arithmetic: One object plus one tone. *Cognition*, **91**, B23–B34.

Köhler, W. (1927). *The Mentality of Apes*. New York, NY: Harcourt, Brace, & Company.

Kolb, A. (1977). How do mother and young of the bat (*Myotis myotis*) recognize each other after mother's return from a hunting flight? *Zi Tierpsychologie*, **44**, 324–431.

Kolling, N., Behrens, T. E. J., Mars, R. B., & Rushworth, M. F. S. (2012). Neural mechanisms of foraging. *Science*, **336**, 95–98.

Komdeur, J., & Hatchwell, B. J. (1999). Kin recognition: Function and mechanisms in avian species. *Trends Ecol Evol*, **14**, 237–241.

Konishi, M. (1965). The role of auditory feedback in the control of vocalization in the white-crowned sparrow. *Z Tierpsychol*, **22**, 770–783.

Kosfeld, M., Heinrichs, M., Zak, P. J., Fischbacher, U., & Fehr, E. (2005). Oxytocin increases trust in humans. *Nature*, **435**, 673–676.

Koski, S. E., & Sterck, E. H. (2007). Triadic postconflict affiliation in captive chimpanzees: Does consolation console? *Anim Behav*, **73**, 133–142.

Kramer, G. (1952). Experiments on bird orientation. *Ibis*, **94**, 265–285.

Krause, J., & Godin, J. G. J. (1996). Influence of prey foraging posture on flight behavior and predation risk: Predators take advantage of unwary prey. *Behav Ecol*, **7**, 264–271.

Krause, J., Orlando, L., Serre, D., *et al.* (2007). Neanderthals in central Asia and Siberia. *Nature*, **449**, 902–904.

Krebs, J. R., & Biebach, H. (1989). Time-place learning by garden warblers (*Sylvia borin*): Route or map? *Ethology*, **83**, 248–256.

Krebs, J. R. & Dawkins, R. 1984. Animal signals: Mind-reading and manipulation. In J. R. Krebs & N. B. Davies (eds.). *Behavioural Ecology: An Evolutionary Approach* (2nd Edn.) (pp. 380–402). Oxford, UK: Oxford University Press.

Krebs, J. R., Erichsen, F. T., Webber, M. I., & Charnov, E. L. (1977). Optimal prey selection by the great tit (*Parus major*). *Anim Behav*, **25**, 30–38.

Krebs, J. R., Kacelnik, A., & Taylor, P. J. (1978). Test of optimal sampling by foraging great tits. *Nature*, **275**, 27–31.

Krubitzer, L. A., Manger, P., Pettigrew, J. D., & Calford, M. B. (1995). Organization of somatosensory cortex in monotremes: In search of the prototypical plan. *J Comp Neurol*, **351**, 261–306.

Krützen, M., Mann, J., Heithaus, M. R., *et al.* (2005). Cultural transmission of tool use in bottlenose dolphins. *Proc Natl Acad Sci USA*, **102**, 8939–8943.

Kruuk, H. (1972). *The Spotted Hyena: A Study of Predation and Social Behavior.* Chicago, IL: University of Chicago Press.

Kuhl, P. K., Stevens, E., Hayashi, A., *et al.* (2006). Infants show a facilitation effect for native language phonetic perception between 6 and 12 months. *Dev Sci*, **9**, F13–F21.

Kuhl, P. K., Williams, K. A., Lacerda, F., Stevens, K. N., & Lindblom, B. (1992). Linguistic experience alters phonetic perception in infants by 6 months of age. *Science*, **255**, 606–608.

Kummer, H., & Goodall, J. (1985). Conditions of innovative behaviour in primates. *Philos Trans Roy Soc Lond B Biol Sci*, **308**, 203–214.

Kutsukake, N., & Castles, D. L. (2004). Reconciliation and post-conflict third-party affiliation among wild chimpanzees in the Mahale Mountains, Tanzania. *Primates*, **45**, 157–165.

Lai, C. S., Fisher, S. E., Hurst, J. A., Vargha-Khadem, F., & Monaco, A. P. (2001). A forkhead-domain gene is mutated in a severe speech and language disorder. *Nature*, **413**, 519–523.

Laland, K. N. (2004). Social learning strategies. *Anim Learn Behav*, **32**, 4–14.

Laland, K. N., & Janik, V. M. (2006). The animal cultures debate. *Trends Ecol & Evol*, **21**, 542–547.

Lane, H. (1976). *The Wild Boy of Aveyron* (Vol. 149). Cambridge, MA: Harvard University Press.

Langley, C. M., Riley, D. A., Bond, A. B., & Goel, N. (1996). Visual search for natural grains in pigeons (*Columba livia*): Search images and selective attention. *J Exp Psychol Anim Behav Process*, **14**, 96–104.

Laude, J. R., Pattison, K. F., & Zentall, T. R. (2012). Hungry pigeons make suboptimal choices, less hungry pigeons do not. *Psychon Bull Rev*, **19**, 884–891.

Laverty, T. M. (1980). Bumblebee foraging: Floral complexity and learning. *Can J Zool*, **58**, 1324–1335.

Leahey, T. H. (1992). The mythical revolution of American psychology. *Am Psychol*, **47**, 308–318.

Leahey, T. H., & Harris, R. J. (2001). *Learning and Cognition* (5th Edn.). Upper Saddle River, NJ: Prentice Hall.

LeDoux, J. (2007). The amygdala. *Curr Biol*, **17**, R868–874.

Lee, V., & Kuhlmeier, V. A. (2013). Young children show a dissociation in looking and pointing behavior in falling events. *Cogn Dev*, **28**, 21–30.

Lee, W. Y., Lee, S., Choe, J. C., & Jablonski, P. G. (2011). Wild birds recognize individual humans: Experiments on magpies, *Pica pica*. *Anim Cogn*, **14**, 817–825.

Lefebvre, L., & Palameta, B. (1988). Mechanisms, ecology, and population diffusion of socially learned, food-finding behavior in feral pigeons. In T. R. Zentall & B. G. Galef Jr. (eds.). *Social Learning: Psychological and Biological Perspectives* (pp. 141–164). Hillsdale, NJ: Lawrence Erlbaum Associates, Inc.

Lejeune, H., & Wearden, J. H. (1991). The comparative psychology of fixed-interval responding: Some quantitative analyses. *Learn Motiv*, **22**, 84–111.

Lemon, W. C., & Barth, R. H. (1992). The effects of feeding rate on reproductive success in the zebra finch, *Taeniopyga guttata*. *Anim Behav*, **44**, 851–857.

Leslie, A. M., & Keeble, S. (1987). Do six-month-old infants perceive causality? *Cognition*, **25**, 265–288.

Lessard, N., Pare, M., Lepore, F., & Lassonde, M. (1998). Early-blind human subjects localize sound sources better than sighted individuals. *Nature*, **395**, 278–280.

Levey D.J., Londoño, G.A., Ungvari-Martin J., *et al.* (2009). Urban mockingbirds quickly learn to identify individual humans. *Proc Natl Acad Sci USA*, **106**, 8959–8962.

Lewens, T. (2007). Adaptation. In D.L. Hull & M. Ruse (eds.). *The Cambridge Companion to the Philosophy of Biology* (pp. 1–21). Cambridge: Cambridge University Press.

Lewis, J. W. (2006). Cortical networks related to human use of tools. *Neuroscientist*, **12**, 211–231.

Lewis, K. P., Jaffe, S., & Brannon, E. M. (2005). Analog number representations in mongoose lemurs (*Eulemur mongoz*): Evidence from a search task. *Anim Cogn*, **8**, 247–252.

Liberman, A. M., Harris, K. S., Hoffman, H. S., & Griffiths, B. C. (1957). The discrimination of speech sounds within and across phoneme boundaries. *J Exp Psychol*, **54**, 358–368.

Libertus, M. E., & Brannon, E. M. (2010). Stable individual differences in number discrimination in infancy. *Dev Sci*, **13**, 900–906.

Libertus, M. E., Feigenson, L., & Halberda, J. (2013). Is approximate number precision a stable predictor of math ability? *Learn Indiv Differ*, **25**, 126–133.

Libertus, M. E., Odic, D., & Halberda, J. (2012). Intuitive sense of number correlates with math scores on college-entrance examination. *Acta Psychol*, **141**, 373–379.

Libet, B., Gleason, C. A., Wright, E. W., & Pearl, D. K. (1983). Time of conscious intention to act in relation to onset of cerebral activity (readiness-potential). The unconscious initiation of a freely voluntary act. *Brain*, **106**, 632–642.

Lima, S. L. (1984). Downy woodpecker foraging behaviour: Efficient sampling in simple stochastic environments. *Ecology*, **65**, 166–174.

Lima, S. L. (1985). Maximizing feeding efficiency and minimizing time exposed to predators: A trade-off in the black-capped chickadee. *Oecologia*, **66**, 60–67.

Lima, S. L. (1994). Collective detection of predatory attack by birds in the absence of alarm signals. *J Avian Biol*, **25**, 319–326.

Lima, S. L., Valone, T. J., & Caraco, T. (1985). Foraging efficiency-predation risk tradeoff in the grey squirrel. *Anim Behav*, **33**, 155–165.

Limongelli, L., Boysen, S. T., & Visalberghi, E. (1995). Comprehension of cause-effect relations in a tool-using task by chimpanzees (*Pan troglodytes*). *J Comp Psychol*, **109**, 18–26.

Lipton, J. S., & Spelke, E. S. (2003). Origins of number sense: Large-number discrimination in human infants. *Psychol Sci*, **14**, 396–401.

Lipton, J. S., & Spelke, E. S. (2004). Discrimination of large and small numerosities by human infants. *Infancy*, **5**, 271–290.

Liu, D., Sabbagh, M. A., Gehring, W. J., & Wellman, H. M. (2009). Neural correlates of children's theory of mind development. *Child Dev*, **80**, 318–326.

Loftus, E. F., & Palmer, J. C. (1974). Reconstruction of automobile destruction: An example of the interaction between language and memory. *J Verbal Learn Verbal Behav*, **13**, 585–589.

Lohmann, K. J., Lohmann, C. M. F., Ehrhart, L. M., Bagley, D. A., & Swing, T. (2004). Geomagnetic map used in sea-turtle navigation. *Nature*, **428**, 909–910.

Lohmann, K. J., Lohmann, C. M. F., & Endres, C. S. (2008). The sensory ecology of ocean navigation. *J Exp Biol*, **211**, 1719–1728.

Lolordo, V. M., Jacobs, W. J., & Foree, D. D. (1982). Failure to block control by a relevant stimulus. *Anim Learn Behav*, **10**, 183–193.

Lopez, J. C., Rodriguez, F., Gomez, Y., *et al.* (2000). Place and cue learning in turtles. *Anim Learn Behav*, **28**, 360–372.

Lorenz, K. Z. (1950). The comparative method in studying innate behaviour patterns. *Symp Soc Exp Biol*, **4**, 221–268.

Lorenz, K. Z. (1952). *King Solomon's Ring*. (Wilson, M. K., trans.). New York: Thomas Y. Cromwell.

Lorenzetti, F. D., Baxter, D. A., & Byrne, J. H. (2008). Molecular mechanisms underlying a cellular analog of operant reward learning. *Neuron*, **59**, 815–828.

Lubyk, D. M., & Spetch, M. L. (2012). Finding the best angle: Pigeons (*Columbia livia*) weight angular information more heavily than relative wall length in an open-field geometry task. *Anim Cogn*, **15**, 305–312.

Luhmann, C. C. (2009). Temporal decision-making; Insights from cognitive neuroscience. *Front Behav Neurosci*, **3**, 1–9.

Lundberg, P. (1988). The evolution of partial migration in birds. *Trends Ecol Evol*, **3**, 172–176.

Lupfer, G., Frieman, J., & Coonfield, D. (2003). Social transmission of flavor preferences in two species of hamsters (*Mesocricetus auratus* and *Phodopus campbelli*). *J Comp Psychol*, **117**, 449–455.

Luria, A. R. (1968). *The Mind of a Mnemonist*. New York, NY: Basic Books Inc.

Lyons, D. E., Young, A. G., & Keil, F. C. (2007). The hidden structure of overimitation. *Proc Natl Acad Sci USA*, **104**, 19751–19756.

Lythgoe, J. N. (1979). *The Ecology of Vision*. Oxford, UK: Clarendon Press.

MacArthur, R. H., & Pianka, E. R. (1966). On optimal use of a patchy environment. *Am Nat*, **100**, 603–609.

Macaskill, A. C., & Hackenberg, T. D. (2012). The sunk cost effect with pigeons: Some determinants of decisions about persistence. *J Exp Anal Behav*, **97**, 85–100.

MacDonald, S. E., Spetch, M. L., Kelly, D. M., & Cheng, K. (2004). Strategies in landmark use by children, adults, and marmoset monkeys. *Learn Motiv*, **35**, 322–347.

Machado, A., & Keen, R. (1999). Learning to time (LET) or scalar expectancy theory (SET)? A critical test of two models of timing. *Psychol Sci*, **10**, 285–290.

MacKillop, J. (2013). Integrating behavioral economics and behavioral genetics: Delayed reward discounting as an endophenotype of addictive disorders. *J Exp Anal Behav*, **99**, 14–31.

Mackintosh, N. J. (1975). A theory of attention: Variations in the associability of stimuli with reinforcement. *Psychol Rev*, **82**, 276–298.

MacLean, E. L., Matthews, L. J., Hare, B. A., *et al.*, (2012). How does cognition evolve? Phylogenetic comparative psychology. *Anim Cogn*, **15**, 223–238.

MacPhail, E. M. (1996). Cognitive function in mammals: The evolutionary perspective. *Cogn Brain Res*, **3**, 279–280.

Madden, G. J., Smethells, J. R., Ewan, E. E., & Hursh, S. R. (2007). Tests of behavioral-economic assessment of relative reinforcer efficacy: Economic substitutes. *J Exp Anal Behav*, **87**, 219–240.

Madsen, E.A., Tunney, R.J., Fieldman, G., Plotkin, H.C., & Dunbar, R.I. (2007). Kinship and altruism: A cross-cultural experimental study. *Br J Psychol*, **98**, 339–359.

Maguire, E. A., Burgess, N., Donnett, J. G., *et al.* (1998). Knowing where and getting there: A human navigation system. *Science*, **280**, 921–934.

Maguire, E. A., Gadian, D. G., Johnsrude, I. S., *et al.* (2000). Navigation-related structural change in the hippocampi of taxi drivers. *Proc Natl Acad Sci USA*, **97**, 4398–4403.

Mahon, B. Z., & Caramazza, A. (2009). Concepts and categories: A cognitive neuropsychological perspective. *Ann Rev Psychol*, **60**, 27–51.

Mahon, B. Z., & Caramazza, A. (2011). What drives the organization of object knowledge in the brain? *Trends Cogn Neurosci*, **15**, 97–103.

Malapani, C., Rakitin, B. C., Levy, R., *et al.* (1998). Coupled temporal memories in Parkinson's disease: A dopamine-related dysfunction. *J Cogn Neurosci*, **10**, 316–331.

Malenka, R. C., & Bear, M. F. (2004). LTP and LTD: An embarrassment of riches. *Neuron*, **4**, 5–21.

Manser, M. B., & Bell, M. B. (2004). Spatial representation of shelter locations in meerkats, *Suricata suricatta*. *Anim Behav*, **68**, 151–157.

Marchette, S. A., Bakker, A., & Shelton, A. M. (2011). Cognitive mappers to creatures of habit: Differential engagement of place and response learning mechanisms predicts human navigational behavior. *J Neurosci*, **31**, 15264–15268.

Mardon, J., Saunders, S. M., Anderson, M. J., Couchoux, C., & Bonadonna, F. (2010). Species, gender, and identity: Cracking petrels' sociochemical code. *Chem Senses*, **35**, 309–321.

Marhold, S., & Wiltschko, W. (1997). Magnetic polarity compass or direction finding in a subterranean mammal. *Naturwissenchaften*, **84**, 421–423.

Marino, L. (2002). Convergence of complex cognitive abilities in cetaceans and primates. *Brain Behav Evol*, **59**, 21–23.

Marino, L., Connor, R. C., Fordyce, R. E., *et al.* (2007). Cetaceans have complex brains for complex cognition. *PLoS Biol*, **5**, e139.

Markman, E. M., & Abelev, M. (2004). Word learning in dogs? *Trends Cogn Sci*, **8**, 479–481.

Markson, L., & Bloom, P. (1997). Evidence against a dedicated system for word learning in children. *Nature*, **385**, 813–815.

Marler, P. (1967). Animal communication signals: We are beginning to understand how the structure of animal signals relates to the function they serve. *Science*, **157**, 769–774.

Marler, P. (1968). Aggregation and dispersal: Two functions in primate communication. In P. Jay (ed.). *Primates* (pp. 420–438). New York, NY: Holt, Rinehart, & Winston.

Marler, P. (1970). A comparative approach to vocal learning: Song development in white-crowned sparrows. *J Comp Physiol Psychol*, **71**, 1–25.

Marler, P., Dufty, A., & Pickert, R. (1986). Vocal communication in the domestic chicken: II. Is a sender sensitive to the presence and nature of a receiver? *Anim Behav*, **34**, 194–198.

Marler, P., & Slabbekoorn, H. (eds.) (2004). *Nature's Music*. San Diego, CA: Elsevier Academic Press.

Marler, P., & Tamura, M. (1964). Culturally transmitted patterns of vocal behavior in sparrows. *Science*, **146**, 1483–1486.

Marsh, H. L., & MacDonald, S. E. (2008). The use of perceptual features in categorization by orangutans (*Pongo abelii*). *Anim Cogn*, **11**, 569–585.

Marsh, H. L., Spetch, M. L., & MacDonald, S. E. (2011). Strategies in landmark use by orangutans and human children. *Anim Cogn*, **14**, 487–502.

Martin, S. J., Vitikainen, E., Drijfhout, F. P., & Jackson, D. (2012). Conspecific ant aggression is correlated with chemical distance, but not with genetic or spatial distance. *Behav Genet*, **42**, 323–331.

Martin-Ordas, G., Call, J., & Colmenares, F. (2008). Tubes, tables and traps: Great apes solve two functionally equivalent trap tasks but show no evidence of transfer across tasks. *Anim Cogn*, **11**, 423–430.

Marzluff, J.M., Walls, J., Cornell, H.N., Withey, J.C., & Craig, D.P. (2010) Lasting recognition of threatening people by wild American crows. *Anim Behav*, **79**, 699–707.

Mateo, J. M. (2002). Kin recognition abilities and nepotism as a function of sociality. *Proc Roy Soc Lond B*, **269**, 721–727.

Mateo, J. M. (2003). Kin recognition in ground squirrels and other rodents. *J Mammal*, **84**, 1163–1181.

Mateo, J. M., & Holmes, W. G. (2004). Cross-fostering as a means to study kin recognition. *Anim Behav*, **68**, 1451–1459.

Mateo, J. M., & Johnson, R. E. (2000). Kin recognition and the 'armpit' effect: Evidence of self-referent phenotype matching. *Proc Roy Soc Lond B*, **267**, 695–700.

Matsuzawa, T. (1985). Colour naming and classification in a chimpanzee (*Pan troglodytes*). *J Hum Evol*, **14**, 283–291.

Matsuzawa, T. (2009). Symbolic representation of number in chimpanzees. *Curr Opin Neurobiol*, **19**, 92–98.

Matyjasiak, P. (2004). Birds associate species-specific acoustic and visual cues: Recognition of heterospecific rivals by male blackcaps. *Behav Ecol*, **16**, 467–471.

Maynard Smith, J., & Harper, D. (2003). *Animal Signals*. Oxford, UK: Oxford University Press.

Mazzocco, M. M., Feigenson, L., & Halberda, J. (2011). Preschoolers' precision of the approximate number system predicts later school mathematics performance. *PLoS One*, **6**, e23749.

McCarthy, R. A., & Warrington, E. A. (1994). Disorders of semantic memory. *Philos Trans Roy Soc Lond B*, **346**, 89–96.

McCloskey, M., Wible, C. G., & Cohen, N. J. (1988). Is there a special flashbulb memory mechanism? *J Exp Psychol General*, **117**, 171–181.

McComb, K., Packer, C., & Pusey, A. (1994). Roaring and numerical assessment in contests between groups of female lions, *Panthera leo*. *Anim Behav*, **47**, 379–387.

McCowan, B., & Reiss, D. (2001). The fallacy of 'signature whistles' in bottlenose dolphins: A comparative perspective of 'signature information' in animal vocalizations. *Anim Behav*, **62**, 1151–1162.

McCrink, K., & Wynn, K. (2004). Large-number addition and subtraction by 9-month-old infants. *Psychol Sci*, **15**, 776–781.

McDonald, R. J., & White, N. M. (1993). A triple dissociation of memory systems: Hippocampus, amygdala, and dorsal striatum. *Behav Neurosci*, **107**, 3–22.

McGaugh, J. L. (2004). The amygdala modulates consolidation of emotionally arousing experiences. *Ann Rev Neurosci*, **27**, 1–28.

McGaugh, J. L., & Krivanek, J. A. (1970). Strychnine effects on discrimination learning in mice: Effects of dose and time of administration. *Physiol Behav*, **5**, 1437–1442.

McGaugh, J. L., & Roozendaal, B. (2009). Drug enhancement of memory consolidation: Historical perspective and neurobiological implication. *Psychopharmacology*, **202**, 3–14.

McGregor, P. K. (1993). Signalling in territorial systems: A context for individual identification, ranging and eavesdropping. *Philos Trans Roy Soc Lond B Biol Sci*, **340**, 237–244.

McGuigan, N., Makinson, J., & Whiten, A. (2011). From over-imitation to super-copying: Adults imitate causally irrelevant aspects of tool use with higher fidelity than young children. *Br J Psychol*, **102**, 1–18.

McNaughton, B. L., Barnes, C. A., Gerrard, J. L., *et al.* (1996). Deciphering the hippocampal polyglot: The hippocampus as a path integration system. *J Exp Biol*, **199**, 173–185.

McNaughton, B. L., Battaglia, F. P., Jensen, O., Moser, E., & Moser, M.-B. (2006). Path integration and the neural basis of the 'cognitive map'. *Nat Rev Neurosci*, **7**, 663–678.

Mech, D.L. (1970). *The Wolf*. New York, NY: Natural History Press.

Meck, W. H., & Church, R. M. (1983). A mode control model of counting and timing processes. *J Exp Psychol Anim Behav Process*, **9**, 320–334.

Mehlhorn, J., & Rehkamper, G. (2009). Neurobiology of the homing pigeon – a review. *Naturwissenschaften*, **96**, 1011–1025.

Melis, A. P., Hare, B., & Tomasello, M. (2006). Engineering cooperation in chimpanzees: Tolerance constraints on cooperation. *Anim Behav*, **72**, 275–286.

Melis, A. P., & Seemann, D. (2010). How is human cooperation different? *Philos Trans Roy Soc B Biol Sci*, **365**, 2663–2674.

Meltzoff, A. N. (2007). 'Like me': A foundation for social cognition. *Developmental Science*, **10**, 126–134.

Mendez, I A., Simon, N. W., Hart, N., *et al.* (2010). Self-administered cocaine causes long-lasting increases in impulsive choice in a delay discounting task. *Behav Neurosci*, **124**, 470–477.

Mendoza-Granados, D., & Sommer, V. (1995). Play in chimpanzees of the Arnhem zoo: Self-serving compromises. *Primates*, **36**, 57–68.

Mendres, K. A., & de Waal, F. (2000). Capuchins do cooperate: The advantage of an intuitive task. *Anim Behav*, **60**, 523–529.

Mennella, J. A., Jagnow, C. P., & Beauchamp, G. K. (2001). Prenatal and postnatal flavor learning by human infants. *Pediatrics*, **107**, 1–6.

Menzel, E. W. (1973). Chimpanzee spatial memory organization. *Science*, **182**, 943–945.

Menzel, R. (2009). Working memory in bees: Also in flies? *J Neurogenet*, **23**, 92–99.

Menzel, R., De Marco, R.J., & Greggers, U. (2006). Spatial memory, navigation, and dance behaviour in *Apis mellifera*. *J Comp Physiol A*, **192**, 889–903.

Menzel, R., Greggers, U., Smith, A., *et al.* (2005). Honey bees navigate according to a map-like spatial memory. *Proc Natl Acad Sci USA*, **102**, 3040–3045.

Mervis, C. B., & Rosch, E. (1981). Categorization of natural objects. *Ann Rev Psychol*, **32**, 89–115.

Mery, F., & Kawecki, T. J. (2002). Experimental evolution of learning abilities in fruit flies. *Proc Natl Acad Sci USA*, **99**, 14274–14279.

Mettke-Hofmann, C., & Gwinner, E. (2003). Long-term memory for a life on the move. *Proc Natl Acad Sci USA*, **100**, 5863–5866.

Michotte, A. (1963). *The Perception of Causality*. Oxford: Basic Books.

Milinski, M. (1979). An evolutionary stable feeding strategy in sticklebacks. *Z Tierpsychol*, **51**, 36–40.

Milinski, M., & Heller, R. (1978). Influence of a predator on the optimal foraging behaviour of sticklebacks (*Gasterosteus aculeatus*). *Nature*, **344**, 330–332.

Milinski, M., Semmann, D., & Krambeck, H. J. (2002). Reputation helps solve the 'tragedy of the commons'. *Nature*, **415**, 424–426.

Miller, G. A. (1956). The magical number seven plus or minus two: Some limits on our capacity for processing information. *Psychol Rev*, **63**, 81–96.

Milner, B. L., Corkin, S., & Teuber, H. L. (1968). Further analysis of the hippocampal amnesic syndrome: 14 year follow up study of H.M. *Neuropsychologia*, **6**, 215–234.

Milner, B. L., Squire, R., & Kandel, E. R. (1998). Cognitive neuroscience and the study of memory. *Neuron*, **20**, 445–468.

Mineka, S., & Cook, M. (1989). Observational conditioning of fear to fear-relevant versus fear-irrelevant stimuli in rhesus monkeys. *J Abnorm Psychol*, **98**, 448–459.

Mischel, W., Shoda, Y., & Rodriguez, M. I. (1989). Delay of gratification in children. *Science*, **244**, 933–938.

Mitchell, C. J., Heyes, C. M., Gardner, M. R., & Dawson, G. R. (1999). Limitations of a bidirectional control procedure for the investigation of imitation in rats: Odour cues on the manipulandum. *Q J Exp Psychol B*, **52**, 193–202.

Mitchell R. W. (2012). Self-recognition in animals. In: M. R. Leary & J. P. Tangney (eds.). *Handbook of Self and Identity* (2nd Edn.) (pp. 656–679). New York, NY: Guilford.

Mittelstaedt, M. L., & Mittelstaedt, H. (1980). Homing by path integration in a mammal. *Naturwissenschaften*, **67**, 566–567.

Mix, K. S., Huttenlocher, J., & Levine, S. C. (2002). Multiple cues for quantification in infancy: Is number one of them? *Psychol Bull*, **128**, 278–294.

Mizumori, S., & Smith, D. (2006). Directing neural representation of space. In M. F. Brown & R. G. Cook (eds.). *Animal Spatial Cognition: Comparative, Neural, and Computational Approaches* [On-line]. Available at: www.pigeon.psy.tufts.edu/asc/balda/

Molet, M., Miller, H. C., Laude, J. R., *et al.* (2012). Decision making by humans in a behavioural task: Do humans, like pigeons, show suboptimal choice? *Learn Behav*, **40**, 439–447.

Morell, V. (2006). Migration and dispersal: Arduous journeys. *Science*, **11**, 783–784.

Morgan, C. A., Grillon, C., Southwick, S. A., Davis, M., & Charney, D. S. (1996). Exaggerated acoustic startle reflex in Gulf War veterans with posttraumatic stress disorder. *Am J Psychiatry*, **153**, 64–68.

Morgan, C. L. (1894). *An Introduction to Comparative Psychology*. London: Walter Scott.

Morris, R. G. M. (1981). Spatial localization does not require the presence of local cues. *Learn Motiv*, **12**, 239–260.

Morris, R. G. M., Anderson, E., Lynch, G. S., & Baudry, M. (1986). Selective impairment of learning and blockade of long-term potentiation by an N-methyl-D-aspartate receptor antagonist, AP5. *Nature*, **319**, 774–776.

Morton, J., & Johnson, M. H. (1991). CONSPEC and CONLERN: A two-process theory of infant face recognition. *Psychol Rev*, **98**, 164.

Mouritsen, H., & Hore, P. J. (2012). The magnetic retina: Light-dependent and trigeminal magnetoreception in migratory birds. *Curr Opin Neurobiol*, **22**, 343–352.

Moyer, R. S., & Landauer, T. K. (1967). Time required for judgements of numerical inequality. *Nature*, **215**, 1519–1520.

Mulcahy, N. J., & Call, J. (2006a). How great apes perform on a modified trap-tube task. *Anim Cogn*, **9**, 193–199.

Mulcahy, N. J., & Call, J. (2006b). Apes save tools for future use. *Science*, **312**, 1038–1040.

Muller, M.N., & Mitani, J.C. (2005). Conflict and cooperation in wild chimpanzees. *Adv Study Behav*, **35**, 275–331.

Munn, C. A. (1986). Birds that 'cry wolf'. *Nature*, **319**, 143–145.

Munz, F. W. (1958). Photosensitive pigments from the retinae of certain deep sea fishes. *J Physiol*, **140**, 220–225.

Murphy, J. G., MacKillop, J., Skidmore, J. R., & Pederson, A. A. (2009). Reliability and validity of a demand curve measure of alcohol dependence. *Exp Clin Psychopharmacol*, **17**, 396–404.

Nader, K., & Hardt, O. (2009). A single standard for memory: The case for reconsolidation. *Nat Rev Neurosci*, **10**, 224–234.

Nader, K., Schafe, G. E., & Le Doux, J. E. (2000). Fear memories require protein synthesis in the amygdala for reconsolidation after retrieval. *Neuron*, **40**, 695–701.

Naef-Daenzer, B. (2000). Patch time allocation and patch sampling by foraging great and blue tits. *Anim Behav*, **59**, 989–999.

Nagel, T. (1974). What is it like to be a bat? *Philos Rev*, **83**, 435–450.

Nairne, J. S., & Pandeirada, J. N. S. (2010). Adaptive memory: Ancestral priorities and the mnemonic value of survival processing. *Cogn Psychol*, **61**, 1–22.

Nairne, J. S., Pandeirada, J. N. S., & Thompson, S. R. (2008). Adaptive memory: The comparative value of survival processing. *Psychol Sci*, **20**, 740–746.

Nasar, S. (1998). *A Beautiful Mind*. New York, NY: Simon & Schuster.

Nash, S., & Domjan, M. (1991). Learning to discriminate the sex of conspecifics in male Japanese quail (*Coturnix coturnix japonica*): Tests of "biological constraints". *J Exp Psychol Anim Behav Process*, **17**, 342–353.

Nelson, D. A., Hallberg, K. I., & Soha, J. A. (2004). Cultural evolution of Puget Sound white-crowned sparrow song dialects. *Ethology*, **110**, 879–908.

Nelson, D. A., & Marler, P. (1989). Categorical perception of a natural stimulus continuum: Birdsong. *Science*, **244**, 976–978.

Nelson, D. A., & Marler, P. (1990). The perception of birdsong and an ecological concept of signal space. In W. C. Stebbins & M. A. Berkley (eds.). *Comparative Perception: Complex Signals* (Vol. 2) (pp. 443–447). New York, NY: John Wiley & Sons.

Nelson, D. A., & Marler, P. (1993). Innate recognition of song in white-crowned sparrows: A role in selective vocal learning? *Anim Behav*, **46**, 806–808.

Newman, G. E., Choi, H., Wynn, K., & Scholl, B. J. (2008). The origins of causal perception: Evidence from postdictive processing in infancy. *Cogn Psychol*, **57**, 262–291.

Nicholls, H. (2007). The royal raccoon from Swedesboro. *Nature*, **446**, 255–256.

Nicol, C. J., & Pope, S. J. (1999). The effects of demonstrator social status and prior foraging success on social learning in laying hens. *Anim Behav*, **57**, 163–171.

Nielsen, M. (2006). Copying actions and copying outcomes: Social learning through the second year. *Dev Psychol*, **42**, 555–565.

Nielsen, M., & Tomaselli, K. (2010). Overimitation in Kalahari Bushman children and the origins of human cultural cognition. *Psychol Sci*, **21**, 729–736.

Niven, J. E., & Laughlin, S. B. (2008). Energy limitation as a selective pressure on the evolution of sensory systems. *J Exp Biol*, **211**, 1792–1804.

Nonacs, P. (2001). State dependent behavior and the Marginal Value Theorem. *Behav Ecol*, **12**, 71–83.

Nowak, M.A., & Sigmund, K. (1998). Evolution of indirect reciprocity by image scoring. *Nature*, **393**, 573–577.

Nowak, M.A., Tarnita, C.E, & Wilson, E.O. (2010). The evolution of eusociality. *Nature*, **466**, 1057–1062.

O'Connell-Rodwell, C. E., Arnason, B. T., & Hart, L. A. (2000). Seismic properties of Asian elephant (*Elephas maximus*) vocalizations and locomotion. *J Acoust Soc Am*, **108**, 3066–3072.

Öhman, A., Dimberg, U., & Ost, L. G. (1985). Animal and social phobias: Biological constraints on learned fear responses. In S. Reiss & R. R. Bootzin (eds.). *Theoretical Issues in Behavior Therapy* (pp. 123–175). Orlando, FL: Academic Press.

Öhman, A., & Mineka, S. (2001). Fears, phobias, and preparedness: Toward an evolved module of fear and fear learning. *Psychol Rev*, **108**, 483–522.

O'Keefe, J., & Conway, D. H. (1978). Hippocampal place units in the freely moving rat: Why they fire where they fire. *Exp Brain Res*, **31**, 573–590.

O'Keefe, J., & Dostrovsky, J. (1971). The hippocampus as a spatial map: Preliminary evidence from unit activity in the freely-moving rat. *Brain Res*, **34**, 171–174.

O'Keefe, J., & Nadel, L. (1978). *The Hippocampus as a Cognitive Map*. London, UK: Oxford University Press.

O'Keefe, J., & Speakman, A. (1987). Single unit activity in the rat hippocampus during a spatial memory task. *Exp Brain Res*, **68**, 1–27.

Olson, D. J, Kamil, A. C., Balda, R. P., & Nims, P. J. (1995). Performance of four seed-caching corvid species in operant tests of nonspatial and spatial memory. *J Comp Psychol*, **109**, 173–181.

Olthof, A., Iden, C.M., & Roberts, W.A. (1997). Judgments of ordinality and summation of number symbols by squirrel monkeys (*Saimiri sciureus*). *J Exp Psychol Anim Behav Process*, **23**, 325–339.

Olton, D. S., Branch, M., & Best, P. F. (1978). Spatial correlates of hippocampal unit activity. *Exp Neurol*, **58**, 387–409.

Olton, D. S., & Samuelson, R. J. (1976). Remembrance of places passed: Spatial memory in rats. *J Exp Psychol Anim Behav Process*, **2**, 97–116.

Onishi, K. H., & Baillargeon, R. (2005). Do 15-month-old infants understand false beliefs? *Science*, **308**, 255–258.

Ophir, E., Nass, C., & Wagner, A. (2011). Cognitive control in media multitaskers. *Proc Natl Acad Sci USA*, **106**, 15583–15587.

Ord, T.J., King, L., & Young, A. R. (2011). Contrasting theory with the empirical data of species recognition. *Evolution*, **65**, 2572–2591.

Orr, M. R. (1992). Parasitic flies (Diptera: Phoridae) influence foraging rhythms and caste division of labor in the leaf-cutter ant, *Atta cephalotes* (Hymenoptera: Formicidae). *Behav Ecol Sociobiol*, **30**, 395–402.

Osthaus, B., Slater, A. M., & Lea, S. E. (2003). Can dogs defy gravity? A comparison with the human infant and a non-human primate. *Dev Sci*, **6**, 489–497.

Ostlund, S. B., & Balleine, B. W. (2007). Orbitofrontal cortex mediates outcome encoding in Pavlovian but not instrumental conditioning. *J Neurosci*, **27**, 4819–4825.

Ottoni, E. B., & Izar, P. (2008). Capuchin monkey tool use: Overview and implications. *Evolutionary Anthropology: Issues, News, and Reviews*, **17**, 171–178.

Packard, M. G., & McGaugh, J. L. (1996). Inactivation of hippocampus or caudate nucleus with lidocaine differentially affects expression of place and response learning. *Neurobiol Learn Mem*, **65**, 65–72.

Packer, C. (1977) Reciprocal altruism in *Papio anubis*. *Nature*, **246**, 441–443.

Paine, T. A., Dringenberg, H. C., & Olmstead, M. C. (2003). Effects of chronic cocaine on impulsivity: Relation to cortical serotonin mechanisms. *Behav Brain Res*, **147**, 135–147.

Palagi, E., Leone, A., Mancini, G., & Ferrari, P. F. (2009). Contagious yawning in gelada baboons as a possible expression of empathy. *Proc Natl Acad Sci USA*, **106**, 19262–19267.

Panksepp, J. (1998). Attention deficit hyperactivity disorders, psychostimulants, and intolerance of childhood playfulness: A tragedy in the making? *Curr Direct Psychol Sci*, **7**, 91–98.

Papi, F., Fiore, L., Fiaschi, V., Benvenuti, S., & Baldaccini, N. E. (1972). Olfaction and homing in pigeons. *Monitore Zoologico Italinao (N.S.)*, **6**, 85–95.

Papi, F., Mariotti, G., Foa, A., & Fiaschi, V. (1980). Orientation and anosmatic pigeons. *J Comp Physiol*, **135**, 227–232.

Park, J. & Brannon, E.M. (2013). Training the approximate number system improves math proficiency. *Psychol Sci*, **24**, 2013–2019.

Parker, E. S., Cahill, L., & McGaugh, J. L. (2006). A case of unusual autobiographical remembering. *Neurocase*, **12**, 35–49.

Parr, L. A. (2001). Cognitive and physiological markers of emotional awareness in chimpanzees (*Pan troglodytes*). *Anim Cogn*, **4**, 223–229.

Parr, L. A. (2003). The discrimination of faces and their emotional content by chimpanzees (*Pan troglodytes*). *Ann N Y Acad Sci*, **1000**, 56–78.

Pascalis, O., de Haan, M., & Nelson, C. A. (2002). Is face processing species-specific during the first year of life? *Science*, **296**, 1321–1323.

Paukner, A., & Anderson, J. R. (2006). Video-induced yawning in stumptail macaques (*Macaca arctoides*). *Biol Lett*, **2**, 36–38.

Payne, K. B., Langbauer Jr, W. R., & Thomas, E. M. (1986). Infrasonic calls of the Asian elephant (*Elephas maximus*). *Behav Ecol Sociobiol*, **18**, 297–301.

Pearce, J. M., Good, M. A., Jones, P. M., & McGregor, A. (2004). Transfer of spatial behaviour between different environments: Implications for theories of spatial learning and for the role of the hippocampus in spatial learning. *J Exp Psychol Anim Behav Process*, **30**, 135–147.

Pearce, J. M., & Hall, G. (1980). A model of Pavlovian learning: Variations in the effectiveness of conditioned but not of unconditioned stimuli. *Psychol Rev*, **87**, 532–552.

Pedersen, E.J., Kurzban, R., & McCullough, M.E. (2013). Do humans *really* punish altruistically? A closer look. *Proc Roy Soc B*, **280**, 2012–2723.

Peeke, H. V. S., & Veno, A. (1973). Stimulus specificity of habituated aggression in three-spined sticklebacks (*Gasterosteus aculeatus*). *Behav Biol*, **8**, 427–432.

Peeters, R., Simone, L., Nelissen, K., *et al.* (2009). The representation of tool use in humans and monkeys: Common and uniquely human features. *J Neurosci*, **29**, 11523–11539.

Pellegrini, A. D. (ed.). (2011). *The Oxford Handbook of the Development of Play*. New York, NY: Oxford University Press.

Pellis, S. M., & Pellis, V. C. (1996). On knowing it's only play: The role of play signals in play fighting. *Aggress Violent Beh*, **1**, 249–268.

Pelphrey, K. A., Sasson, N. J., Reznick, J. S., *et al.* (2002). Visual scanning of faces in autism. *J Autism Devel Disord*, **32**, 49–261.

Pelphrey, K. A., Singerman, J. D., Allison, T., & McCarthy, G. (2003). Brain activation evoked by perception of gaze shifts: The influence of context. *Neuropsychologia*, **41**, 156–170.

Penfield, W., & Evans, J. (1935). The frontal lobe in man: A clinical study of maximum removals. *Brain*, **58**, 115–133.

Penn, D. C., Holyoak, K. J., & Povinelli, D. J. (2008). Darwin's mistake: Explaining the discontinuity between human and nonhuman minds. *Behav Brain Sci*, **31**, 109–178.

Penn, D. C., & Povinelli, D. J. (2007). Causal cognition in human and nonhuman animals: A comparative, critical review. *Ann Rev Psychol*, **58**, 97–118.

Penn, D. C., & Povinelli, D.J. (2013). The comparative delusion: The 'behavioristic'/'mentalistic' dichotomy in comparative theory of mind research. In H. S. Terrace & J. Metcalfe (eds.). *Agency and Joint Attention*. New York, NY: Oxford University Press.

Pennisi, E. (2009). On the origin of cooperation. *Science*, **325**, 1196–1199.

Pepperberg, I. M. (1999). *The Alex Studies: Cognitive and Communicative Abilities of Grey Parrots*. Cambridge, MA: Harvard University Press.

Pepperberg, I. M. (2006). Grey parrot numerical competence: A review. *Anim Cogn*, **9**, 377–391.

Pepperberg, I. M., Willner, M. R., & Gravitz, L. B. (1997). Development of Piagetian object permanence in grey parrot (*Psittacus erithacus*). *J Comp Psychol*, **111**, 63–75.

Perdeck, A. C. (1958). Two types of orientation in migrating starlings, *Sturnus vulgaris* L. and chaffinches, *Fringilla coelebs*, as revealed by displacement experiments. *Ardea*, **46**, 1–37.

Peters, J., & Buchel, C. (2011). The neural mechanisms of intertemporal decision-making: Understanding variability. *Trends Cogn Sci*, **15**, 227–239.

Petrie, M., & Halliday, T. (1994). Experimental and natural changes in the peacock's (*Pavo cristatus*) train can affect mating success. *Behav Ecol Sociobiol*, **35**, 213–217.

Phelps, E. A., & LeDoux, J. E. (2005). Contributions of the amygdala to emotion processing: From animal models to human behavior. *Neuron*, **48**, 175–187.

Phillips, J.B., Schmidt-Koenig, K., & Muheim, R. (2006). True navigation: Sensory bases of gradient maps. In M. F. Brown & R. G. Cook (eds.). *Animal Spatial Cognition: Comparative, Neural, and Computational Approaches*. [On-line]. Available at: www.pigeon.psy.tufts.edu/asc/phillips/

Pica, P., Lemer, C., Izard, V., & Dehaene, S. (2004). Exact and approximate arithmetic in an Amazonian indigene group. *Science*, **306**, 499–503.

Pietrewicz, A. T., & Kamil, A. C. (1981). Search images and the detection of cryptic prey: An operant approach. In A. C. Kamil & T. D. Sargent (eds.). *Foraging Behavior: Ecological, Ethological and Psychological Approaches* (pp. 311–331). New York, NY: Garland STPM Press.

Piffer, L., Agrillo, C., & Hyde, D.C. (2012). Small and large number discrimination in guppies. *Anim Cogn*, **15**, 215–221.

Pilley, J. W., & Reid, A. K. (2011). Border collie comprehends object names as verbal referents. *Behav Proc*, **86**, 184–195.

Pinel, P., Piazza, M., Le Bihan, D., & Dehaene, S. (2004). Distributed and overlapping cerebral representations of number, size, and luminance during comparative judgments. *Neuron*, **41**, 983–993.

Pinker, S., & Jackendoff, R. (2005). The faculty of language: What's special about it? *Cognition*, **95**, 201–236.

Platek, S. M., Critton, S. R., Myers, T. E., & Gallup, G. G. (2003). Contagious yawning: The role of self-awareness and mental state attribution. *Cogn Brain Res*, **17**, 223–227.

Platt, J. R., & Johnson, D. M. (1971). Localization of position within a homogeneous behavior chain: Effects of error contingencies. *Learn Motiv*, **2**, 386–414.

Pollard, K. A., & Blumstein, D. T. (2011). Social group size predicts the evolution of individuality. *Curr Biol*, **21**, 413–417.

Poole, J. H., Payne, K., Langbauer, W. R. Jr., & Moss, C. J. (1988). The social contexts of some very low frequency calls of African elephants. *Behav Ecol Sociobiol*, **22**, 385–392.

Pothuizen, H. H., Aggleton, J. P., & Vann, S. D. (2008). Do rats with retrosplenial cortex lesions lack direction? *Eur J Neurosci*, **28**, 2486–2498.

Poulin, M. (2012). Our genes want us to be altruists. *Observer, 25*, No. 10.

Poulin, M. J., Holman, E. A., & Buffone, A. (2012). The neurogenetics of nice receptor genes for oxytocin and vasopressin interact with threat to predict prosocial behavior. *Psychol Sci*, **23**, 446–452.

Povinelli, D.J. (2000). *Folk Physics for Apes: The Chimpanzee's Theory of How the World Works*. Oxford, UK: Oxford University Press.

Povinelli, D. J., & Eddy, T. J. (1996). Factors influencing young chimpanzees' (*Pan troglodytes*) recognition of attention. *J Comp Psychol*, **110**, 336–345.

Povinelli, D. J., & Preuss, T. M. (1995). Theory of mind: Evolutionary history of a cognitive specialization. *Trends Neurosci*, **18**, 418–424.

Povinelli, D. J., & Vonk, J. (2004). We don't need a microscope to explore the chimpanzee's mind. *Mind Lang*, **19**, 1–28.

Povinelli, D. J., Nelson, K. E., & Boysen, S. T. (1990). Inferences about guessing and knowing by chimpanzees (*Pan troglodytes*). *J Comp Psychol*, **104**, 203–210.

Povinelli, D. J., Gallup, G. G. Jr., Eddy, T., Bierschwale, D. T., Engstrom, M. C., Perilloux, H. K., & Toxopeus, I. B. (1997). Chimpanzees recognize themselves in mirrors. *Anim Behav*, **53**, 1083–1088.

Prather, J. F., Peters, S., Nowicki, S., & Mooney, R. (2008). Precise auditory–vocal mirroring in neurons for learned vocal communication. *Nature*, **451**, 305–310.

Pravosudov, V. V., & Clayton, N. S. (2002). A test of the adaptive specialization hypothesis: Population differences in caching, memory, and the hippocampus of black-capped chickadees (*Poecile atricapilla*). *Behav Neurosci*, **116**, 515–522.

Premack, D. (1971). Language in chimpanzees. *Science*, **172**, 808–822.

Premack, D. (1995). Cause/induced motion: intention/spontaneous motion. In J.-P. Changeux and J. Chavaillion (eds.). *The Origins of the Human Brain*. Oxford, UK: Claredon Press.

Premack, D., & Premack, A. J. (1983). *The Mind of an Ape*. New York, NY: Norton.

Premack, D., & Woodruff, G. (1978). Does the chimpanzee have a theory of mind? *Behav Brain Sci*, **1**, 515–526.

Pritchard, A. E., Nigro, C. A., Jacobson, L. A., & Mahone, E. M. (2012). The role of neuropsychological assessment in the functional outcomes of children with ADHD. *Neuropsychol Rev*, **22**, 54–68.

Proctor, C. J., Broom, M., & Ruxton, G. D. (2001). Modeling antipredator vigilance and flight response in group foragers when warning signals are ambiguous. *J Theor Biol*, **211**, 409–417.

Provine, R. R. (1986). Yawning as a stereotyped action pattern and releasing stimulus. *Ethology*, **72**, 109–122.

Provine, R. R. (1989). Faces as releasers of contagious yawning: An approach to face detection using normal human subjects. *Bull Psychon Soc*, **27**, 211–214.

Pulido, F., Berthold, P., Mohr, G., & Querner, U. (2001). Heritability of the timing of autumn migration in a natural bird population. *Proc Roy Soc Lond B*, **268**, 953–959.

Pyke, G. H. (1984). Optimal foraging theory: A critical review. *Ann Rev Ecol Syst*, **15**, 523–575.

Pylyshyn, Z. W., & Storm, R. W. (1988). Tracking multiple independent targets: Evidence for a parallel tracking mechanism. *Spatial Vision*, **3**, 179–197.

Quick, N. J., & Janik, V. M. (2012). Bottlenose dolphins exchange signature whistles when meeting at sea. *Proc Roy Soc B Biol Sci*, **279**, 2539–2545.

Quinn, T. P., & Busack, C. A. (1985). Chemosensory recognition of siblings in juvenile coho salmon (*Oncorhynchus kisutch*). *Anim Behav*, **33**, 51–56.

Quirk, G. J., Muller, R. U., & Kubie, J. L. (1990). The firing of hippocampal place cells in the dark depends on the rat's recent experience. *J Neurosci*, **10**, 2008–2017.

Rakison, D. H., & Poulin-Dubois, D. (2001). Developmental origin of the animate-inanimate distinction. *Psychol Bull*, **127**, 49–62.

Rakoczy, H., Warneken, F., & Tomasello, M. (2008). The sources of normativity: Young children's awareness of the normative structure of games. *Dev Psychol*, **44**, 875–881.

Rand, A., & Peikoff, L. (1979). *Introduction to Objectivist Epistemology*. New York, NY: New American Library.

Range, F., Aust, U., Steurer, M., & Huber, L. (2008). Visual categorization of natural stimuli by domestic dogs. *Anim Cogn*, **11**, 339–347.

Rasmussen, K., Palacios, D. M., Calambokidis, J., *et al.* (2007). Southern Hemisphere humpback whales wintering off Central America: Insights from water temperature into the longest mammalian migration. *Biol Lett*, **3**, 302–305.

Rauschecker, J. P., & Kniepert, U. (1994). Auditory localization behavior in visually deprived cats. *Eur J Neurosci*, **6**, 149–160.

Reed, D. D., Critchfield, T. S., & Martins, B. K. (2006). The generalized matching law in elite sports competition: Football play calling as operant choice. *J Appl Behav Anal*, **39**, 281–297.

Rescorla, R. A. (1988). Classical conditioning: It's not what you think it is. *Am Psychol*, **43**, 151–160.

Rescorla, R. A., & Wagner, A. R. (1972). A theory of Pavlovian conditioning: Variations in the effectiveness of reinforcement and nonreinforcement. In A. H. Black & W. F. Profasy (eds.). *Classical Conditioning II* (pp. 64–69). New York, NY: Appleton-Century-Crofts.

Révész, G. (1924). Experiments on animal perception. *Br J Psychol*, **14**, 387–414.

Reznikova, Z. (2007). *Animal Intelligence: From Individual to Social Cognition*. Cambridge, UK: Cambridge University Press.

Richerson, P. J., & Boyd, R. (2005). *Not by Genes Alone: How Culture Transformed Human Evolution*. Chicago, IL: University of Chicago Press.

Rijnsdorp, A., Daan, S., & Dijkstra, C. (1981). Hunting in the kestrel, *Falco tinnunculus*, and the adaptive significance of daily habits. *Oecologia*, **50**, 391–406.

Rilling, J. K. (2006). Human and nonhuman primate brains: Are they allometrically scaled versions of the same design? *Evol Anthropol*, **15**, 65–77.

Rilling, J.K., Gutman, D.A., Zeh, T.R., *et al.* (2002). A neural basis for social cooperation. *Neuron*, **35**, 395–405.

Rinkevich, B., Porat, R., & Goren, M. (1995). Allorecognition elements on a urochordate histocompatibility locus indicate unprecedented extensive polymorphism. *Proc Roy Soc Lond B*, **260**, 319–324.

Rivas, E. (2005). Recent use of signs by chimpanzees (*Pan troglodytes*) in interactions with humans. *J Comp Psychol*, **119**, 404–417.

Rivera-Gaxiola, M., Silva-Pereyra, J., & Kuhl, P.K. (2005). Brain potentials to native and non-native speech contrasts in 7- and 11-month-old American infants. *Dev Sci*, **8**, 162–172.

Rizley, R.C., & Rescorla, R.A. (1972). Associations in second-order conditioning and sensory preconditioning. *J Comp Physiol Psychol*, **81**, 1–11.

Roberts, D. (1941). Imitation and suggestion in animals. *Bull Anim Behav*, **1**, 11–19.

Roberts, S.K. de F. (1965). Photoreception and entrainment of cockroach activity rhythms. *Science*, **148**, 958–959.

Roberts, S. (1981). Isolation of an internal clock. *J Exp Psychol Anim Behav Process*, **7**, 242–268.

Roberts, W.A. (1979). Spatial memory in the rat on the hierarchical maze. *Learn Motiv*, **10**, 117–140.

Roberts, W.A. (1998). *Principles of Animal Cognition*. Boston: The McGraw-Hill Companies Inc.

Roberts, W.A., & Mazmanian, D.S. (1988). Concept learning at different levels of abstraction by pigeons, monkeys, and people. *J Exp Psychol Anim Behav Process*, **14**, 247–260.

Roberts, W.A., & Mitchell, S. (1994). Can a pigeon simultaneously process temporal and numerical information? *J Exp Psychol Anim Behav Process*, **20**, 66–78.

Roberts, W.A., & Pearce, J.M. (1998). Control of spatial behavior by an unstable landmark. *J Exp Psychol Anim Behav Process*, **24**, 172–184.

Roitblat, H.L. (1987). *Introduction to Comparative Cognition*. New York, NY: W.H. Freeman & Co.

Rolls, E.T. (2004). The functions of the orbitofrontal cortex. *Brain Cogn*, **55**, 11–29.

Rolls, E.T., & Grabenhorst, F. (2008). The orbitofrontal cortex and beyond: From affect to decision making. *Prog Neurobiol*, **86**, 216–244.

Romanes, G.J. (1883). *Animal Intelligence*. New York, NY: D. Appleton and Company.

Rooney, N.J. (1999). *Play behaviour of the domestic dog Canis familiaris, and its effect upon the dog-human relationship* (Doctoral dissertation, University of Southampton, Southampton, England). Retrieved from http://ethos.bl.uk/OrderDetails.do?uin=uk.bl.ethos.298116

Rosati, A.G., Stevens, J.R., Hare, B., & Hauser, M.D. (2007). The evolutionary origins of human patience: Temporal preferences in chimpanzees, bonobos, and human adults. *Curr Biol*, **17**, 1663–1668.

Rosch, E. (1973). Natural categories. *Cogn Psychol*, **4**, 328–350.

Rouder, J.N., & Ratcliff, R. (2006). Comparing exemplar- and rule-based theories of categorization. *Curr Direct Psychol Sci*, **15**, 9–13.

Rovee-Collier, C. (1999). The development of infant memory. *Curr Direct Psychol Sci*, **8**, 80–85.

Rozin, P., & Kalat, J.W. (1971). Specific hungers and poison avoidance as adaptive specializations of learning. *Psychol Rev*, **78**, 459–486.

Rugani, R., Fontanari, L., Simoni, E., Regolin, L., & Vallortigara, G. (2009). Arithmetic in newborn chicks. *Proc Roy Soc B Biol Sci*, **276**, 2451–2460.

Rugani, R., Regolin, L., & Vallortigara, G. (2008). Discrimination of small numerosities in young chicks. *J Exp Psychol Anim Behav Process*, **34**, 388–399.

Rumbaugh, D.M. (ed.) (1977). *Language Learning by a Chimpanzee: The Lana Project*. New York, NY: Academic Press.

Rumbaugh, D.M., & Savage-Rumbaugh, E.S. (1994). Language in comparative perspective. In N.J. Mackintosh (ed.). *Animal Learning and Cognition. Handbook of Perception and Cognition Series* (2nd Edn.) (pp. 307–333). San Diego, CA: Academic Press.

Rumbaugh, D.M., Savage-Rumbaugh, E.S., & Sevcik, R.A. (1994). Biobehavioral roots of language. In R.W. Wrangham, W.C. McGrew, & P.G. Heltne (eds.). *Chimpanzee Cultures* (pp. 319–334). Cambridge, MA: Harvard University Press.

Ruxton, G.D., Sherratt, T.N., & Speed, M.P. (2004). *Avoiding Attack: The Evolutionary Ecology of Crypsis, Warning Signals, and Mimicry* (Vol. 249). Oxford, UK: Oxford University Press.

Sablin, M.V., & Khlopachev, G.A. (2002). The earliest ice age dogs: Evidence from Eliseevichi I. *Curr Anthropol*, **43**, 795–799.

Sagi, A., & Hoffman, M.L. (1976). Empathic distress in the newborn. *Dev Psychol*, **12**, 175.

Saksida, L.M., & Wilkie, D.M. (1994). Time-of-day discrimination by pigeons, *Columba livia*. *Anim Learn Behav*, **22**, 143–154.

Salwiczek, L.H., & Bshary, R. (2011). Cleaner wrasses keep track of the 'when' and 'what' in a foraging task. *Ethology*, **117**, 939–948.

Sandstrom, N.J., Kaufman, J., & Huettel, S.A. (1998). Males and females use different distal cues in a virtual environment navigation task. *Cogn Brain Res*, **6**, 351–360.

Santorelli, L.A., Christopher, R.L., Thompson, E.V., *et al.* (2008). Facultative cheater mutants reveal the genetic complexity of cooperation in social amoebae. *Nature*, **451**, 1107–1110.

Santos, L.R., Barnes, J.L., & Mahajan, N. (2005). Expectations about numerical events in four lemur species (*Eulemur fulvus*, *Eulemur mongoz*, *Lemur catta* and *Varecia rubra*). *Anim Cogn*, **8**, 253–262.

Santos, L.R., & Hauser, M.D. (2002). A non-human primate's understanding of solidity: Dissociations between seeing and acting. *Dev Sci*, **5**, F1-F7.

Santos, L.R., & Hughes, K.D. (2009). Economic cognition in humans and animals: The search for core mechanisms. *Curr Opin Neurobiol*, **19**, 63–66.

Santos, L.R., Pearson, H.M., Spaepen, G.M., Tsao, F., & Hauser, M.D. (2006). Probing the limits of tool competence: Experiments with two non-tool-using species (*Cercopithecus aethiops* and *Saquinus oedipus*). *Anim Cogn*, **9**, 94–109.

Sanz, C., Boesch, C. & Call, J. (eds.) (2013). *Tool-use: Cognitive Requirements and Ecological Determinants*. Cambridge, UK: Cambridge University Press.

Saucier, D., & Cain, D.P. (1995). Spatial learning without NMDA-dependent long-term potentiation. *Nature*, **9**, 186–189.

Saucier, D., Lisoway, A., Green, S., & Elias, L. (2007). Female advantage for object location memory in peripersonal but not extrapersonal space. *J Int Neuropsychol Soc*, **13**, 683–686.

Saura, C. A., & Valero, J. (2011). The role of CREB signaling in Alzheimer's disease and other cognitive disorders. *Rev Neurosci*, **22**, 153–169.

Savage-Rumbaugh, E. S., & Lewin, R. (1984). *Kanzi, the Ape at the Brink of the Human Mind*. New York, NY: John Wiley & Sons, Inc.

Savage-Rumbaugh, E. S., McDonald, K., Sevcik, R. A., Hopkins, W. D., & Rubert, E. (1986). Spontaneous symbol acquisition and communicative use by pygmy chimpanzees (*Pan paniscus*). *J Exp Psychol Gen*, **115**, 211–235.

Savage-Rumbaugh, E. S., Murphy, J., Sevcik, R. A., *et al.* (1993). Language comprehension in ape and child. *Monogr Soc Res Child Dev*, **58**, 1–256.

Savage-Rumbaugh, E. S., Rumbaugh, D. M., Smith, S. T., & Lawson, J. (1980). Reference: The linguistic essential. *Science*, **210**, 922–925.

Save, E., Cressant, A., Thinus-Blanc, C., & Poucet, B. (1998). Spatial firing of hippocampal place cells in blind rats. *J Neurosci*, **18**, 1818–1826.

Saviola, A. J., Chiszar, D., & MacKessy, S. P. (2012). Ontogenetic shift in response to prey-derived chemical cues in prairie rattlesnakes *Crotalus viridis viridis*. *Curr Zool*, **58**, 549–555.

Saviola, A. J., McKenzie, V. J., & Chiszar, D. (2012). Chemosensory responses to chemical and visual stimuli in five species of colubrid snakes. *Acta Herpetol*, **7**, 91–103.

Saxe, R., & Kanwisher, N. (2003). People thinking about thinking people: The role of the temporo-parietal junction in "theory of mind". *NeuroImage*, **19**, 1835–1842.

Saxe, R., & Powell, L. J. (2006). It's the thought that counts: Specific brain regions for one component theory of mind. *Psychol Sci*, **17**, 692–699.

Sayigh, L. S., Esch, H. C., Wells, R. S., & Janik, V. M. (2007). Facts about signature whistles of bottlenose dolphins, *Tursiops truncatus*. *Anim Behav*, **74**, 1631–1642.

Schacter, D. L. (1999). The seven sins of memory: Insights from psychology and cognitive neuroscience. *Am Psychol*, **54**, 182–203.

Scheel, D., & Packer, C. (1991). Group hunting behaviour of lions: A search for cooperation. *Anim Behav*, **41**, 697–709.

Scheer, F. A., Wright, K. P. Jr., Kronauer, R. E., & Czeisler, C. A. (2007). Plasticity of the intrinsic period of the human circadian timing system. *PLoS One*, **2**, e721.

Schelling, T. (1980). *The Strategy of Conflict*. Boston, MA: Harvard University Press.

Schiff, W. (1965). Perception of impending collision: A study of visually directed avoidant behavior. *Psychol Monogr Gen Appl*, **79**, 1–26.

Schiff, W., Caviness, J. A., & Gibson, J. J. (1962). Persistent fear responses in rhesus monkeys to the optical stimulus of 'looming'. *Science*, **136**, 182–183.

Schiller, P. (1952). Innate constituents of complex responses in primates. *Psychol Rev*, **59**, 177–191.

Schino, G., di Sorrentino, E.P., & Tiddi, B. (2007). Grooming and coalitions in Japanese macaques (*Macaca fuscata*): Partner choice and the time frame reciprocation. *J Comp Psychol*, **121**, 181–188.

Schippers, M. C., Binnekade, R., Schofelmeer, A. N., Pattij, T., & De Vries, T. J. (2012). Unidirectional relationship between heroin self-administration and impulsive decision-making in rats. *Psychopharmacology*, **219**, 443–452.

Schneider, G. E. (1969). Two visual systems. *Science*, **163**, 895–902.

Scholl, B. J. (2001). Objects and attention: The state of the art. *Cognition*, **80**, 1–46.

Scholz, A. T., Horrall, R. M., Cooper, J. C., & Hassler, D. (1976). Imprinting to chemical cues: the basis for home stream selection in salmon. *Science*, **192**, 1247–1249.

Schusterman, R. J., & Kastak, D. (1998). Functional equivalence in a California sea lion: Relevance to animal social and communicative interactions. *Anim Behav*, **55**, 1087–1095.

Searle, J. R. (1995). *The Construction of Social Reality*. New York, NY: Free Press.

Seed, A. M., Clayton, N.S., & Emery, N.J. (2007). Postconflict third-party affiliation in rooks, *Corvus frugilegus*. *Curr Biol*, **17**, 152–158.

Seed, A. M., Tebbich, S., Emery, N. J., & Clayton, N. S. (2006). Investigating physical cognition in rooks, *Corvus frugilegus*. *Curr Biol*, **16**, 697–701.

Seeley, T. D. (1995). *The Wisdom of the Hive*. Cambridge, MA: Harvard University Press.

Seghers, B. H. (1974). Schooling behaviour in the guppy (*Poecilia reticulata*): An evolutionary response to predation. *Evolution*, **28**, 486–489.

Senju, A., Maeda, M., Kikuchi, Y., *et al.* (2007). Absence of contagious yawning in children with autism spectrum disorder. *Biol Lett*, **3**, 706–708.

Sevenster, P. (1973). Incompatibility of response and reward. In R.A. Hinde & J. Stevenson-Hinde (eds.). *Constraints on Learning, Limitations and Predispositions* (pp. 488). Oxford: Academic Press.

Seyfarth, E. A., Hergernroder, R., Ebbes, H., & Barth, F. G. (1982). Idiothetic orientation of a wandering spider: Compensations of detours and estimates of goal distance. *Behav Ecol Sociobiol*, **11**, 139–148.

Seyfarth, R. M. (1980). The distribution of grooming and related behaviours among adult female vervet monkeys. *Anim Behav*, **28**, 798–813.

Seyfarth, R. M., & Cheney, D. L. (1986). Vocal development in vervet monkeys. *Anim Behav*, **34**, 1640–1658.

Seyfarth, R. M., & Cheney, D. L. (2003a). Meaning and emotion in animal vocalizations. *Ann N Y Acad Sci*, **1000**, 32–55.

Seyfarth, R. M., & Cheney, D. L. (2003b). The structure of social knowledge in monkeys. In F. B. M. de Waal & P. L. Tyak (eds.). *Animal Social Complexity: Intelligence, Culture, and Individualized Societies* (pp. 207–229). Cambridge, MA: Harvard University Press.

Seyfarth, R. M., Cheney, D. L., & Marler, P. (1980). Monkey responses to three different alarm calls: Evidence of predator classification and semantic communication. *Science*, **210**, 801–803.

Shellmann, J. S., & Gamboa, G. J. (1982). Nestmate discrimination in social wasps: The role of exposure to nest and nestmates (*Polistes fuscatus*, Hymenoptera: Vespidae). *Behav Ecol Sociobiol*, **11**, 51–53.

Sherman, P. W., Reeve, H. K., & Pfennig, D. W. (1997). Recognition systems. In J. R. Krebs & N. B. Davies (eds.). *Behavioural Ecology: An Evolutionary Approach* (4th Edn.) (pp. 69–96). Oxford, UK: Blackwell Scientific.

Sherry, D. F., & Galef, B. G. (1984). Cultural transmission without imitation: Milk bottle opening by birds. *Anim Behav*, **32**, 937–938.

Sherry, D. F., & Schacter, D. L. (1987). The evolution of multiple memory systems. *Psychol Rev*, **94**, 439–454.

Sherry, D. F., Vaccarino, A. L., Buckenham, K., & Herz, R. S. (1989). The hippocampal complex of food-storing birds. *Brain Behav Evol*, **34**, 308–317.

Shettleworth, S. J. (1975). Reinforcement and the organization of behavior in golden hamsters: Hunger, environment, and food reinforcement. *J Exp Psychol Anim Behav Process*, **1**, 56–87.

Shettleworth, S. J. (1978). Reinforcement and the organization of behavior in golden hamsters: Pavlovian conditioning with food and shock unconditioned stimuli. *J Exp Psychol Anim Behav Process*, **4**, 152–169.

Shettleworth, S. J. (2009). The evolution of comparative cognition: Is the snark still a boojum? *Behav Process*, **80**, 210–217.

Shettleworth, S. J. (2010). *Cognition, Evolution, and Behavior.* New York, NY: Oxford University Press.

Shettleworth, S. J., & Juergensen, M. R. (1980). Reinforcement and the organization of behavior in golden hamsters: Brain stimulation reinforcement for several action patterns. *J Exp Psychol Anim Behav Process*, **6**, 352–375.

Shettleworth, S. J., & Krebs, J. R. (1982). How marsh tits find their hoards: The roles of site preference and spatial memory. *J Exp Psychol Anim Behav Process*, **36**, 87–105.

Shimizu, T., Bowers, A. N., Budzynski, C., Kahn, M. C., & Bingman, V. P. (2004). What does a pigeon brain look like during homing? Selective examination of ZENK expression in the telencephalon of pigeons navigating home. *Behav Neurosci*, **118**, 845–851.

Shizgal, P. (1997). Neural basis of utility estimation. *Curr Opin Neurobiol*, **7**, 198–208.

Shoda, Y., Mischel, W., & Peake, P. K. (1990). Predicting adolescent cognitive and self-regulatory competencies from preschool delay of gratification: Identifying diagnostic conditions. *Dev Psychol*, **26**, 978–986.

Shumaker, R. W., Walkup, K. R., & Beck, B. B. (2011). *Animal Tool Behavior: The Use and Manufacture of Tools by Animals* (Revised and updated ed.). Baltimore, MA: The Johns Hopkins University Press.

Sibly, R. M. (1975). How incentive and deficit determine feeding tendency. *Anim Behav*, **23**, 437–446.

Sibly, R. M., & McFarland, D. J. (1976). On the fitness of behaviour sequences. *Am Nat*, **110**, 601–617.

Sidman, M. (2000). Equivalence relations and the reinforcement contingency. *J Exp Anal Behav*, **74**, 296–300.

Siegel, S. (1983). Classical conditioning, drug tolerance, and drug dependence. In Y. Israel, F. T. Glaser, H. Kalant, R. E. Popham, W. Schmidt, & R. G. Smart (eds.). *Research Advances in Alcohol and Drug Problems* (Vol. 2) (pp. 207–246). New York, NY: Plenum.

Siegel, S. (1984). Pavlovian conditioning and heroin overdose: Reports by overdose victims. *Bull Psychon Soc*, **22**, 428–430.

Siegel, S., Hinson, R. E., Krank, M.D., & McCully, J. (1982). Heroin "overdose" death: The contribution of drug-associated environmental cues. *Science*, **216**, 436–437.

Silk, J.B. (2007a). Empathy, sympathy, and prosocial preferences in primates. In R. I. M. Dunbar & L. Barrett (eds.). *The Oxford Handbook of Evolutionary Psychology* (pp. 115–126). Oxford, UK: Oxford University Press.

Silk, J. B. (2007b). Animal behavior: Conflict management is for the birds. *Curr Biol*, **17**, R50-R51.

Silk, J.B., Brosnan, S.F., Vonk, J., *et al.* (2005). Chimpanzees are indifferent to the welfare of unrelated group members. *Nature*, **437**, 1357–1359.

Silk, J. B., & House, B. R. (2011). Evolutionary foundations of human prosocial sentiments. *Proc Natl Acad Sci USA*, **108**, 10910–10917.

Silva, F. J., Page, D. M., & Silva, K. M. (2005). Methodological-conceptual problems in the study of chimpanzees' folk physics: How studies with adult humans can help. *Anim Learn Behav*, **33**, 47–58.

Silverman, I., Choi, J., & Peters, M. (2007). The hunter-gatherer theory of sex differences in spatial abilities: Data from 40 countries. *Arch Sex Behav*, **36**, 261–268.

Simion, F., Regolin, L., & Bulf, H. (2008). A predisposition for biological motion in the newborn baby. *Proc Natl Acad Sci USA*, **105**, 809.

Simon, T., Hespos, S.J., & Rochat, P. (1995). Do infants understand simple arithmetic? A replication of Wynn (1992). *Cogn Dev*, **10**, 253–269.

Skinner, B. F. (1948). "Superstition" in the pigeon. *J Exp Psychol*, **38**, 168–172.

Skinner, B.F. (1957). *Verbal Behavior.* New York: Appleton-Century-Crofts.

Slocombe, K. E., & Zuberbühler, K. (2005). Functionally referential communication in a chimpanzee. *Curr Biol*, **15**, 1779–1784.

Smith, J. D., Minda, J. P., & Washburn, D. A. (2004). Category learning in rhesus monkeys: A study of the Shepherd, Hovland and Jenkins (1961) tasks. *J Exp Psychol Gen*, **133**, 398–414.

Smith, J. D., Redford, J. S., & Hass, S. M. (2008). Prototype abstraction by monkeys (*Macaca mulatta*). *J Exp Psychol Gen*, **137**, 390–401.

Smulders, T. V., & DeVoogd, T. J. (2000). Expression of immediate early genes in the hippocampal formation of the black-capped chickadee (*Poecile atricapillus*) during a food-hoarding task. *Behav Brain Res*, **114**, 39–49.

Smulders, T. V., Gould, K. L., & Leaver, L. A. (2010). Using ecology to guide the study of cognitive and neural mechanisms of different aspects of spatial memory in food-hording animals. *Philos Trans Roy Soc Lond B*, **365**, 883–900.

Smuts, B. B. (1985). *Sex and Friendship in Baboons*. Hawthorn, NY: Aldine.

Snyder, J. B., Nelson, M. E., Burdick, J. W., & MacIver, M. A. (2007). Omnidirectional sensory and motor volumes in electric fish. *PLoS Biol*, **11**, e301.

Sobel, D. M., & Kirkham, N. Z. (2007). Interactions between causal and statistical learning. In A. Gopnik & L. Schulz (eds.). *Causal Learning: Psychology, Philosophy, and Computation* (pp. 139–153). New York, NY: Oxford University Press.

Sobel, D. M., Tenenbaum, J. B., & Gopnik, A. (2004). Children's causal inferences from indirect evidence: Backwards blocking and Bayesian reasoning in preschoolers. *Cogn Sci*, **28**, 303–333.

Sokolowski, M. B., Tonneau, F., Feixz, I., & Baqué, E. (1999). The ideal free distribution in humans: An experimental test. *Psychon Bull Rev*, **6**, 157–161.

Soon, C., Brass, M., Heinze, H., & Haynes, J. (2008). Unconscious determinants of free decisions in the human brain. *Nat Neurosci*, **11**, 543–545.

Spelke, E. S., Breinlinger, K., Macomber, J., & Jacobson, K. (1992). Origins of knowledge. *Psychol Rev*, **99**, 605–632.

Spelke, E. S., Kestenbaum, R., Simons, D. J., & Wein, D. (1995). Spatiotemporal continuity, smoothness of motion and object identity in infancy. *Br J Dev Psychol*, **13**, 113–142.

Spence, K. W. (1937). Experimental studies of learning and the higher mental processes in infra-human primates. *Psychol Bull*, **34**, 806–850.

Spence, K. W. (1938). Gradual versus sudden solution of discrimination problems by chimpanzees. *J Comp Psychol*, **1**, 213–224.

Sperber, D., Premack, D., & Premack, A.J. (eds.) (1995). *Causal Cognition: A Multidisciplinary Debate*. Oxford, UK: Oxford University Press.

Squire, L. R. (2009). Memory and brain systems: 1969–2009. *J Neurosci*, **29**, 12711–12716.

Squire, L. R., & Wixted, J. T. (2011). The cognitive neuroscience of human memory since H.M. *Ann Rev Neurosci*, **34**, 259–288.

Squire, L. R., & Zola-Morgan, S. (1988). Memory: Brain systems and behavior. *Trends Neurosci*, **11**, 170–175.

Squire, L. R., & Zola-Morgan, S. (1991). The medial temporal lobe memory system. *Science*, **253**, 1380–1386)

Srinivasan, M. V., Zhang, S. W., Lehrer, M., & Collett, T. S. (1996). Honeybee navigation *en route* to the goal: Visual flight control and odometry. *J Exp Biol*, **199**, 237–244.

Staddon, J. E. R. (2005). Interval timing: Memory, not a clock. *Trends Cogn Sci*, **9**, 312–314.

Staddon, J. E. R., & Cerutti, D. T. (2003). Operant conditioning. *Ann Rev Psychol*, **54**, 115–144.

Staddon, J. E. R., & Higa, J. J. (1999). Time and memory: Towards a pacemaker-free theory of interval timing. *J Exp Anal Behav*, **71**, 215–251.

Staddon, J. E. R., & Higa, J. J. (2006). Interval timing. *Nat Rev Neurosci*, **7**, 1–2.

Staddon, J. E. R., & Simmelhag, V. L. (1971). The "superstitious experiment": A reexamination of its implications for the principle of adaptive behavior. *Psychol Rev*, **78**, 3–43.

St Amant, R., & Horton, T. E. (2008). Revisiting the definition of animal tool use. *Anim Behav*, **75**, 1199–1208.

Stanford, C.B. (1998). *Chimpanzee and Red Colobus: The Ecology of Predator and Prey.* Cambridge, MA: Harvard University Press.

Steiniger, F. (1950). Beiträge zur Soziologie und sonstigen Biologie der Wanderratte. *Z Tierpsychol*, **7**, 356–379.

Stephens, D. W. (1981). The logic of risk-sensitive foraging preferences. *Anim Behav*, **29**, 628–629.

Stephens, D. W. (1991). Change, regularity and value in the evolution of animal learning. *Behav Ecol*, **2**, 77–89.

Stephens, D. W. (2008). Decision ecology: Foraging and the ecology of animal decision making. *Cogn Affect Behav Neurosci*, **8**, 475–484.

Stephens, D. W., & Anderson, D. (2001). The adaptive value of preference for immediacy: When shortsighted rules have farsighted consequences. *Behav Ecol*, **12**, 330–339.

Stephens, D. W., & Krebs, J. R. (1986). *Foraging Theory.* Princeton, NJ: Princeton University Press.

Stuss, D. T., & Benson, D. F. (1986). *The Frontal Lobes*. New York, NY: Raven Press.

Suga, N. (1989). Principles of auditory information-processing derived from neuroethology. *J Exp Biol*, **146**, 277–286.

Sugita, Y. (2008). Face perception in monkeys reared with no exposure to faces. *Proc Natl Acad Sci USA*, **105**, 394–398.

Sulin, R. A., & Dooling, D. J. (1974). Intrusion of a thematic idea in retention of prose. *J Exp Psychol*, **103**, 255–262.

Sumpter, D. J. T. (2006). The principles of collective animal behaviour. *Philos Trans Roy Soc B*, **361**, 5–22.

Surian, L., Caldi, S., & Sperber, D. (2008). Attribution of beliefs by 13-month-old infants. *Psychol Sci*, **18**, 580–586.

Surlykke, A., & Fullard, J. H. (1989). Hearing of the Australian whistling moth, *Hecatesia thyridion. Naturwissenschaften*, **76**, 132–134.

Sutherland, W. J., Townsend, C. R., & Patmore, J. M. (1988). A test of ideal free distribution with unequal competitors. *Behav Ecol Sociobiol*, **23**, 51–53.

Sutton, J. E., Olthof, A., & Roberts, W. A. (2000). Landmark use by squirrel monkeys (*Saimiri sciurues*). *Anim Learn Behav*, **28**, 28–42.

Svensson, E. I., Eroukhmanoff, F., Karlsson, K., Runemark, A., & Brodin, A. (2010). A role for learning in population divergence of mate preferences. *Evolution*, **64**, 3101–3113.

Swanson, J., Baler, R. D., & Volkow, N. D. (2011). Understanding the effects of stimulant medications on cognition in individuals with attention-deficit hyperactivity disorder: A decade of progress. *Neuropsychopharmacology*, **36**, 207–226.

Tan, S., Amos, W., & Laughlin, S. B. (2005). Captivity selects for smaller eyes. *Curr Biol*, **15**, R540-R542.

Tanaka, S. C., Balleine, B. W., & O'Doherty, J. P. (2008). Calculating consequences: Brain systems that calculate the causal effects of actions. *J Neurosci*, **28**, 6750–6755.

Tarsitano, M. S., & Andrew, R. (1999). Scanning and route selection in the jumping spider *Portia labiate. Anim Behav*, **58**, 255–265.

Tautz, J., Casas, J., & Sandeman, D. (2001). Phase reversal of vibratory signals in honeycomb may assist dancing honeybees to attract their audience. *J Exp Biol*, **204**, 3737–3746.

Taylor, A. H., Hunt, G. R., Medina, F. S., & Gray, R. D. (2009). Do New Caledonian crows solve physical problems through causal reasoning? *Proc Roy Soc B Biol Sci*, **276**, 247–254.

Tebbich, S., & Bshary, R. (2004). Cognitive abilities related to tool use in the woodpecker finch, *Cactospiza pallida. Anim Behav*, **67**, 689–697.

Tenenbaum, J. B., & Griffiths, T. L. (2003). Theory-based causal inference. In S. Becker, S. Thrun, & K. Obermayer (eds.).

Advances in Neural Information Processing Systems (pp. 43–50). Cambridge, MA: MIT Press.

Tenenbaum, J. B., Griffiths, T. L., & Niyogi, S. (2007). Intuitive theories as grammars for causal inference. In A. Gopnik & L. Shulz (eds.). *Causal Learning: Psychology, Philosophy, and Computation* (pp. 301–322). New York, NY: Oxford University Press.

Tennie, C., Call, J., & Tomasello, M. (2009). Ratcheting up the ratchet: On the evolution of cumulative culture. *Philos Trans Roy Soc B Biol Sci*, **364**, 2405–2415.

Terrace, H. S. (1991). Chunking during serial learning by a pigeon: 1. Basic evidence. *J Exp Psychol Anim Behav Process*, **17**, 81–93.

Terrace, H. S., Petitto, L. A., Sanders, R. J., & Bever, T. G. (1979). Can an ape create a sentence? *Science*, **206**, 891–902.

Thompson, R. F., & Krupa, D. J. (1994). Organization of memory traces in the mammalian brain. *Ann Rev Neurosci*, **17**, 519–549.

Thompson, R. F., & Steinmetz, J. E. (2009). The role of the cerebellum in classical conditioning of discrete behavioral responses. *Neuroscience*, **162**, 732–755.

Thompson, R. K. R. (1995). Natural and relational concepts in animals. In H. Roitblat & J. A. Meyer (eds.). *Comparative Approaches to Cognitive Science* (pp. 175–224). Cambridge, MA: MIT Press.

Thompson, R. K. R., & Oden, D. L. (2000). Categorical perception and conceptual judgments by nonhuman primates: The paleological monkey and the analogical ape. *Cogn Sci*, **24**, 363–396.

Thompson, R. K. R., Oden, D. L., & Boysen, S. T. (1997). Language-naïve chimpanzees (*Pan troglodytes*) judge relations between relations in a conceptual matching-to-sample task. *J Exp Psychol Anim Behav Process*, **23**, 31–43.

Thorndike, E. L. (1911). *Animal Intelligence: Experimental Studies*. New York: Macmillan.

Thornton, A., & McAuliffe, K. (2006). Teaching in wild meerkats. *Science*, **313**, 227–229.

Thorpe, W. H. (1956). *Learning and Instinct in Animals*. London, UK: Methuen.

Thorpe, W. H. (1963). *Learning and Instinct in Animals* (2nd Edn.). Cambridge, MA: Harvard University Press.

Thorup, K., Holland, R. A., Tettrup, A. P., & Wikelski, M. (2010). Understanding the migratory orientation program of birds: Extending laboratory studies to study free-flying migrants in a natural setting. *Integr Comp Biol*, **50**, 315–322.

Thorup, K., Ortvad, T. E., Rabol, J., *et al.* (2011). Juvenile songbirds compensate for displacement to oceanic islands during autumn migration. *PLoS One*, **6**, e17903.

Tibbets, E. A. (2002). Visual signals of individual identify in the wasp *Polistes fuscatus*. *Proc Roy Soc Lond B*, **269**, 1423–1428.

Tibbets, E. A., & Dale, J. (2007). Individual recognition: It is good to be different. *Trends Ecol Evol*, **22**, 529–537.

Timberlake, W. (1984). Behavior regulation and learned performance: Some misapprehensions and disagreements. *J Exp Anal Behav*, **41**, 355–375.

Tinbergen, N. (1951). *The Study of Instinct*. Oxford, UK: Clarendon Press.

Tinbergen, N. (1958). *Curious Naturalist*. Garden City, New York, NY: Doubleday.

Tinbergen, N. (1960). The natural control of insects in pinewoods. I. Factors influencing the intensity of predation by song birds. *Arch Neerlandaises Zool*, **13**, 265–343.

Tinbergen, N. (1963). On aims and methods of ethology. *Z Tierpsychol*, **20**, 410–433.

Tinbergen, N. (1964). The evolution of signalling devices. In W. Etkin (ed.). *Social Behavior and Organization among Vertebrates* (pp. 206–230). Chicago, IL: The University of Chicago Press.

Tobin, H., & Logue, A. W. (1994). Self-control across species (*Columba livia, Homo sapiens* and *Rattus norvegicus*). *J Comp Psychol*, **108**, 126–133.

Todd, P. M., & Gigerenzer, G. (2000). Précis of simple heuristics that make us smart. *Behav Brain Sci*, **23**, 727–741.

Todd, P. M., & Gigerenzer, G. (2007). Environments that make us smart: Ecological rationality. *Curr Direct Psychol Sci*, **16**, 167–171.

Tolman, C. W. (1964). Social facilitation of feeding behaviour in the domestic chick. *Anim Behav*, **12**, 245–251.

Tolman, E. C. (1932). *Purposive Behavior in Animals and Men*. New York, NY: Appleton Century-Crofts.

Tolman, E. C. (1948). Cognitive maps in rats and men. *Psychol Rev*, **55**, 189–208.

Tolman, E. C., & Honzik, C. H. (1930). 'Insight' in rats. *Publications in Psychology*, **4**, 215–232.

Tomasello, M. (1990). Cultural transmission in the tool use and communicatory signalling of chimpanzees? In S. Parker & K. Gibson (eds.). *"Language" and Intelligence in Monkeys and Apes: Comparative Developmental Perspectives* (pp. 274–311). Cambridge, UK: Cambridge University Press.

Tomasello, M. (1998). Emulation learning and cultural learning. *Behav Brain Sci*, **21**, 703–704.

Tomasello, M. (2009). *Why We Cooperate*. Cambridge, MA: The MIT Press.

Tomasello, M., & Carpenter, M. (2005). Intention reading and imitative learning. In S. Hurley & N. Chater (eds.). *Perspectives on Imitation: Imitation, Human Development, and Culture* (pp. 133–148). Cambridge, MA: MIT Press.

Tomasello, M., Carpenter, M., Call, J., Behne, T., & Moll, H. (2005). Understanding and sharing intentions: The origins of cultural cognition. *Behav Brain Sci*, **28**, 675–735.

Tomasello, M., Kruger, A. C., & Ratner, H. H. (1993). Cultural learning. *Behav Brain Sci*, **16**, 495–511.

Tomonaga, M., Imura, T., Mizuno, Y., & Tanaka, M. (2007). Gravity bias in young and adult chimpanzees (*Pan troglodytes*): Tests with a modified opaque-tube task. *Dev Sci*, **10**, 411–421.

Topál, J., Gácsi, M., Miklósi, Á., *et al.* (2005). Attachment to humans: A comparative study on hand-reared wolves and differently socialized dog puppies. *Anim Behav*, **70**, 1367–1375.

Topál, J., Miklósi, Á., Csányi, V., & Dóka, A. (1998). Attachment behavior in dogs (*Canis familiaris*): A new application of Ainsworth's (1969) Strange Situation Test. *J Comp Psychol*, **112**, 219–229

Treiber, C. D., Salzer, M. C., Riegler, J., *et al.* (2012). Clusters of iron-rich cells in the upper beak of pigeons are macrophages not magnetosensitive neurons. *Nature*, **484**, 367–370.

Treisman, A. (1988). Features and objects: The fourteenth Bartlett Memorial lecture. *Q J Exp Psychol*, **40B**, 201–223.

Trewavas, A. J. (1999). How plants learn. *Proc Natl Acad Sci USA*, **96**, 4216–4218.

Trimmer, P. C., Houston, A. I., Marshall, J. A. R., *et al.* (2011). Decision-making under uncertainty: Biases and Bayesians. *Animal Cognition*, **14**, 465–476.

Trivers, R. (1971). The evolution of reciprocal altruism. *Q Rev Biol*, **46**, 35–57.

Troje, N. F. (2002). Decomposing biological motion: A framework for analysis and synthesis of human gait patterns. *J Vision*, **2**, 371–387.

Troje, N. F., Huber, L., Loidolt, M., Aust, U., & Fieder, M. (1999). Categorical learning in pigeons: The role of texture and shape in complex static stimuli. *Vision Res*, **39**, 353–366.

Tudusciuc, O., & Nieder, A. (2007). Neuronal population coding of continuous and discrete quantity in the primate posterior parietal cortex. *Proc Natl Acad Sci USA*, **104**, 14513–14518.

Tulving E. (1972). Episodic and semantic memory. In E. Tulving & W. Donaldson (eds.). *Organization of Memory* (pp. 381–402). New York, NY: Academic Press.

Tulving, E. (1989). Remembering and knowing the past. *American Scientist*, **40**, 385–398.

Tulving, E., & Schacter, D. L. (1990). Priming and human memory systems. *Science*, **247**, 301–306.

Turing, A. M. (1950). Computing machinery and intelligence. *Mind*, **59**, 433–460.

Tversky, A., & Kahneman, D. (1974). Judgment under uncertainty: Heuristics and biases. *Science*, **185**, 1124–1131.

Tyack, P. & Sayigh, L. (1997). Vocal learning in cetaceans. In C. Snowdon & M. Hausburger (eds.). *Social Influences on Vocal Development* (pp. 208–233). Cambridge, UK: Cambridge University Press.

Tyler, L. K., & Moss, H. E. (2001). Towards a distributed account of conceptual knowledge. *Trends Cogn Sci*, **5**, 244–252.

Uller, C. (2004). Disposition to recognize goals in infant chimpanzees. *Anim Cogn*, **7**, 154–161.

Uller, C., Jaeger, R., Guidry, G., & Martin, C. (2003). Salamanders (*Plethodon cinereus*) go for more: Rudiments of number in an amphibian. *Anim Cogn*, **6**, 105–112.

Urcelay, G. P., & Miller, R. R. (2010). On the generality and limits of abstraction in rats and humans. *Anim Cogn*, **13**, 21–32.

Urcuioli, P. J. (2006). Responses and acquired equivalence classes. In E. A. Wasserman & T. R. Zentall (eds.). *Comparative Cognition: Experimental Explorations of Animal Intelligence* (pp. 405–421). New York, NY: Oxford University Press.

Vacha, M. (2006). Laboratory behavioural assay of insect magnetoreception: Magnetosensitivity of *Periplaneta americana*. *J Exp Biol*, **209**, 3882–3889.

Valenza, E., Simion, F., Cassia, V. M., & Umilta, C. (1996). Face preference at birth. *J Exp Psychol Hum Percept Perform*, **22**, 892–903.

Vallortigara, G., Regolin, L., & Marconato, F. (2005). Visually inexperienced chicks exhibit spontaneous preference for biological motion patterns. *PLoS Biol*, **3**, e208.

Vallortigara, G., Zanforlin, M., & Pasti, G. (1990). Geometric modules in animals' spatial representations: A test with chicks (*Gallus gallus domesticus*). *J Comp Psychol*, **104**, 248–254.

Van Baaren, R. B., Holland, R. W., Kawakami, K., & Van Knippenberg, A. (2004). Mimicry and prosocial behavior. *Psychol Sci*, **15**, 71–74.

Van Bergen, Y., Coolen, I., & Laland, K. N. (2004). Nine-spined sticklebacks exploit the most reliable source when public and private information conflict. *Proc Roy Soc Lond B Biol Sci*, **271**, 957–962.

Vander Wyk, B. C., Hudac, C. M., Carter, E. J., Sobel, D. M., & Pelphrey, K. A. (2009). Action understanding in the superior temporal sulcus region. *Psychol Sci*, **20**, 771–777.

Van Lawick-Goodall, J. (1970). Tool-using in primates and other vertebrates. In D. S. Lehrman (ed.). *Advances in the Study of Behavior* (Vol. 3) (pp.195–249). New York, NY: Academic Press.

vanMarle, K. (2013). Infants use different mechanisms to make small and large number ordinal judgments. *J Exp Child Psychol*, **114**, 102–110.

vanMarle, K., & Wynn, K. (2006). Six-month-old infants use analog magnitudes to represent duration. *Dev Sci*, **9**, F41–F49.

vanMarle, K., & Wynn, K. (2011). Tracking and quantifying objects and non-cohesive substances. *Dev Sci*, **14**, 502–515.

Vann, S. D., Aggleton, J. P., & Maguire, E. A. (2009). What does the retrosplenial cortex do? *Nat Rev Neurosci*, **10**, 792–802.

van Schaik, C. P., Deaner, R. O., & Merrill, M. Y. (1999). The conditions for tool use in primates: Implications for the evolution of material culture. *J Hum Evol*, **36**, 719–741.

Vardanis, Y., Klaassen, R. H., Strandberg, R., & Alerstam, T. (2011). Individuality in bird migration: Routes and timing. *Biol Lett*, **7**, 502–505.

Vargha-Khadem, F., Gadian, D. G., Copp, A., & Mishkin, M. (2005). FOXP2 and the neuroanatomy of speech and language. *Nat Rev Neurosci*, **6**, 131–138.

Vasconcelos, M., Monteiro, T., & Kacelnik, A. (2013). Context-dependent preferences in starlings: Linking ecology, foraging and choice. *Pub Libr Sci*, **8**, e64934.

Veblen, T. B. (1899). *The Theory of the Leisure Class: An Economic Study in the Evolution of Insitutions*. New York, NY: Macmillan.

Vickers, N. J. (2006). Winging it: Moth flight behavior and responses of olfactory neurons are shaped by pheromone plume dynamics. *Chem Senses*, **31**, 155–166.

Visalberghi, E., & Fragaszy, D. M. (1990). Food-washing behaviour in tufted capuchin monkeys, *Cebus apella*, and crabeating macaques, *Macaca fascicularis*. *Anim Behav*, **40**, 829–836.

Visalberghi, E., Fragaszy, D. M., & Savage-Rumbaugh, S. (1995). Performance in a tool-using task by common chimpanzees (*Pan troglodytes*), bonobos (*Pan paniscus*), an orangutan (*Pongo pygmaeus*), and capuchin monkeys (*Cebus apella*). *J Comp Psychol*, **109**, 52–60.

Visalberghi, E., & Limongelli, L. (1994). Lack of comprehension of cause-effect relations in tool-using capuchin monkeys (*Cebus apella*). *J Comp Psychol*, **108**, 15–22.

Visalberghi, E., Quarantotti, B. P., & Tranchida, F. (2000). Solving a cooperation task without taking into account the partner's behavior: The case of capuchin monkeys (*Cebus apella*). *J Comp Psychol*, **114**, 297.

Vlamings, P. H., Uher, J., & Call, J. (2006). How the great apes (*Pan troglodytes*, *Pongo pygmaeus*, *Pan paniscus*, and *Gorilla gorilla*) perform on the reversed contingency task: The effects of food quantity and food visibility. *J Exp Psychol Anim Behav Process*, **32**, 60–70.

Volkow, N. D., Wang, G. J., Tomasi, D., & Baler, R. D. (2013). Obesity and addiction: Neurobiological overlaps. *Obes Rev*, **14**, 2–18.

Von Frijtag, J. C., Schot, M., van den Bos, R., & Spruijt, B. M. (2002). Individual housing during the play period results in changed responses to and consequences of a psychosocial stress situation in rats. *Dev Psychobiol*, **41**, 58–69.

von Frisch, K. (1953). *The Dancing Bees*. New York, NY: Harcourt Brace.

von Frisch, K. (1967). *The Dance Language and Orientation of Bees*. Cambridge, MA: Harvard University Press.

Vonk, J., Brosnan, S.F., Silk, J.B., *et al*. (2008). Chimpanzees do not take advantage of very low cost opportunities to deliver food to unrelated group members. *Anim Behav*, **75**, 1757–1770.

Vonk, J., & MacDonald, S. E. (2004). Levels of abstraction in orangutan (*Pongo abelii*) categorization. *J Comp Psychol*, **118**, 3–13.

Vuchinich, R. E. (1999). Behavioral economics as a framework for organizing the expanded range of substance abuse interventions. In: J. A. Tucker, D. M. Donovan, & G. A. Marlatt (eds.). *Changing Addictive Behavior: Bridging Clinical and Public Health Strategies* (pp.191–218). New York, NY: Guildford Press.

Vyse, S. A. (1997). *Believing in Magic: The Psychology of Superstition*. New York, NY: Oxford University Press.

Waga, I. C., Dacier, A. K., Pinha, P. S., & Tavares, M. C. H. (2006). Spontaneous tool use by wild capuchin monkeys (*Cebus libidinosus*) in the cerrado. *Folia Primatol*, **77**, 337–344.

Walcott, C., & Green, P. R. (1974). Orientation of homing pigeons altered by a change in the direction of an applied magnetic field. *Science*, **184**, 180–182.

Waldmann, M. R. (1996). Knowledge-based causal induction. *Psychol Learn Motiv*, **34**, 47–88.

Waldmann, M. R., & Hagmayer, Y. (2005). Seeing versus doing: Two modes of accessing causal knowledge. *J Exp Psychol Learn Mem Cogn*, **31**, 216–227.

Wallace, A. R. (1870). *Contributions to the Theory of Natural Selection: A Series of Essays*. New York, NY: Macmillan & Co.

Wallace, D. F., Hines, D. J., Pellis, S. M., & Whishaw, I. Q. (2002). Vestibular information is required for dead reckoning in the rat. *J Neurosci*, **22**, 10009–10017.

Wallis, J. D. (2011). Cross-species studies of orbitofrontal cortex and value-based decision-making. *Nat Neurosci*, **15**, 13–19.

Wallis, J. D., Anderson, K. C., & Miller, E. K. (2001). Single neurons in prefrontal cortex encode abstract rules. *Nature*, **411**, 953–956.

Wallman, J. (1992). *Aping Language*. Cambridge, UK: Cambridge University Press.

Wallraff, H. G. (2004). Avian olfactory navigation: Its empirical foundation and conceptual state. *Anim Behav*, **67**, 189–204.

Walsh, V. (2003). A theory of magnitude: Common cortical metrics of time, space and quantity. *Trends Cogn Sci*, **7**, 483–488.

Wang, R. F., & Brockmole, J. R. (2003). Simultaneous spatial updating in nested environments. *Psychon Bull Rev*, **10**, 981–986.

Wang, R. F., & Spelke, E. S. (2002). Human spatial representation: Insights from animals. *Trends Cogn Sci*, **6**, 376–382.

Wang, Y., & Frost, B. J. (1992). Time to collision is signaled by neurons in the nucleus rotundus of pigeons. *Nature*, **356**, 236–238.

Want, S. C., & Harris, P. L. (2002). How do children ape? Applying concepts from the study of non-human primates to the developmental study of 'imitation'in children. *Dev Sci*, **5**, 1–13.

Ward, A. J., Herbert-Read, J. E., Jordan, L.A., *et al*. (2013). Initiators, leaders, and recruitment mechanisms in the collective movements of damselfish. *Am Nat*, **181**, 748–760.

Ward, A. J., Krause, J., & Sumpter, D. J. (2012). Quorum decision-making in foraging fish shoals. *PLoS One*, **7**, e32411.

Warden, C. J., & Jackson, T. A. (1935). Imitative behavior in the rhesus monkey. *J Genet Psychol*, **46**, 103–125.

Warneken, F., Hare, B., Melis, A.P., Hanus, D., & Tomasello, M. (2007). Spontaneous altruism by chimpanzees and young children. *PLoS Biol*, **5**, e184.

Warneken, F., Lohse, K., Melis, A. P., & Tomasello, M. (2011). Young children share the spoils after collaboration. *Psychol Sci*, **22**, 267–273.

Warneken, F., & Tomasello, M. (2006). Altruistic helping in human infants and young chimpanzees. *Science*, **311**, 1301–1303.

Warneken, F., & Tomasello, M. (2009). Varieties of altruism in children and chimpanzees. *Trends Cogn Sci*, **13**, 397.

Warrant, E., Kelber, A., Gislén, A., *et al*. (2004). Nocturnal vision and landmark orientation in a tropical halictid bee. *Curr Biol*, **14**, 1309–1318.

Wasserman, E. (2007). Autobiography of Division 3 President-Elect. *Exp Psychol Bull*, **11**.

Wasserman, E. A., & Castro, L. (2005). Surprise and change: Variations in the strength of present and absent cues in causal learning. *Learn Behav*, **33**, 131–146.

Wasserman, E. A., & Miller, R. R. (1997). What's elementary about associative learning? *Ann Rev Psychol*, **48**, 573–607.

Wasserman, E. A., Kleidinger, R. E., & Bhatt, R. S. (1988). Conceptual behavior in pigeons: categories, subcategories and pseudocategories. *J Exp Psychol Anim Behav Process*, **14**, 235–246.

Watanabe, S. (1993). Object-picture equivalence in the pigeon: An analysis with natural concept and pseudocategory discriminations. *Behav Process*, **30**, 225–231.

Watanabe, S., & Ono, K. (1986). An experimental analysis of "empathic" response: Effects of pain reactions of pigeon upon other pigeon's operant behavior. *Behav Process*, **13**, 269–277.

Watson, J. B. (1913). Psychology as the behaviorist views it. *Psychol Rev*, **72**, 89–104.

Weary, D. M., & Krebs, J. R. (1992). Great tits classify songs by individual voice characteristics. *Anim Behav*, **43**, 283–287.

Weber, E. U., Shafir, S., & Blais, A.-R. (2004). Predicting risk sensitivity in humans and lower animals: Risk as variance or coefficient of variation. *Psychol Rev*, **111**, 430–445.

Wehner, R., & Muller, M. (2006). The significance of direct sunlight and polarized skylight in the ant's celestial system of navigation. *Proc Natl Acad Sci USA*, **103**, 12575–12579.

Wehner, R., & Srinivasan, M.V. (1981). Searching behavior of desert ants, genus *Cataglyphyis* (Formicidae, Hymenoptera). *J Comp Physiol*, **142**, 315–338.

Wiener, S.I. (1993). Spatial and behavioral correlates of striatal neurons in rats performing a self-initiated navigation task. *J Neurosci*, **13**, 3802–3817.

Weir, A.A., Chappell, J., & Kacelnik, A. (2002). Shaping of hooks in New Caledonian crows. *Science*, **297**, 981.

Weir, A.A., & Kacelnik, A. (2006). A New Caledonian crow (*Corvus moneduloides*) creatively re-designs tools by bending or unbending aluminium strips. *Anim Cogn*, **9**, 317–334.

Weiskrantz, L. (1976). *Blindsight – A Case Study and Implications*. Oxford, UK: Clarendon Press.

Weisman, R.G., Balkwill, L.-L., Hoeschele, M., *et al.* (2010). Absolute pitch in boreal chickadees and humans: Exceptions that test a phylogenetic rule. *Learn Motiv*, **41**, 156–173.

Weisman, R.G., Njegovan, M.G., Williams, M.T., Cohen, J.S., & Sturdy, C.B. (2004). A behavior analysis of absolute pitch: Sex, experience, and species. *Behav Process*, **66**, 289–307.

Weissburg, M.J., Doall, M.H., & Yen, J. (1998). Following the invisible trail: Kinematic analysis of mate-tracking in the copepod *Temora longicornis*. *Philos Trans Roy Soc B*, **29**, 701–712.

Wellman, H.M., & Liu, D. (2004). Scaling of theory-of-mind tasks. *Child Dev*, **75**, 523–541.

Wenner, A.M. (1967). Honey bees: Do they use the distance information contained in their dance maneuver? *Science*, **155**, 847–849.

Wenner, A.M., Wells, P.H., & Rohlf, F.J. (1967). An analysis of the waggle dance and recruitment in honey bees. *Physiol Zool*, **40**, 317–344.

Werker, J.F., & Tees, R.C. (1984). Cross-language speech perception: Evidence for perceptual reorganization during the first year of life. *Infant Behav Dev*, **7**, 49–63.

West, M.J., King, A.P., & White, D.J. (2003). Discovering culture in birds: the role of learning and development. In F.B.M. de Waal & P.L. Tyack (eds.). *Animal Social Complexity: Intelligence, Culture, and Individualized Societies* (pp. 470–492). Cambridge, MA: Harvard University Press.

Whishaw, I.Q., Gorney, B.P., & Dringenberg, H.C. (1991). The defensive strategies of foraging rats: A review and synthesis. *Psychol Rev*, **41**, 185–205.

Whishaw, I.Q., & Tomie, J. (1989). Food-pellet size modifies the hoarding behavior of foraging rats. *Psychobiology*, **17**, 93–101.

White, D.J., & Galef, B.G. (1999). Affiliative preferences are stable and predict mate choices in both sexes of Japanese quail, *Coturnix japonica*. *Anim Behav*, **58**, 865–871.

White, D.J., & Galef, B.G., Jr. (2000). Differences between the sexes in direction and duration of response to seeing a potential sex partner mate with another. *Anim Behav*, **59**, 1235–1240.

Whiten, A. (2006). The dissection of imitation and its "cognitive kin" in comparative and developmental psychology. In S.J. Rogers & J.H.G. Williams (eds.). *Imitation and the Social Mind: Autism and Typical Development* (pp. 227–250). New York, NY: The Guilford Press.

Whiten, A., Custance, D.M., Gomez, J.C., Teixidor, P., & Bard, K.A. (1996). Imitative learning of artificial fruit processing in children (*Homo sapiens*) and chimpanzees (*Pan troglodytes*). *J Comp Psychol*, **110**, 3–14.

Whiten, A., Goodall, J., McGrew, W.C., *et al.* (1999). Cultures in chimpanzees. *Nature*, **399**, 682–685.

Whiten, A., & Ham, R. (1992). On the nature and evolution of imitation in the animal kingdom: Reappraisal of a century of research. *Adv Study of Behav*, **21**, 239–283.

Whiten, A., Horner, V., & De Waal, F.B. (2005). Conformity to cultural norms of tool use in chimpanzees. *Nature*, **437**, 737–740.

Whiten, A., Horner, V., Litchfield, C.A., & Marshall-Pescini, S. (2004). How do apes ape? *Anim Learn Behav*, **32**, 36–52.

Whitlock, J.R., Heynen, A.J., Shuler, M.G., & Bear, M.F. (2006). Learning induced long-term potentiation in the hippocampus. *Science*, **313**, 1093–1097.

Widdig, A. (2007). Paternal kin discrimination: The evidence and likely mechanisms. *Biol Rev*, **82**, 319–334.

Wiener, S.I. (1993). Spatial and behavioral correlates of striatal neurons in rats performing a self-initiated navigation task. *J Neurosci*, **13**, 3802–3817.

Wilhelm, C.J., & Mitchell, S.H. (2009). Strain differences in delay discounting using inbred rats. *Genes Brain Behav*, **8**, 426–434.

Wilkinson, G.S. (1984). Reciprocal food sharing in the vampire bat. *Nature*, **308**, 181–184.

Wilkinson, G.S. (1985). The social organization of the common vampire bat. *Behav Ecol Sociobiol*, **17**, 123–134.

Wilkinson, G.S. (1992). Communal nursing in the evening bat, *Nycticeius humeralis*. *Behav Ecol Sociobiol*, **31**, 225–235.

Williams, C.L., Barnett, A.M., & Meck, W.H. (1990). Organization effects of early gonadal secretions on sexual-differentiation in spatial memory. *Behav Neurosci*, **104**, 84–97.

Williams, G.C. (1966). *Adaptation and Natural Selection*. Princeton, NJ: Princeton University Press.

Williams, H. (2008). Birdsong and singing behavior. In H.P. Ziegler & P. Marler (eds.). *Neuroscience of Birdsong* (pp. 32–49). Cambridge, UK: Cambridge University Press.

Williams, H., Levin, I.I., Norris, D.R., Newman, A.E.M., Wheelwright, N.T. (2013). Three decades of cultural evolution in Savannah sparrow songs. *Anim Behav*, **85**, 213–223.

Williamson, R.A., & Markman, E.M. (2006). Precision of imitation as a function of preschoolers' understanding of the goal of the demonstration. *Dev Psychol*, **42**, 723–731.

Wills, T.A., Vaccaro, D., & McNamara, G. (1994). Novelty seeking, risk taking, and related constructs as predictors of adolescent substance use: An application of Cloninger's theory. *J Subst Abuse*, **6**, 1–20.

Wilmer, J.B., Germine, L., Chabris, C.F., *et al.* (2010). Human face recognition ability is specific and highly heritable. *Proc Natl Acad Sci USA*, **107**, 5238–5241.

Wilson, B., Mackintosh, N.J., & Boakes, R.A. (1985). Transfer of relational rules in matching and oddity learning by pigeons and corvids. *Q J Exp Psychol*, **37B**, 313–332.

Wilson, E.O. (1975). *Sociobiology: The New Synthesis*. Cambridge, MA: Harvard University Press.

Wiltschko, R., Stapput, K., Thalau, P., & Wiltschko, W. (2010). Directional orientation of birds by the magnetic field under different light conditions. *J Roy Soc Interface*, **7**, S163-S177.

Wiltschko, W., & Wiltschko, R. (1995). *Magnetic Orientation in Animals*. Berlin: Springer-Verlag.

Wiltschko, W., & Wiltschko, R. (2012). Global navigation in migratory birds: Tracks, strategies, and interactions between mechanisms. *Curr Opin Neurobiol*, **22**, 328–335.

Wiltschko, W., Wiltschko, R., Grueter, M., & Kowalski, U. (1987). Pigeon homing: Early experience determines what factors are used for navigation. *Naturwissenchaften*, **74**, 196–198.

Wise, R. A., & Rompré, P. P. (1989). Brain dopamine and reward. *Ann Rev Psychol*, **40**, 191–225.

Witschko, R., & Witschko, W. (2003). Avian navigation: From historical to modern concepts. *Anim Behav*, **65**, 257–272.

Wittig, R. M., & Boesch, C. (2003). Food competition and linear dominance hierarchy among female chimpanzees of the Tai National Park. *Int J Primatol*, **24**, 847–867.

Wittlinger, M., Wenher, R., & Wolf, H. (2006). The ant odometer: Stepping on stilts and stumps. *Science*, **312**, 1965–1967.

Wolpert, D. M., Goodbody, S. J., & Husain, M. (1998). Maintaining internal representations: The role of the human superior parietal lobe. *Nat Neurosci*, **1**, 529–533.

Wood, J. N., & Hauser, M. D. (2011). Replication of "The perception of rational, goal-directed action in nonhuman primates". *Science*, posted 29 April 2011.

Wood, J. N., Glynn, D. D., Phillips, B. C., & Hauser, M. D. (2007). The perception of rational, goal-directed action in nonhuman primates. *Science*, **317**, 1402–1405.

Woodward, A. L. (1998). Infants selectively encode the goal object of an actor's reach. *Cognition*, **69**, 1–34.

Woollett, K., & Maguire, E. A. (2011). Acquiring 'the knowledge" on London's layout drives structural brain changes. *Curr Biol*, **21**, 2109–2114.

Wrangham, R. W. (1980). An ecological model of female-bonded primate groups. *Behaviour*, **75**, 262–299.

Wray, M. K., Klein, B. A., Mattila, H. R., & Seeley, T. D. (2008). Honeybees do not reject dances for 'implausible' locations: Reconsidering the evidence for cognitive maps in insects. *Anim Behav*, **76**, 261–269.

Wynn, K. (1992). Addition and subtraction by human infants. *Nature*, **358**, 749–750.

Wynn, T., & Coolidge, F. L. (2011). The implications of working memory for the evolution of modern cognition. *Int J Evol Biol*, **2011**, 741357.

Wynne, C. D. L., Udell, M. A. R., & Lord, K. A. (2008). Ontogeny's impacts on human-dog communication. *Anim Behav*, **76**, e1-e4.

Wyttenbach, R. A., May, M. L., & Hoy, R. R. (1996). Categorical perception of sound frequency by crickets. *Science*, **273**, 1542–1544.

Xu, F. (2003). Numerosity discrimination in infants: Evidence for two systems of representations. *Cognition*, **89**, B15-B25.

Xu, F., & Spelke, E. S. (2000). Large number discrimination in 6-month-old infants. *Cognition*, **74**, B1–B11.

Yack, J. E. (2004). The structure and function of auditory chordotonal organs in insects. *Microsc Res Tech*, **63**, 315–337.

Yamamoto, S., Humle, T., & Tanaka, M. (2009). Chimpanzees help each other upon request. *PLoS One*, **4**, e7416.

Yamamoto, S., & Tanaka, M. (2010). The influence of kin relationship and reciprocal context on chimpanzees' other-regarding preferences. *Anim Behav*, **79**, 595–602.

Yerkes, R. M., & Spragg, S. D. S. (1937). La mésure du comportement adapté chez les chimpanzes. *Journal de Psycholgie Normale et Pathologique*, **34**, 449–474.

Young, M. E., Beckmann, J. S., & Wasserman, E. A. (2006). Pigeons' discrimination of Michotte's Launching Effect. *J Exp Anal Behav*, **86**, 223–237.

Zach, R. (1979). Shell dropping: Decision-making and optimal foraging in Northwestern crows. *Behaviour*, **68**, 106–117.

Zahavi, A. (1975). Mate selection: A selection for a handicap. *J Theor Biol*, **53**, 205–214.

Zahn-Waxler C., Hollenbeck B., & Radke-Yarrow, M. (1984). The origins of empathy and altruism. In M. W. Fox & L. D. Mickley (eds.). *Advances in Animal Welfare Science* (pp. 21–41). Netherlands: Springer.

Zajonc, R. B. (1965). Social facilitation. *Science*, **149**, 269–274.

Zak, P. J., Stanton, A. A., & Ahmadi, S. (2007). Oxytocin increases generosity in humans. *PLoS One*, **2**, e1128.

Zapka, M., Heyers, D., Hein, C. M., *et al.* (2009). Visual but not trigeminal mediation of magnetic compass information in a migratory bird. *Nature*, **461**, 1274–1278.

Zatorre, R. J. (2003). Absolute pitch: A model for understanding the influence of genes and development on neural and cognitive function. *Nat Neurosci*, **6**, 692–695.

Zeeb, F. D., Robbins, T. W., & Winstanley, C. A. (2009). Serotonergic and dopaminergic modulation of gambling behavior as assessed using a novel rat gambling task. *Neuropsychopharmacology*, **34**, 2329–2343.

Zentall, T. R. (2006). Imitation: Definitions, evidence, and mechanisms. *Anim Cogn*, **9**, 335–353.

Zentall, T. R. (1993). Animal cognition: An approach to the study of animal behavior. In T. R. Zentall (ed.), *Animal Cognition: A Tribute to Donald A. Riley*. Hillsdale, NJ: Lawrence Erlbaum Associates.

Zentall, T. R. (2011). Social learning mechanisms: Implications for a cognitive theory of imitation. *Interact Stud*, **12**, 233–261.

Zentall, T. R. (2012). Perspectives on observational learning in animals. *J Comp Psychol*, **126**, 114–128.

Zentall, T. R., & Levine, J. M. (1972). Observational learning and social facilitation in the rat. *Science*, **178**, 1220–1221.

Zentall, T. R., & Stagner, J. P. (2011). Maladaptive choice behaviour by pigeons: An animal analogue and possible mechanisms for gambling (sub-optimal human decision making behaviour). *Proc Roy Soc B*, **278**, 1203–1208.

Zentall, T. R., Sutton, J. E., & Sherburne, L. M. (1996). True imitative learning in pigeons. *Psychol Sci*, **7**, 343–346.

Zentall, T. R., Wasserman, E. A., Lazareva, O. F., Thompson, R. R. K., & Rattermann, M. J. (2008). Concept learning in animals. *Comp Cogn Behav Rev*, **3**, 13–45.

Zhou, S., & Crystal, J. D. (2011). Validation of a rodent model of episodic memory. *Anim Cogn*, **14**, 325–340.

Zuberbühler, K. (2003). Referential signaling in non-human primates: Cognitive precursors and limitations for the evolution of language. *Adv Study Behav*, **33**, 265–307.

Zuberbühler, K., Cheney, D. L., & Seyfarth, R. M. (1999). Conceptual semantics in a nonhuman primate. *J Comp Psychol*, **113**, 33–42.

FIGURE CREDITS

Chapter 1

Opening image: © Nigel Cattlin/Alamy.

1.1 Courtesy of Dr. Bennett Galef.
1.2 © Stubblefield Photography/Shutterstock.com.
1.3 © Classic Image/Alamy.
1.4 Courtesy of Wikimedia Commons.
1.5 Illustration by Simon Tegg, redrawn from Köhler, W. (1927). *The Mentality of Apes*, 2nd Rev. Edn. (Winter, E., trans.). London, UK: Routledge & Kegan Paul.
1.6 Illustration by Simon Tegg.
1.7 © INTERFOTO/Alamy.
1.8 Illustration by Simon Tegg.
1.9 Illustration by Simon Tegg.
1.10 Illustration by Simon Tegg.

Chapter 2

Opening image: © Teguh Tirtaputra/Shutterstock.com

2.1 © Thomas O'Neil/Shutterstock.com
2.2 Illustration by Simon Tegg.
2.3 Illustration by Simon Tegg.
2.4 (A) © iStockphoto.com/k1uk; (B) © iStockphoto.com/Zzvet.
2.5 © iStockphoto.com/mvanhoutte.
2.6 Illustration by Simon Tegg.
2.7 Reprinted with permission from Burley & Symanski (1998).
2.8 Illustration by Simon Tegg.
2.9 Illustration by Simon Tegg.
2.10 Reprinted with permission from Kruitzer, L. A., & Seelke, A. M. H. (2012). Cortical evolution in mammals: The bane and beauty of phenotypic variability. *Proc Natl Acad Sci USA*, **109**, 10647–10654.
2.11 Illustration by Simon Tegg, redrawn with permission from Rosenzweig, M. R., Breedlove, A. M., & Leiman, A. L. (2002). *Biological Psychology. An Introduction to Behavioral, Cognitive, and Clinical Neuroscience*. Sunderland, MA: Sinauer Associates Inc.
2.12 Credit: Ronald C. James, used with permission.
2.13 Illustration by Simon Tegg.
2.14 Illustration by Simon Tegg.
2.15 © iStockphoto.com/Greyfebruary.
2.16 Illustration by Simon Tegg.
2.17 Courtesy of Dr. Barrie Frost.
2.18 © artpartner-images.com/Alamy.
2.19 Courtesy of Dr. Alan Bond.
2.20 Courtesy of Dennis Paulson.
2.21 © iStockphoto.com/ChrisGorgio.

Chapter 3

Opening image: courtesy of Susan Magwood.
3.1 Illustration by Simon Tegg.
3.2 Reprinted with permission from Nairne & Pandeirada (2010).
3.3 © littleny/Shutterstock.com
3.4 Illustration by Simon Tegg.
3.5 Illustration by Simon Tegg.
3.6 Illustration by Simon Tegg.
3.7 Illustration by Simon Tegg.
3.8 Illustration by Simon Tegg.
3.9 Illustration by Simon Tegg.
3.10 Illustration by Simon Tegg.
3.11 Illustration by Simon Tegg.
3.12 Reprinted with permission from Knowlton, B. J., Mangels, J. A., & Squire, L. R. (1996). Neostriatal habit learning system in humans. *Science*, **273**, 1399–1402.
3.13 Illustration by Simon Tegg.
3.14 Courtesy of Dr. Nicole Clayton.
3.15 Illustration by Simon Tegg.
3.16 Illustration by Simon Tegg.
3.17 Illustration by Simon Tegg.

Chapter 4

Opening image: courtesy of Ingrid Taylar. Source – Flickr: Mikiko the Quail (made available under http://creativecommons.org/licenses/by/2.0/legalcode).
4.1 © RIA Novosti/Alamy.
4.2 AUTHOR – provide a credit for ths figure.
4.3 © Gita Memmena/Shutterstock.com.
4.4 Illustration by Simon Tegg.
4.5 Reprinted with permission from Huff, N. C., Zielinski, D. J., Fecteaau, M. E., Brady, R., & LaBar, K. S. (2010). Human fear conditioning conducted in full immersion 3-dimensional virtual reality. *J Visual Exp,* **42**, published on-line doi: 10.3791/1993.
4.6 Reprinted with permission from Gass *et al.* (2009).
4.7 Illustration by Simon Tegg.
4.8 Illustration by Simon Tegg.
4.9 Courtesy of Dr. Karen Hollis.
4.10 Illustration by Simon Tegg.
4.11 Illustration by Simon Tegg.
4.12 Illustration by Simon Tegg.
4.13 Illustration by Simon Tegg.
4.14 Illustration by Simon Tegg.
4.15 Reprinted with permission from Cusato & Domjan (1998).
4.16 Illustration by Simon Tegg.
4.17 Reprinted with permission from Jenkins & Moore (1973).
4.18 Illustration by Simon Tegg.
4.19 Illustration by Simon Tegg.
4.20 Illustration by Simon Tegg.
4.21 Illustration by Simon Tegg.

4.22 Illustration by Simon Tegg, redrawn with permission from de Wit, S., & Dickinson, A. (2009). Associative theories of goal-directed behavior: A case for animal-human translational models. *Psychol Res*, **73**, 463–476.
4.23 Illustration by Simon Tegg.
4.24 Illustration by Simon Tegg.
4.25 Illustration by Simon Tegg.
4.26 Reprinted with permission from Baxter & Byrne (2006).

Chapter 5

Opening image: © AlexAranda/Shutterstock.com.
5.1 Illustration by Simon Tegg, redrawn with permission from Olson, D. J., Kamil, A. C., Balda, R. P., & Nims, P. J. (1995) Performance of four seed-caching corvid species in operant tests of non-spatial and spatial memory. *J Comp Psychol*, **109**, 173–181.
5.2 Illustration by Simon Tegg.
5.3 © blickwinkel/Alamy.
5.4 Illustration by Simon Tegg.
5.5 Illustration by Simon Tegg.
5.6 Courtesy of Dr. Sue Healy.
5.7 Reprinted with permission from Muller & Wehner (1988). Path integration in desert ants, *Cataglyphis fortis*. *Proc Natl Acad Sci USA*, **85**, 5287–5290.
5.8 Reprinted with permission from Cheng, K. (1986). A purely geometric module in the rat's spatial representation. *Cognition*, **23**, 149–178.
5.9 Image courtesy of TreasuryTag at en.wikipedia (http://creativecommons.org/licenses/by-sa/3.0/legalcode).
5.10 Illustrations by Simon Tegg.
5.11 © Sekar B/Shutterstock.com.
5.12 Illustration by Simon Tegg.
5.13 Reprinted with permission from Bingman *et al.* (2006).
5.14 Illustration by Simon Tegg, redrawn with permission from Perdeck (1958).
5.15 Illustration by Simon Tegg.
5.16 Reprinted with permission from Packard & McGaugh (1996).

Chapter 6

Opening image: © EBFoto/Shutterstock.com.
6.1 Illustration by Simon Tegg, after Biebach *et al.* (1998) with permission.
6.2 Illustration by Simon Tegg, adapted with permission.
6.3 Illustration by Simon Tegg, redrawn from Henderson *et al.* (2006) with permission.
6.4 Illustration by Simon Tegg, redrawn with permission.
6.5 Illustration by Simon Tegg, redrawn from Meck and Church (1983) with permission.
6.6 Illustration by Simon Tegg, adapted from Honig and Stewart (1989) with permission.
6.7 Illustration by Simon Tegg, redrawn with permission from Cantlon and Brannon (2007).
6.8 Illustration by Simon Tegg, redrawn with permission from Cantlon *et al.* (2009).
6.9 Courtesy of Dr. Elizabeth Brannon
6.10 Illustration by Simon Tegg, redrawn with permission from Barner *et al.* (2008).
6.11 Illustration by Simon Tegg, adapted with permission from Boysen and Berntson (1989).
6.12 Illustration by Simon Tegg, redrawn with permission.

Chapter 7

Opening image: © Feng Yu/Shutterstock.com.

7.1 (A) © tolmachevr/Shutterstock.com; (B) Illustration by Simon Tegg.
7.2 Illustration by Simon Tegg.
7.3 Reprinted with permission from Lima (1984).
7.4 Illustration by Simon Tegg.
7.5 © Dray van Beeck/Shutterstock.com.
7.6 Illustration by Simon Tegg.
7.7 Illustration by Simon Tegg.
7.8 Illustration by Simon Tegg.
7.9 Illustration by Simon Tegg.
7.10 Courtesy of Dr. Alex Kacelnik.
7.11 Reprinted with permission from Stephens, D. W., & Anderson, D. (2001). The adaptive value of preference for immediacy: When shortsighted rules have farsighted consequences. *Behav Ecol*, **12**, 330–339.
7.12 Illustration by Simon Tegg.
7.13 Reprinted with permission from Bigelow, H. J. (1850). Dr. Harlow's case of recovery from the passage of an iron bar through the head. *Am J Med Sci*, n.s. v.20: 13–22.
7.14 Illustration by Simon Tegg.
7.15 Reprinted with permission from Peters & Buchel (2011).

Chapter 8

Opening image: © Mint Images/SuperStock.

8.1 Illustration by Simon Tegg, redrawn from Leslie and Keeble (1987) with permission.
8.2 Illustration by Simon Tegg.
8.3 Courtesy of Dr. Ed Wasserman.
8.4 Illustration by Simon Tegg, adapted from Spelke *et al.* (1992) with permission.
8.5 Photo: Ewa Krzyszczyk for Mann *et al.* (2008).
8.6 © iStockphoto.com/MikeLane45.
8.7 Illustration by Simon Tegg, adapted from Visalberghi and Limongelli (1994).
8.8 From Taylor *et al.* (2009) with permission.
8.9 Photo: Laurie Santos for Santos *et al.* (2006).

Chapter 9

Opening image: © Georgios Kollidas/Shutterstock.com. Engraving by C. E.Wagstaff.

9.1 Reprinted from Sturdy, C. B., Phillmore, L. S., & Weisman, R. G. (2000). Call-note discriminations in black-capped chickadees (*Poecile atricapillus*). *J Comp Psychol*, **114**, 357–364.
9.2 Illustration by Simon Tegg.
9.3 Illustration by Simon Tegg.
9.4 Illustration by Simon Tegg.
9.5 Reprinted with permission from Weisman *et al.* (2010).
9.6 Illustration by Simon Tegg.
9.7 Illustration by Simon Tegg.
9.8 Reprinted with permission from Flemming & Jennedy (2011).
9.9 © Circumnavigation/Shutterstock.com.
9.10 © Rolf Nussbaumer Photography/Alamy.

9.11 Courtesy of Ana Sanz Aguilar.
9.12 Courtesy of Dr. Francesco Bonadonna.
9.13 © mooinblack/Shutterstock.com.
9.14 © Taiftin/Shutterstock.com.
9.15 Illustration by Simon Tegg.
9.16 Illustration by Simon Tegg.

Chapter 10

Opening image: © Andrew King/ZSL Tsaobis Baboon Project.
10.1 Used with permission from Simion *et al.* (2008).
10.2 © Minden Pictures/SuperStock.
10.3 Reprinted with permission from Sugita (2008).
10.4 Illustration by Simon Tegg, based on description in Parr (2003).
10.5 Illustration by Simon Tegg, adapted from Povinelli and Preuss (1995) and Povinelli *et al.* (1997) with permission.
10.6 Redrawn from a frame of the movie by Heider and Simmel (1944).
10.7 Adapted with permission from Wood *et al.* (2007).
10.8 Illustration by Simon Tegg.
10.9 Illustration by Simon Tegg, drawn based on description in Povinelli, Nelson, and Boysen (1990).
10.10 Courtesy of Rodney Birch.
10.11 Illustration by Simon Tegg, adapted from Hare *et al.* (2000) with permission.
10.12 Illustration by Simon Tegg.
10.13 Illustration by Simon Tegg.
10.14 © Fotosearch/SuperStock.
10.15 Illustration by Simon Tegg, redrawn with permission from Cheney and Seyfarth (1990a).

Chapter 11

Opening image: © iStockphoto.com/Dervical.
11.1 Illustration by Simon Tegg.
11.2 Photograph courtesy of Walter Siegmund/Wikipedia (http://creativecommons.org/licenses/by/2.5/legalcode).
11.3 Photograph courtesy of Richard Ling/Wikipedia (http://creativecommons.org/licenses/by-sa/2.0/legalcode).
11.4 Illustration by Simon Tegg, based on the description in Warneken and Tomasello (2006).
11.5 © Minden Pictures/SuperStock.
11.6 Illustration by Simon Tegg, redrawn with permission from Jensen *et al.* (2007).
11.7 Courtesy of Dr. Joan Silk.
11.8 Illustration by Simon Tegg, redrawn with permission.
11.9 Illustration by Simon Tegg.
11.10 Used with permission from Melis *et al.* (2006).

Chapter 12

Opening image: Courtesy of Rodney Birch.
12.1 (A) © AlessandroZocc/Shutterstock.com; (B) © iStockphoto.com/Coleman515.
12.2 © Johan Swanepoel/Shutterstock.com.
12.3 Illustration by Simon Tegg.

12.4 Illustration based on the description in von Frisch (1953).
12.5 Sonogram illustration by Simon Tegg, redrawn with permission from Seyfarth *et al.* (1980).
12.6 Photograph courtesy of Dick Daniels (http://carolinabirds.org/)/Wikipedia (http://creativecommons.org/licenses/by-sa/3.0/legalcode).
12.7 Courtesy of Dr. Dorothy Cheney.
12.8 Illustration by Simon Tegg.
12.9 Illustration by Simon Tegg, drawn based on description in Premack and Premack (1983).
12.10 Illustration by Simon Tegg, drawn based on description in Savage-Rumbaugh *et al.* (1986).

Chapter 13

Opening image: © Minden Pictures/SuperStock.
13.1 Adapted with permission from Haun *et al.* (2012).
13.2 Illustration by Simon Tegg.
13.3 Illustration by Simon Tegg.
13.4 Redrawn from Brainard and Doupe (2002) with permission.
13.5 Illustration by Simon Tegg, adapted with permission from Klein & Zentall (2003).
13.6 Illustration by Simon Tegg, adapted with permission from Whiten *et al.* (1996).
13.7 Courtesy of Dr. Thomas Zentall.
13.8 Illustration by Simon Tegg, adapted from Horner and Whiten (2005) with permission.

INDEX